ECOLOGY OF EUROPEAN RIVERS

Frontispiece: The Rhine valley near Bacharach—a part of the river well known to tourists. The many physical and chemical changes which have taken place in the Rhine are described in Chapter 10.

Ecology of European Rivers

EDITED BY

B. A. WHITTON
MA, DSc
Department of Botany
University of Durham
England

BLACKWELL SCIENTIFIC PUBLICATIONS
OXFORD LONDON EDINBURGH
BOSTON PALO ALTO MELBOURNE

QH 97

© 1984 by
Blackwell Scientific Publications
Editorial offices:
Osney Mead, Oxford, OX2 0EL
8 John Stret, London WC1N 2ES
9 Forrest Road, Edinburgh, EH1 2QH
52 Beacon Street, Boston, Massachusetts 02108,
    USA
706 Cowper Street, Palo Alto, California 94301,
    USA
99 Barry Street, Carlton, Victoria 3053,
    Australia

First published 1984

Set by Oxprint Ltd, Oxford

Printed and bound in Great Britain by
Butler & Tanner Ltd, Frome and London

DISTRIBUTORS

USA and Canada
    Blackwell Scientific Publications Inc
    PO Box 50009, Palo Alto
    California 94303

Australia
    Blackwell Scientific Book Distributors
    31 Advantage Road, Highett
    Victoria 3190

British Library
Cataloguing in Publication Data

Ecology of European rivers.
    1. Stream ecology—Europe
    I. Whitton, B.A.
    574.5′26323′094   QH135
    ISBN 0-632-00816-4

Contents

Contributors

A. ALMESTRAND *Scandiaconsult AB, P.O. Box 17040, S-200 10 Malmo, Sweden*

A. ALMESTRAND *Department of Plant Physiology, University of Lund, P.O. Box 7007, S-220 07 Lund, Sweden*

D. BACKHAUS *Landesanstalt für Umweltschutz Baden-Württemberg, Institut für Wasser- und Abfallwirtschaft, Postfach 210752, 7500 Karlsruhe 21, Germany*

A. D. BERRIE *Freshwater Biological Association River Laboratory, East Stoke, Wareham, Dorset BH20 6BB, U.K.*

M. BROOKER *Welsh Water Authority, Directorate of Scientific Services, Nash Area Laboratory, West Nash, Newport, Gwent NP6 2YH, U.K.*

J. CAPBLANCQ *Laboratoire d'Hydrobiologie, Université Paul-Sabatier, 118 Route Narbonne, 31062 Toulouse Cedex, France*

G. CHIAUDANI *Dipartimento di Biologia (Sezione di Ecologia), Università degli Studi Milano, Via Celoria 26, 20133 Milano, Italy*

D. T. CRISP *Teesdale Unit, c/o Northumbrian Water Authority, Lartington Treatment Plant, Lartington, Barnard Castle, Co. Durham, DL12 9DW, U.K.*

H. DÉCAMPS *Centre d'Ecologie des Resources Renouvelables, 29 rue Jeanne-Marvig, 31055 Toulouse Cedex, France*

J-P. DESCY *Département de Botanique, Université de Liège, Sart Tilman, B 4000 Liège, Belgium*

R. W. EDWARDS *Department of Applied Biology, University of Wales Institute of Science and Technology, King Edward VII Avenue, Cardiff CF1 3NU, U.K.*

A. EMPAIN *Jardin Botanique de l'État, Domaine de Bouchout, B1860 Meise, Belgium*

M. ESTRADA *Instituto de Investigaciones Pesqueras, Paseo Nacional s/n, Barcelona 3, Spain*

G. FRIEDRICH *Landesamt für Wasser und Abfall Nordrhein-Westfalen, Auf dem Draap 25, 4000 Düsseldorf 1, Germany*

G. GONZALEZ *Departament d'Ecologia, Facultat de Biologia, Universitat de Barcelona, Avda. Diagonal 645, Barcelona 28, Spain*

P. B. HEISE *The Gudenå Committee, Teknisk Forvaltning, Århus Amtskommune, Lyseng Allé 1, DK-8270 Højbjerg, Denmark*

H. HEUFF *Department of Botany, University College, Dublin, Ireland (but present address c/o Forest and Wildlife Service, 2 Sidmonton Place, Bray, Co. Wicklow, Ireland)*

D. G. HOLLAND *North West Water Authority, Rivers Division, Liverpool Road, Great Sankey, Warrington WA5 3LP, U.K.*

K. HORKAN *An Foras Forbartha Teoranta, Laboratories, Pottery Road, Kill of the Grange, Dun Laoghaire, Co. Dublin, Ireland*

J. P. C. HARDING *Wanlip Water Reclamation Works, Severn-Trent Water Authority, Soar Division, near Leicester, U.K. (article written during employment with North West Water Authority)*

B. KAWECKA *Laboratory of Water Biology, Polish Academy of Science, Sławkowska 17, Kraków, Poland*

CONTRIBUTORS

P. LAVANDIER *Laboratoire d'Hydrobiologie, Université Paul-Sabatier, 118 Route Narbonne, 31062 Toulouse Cedex, France*

A. LILLEHAMMER *Zoological Museum, University of Oslo, Toyen, Norway*

R. MARCHETTI *Consiglio Nazionale della Ricerche, Istituto di Ricerca sulle Acque, 20047 Brugherio (Milano), Italy; (also at Departimento di Biologia (Sezione di Ecologia), Università degli Studi di Milano, Via Celoria 26, 20133 Milano, Italy)*

D. MÜLLER *Bundesanstalt für Gewässerkunde, Postfach 309, 5400 Koblenz, Germany*

I. PINTÉR *Landesanstalt für Umweltschutz Baden-Württemberg, Institut für Wasser- und Abfallwirtschaft, Postfach 210752, 7500 Karlsruhe 21, Germany*

N. PRAT *Department d'Ecologia, Facultat de Biologia, Universitat de Barcelona, Avda. Diagonal 645, Barcelona 28, Spain*

M. A. PUIG *Department d'Ecologia, Facultat de Biologia, Universitat de Barcelona, Avda. Diagonal 645, Barcelona 28, Spain*

B. SZCZĘSNY *Laboratory of Water Biology, Polish Academy of Science, Sławkowska 17, Kraków, Poland*

O. M. SKULBERG *Norwegian Institute for Water Research, P.O. Box 333, Blindern, Oslo 3, Norway*

M. F. TORT *Department d'Ecologia, Facultat de Biologia, Universitat de Barcelona, Avda. Diagonal 645, Barcelona 28, Spain*

J. N. TOURENQ *Laboratoire d'Hydrobiologie, Université Paul-Sabatier, 118 Route Narbonne, 31062 Toulouse Cedex, France*

G. VAN URK *Rijksinstituut voor Zuivering van Afvalwater, Postbus 17, Maerlant 6, 8200 AA Lelystad, The Netherlands*

B. A. WHITTON *Department of Botany, University of Durham, Co. Durham DH1 3LE, U.K.*

P. F. WILLIAMS *Nature Conservancy Council, 44 The Parade, Roath, Cardiff CF2 3AB, U.K.*

R. WILLIAMS *Department of Applied Biology, University of Wales Institute of Science and Technology, King Edward VII Avenue, Cardiff CF1 3NU, U.K.*

J. F. WRIGHT *Freshwater Biological Association River Laboratory, East Stoke, Wareham, Dorset BH20 6BB, U.K.*

Preface

The aim of this book is to give an overall picture of the ecology of selected rivers in Europe. What are the key features of the environment? What generalizations can be made about their bacteria, plants and animals? What is known about the productivity in different parts of the river? How has man changed the river from its natural state and how is the river managed nowadays? The chapters have been written both to interest the student or researcher with a general interest in river ecology and also to provide more detailed information for those wanting to know about a particular river.

An overview of the chapters

The twenty chapters include examples from twelve countries. The examples range in size from the Rhine to two tiny streams which have been the subject of intensive research and show features of particular interest. Most of the rivers might be described as medium-sized by European standards, running for distances of several hundred kilometres. No doubt they are small by world standards, but this has the advantage that in at least some cases it is possible to generalize about the whole river. Europe's largest river, the Danube, was not included because it received monographic treatment during the 1960s. (R. Liepolt, 1967, *Limnologie der Donau*, Schweizerbart, Stuttgart.)

Chapter 1 deals with the Meuse in Belgium, where the river is subject to a particularly wide variety of pollutants: heavy metals, radionuclides, organic and industrial wastes and thermal pollution. In response to this, researchers at Liège have developed a variety of biologically orientated methods for monitoring these pollutants. The chapter shows how these methods can provide general information of the ecology of the river. The Meuse is discussed again in Chapter 16.

The Danish, Norwegian and Swedish rivers (Gudenå, Chapter 2; Glåma, Chapter 17; Rönnea, Chapter 20) resemble each other in that they have relatively long lowland stretches with many lakes and reservoirs in the catchment which have a profound influence on their ecology. All three rivers are clearly very important for recreational purposes and subject to intensive management.

The inclusion of five rivers from the United Kingdom is a reflection of editorial bias rather than a lack of well-studied rivers elsewhere in Europe. It seemed at one time that the Wye (Chapter 3), which rises in Wales but flows into the Severn in England, would eventually be influenced by the building of a much increased reservoir at an upstream site. At the same time this river has been considered particularly important for conservation purposes. The possible conflict of interests has led to a wide range of studies on the Wye and it is possible to generalize in a way which is difficult for most rivers. The Ebbw—in Wales (Chapter 4)—provides a good example of a river with an upstream clean stretch and a downstream one heavily polluted by coal-mining and the effluent from a major steel-works. It is much more difficult to generalize on the Mersey system (Chapter 5), because of the variety of types of pollution. At one time this system included some of the most notorious stretches of river in Britain, but many have been improved greatly during the past two decades. That we know as much as we do about the ecology of the Mersey system, and the ecological changes which have taken place, is due to the exceptional enthusiasm of the biological staff of the local water authority.

The Tees (Chapter 6) is similar to the Wye in that it is fast-moving, subject to rapid changes in flow at all times of year and relatively unpolluted for the majority of its length. A chemical and biological survey made on this river in the 1930s was, at the time, the most detailed study which had been made on any British river and one of the most detailed for any river in Europe. It is therefore possible to make some comment on long-term changes.

A winterbourne is the name given to a stream rising on the chalk of southern England and which flows for only part of the year. The example chosen (Chapter 7), which is actually known as the Winterbourne Stream, dries out in the upper, but not the lower part of its course each summer. The Estaragne (Chapter 9) is another small stream, which drains a permanent field of consolidated snow in the High Pyrenees. There is again a marked seasonal pattern of flow changes, but in this case minimum flows are in winter, when the remaining water passes under the thick snow cover, which lasts for six to eight months near the top of the stream.

Movement of water in the Lot (Chapter 8), a major tributary of the Garonne in SW France, has been sufficiently influenced by on-stream reservoirs and weirs to permit dense phytoplankton populations to develop when summer flows are low. Although many features of this river still await study, some aspects of chemical and phytoplankton dynamics in the stretch below the junction of the Haut-Lot and the Truyère have been quantified to a much greater extent than on most rivers. The chapter summarizes this work, which includes modelling and behaviour of phytoplankton under different flow conditions.

The Rhine deservedly gets the longest treatment. Chapter 10 takes the river down to the German–Dutch border, while the downstream parts are included in Chapter 16. Some of the oldest accounts of river biology anywhere are for the Rhine and Chapter 10 refers to many older papers, as well as 1983 reports, in order to give a broad picture of changes which have taken place over the past century. Although the middle and lower stretches of the Rhine are still heavily polluted by wastes from the potash mines in Alsace, in general the authors' verdict on the state of pollution in the river is reasonably favourable. Improved conditions since the early 1970s are due largely to control measures, though also slightly to decreased economic activity. The pollution status of the Neckar (Chapter 11), one of the Rhine's main tributaries, has again much improved in recent years, but the condition of the effluents entering the river in the region of Stuttgart is still most unsatisfactory. Problems due to organic pollution on the Neckar are compounded by the heat load added by a series of power stations. The chapter has details of measures which have been adopted to minimize the effects of thermal pollution.

The three Irish rivers include Ireland's longest river, the Shannon (Chapter 12), one of its least polluted rivers, the Caragh (Chapter 13), and one of its rivers most subject to nutrient enrichment, the Suir (Chapter 14). The Caragh has no doubt changed considerably since pre-historic times because of the much reduced tree cover; nevertheless its soft water, low in inorganic nutrients but rich in humic materials, provides an interesting environment which deserves much more detailed study.

The Po (Chapter 15) has the second largest catchment of the rivers in this book. Many of its tributaries derive ultimately from mountain streams, but their influence is partly buffered by the presence of large lakes on the tributaries draining the Alps to the north. Much of the chapter is concerned with the middle and lower parts of the main river.

The Polish river, the Dunajec (Chapter 18), also rises in a high mountain region, but in this case the more upstream parts of the tributaries have received the most study. Its uppermost streams have some resemblance to the Estaragne, though with a much shorter period of snow cover. The river flows through some of the most beautiful scenery of any river described here, but some of the more downstream stretches receive heavy loads of poorly treated organic effluent. Two on-stream reservoirs also have a marked effect on the downstream river.

Chapter 16 (lower Meuse–Rhine) deals with the large river system flowing through the Netherlands. This is complicated not only because of the branching and coming together of different rivers, but also because of the huge changes which man has made to the behaviour of the rivers and the location of their channels. In particular a major dam, which was built to protect the south-western part of the country and completed in 1970, has had profound ecological effects. The chapter includes an account of the unique freshwater tidal area, which was largely destroyed by the dam, and of events since then.

Up to now there has been less research on rivers in Spain than in any other country covered in this book. The team of river ecologists at Barcelona are, however, clearly determined to change this situation and have assembled much information about the Llobregat within just a few years. This river (Chapter 19) arises on the other side of the Pyrenees to the Estaragne. It is a typical Mediterranean river in that flows are very high in spring at the time of snow melt, but low in late summer. The flow regime is somewhat modified by the presence of an on-stream reservoir on the main river and the chemistry of the water is changed downstream by the influence of various salt beds.

An understanding of the ecological processes operating in a river is essential if all the varying demands on the water are to be resolved, yet at the same time the river is to retain something of its original character. Progress can come partly from generalization based on observations of a wide variety of rivers and partly from intensive quantitative studies on a few selected ones. It is hoped that *Ecology of European Rivers* aids in both ways.

Editorial comment

With such a variety of countries and authors, it is not surprising that the original versions of the chapters had a great variety of names and terms for the same thing. I have gone some way towards standardizing these, but it seemed both impractical and unnecessary to bring about complete uniformity. Local names for towns and rivers have mostly been retained, except where the English equivalent is especially familiar, as with the Rhine. Terms and units for most physical and chemical variables are the same throughout the book. Among the exceptions are 'flow' and 'discharge', words which are in most cases used interchangeably. Concentrations of ions such as nitrate and phosphate are almost always expressed as the element.

As far as possible the same Latin binomial has always been used for a particular species, though synonyms are sometimes added in brackets. No doubt I have overlooked some examples of synonymy, especially among the invertebrates. In the case of common flowering plants and fish, authors differ as to whether they use English, Latin or both names. A list of equivalents (Appendix B) should prevent any confusion.

It will be clear from the references how much information is available only in local journals and reports. An English translation of the title of a book or paper is added after each reference in Danish, Dutch, Italian, Norwegian, Polish, Spanish and Swedish, but not French or German. As this is based on the assumption that the last two languages are known more widely than the others, I hope it gives no offence to readers in other countries.

I thank all the contributors for their patience with the editor and his editing. Because of a delay in processing chapters which were submitted first, several chapters had to be modified considerably to bring them up-to-date at proof stage. The authors of Chapter 4 have now published a book on the Wye (*The Ecology of the Wye*, 1982, Junk), which gives a much fuller account of this river. I am also most grateful to the many people who have aided with advice, to Audrey Richardson for typing various early versions of chapters, to Margaret Creighton for typing the camera-ready version, with all the difficulties due to accents in the various languages, and to Penny Baker of Blackwell Scientific Publications for dealing with a range of practical problems.

Brian Whitton

1 · Meuse

J-P. DESCY & A. EMPAIN

1.1   INTRODUCTION

This chapter deals with the 38.7% of the Meuse catchment which lies
within Belgium, while the more downstream part of the river is discussed
in Chapter 16.   The river basin is of considerable importance for the
south of Belgium, providing drinking water and also water for industry
and power stations; the river itself is also important for navigation
and, together with its tributaries, for recreation.   Although a wide
range of ecological studies have been made on the Meuse basin, few deal
in depth with fundamental ecological questions.   Nevertheless ecologists
have contributed by focusing their attention to the use of biological
methods for assessing the numerous pollutants which reach the rivers and
it is this aspect which the present chapters treat in most detail.   Part
of the work has come from our Laboratory of Hydrobiology at Liège
University; for the remainder we have included ecological results with a
special interest at the scale of the whole basin.

Fig. 1.1 Meuse at Liège
(200 m wide) at confluence
with R.Ourthe (right).   The
main industrial zone is
located upstream of this site.

An indication of the importance of the Meuse basin may be provided
by some simple statistics.   It supplies most of the drinking water for
the southern part of the country and also Bruxelles and the great cities
of the north (Gent, Antwerpen); 80% of the water is collected underground and 20% by surface water treatment.   In addition artificial lakes
provide a total storage capacity of 58.3×10^6 m^3 per year.   Surface
waters are also needed for industry (455×10^6 m^3 per year), agriculture
(235×10^6 m^3 per year) and navigation (1800×10^6 per year)
(Vereerstraeten, 1971).   As the supply of underground water decreases,
the use of surface waters will rise, so the preservation of river water
quality is going to become more and more essential.

Fig. 1.2 (opposite)   Meuse and its tributaries: A) whole basin;
Belgian part lies in shaded area ; B) schematic geological features;
C) Belgian part of basin, showing river system and main reservoirs.

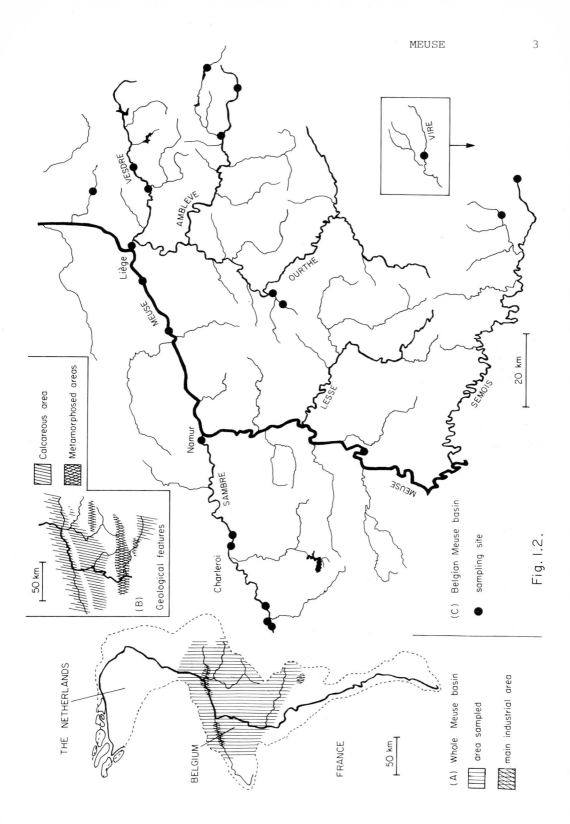

Fig. 1.2.

(A) Whole Meuse basin

▨ area sampled

▨ main industrial area

(B) Geological features

▨ Calcareous area

▨ Metamorphosed areas

(C) Belgian Meuse basin

● sampling site

1.2   GENERAL DATA

DIMENSIONS

Area of catchment   13489 km^2 in Belgium (38.7% of whole basin)

CLIMATE

Precipitation   800-1400 mm yr^{-1} (over 1000 mm yr^{-1} in the Ardenne)

Temperature (mean annual)   7-9 $^\circ$C

GEOLOGY (see Fig. 1.2)

schistose and siliceous rocks in the east and south (Lower Devonian and Cambro-Silurian)
limestone and calcareous rocks in remainder (middle and upper Devonian, Carboniferous and secondary rocks)

LAND USE

Ardennes   forestry, agriculture, livestock
Other rural areas   agriculture, livestock
Industrial development
   iron-steel industry   mostly around Charleroi and Liège
   textile industries   Verviers
   power stations   include two nuclear plants, one in France, near the border (Chooz, 270 MWe) and one upstream of Liège (Tihange I, 879 MWe)

POPULATION

Ardennes   10-50 people km^{-2} Typical towns   3000 people km^{-2}
Other rural   Large cities along the   25000 people km^{-2}
   areas   50-300 people km^{-2} "iron-coal" axis

PHYSICAL AND CHEMICAL FEATURES OF RIVER

(Figures for lengths given in brackets refer to the whole river; those for sources in brackets refer to rivers which originate in France. Data for precipitation and flow are based on 1954-1963.

	length (km)	source (m)	mean slope	mean rainfall	flow module	flow (m^3 s^{-1}) minimum monthly average	maximum monthly average
Meuse	183 (890)	(402)	0.25	898.8	244	70	436
Semois	177	413	1.54	1144.6	29	17.5	40
Lesse	83	405	3.7	947.7	24.3	11.6	39.9
Sambre	86 (190)	(212)	0.25	832.8	27.5	7.6	59.7
Ourthe	175	507	2.5	976.8	61	10.3	32

Ranges for physico-chemical variables in the Meuse basin

variable	unit	mean value for 'natural' stations	'natural range'	extremes	
t	°C	-	0 - 23	32	Sambre et Marcinelle and Couillet
pH	-	6.9	3.4 - 9.1	10.0	Warche et Bullange
O_2	% saturation	98.2	74 - 130	min: 0% / max:140%	Sambre et Marcinelle / Marchette at Rabosée
conductivity	$\mu S\ cm^{-1}$ at 25°C	216	11 - 674	3375	Semois at Bohan / Sambre at Couillet
alkalinity	$mg\ l^{-1}$ $CaCO_3$	63	0.0 - 280	383	Marchette at Rabosée
total organic C	$mg\ l^{-1}$	3.0	<0.2 - 13.1	176	Sambre at Couillet
Na	"	6.5	1.0 - 18.0	288	Sambre at Namur
K	"	1.7	<0.1 - 9.2	28.4	Vesdre at Pepinster
Mg	"	4.7	0.7 - 17.4	23.7	Berwinne at Val Dieu
Ca	"	28.6	1.1 - 112	423	Sambre at Couillet
ammonia-N	"	0.134	<0.002 - 0.660	21.9	Semois at Arlon
NO_2-N	"	0.024	<0.002 - 0.214	9.61	Sambre at Marcinelle
NO_3-N	"	2.2	<0.1 - 6.3	10.1	Vire at Latour
total reactive orthophosphate-P	"	0.128	<0.005 - 0.6	16.6	Marchette at Rabosée
reactive-Si	"	2.3	0.5 - 8.4	24	Berwinne at Val Dieu
Cl	"	14.5	2.3 - 43.8	840	Sambre at Couillet
SO_4-S	"	6.7	1.0 - 20.7	119	Berwinne at Val Dieu

The values for 'natural' stations are a selection of 531 results from 1040 (123 stations from 201): all obviously polluted stations are discarded from the pool, but 'natural' reference in Belgium is utopian. The 'natural' range includes the lowest value found, while the upper limit is defined by the largest type value (mean plus two sigma); due to the difficulty of defining true natural conditions, this upper side is often high (see ammonia, nitrite and phosphate).

1.3   PHYSICO-CHEMICAL FEATURES OF THE RIVERS

1.3   NATURAL TYPOLOGY

Geology is a key factor determining the ecological features of running
waters in the Belgian part of the Meuse basin. Differences in litho-
logy lead to a great range in the chemical composition of the waters
and several attempts have been made to make a chemical classification
helpful in interpreting the ecology of rivers in the region.

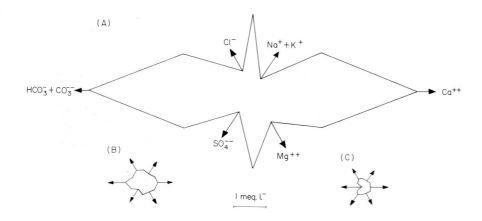

Fig. 1.3 Major elements dissolved in water, according to Symoens
(1957): A, alkaline type;   B, neutral type;   C, acid type.

Symoens (1957) proposed a classification based on the mineral content of the water (Fig. 1.3). As his terminology was derived from
local names, it is replaced here with acid, neutral and alkaline.

  Acid type ('type fagnard') acid water (pH 4 to 6, or lower) with
      low mineral content, especially HCO_3^-; usually rise in Ardennes
      peat-bogs and flow over Cambrian rocks; mountain and boreal
      algae present in flora (Fabri & Leclercq, 1979).
  Neutral type ('type ardennais') pH circumneutral and with a
      slightly higher mineral content than the acid type of water;
      usually flow over lower Devonian or Cambro-Silurian rocks.
  Alkaline type ('type condrusien') pH over 7.5 in natural
      conditions; conductivity in range 250 to 600 µS cm^{-1} at 25 °C
      and mineral   content dominated by Ca^{2+} and HCO_3^-; flow over
      rocks rich in calcium carbonate.

Fig. 1.4 (opposite) Alkalinity in rivers of Belgian Meuse Basin.
(Mean for 1973-1980).

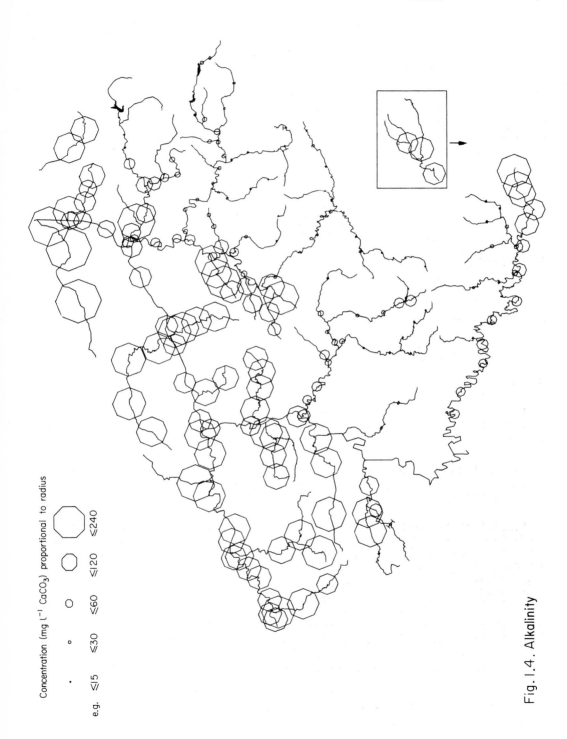

Concentration (mg l^{-1} CaCO$_3$) proportional to radius

e.g.   ≤15   ≤30   ≤60   ≤120   ≤240

Fig. I.4. Alkalinity

In a synthesis we made for a general survey of the Meuse basin (Descy & Empain, 1981), we attempted to redefine this chemical typology statistically. A comparison with the above types is given in Table 1.1.

Table 1.1 Types of water in the Meuse basin. (Alkalinity range is equal to mean ± 2 standard deviations, including about 95% of the natural variation in each type).

General type (Descy & Empain, 1981)	Symoens (1957) type	Alkalinity (mg l^{-1} $CaCO_3$)
1	acid (fagnard)	0 – 5
2	neutral (ardennais)	6 – 20
3		16 – 50
4	alkaline (condrusien)	51 – 110
5		130 – 265

Many rivers belong successively to different types, as they rise in the Ardennes, often in peat-bogs, and then flow over Devonian or Carboniferous limestone; such rivers rarely reach type 5, however; these latter rise on limestone or chalk (Fig. 1.4).

This typology is important when considering the applied ecology of the Meuse basin. It provides a guide to natural productivity at the different trophic levels and the composition and structure of the plant communities (see below); it is also very useful when assessing pollution, since the ecological effects of most pollutants appear to depend on the natural type of the river concerned.

1.32 POLLUTION

1.321 Mineral pollution

Pollution leading to marked changes in the composition of major elements occurs in the Meuse basin, mainly in industrial zones. The impact of such pollution depends on the natural chemistry of the water-course. For example, Na pollution at a particular level might be acceptable for a calcareous river, but three times too high for a stream of type 1. There are also many sites where pollution by heavy metals takes place. The majority occur in the 'iron-coal' axis, along the rivers Sambre and Meuse; others are associated with old mine operations or ore processing (mainly Gueule and Vesdre) or with particular industries like tanneries (Warche, Amblève, Viroin).

Fig. 1.5 (opposite)   Reactive phosphate (as P) in Belgian Meuse basin.   (Mean for 1973-1980).

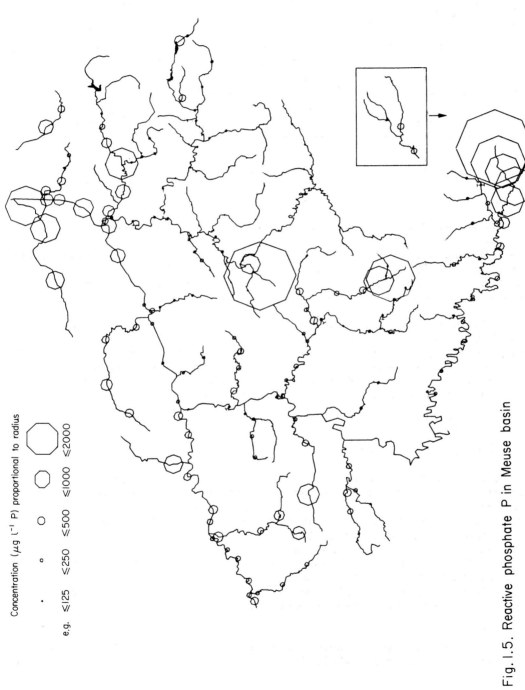

Concentration (μg l^{-1} P) proportional to radius

e.g. $\leqslant$125 $\leqslant$250 $\leqslant$500 $\leqslant$1000 $\leqslant$2000

Fig. I.5. Reactive phosphate P in Meuse basin

Particularly heavy contamination with Cu occurs in the upper Sambre
and with chromium in the Warche; (Mouvet, 1981); Zn, Cd and Pb
pollution are associated with industrial areas and old mining areas.
Waters are also enriched naturally by heavy metals in some areas;
typically these are associated with metamorphism and old geological
faults.

1.322 Organic pollution

Heavy organic pollution occurs mainly where the population density is
high and, in the large rivers, is often combined with industrial
pollution. In some other instances, extreme situations may be
encountered where the sewage flow is much higher than that of the
river itself; two of the most striking examples are the Semois below
Arlon, which is seriously polluted for about 30 km and the Magne,
which receives slaughter-house effluent near the source of the river.
The effects of these effluents are evident in the reduced levels of
dissolved oxygen in many stretches. Quite apart from these situations,
most of the rivers are subject to marked eutrophication (Fig. 1.5), the
only exceptions being the highest points of the basin, where the density
of the population is low. It will be clear that treatment of wastes is
poorly developed in Belgium, which is probably one of the most backward
countries in Europe in this respect; for instance, only about 6% of the
sewage is treated at present, as opposed to 80% in our nearest neigh-
bour, the Grand-Duchy of Luxembourg.

1.323 Thermal pollution

Thermal pollution from the cooling systems of factories and power
stations occurs at some sites in addition to other types of pollution.
The amount of heat discharged is sometimes sufficient to bring about a
marked rise in temperature. For instance, a study carried out in
1973-74 in the Sambre below Charleroi showed that the mean temperature
rise over a stretch of a few kilometres was 9.3°C, with a maximum
recorded of 32°C (Descy & Empain, 1976). At the Tihange I power plant
on the Meuse, the heat discharge typically brings about a 2 to 3°C
increase when mixing is complete (observed maximum 4.2°C). This in
itself is unlikely to be critical for many processes, but the total
rise from a variety of sources may reach 10°C at Liège. As these
conditions occur with low flow (<50 m^3 s^{-1}) and consequently a high
organic content in the water, the oxygen levels may drop below 50%
saturation. Mathematical models have been constructed (Smitz, 1976)
to predict the thermal behaviour of the Meuse in relation to river flow.
Unfortunately the results indicate that temperature rises resulting from
pollution cannot be reduced sharply by using reservoir water to regulate
flows at critical periods. A system of monitoring has been developed
to assess the ecological impact of the Tihange I power station; this
incorporates most of the techniques mentioned in the present chapter
(Lambinon *et al.*, 1977; Kirchmann *et al.*, 1980). A similar programme

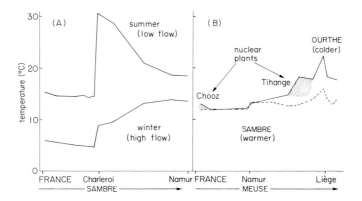

Fig. 1.6   Thermal profiles in Sambre and Meuse.
A)   Summer and winter profiles. Summer has high warming, but good
     atmospherical cooling downstream due to low flow-rate.
B)   Typical profiles before and after commencement of nuclear plant
     at Tihange.

should be put into action now in the Upper Meuse in advance of further
nuclear power stations in order to provide ecological base-lines for
future comparison ('point zéro écologique').

Problems associated with thermal pollution are also likely to arise
with the development of nuclear power stations at Chooz, just upstream
of the French-Belgian border. Here a model of possible thermal changes
in the water has shown that it would be better to control the heat
discharge by using and releasing water at several (minimum of two) sites
along the main river. Such a programme would however require the
construction of new dams and these themselves would undoubtedly cause
other types of damage to the environment.

1.324 Radioactive pollution

Two sites on the Meuse have been the subject of special checks on
radiocontamination of the aquatic ecosystem; these are the Belgian
stretch downstream of the Chooz nuclear plant (270 MWe) and the stretch
below the Tihange I nuclear plant (870 MWe). Most of the ecological
monitoring is based on the measurement of activity in aquatic organisms,
particularly bryophytes (1.42) and fish (1.43). The level of contamination is low under normal conditions, with activities in water and
suspended matter often below the limits of detection; some radio-
elements have however been found to reach detectable levels in various
organisms and in sediments (Kirchmann *et al*., 1980). Experimental
studies included in this programme deal with uptake and decontamination
kinetics, e.g. uptake of ^{3}H by *Scenedesmus obliquus* in culture.

Table 1.2   Radionuclides in water, bryophytes and fish, downstream of the nuclear plant of Chooz (Kirchmann *et al.*, 1979); (pCi 1^{-1}, pCi kg^{-1} radionuclide wet weight; fish are eviscerated).

		1971	1974	1977
^{54}Mn	Water	6.52	0.15	0.03
	Bryophytes	10^4-10^5	10^3-10^4	10^3-10^4
	Fish	10^2-10^3	$<10^2-10^3$	$<10^1-10^2$
^{60}Co	Water	2.19	0.16	0.05
	Bryophytes	10^4-10^5	10^3-10^4	$<10^2-10^4$
	Fish	10^2-10^3	$<10^1-10^2$	$<10^2$
^{134}Cs	Water	1.63	1.31	0.10
	Bryophytes	10^3-10^4	10^2-10^4	10^2-10^3
	Fish	10^3-10^4	10^2-10^3	10^1-10^2
^{137}Cs	Water	0.27	1.56	0.14
	Bryophytes	10^3-10^4	10^2-10^3	10^1-10^3
	Fish	10^2-10^3	10^2-10^3	10^1-10^2

1.4   ECOLOGY OF AQUATIC COMMUNITIES

1.41   ALGAE

The benthos has been studied in more detail than the plankton. A complete study of the Belgian Meuse, Descy (1973, 1976) included both macroscopic and microscopic species. Typically there is a *Cladophora glomerata* community, accompanied under particular conditions by other species such as *Lemanea fluviatilis*, *Vaucheria* sp., *Bangia atropurpurea* and blue-green algae.   If the basin as a whole is considered then the communities dominated by particular macro-algae can be seen to differ according to the natural water quality, as shown by Symoens (1957), who noted an ecological succession of *Microspora, Vaucheria* and *Cladophora* communities. A more precise approach is now required, with statistical and multivariate analysis of quantitative samples taken from river sites over the whole basin. Such an approach has already been adopted since 1976 for diatom communities (Descy, 1979, 1980). This has led to a typology of communities closely related to the chemical types of the rivers. The five 'biotypes', or ecological groups formed by dominant species, are also close to the associations described by Symoens (1957), but the use of quantitative data and data processing by factor analysis enabled the results to be analysed objectively.

Very few data are available on biomass and productivity of the
benthic algae. Local measurements in the Meuse have indicated a
periphyton biomass ranging from 5 to 20 µg cm^{-2} on artificial substrates. The biomass measured in an Ardennes stream was at least five
times smaller (Fabri, 1977).

There have been no comprehensive studies of the phytoplankton of the
Meuse basin, although a number of papers do give records for the Meuse
(Pierre, 1974, for the upper and middle Meuse in France; Symoens, 1957
and Jansen, 1967, for the Belgian part). The French and Belgian studies
of the Meuse both showed a similar flora, with the following dominants:
*Cyclotella meneghiniana, Fragilaria construens, Melosira varians,
M. italica, Stephanodiscus hantzschii, Scenedesmus quadricauda.* Well
developed plankton communities probably only occur in the large
canalized stretches of the Meuse and Sambre. Measurements on phytoplankton biomass upstream of Liège have showed chlorophyll contents
ranging from 3 to 40 mg chlorophyll a m^{-3}, with a maximum in summer
(Declercq-Versele & Kirchmann, 1982).

1.42 BRYOPHYTES

Bryophyte communities occur mainly on rocks and concrete surfaces in
fast flowing water, such as rapids and natural waterfalls or, in
downstream stretches of the Sambre and Meuse, the artificial falls
associated with fluvial traffic movement. The biomass involved varies
greatly, ranging from well-defined strips limited to a few hundred
metres downstream of artificial falls to a relatively continuous cover
in rapids where the depth is not greater than 0.5 m and where there is
no disturbance of the bed by trampling. The populations of the main
gated rivers, the Meuse and Sambre, are dominated by the genus
Cinclidotus, mainly *C. nigricans, C. fontinaloides* and *C. danubicus*,
with sometimes also *C. aquaticus* and *Dialytrichia mucronata* (= *C.
mucronatus*).

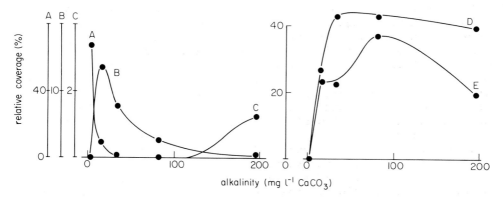

Fig. 1.7 Relationship between cover of bryophytes and alkalinity:
A, *Scapania undulata*; B, *Chiloscyphus polyanthus*; C, *Cinclidotus
danubicus*; D, *Platyhypnidium riparioides*; E, *Fontinalis antipyretica.*

The ecology of individual species has been studied in some detail
(Empain, 1974, 1978; Empain *et al.*, 1980), with data on the influence of
frequency of immersion, resistance to desiccation and solar heating,
photosynthesis and resistance to pollution. Species distribution
within the basin shows a picture closely correlated with alkalinity
(Fig. 1.7). Some species, such as *Scapania undulata, Chiloscyphus
polyanthus* and *Cinclidotus danubicus* are limited to a narrow range of
alkalinity concentrations; others like *Platyhypnidium riparioides
(= Rhynchostegium riparioides)* and *Fontinalis antipyretica* are found
over a wide alkalinity range.

1.43   FISH

According to the system proposed by Huet (1949), the changes in fish
communities on passing downstream depends under natural conditions
mainly on physical factors such as altitude, slope, width and water
temperature. In the Meuse basin, the most frequent communities belong
to the grayling and barbel zones, the trout zone being limited mostly
to upland streams and stretches of the main rivers. The Meuse itself
has communities dominated by still water cyprinids, but theoretically
it belongs to the lower barbel zone. In unpolluted stretches, species
diversity is greatest in the barbel zone (up to 23 species) and least
in the trout zone (3-8 species); the more acid streams provide an
extreme example, with no fish present at all.

Measurements of fish biomass have been made by Huet and Timmermans
(1976) and Philippart (1979, 1980). In natural conditions the biomass
may reach 500 - 700 kg ha^{-1} in small rivers (e.g. upper Amblève,
520 kg ha^{-1}; Eau d'Heure in the Ourthe basin, 670 kg ha^{-1}; lower
Méhaigne, 470 kg ha^{-1}). In the main rivers the mean biomass is lower,
with values of 250 - 350 kg ha^{-1} being recorded.

A very detailed description of the features of walloon rivers has
been made in a guide for anglers in the south of Belgium (Balzat &
Dussart, 1978); even if the data are not always very precise, this book
gives much useful information on not only the fish fauna, but also
geography, geology, chemistry, vegetation and invertebrates. Philippart
(1979) gives an account of the influence on population dynamics of
ecological and sociological factors and the economics of sport fishery;
it also describes the relationship between the growth rate of fish
species and physico-chemical factors.

1.5   WATER QUALITY ASSESSMENT BY COMMUNITIES

1.51   USE OF DIATOMS

A method for estimating water quality has been developed (Descy, 1979,
1980) which is of general applicability and which makes use of diatoms
and data about their response to pollution. It involves a comparison
of communities found at a site with the 'normal' communities that might

be expected in the 'natural' type of the particular watercourse. Briefly this 'diatom index' is based on quantitative data obtained by counting (relative abundance of species in sample) and takes into account the sensitivity to pollution of about 100 species, as well as their 'indicator value' related to their ecological amplitude towards pollution.

The computed diatom index ranges from 1 for heavily polluted sites to 5 for clean water; in practice six classes of index values have been adopted for interpretative purposes (Fig. 1.8). The method has been applied successfully in very diverse situations within the Meuse basin; it usually shows a very close relationship with chemical analysis and has led in some cases to the detection of discontinuous pollution. Use of the method for the Meuse basin survey led to the following conclusions:

5% sampling sites are unpolluted; however 31% may be considered as presenting a good water quality;

44% are subjected to moderate eutrophication, mostly in the rivers with lower mineral content (types 1,2,3) which are more sensitive to pollution;

9% show heavy eutrophication of medium pollution;

16% present heavy pollution.

1.52   USE OF BRYOPHYTES

A somewhat similar approach to that with diatoms has been developed for bryophytes. The first attempts (Empain 1973, 1978) were devoted mainly to alkaline waters (type 5), where a high correlation was found with the diatom index; the occasional differences may be explained by the very different time taken for recovery - weeks for diatoms, years for bryophytes. More recently a generalised approach (Descy & Empain, 1981) has been developed which takes account of the natural type of the water. The key steps needed to compute the generalized bryophyte index are the evaluation of sensitivity coefficient for each species and the estimation of mean value of 'crude bryophyte index' for each type of water. When pollution is too high, no bryophytes are found, such as parts of the Sambre and Semois, and the crude bryophyte index equals 0. The final steps are the calculation of crude bryophyte index for a particular station and the generalization of the result by taking account of the mean value of the water type of the sampling site. The generalized bryophyte index gives results similar to those from the diatom index. Among the 30 sites selected by the former as strongly polluted, 27 were selected by the bryophyte index; the 3 discrepancies are easily explained by local sampling constraints. Some other sites were selected by the bryophyte index because of the discontinuous nature of the pollution, such as rivers polluted by sugar-beet waste (Méhaigne and Geer) or toxic wastes (Hoyoux). The main advantage of the bryophyte index lies in the long-term integration of pollution events, whereas the diatom index reflects more short-term events; it is also more sensitive to eutrophication.

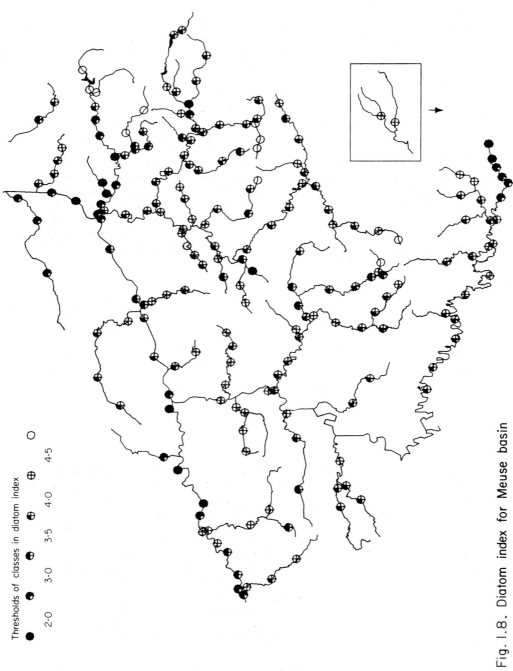

Thresholds of classes in diatom index

2·0   3·0   3·5   4·0   4·5

Fig. I.8. Diatom index for Meuse basin

1.53   USE OF MACROINVERTEBRATES

A survey of Belgian rivers based on the use of macroinvertebrates has
recently been published by the Ministry of Public Health (I.H.E., 1979),
following collaborative work between various institutes and laboratories.
The 'biotic index' method of Verneaux and Tuffery (1967) was employed,
which is itself derived from the Trent River Board index (Woodiwiss,
1964). Unfortunately the results of this survey appear to give an
unrealistic picture of water quality in the Meuse basin. As far as the
sites common with our own studies are concerned, there are some marked
differences. 47% of the sites in the I.H.E. survey were recorded as
having very good water quality (biotic index $\geqslant$ 9) versus 31% for the
diatom index. In contrast, 11% of sites on the biotic index were
heavily polluted versus 19% on the diatom index. However the total
number of sites of acceptable water quality were similar (73% v 75%).
We conclude that the differences lie mainly in the sampling procedure
used for macroinvertebrates, like the restriction to one or two periods
for sampling (autumn 1978, spring 1979), localization of the reaches
studied and difficulties in deep gates stretches. Hawkes (1979) has
pointed out limitations in the use of macroinvertebrates, such as those
resulting from seasonality and how this should be taken into account
when planning sampling programmes.

Our conclusion is that the I.H.E. pollution survey underestimates
some examples of heavy pollution, such as the Meuse between Liège and
Visé and the Vesdre downstream of Verviers. On the other hand it
overestimates water quality in the less polluted parts of the catchment,
such as the Ourthe and Lesse. This emphasizes the risks involved in
using biological methods in monitoring programmes without sufficient
care. Biological and physico-chemical methods are complementary, as
they provide different sorts of information; both are needed for a full
assessment.

1.54   USE OF FISH

Many studies have been made by the Fish Research Unit of Liège University on the state of fish populations in the Meuse basin. The programme
which started with a survey of the Ourthe basin (Philippart, 1979) is
now extended to the whole Meuse catchment. A key aim is to provide
reference data for long-term comparison, but some estimates of change
are already possible; these have been made by comparison either with
any available earlier data or with reference stretches belonging to the
same type. In the Ourthe basin, for example, the loss of biomass has
been estimated as 38.6 tonnes, mostly affecting *Salmo trutta, Chondrostoma nasus* and *Thymallus thymallus*. In some cases (Fig. 1.9) the
change in biomass is minor, but the community structure is affected
greatly by the increase of tolerant species *(Leuciscus cephalus, L.
leuciscus, Gobio gobio, Phoxinus phoxinus)* without recovery of the

Fig. 1.8 (opposite) Diatom index for Belgian Meuse basin, 1973-1980
(Descy 1979, 1980).

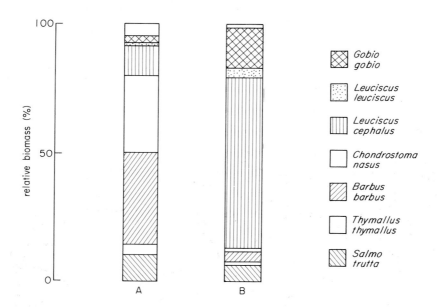

Fig. 1.9 Example of changes in a fish community following recovery after pollution: A, normal structure in the upper part of the gray-ling zone ; B, unbalanced structure in the Amblève ten years after heavy pollution.

normal population. This type of study can provide information on long-term changes long after the causes have disappeared. Moreoever fish are sensitive to alterations not often apparent with the other, monitoring programmes, such as the destruction of biotopes by damming, canal-ization, dredging and water abstraction.

1.6   USE OF BRYOPHYTES TO MONITOR HEAVY METALS AND RADIONUCLIDES

As has already been mentioned, many sites on rivers in the Meuse basin are influenced by heavy metal pollution. The accumulation of these metals in bryophytes has been monitored in the Meuse and Sambre since 1973 and in other rivers since 1977. A semi-popular account of the results has been given by Descy *et al.* (1981). For monitoring purposes, the main advantages are the ease of sampling, the huge increase in levels to be analysed (10^3 to 10^5 times the mean concentration in the water) and the integration of large fluctuations in the water. Metal concentrations in the bryophytes are sometimes very high. For instance chromium has reached as much as 13 mg g^{-1} in the Warche and Amblève downstream of a tannery (Mouvet, 1981). Reference to water type and

Fig. 1.10 (opposite)   Zinc in bryophytes (mg kg^{-1} dry weight, mean concentration) in Belgian Meuse basin, 1973-1980.

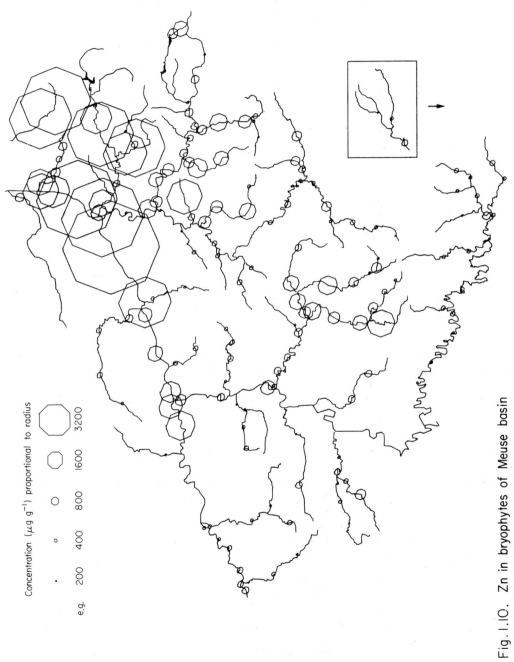

Concentration (μg g^{-1}) proportional to radius

e.g.   200   400   800   1600   3200

Fig. 1.10.   Zn in bryophytes of Meuse basin

the natural range of concentration makes it easy to discriminate between natural enrichment in particular areas and industrial discharges. Copper is found at particularly high concentrations in mosses of the upper Belgian Sambre, downstream of industry producing electrical wires. The mean values in the mosses vary between 5 and 20 times the natural limits, whereas the mean values for the water fit the natural range for this type of water (max. 40 µg l^{-1}); peak values in the water reach 2 to 6 times the natural upper limit, whereas those in the mosses 7 to 25 times their natural upper limit.

Bryophytes have also played an important role in monitoring radionuclides, since they are the most contaminated element of the river ecosystem (Fig. 1.11) (Kirchmann & Lambinon, 1973). They have been used regularly downstream of the Tihange nuclear plant (Kirchmann *et al.*, 1979) and, together with other organisms, every three years on the upper Meuse (Lambinon *et al.*, 1976, 1977).

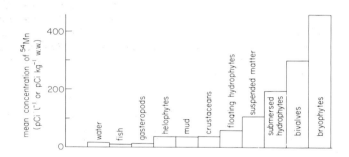

Fig. 1.11 ^{54}Mn in Meuse: mean concentration in water, mud and aquatic organisms downstream of the nuclear plant of Chooz (France, Ardennes).

1.7 DISCUSSION

Much of the knowledge of the ecology of the Meuse basin has been built up from the detailed studies of particular types of organism, often associated with programmes to assess water quality and pollution. There are obvious gaps, such as the lack of detailed research on bacteria, macrophytes and plankton. Among the physico-chemical studies there are very few systematic results for some elements and also pesticides and certain other compounds. We hope that this chapter has made clear the need for integrated studies involving both abiotic parameters and community studies. Reference to natural conditions is important for the assessment of water quality and the ecological impact of pollutants. A general approach involving all the important factors is complicated but needed if ecological information is to be used as an aid to optimal river management, which involves drinking water supply, the needs of industry, navigation, agriculture and angling and, finally, the maintenance of ecological equilibrium.

REFERENCES

Balzat N. H. & Dussart A. (1978) *Guide de la pêche en Ardenne,*384 pp. Duculot, Paris - Gembloux.

Descy J.-P. (1973) La végétation algale benthique de la Meuse belge et ses relations avec la pollution des eaux. *Lejeunia* 66, 62 pp.

Descy J.-P. (1976) Etude quantitative du peuplement algal benthique en vue de l'établissement d'une méthodologie d'estimation biologique de la qualité des eaux courantes : Application au cours belge de la Meuse et de la Sambre. In: *Recherche et Technique au service de l'Environnement.* 382 pp., pp. 159-206. Debedoc, Liège.

Descy J.-P. (1979) A new approach to water quality estimation using diatoms. *Beih. Nova Hedw.* 64, 305-23.

Descy J.-P. (1980) Utilisation des algues benthiques comme indicateurs biologiques de la qualité des eaux courantes. In: Pesson P., *La pollution des eaux continentales. Incidence sur les biocénoses aquatiques,* 2me Edn. 345 pp., pp. 169-94. Gauthier-Villars, Paris.

Descy J.-P. & Empain A. (1976) Analyse physico-chimique de la Meuse et de la Sambre belges, en 1973 et 1974. In : *Recherche et Technique au service de l'Environnement.* 382 pp., pp. 139-57.

Descy J.-P. & Empain A. (1981) *Inventaire de la qualité des eaux courantes en Wallonie (Bassin Wallon de la Meuse).* Rapport de synthèse, 1, 87 pp., 2, 194 pp., 3, 37 pp. Université de Liège (limited circulation).

Descy J.-P, Empain A. & Lambinon J. (1981) *La Qualite des Eaux Courantes en Wallonie - Bassin de la Meuse.* 18 pp. Secrétariat d'etat à l'environnement, a l'aménagement du territoire et à l'eau pour la Wallonie, Square de Meeus, 35, B-1040 Bruxelles.

Empain A. (1973) La végétation bryophytique aquatique et subaquatique de la Sambre belge, son déterminisme écologique et ses relations avec la pollution des eaux. *Lejeunia* 63, 58 pp.

Empain A. (1974) Relations quantitatives entre les bryophytes de la Sambre belge et leur fréquence d'émersion : distribution verticale et influence de la pollution. *Bull. Soc. r. Bot. Belg.* 107, 361-74.

Empain A. (1976) Les bryophytes aquatiques utilisés comme traceurs de la contamination en métaux lourds des eaux douces. *Mém. Soc. r. Bot. Belg.* 2, 201-16..

Empain A. (1978) Relations quantitatives entre les populations de bryophytes aquatiques et la pollution des eaux courantes. Définition d'un indice de qualité des eaux. *Hydrobiologia* 60, 49-74.

Empain A., Lambinon J., Mouvet C. & Kirchmann R. (1980) Utilisation des bryophytes aquatiques et subaquatiques comme indicateurs biologiques de la qualité des eaux courantes. In: *La pollution des eaux continentales. Incidence sur les biocénoses aquatiques,* 2me Edn. 345 pp., pp, 195-223. Gauthier Villars, Paris.

Fabri R. (1977) Végétation, production primaire et caractéristiques physico-chimiques d'une rivière de haute Ardenne (Belgique): la Warche superiéure. *Lejeunia* 87, 43 pp.

Fabri R. & Leclercq L. (1979) Flore et végétation algale des ruisseaux de la Réserve naturelle domaniale des Hautes Fagnes (province de Liège, Belgique). Serv. Conserv. Nat., trav. n° 11, 48 pp.

Hawkes H. A. (1979) Invertebrates as indicators of river water quality.
    In: James A. & Evison L. (Eds) *Biological Indicators of Water
    Quality*. pp. 2.1 - 2.45. John Wiley & Sons, Chichester.
Huet M. (1949) Aperçu des relations entre la pente et les populations
    piscicoles des eaux courantes. *Schweiz. Z. Hydrol.* 11, 332-51.
Huet M. & Timmermans J. A. (1976) Influence sur les populations de
    poissons des aménagements hydrauliques de petits cours d'eau assez
    rapides. *Trav. Stat. Rech. Eaux et Forêts Groenendael*, D, 46, 27 pp.
I. H. E. (1979) Carte de la qualité biologique des cours d'eau en
    Belgique. 61 pp. + 1 map, Institut d'Hygiène et d'Epidémiologie,
    Bruxelles.
Jansen P. (1967) Een onderzoek naar de kwaliteit van het Maaswater in
    Frankrijk en België. *Nat. Hist. Maandbl.* 56, 74-84.
Kirchmann R. & Lambinon J. (1973) Bioindicateurs végétaux de la contam-
    ination d'un cours d'eau par des effluents d'une Centrale Nucléaire à
    eau pressurisée. Evaluation des rejets de la Centrale de la SENA
    (Chooz, Ardennes francaises) au moyen des végétaux aquatiques et
    ripicoles de la Meuse. *Bull. Soc. r. Bot. Belg.* 106, 187-201.
Kirchmann R., Lambinon J., Empain A. & Bonnijns-Van Gelder E. (1980)
    Choix et performances de bioindicateurs pour la surveillance des sites
    d'implantation d'installations nucléaires. In: *IIe Symposium Inter-
    national de Radioécologie*. 751 pp., pp. 557-92. CEA - EDF, Cadarache.
Lambinon J., Kirchmann R., Colard J. (et coll.) (1976) Evolution récente
    de la contamination radioactive des écosystèmes aquatique et ripicole
    de la Meuse par les effluents de la Centrale Nucléaire de la SENA
    (Chooz, Ardennes françaises). *Mém. Soc. r. Bot. Belg.* 7, 157-75.
Lambinon J., Descy J.-P., Empain A., Kirchmann R. & Bonnijns-Van Gelder E.
    (1977) La surveillance écologique des sites d'implantation des centrales
    nucléaires. Effets des rejets d'effluents sur l'écosystème mosan: acquis
    et perspectives. *Ann. Ass. Belge Radiopro.* 2, 201-16.
Mouvet C. (1981) Pollution de l'Amblève par les métaux lourds, en par-
    ticulier le chrome : dosage dans les eaux et les bryophytes aquatiques.
    Trib. Cebedeau 445, 527-38.
Philippart J.-C. (1979) Evaluation des ressources piscicoles et hali-
    eutiques dans les rivières du bassin de la Meuse. Définition d'une
    méthode d'étude. In: Calembert L. (Ed.) *Problématique et Gestion des
    Eaux Intérieures*. 967 pp., pp. 481-495. Derouaux, Liège.
Philippart J.-C. (1980) Incidences de la pollution organique et de
    l'eutrophisation sur la faune ichtyologique de la Semois. *Annls Limnol.*
    16, 77-89.
Pierre J.-F. (1974) Contribution à l'étude hydrobiologique des eaux
    superficielles du bassin Rhin-Meuse. I. Evolution du phytoplancton
    des eaux du cours moyen et supérieur de la Meuse. *Bull. Acad. Soc.
    Lorr. Sciences* 13, 91-108.
Smitz, J. (1976) Modèle thermique de rivière - température naturelle et
    rejets thermiques. In: *Recherche et Technique au service de l'Environne-
    ment*. 382 pp., pp. 117-37, Cebedoc, Liège.
Symoens J-J. (1957) Les eaux douces de l'Ardenne et des régions voisines:
    les milieux et leur végétation algale. *Bull. Soc. r. Bot. Belg.* 89,
    111-314.

Vereerstraeten J. (1971) Le bassin de la Meuse. Etude de géographie
    hydrologique. *Rev. belg. Géogr.* 94, 339 pp.
Verneaux J. & Tuffery G. (1967) Une méthode zoologique pratique de
    détermination de la qualité biologique des eaux courantes. Indices
    biotiques. *Annals scient. Univ. Besançon, Zoologie* 3, 79-80.
Woodiwiss F. S. (1964) The biological system of stream classification
    used by the Trent River Board. *Chem. Ind.*, 443-447.

A shallow stretch of the Meuse below a weir near the France-Belgium
border, showing patches of *Scirpus lacustris*. The red alga *Thorea
ramosissima* is frequent here in late summer.

2 · Gudenå

P. HEISE

2.1 INTRODUCTION AND HISTORY

The Gudenå is Denmark's longest river. Although its 146 km is short in comparison with many of the rivers in this book, nevertheless the river and its catchment have a long history of human use and the river itself has many features of ecological interest. It also provides a good case study for considering the problems of river management in Denmark.

The river rises in springs at Tinnet in the hills of Midjutland close to the maximum extension of the ice during the last (Weichelian) glaciation. Only 200 m from the Gudenå spring area, occur the springs of another major river, the Skjern Å. The Gudenå discharges into the Randers Fjord, a typical Danish estuary, and further into the Kattegat; the combined catchments of the Gudenå and the Randers Fjord constitute about 8% of the area of Denmark. The river runs through a series of lakes moulded during the last glaciation, but man-made impoundments have been added to these. The Tange Lake, for example, was established in 1920 to provide hydro-electric power at Tange. The Gudenå, its tributaries and the lakes, together with surrounding hills and woods, form the beautiful landscape of Midjutland known as 'The Danish Lake District'.

The Gudenå area has a long history of intensive use. The Stone Age settlements on the banks of the river formed the so-called 'Gudenå-culture', while great monasteries were established close to the river and its lakes during the Middle Ages. Hydro-power was used during the latter period for corn and paper mills and the waters also served the monasteries for fishing and transportation. The Benedictines from the town of Vorre dug a canal - three kilometres long - and diverted a stretch of the Gudenå into the 'Klosterkanal' (Monastery Canal), which provided hydro-power for the 'Klostermølle' (Monastery Mill) at Mossø Lake. (The mill is owned by Ministry of Environment and a water-mill museum is planned.)

Some transport of both people and goods took place on the Gudenå before 1850, but following the opening of the paper factory at Silkeborg in this year, the use of barges between the towns of Randers and Silkeborg was systematized. From Randers to Silkeborg the barges were pulled upstream by horses or men ('hired men') from the shore; the flow of water was itself sufficient to carry the barges and their goods on their downstream journey. At that time a great canalization project was fulfilled in the Silkeborg to Randers stretch, especially the part between Silkeborg and Tange, which was dredged to maintain a summer water level above one metre.

When the railroad connected the towns and the hydro-electric power station at Tange was built, the use of barges came to an end. Today it is tourists who sail on the Gudenå and the lakes. The tourist boats take some 120000 passengers a year, while canoes are hired by 25000 persons for tours of at least five days. About 1500 motorboats are located on the lakes and the downstream stretch of the Gudenå. There are 100000 anglers visiting the Gudenå and its tributaries each year. Tourists thus represent an important group of users demanding a high quality of water in the Gudenå system.

Fig. 2.1 Upper Gudenå running through intensively used farm lands.

Fig. 2.2 Middle Gudenå at Tvilum, 10 km upstream of Lake Tange: a place visited by many anglers and nature lovers.

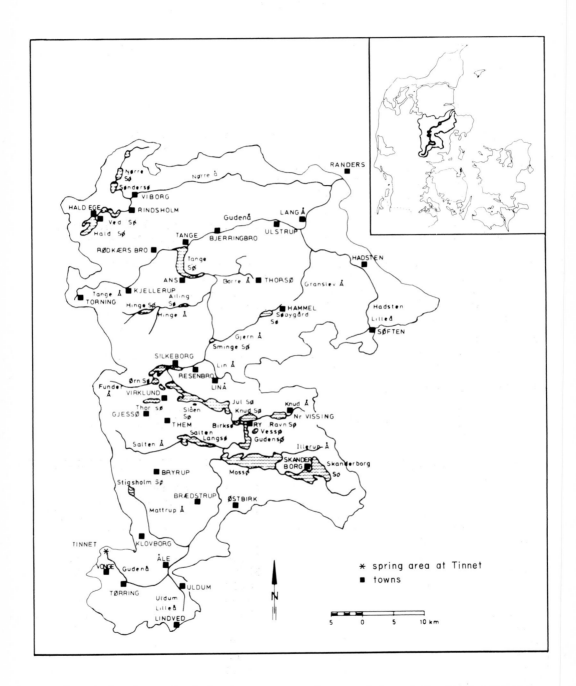

Fig. 2.3 The Gudenå drainage area: towns, lakes and water courses.

2.2   GENERAL DATA

DIMENSIONS
Area of catchment 2600 km^2 (3300 km^2 with Randers Fjord)
Length of river   146 km
Altitude at source 70 m

CLIMATE
Precipitation
mean: 720 mm yr^{-1}
range: 650 (in NE) to 750 mm yr^{-1} (in SW)

LAND USE
agricultural areas 1700 km^2 (65%)
forests and moors 650 km^2 (25%)
towns, paved areas 150 km^2 (6%)
water surfaces* 100 km^2 (4%)
*In the Gudenå drainage area there are about 40 lakes or impoundments greater than 0.1 km^2, and the water surface of these 40 lakes amounts to 75 km^2.
The drainage area of the Gudenå covers three counties (amt):
Vejle amt   (10% of the drainage area)
Viborg amt   (20% of the drainage area)
Århus amt   (70% of the drainage area)
and 30 municipalities

POPULATION
Largest towns are Randers, Silkeborg, Skanderborg and Viborg.
Total = 190000 (= 73 people km^{-2})

PHYSICAL FEATURES OF RIVER

Comparison of hydrological data for Gudenå with other Danish rivers
(Rate of flow given both as annual mean rate and median of annual minimum rate)

river	length	catchment	flow		precipitation
			annual mean	median minimum	
	(km)	(km^2)	(1 s^{-1})	(1 s^{-1})	(mm yr^{-1})
Gudenå	146	2600	31000	20000*	720
Storå	100	1030	13900	6800	740
Skjernå	93	2210	32700	21700	750
Suså	81	815	6400	1000*	650
Karup Å	77	820	9400	7000	690
Odense Å	60	606	5800	1750*	650
Kongeå	60	460	6900	3100	770
Mølleå	40	116	340	50*	620
Nørreå†	40	394	4500	2500*	690
Hadsten Lilleå†	28	292	2600	1250	630
Salten Å†	22	124	1500	1200	730

*flow is influenced by lakes or impoundments with regulation
†important tributaries of Gudenå

Topography of Gudenå

average slope (see Fig. 2.4)   0.047%, but there are 8 dams totalling 30 m
height. (Hydroelectric plants at Vestbirk
and Tange each have a height of fall of about
10 m)

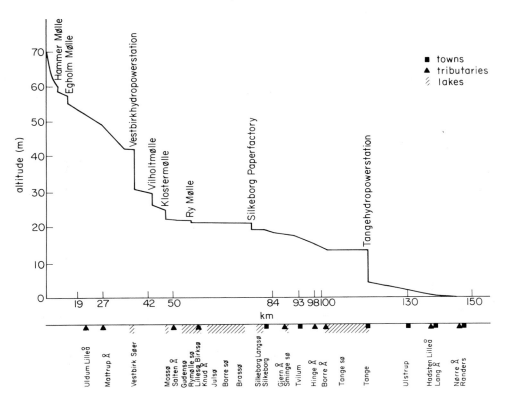

Fig. 2.4 Change in elevation in Gudenå on passing downstream

Zonation

Three stretches can be recognized based on profile and shape:

zone	length	limits	slope
upper	46 km	Tinnet to Mossø	0.03-0.40%
middle	65 km	Mossø to Tange (including 40 km through lakes	0.006-0.06%
lower	35 km	Tange to Randers	0.006-0.02%

Cross-section of river

	width (m)	depth (m)	summer av. velocity (m s^{-1})
km 10 (from source)	6	0.4 (av.)	0.5-2.0
km 100-Randers Fjord	25-40	2.5-4.0	0.1-0.5

Flow per unit area of catchment

average   12.5 1 s^{-1} km^{-2} (corresponds to 54% precipitation).
(Rate varies considerably according to part of catchment.
The tributaries and Salten A, Funder Å and Nørre Å 'steal'
groundwater from outside the watershed and their average
flow is 16-20 1 s^{-1} km^{-1}.)

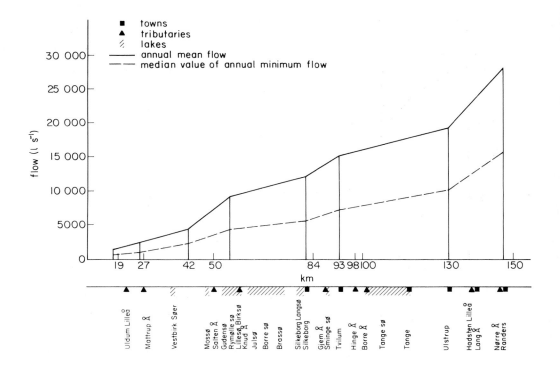

Fig. 2.5 Change in flow in Gudenå on passing downstream. Results
from 'The Gudenå Investigation 1973-1975' (Water Quality Institute,
1976; Hydrological Survey of the Danish Land Development Service).

Sources

c. 550 springs and headwater streams

Main tributaries

Uldum Lilleå	Funder Å	Tange Å
Mattrup Å	Gjern Å	Hadsten Lilleå
Salten Å	Hinge Å	Nørre Å
Knud Å		

2.3 WATER QUALITY

2.31 SPRING AREAS

The natural quality of water in the Gudenå catchment is determined by
the surface of the land and the geology of the area that is drained by
the river. However much of the water is influenced now by sewage discharges, effluents from industry and fish farms and pollution from
rural activities, with consequent enrichment in organic matter, nitrogen
and phosphorus. Of about 550 spring areas which form the sources of
the Gudenå and its tributaries, only some 15% are undisturbed today
(F. Jensen, 1976; Water Quality Institute, 1976). The composition of
water in some of these spring areas has been investigated intensively
(e.g. Warncke, 1980) and typical concentrations for three examples are
given in Table 2.1. Estimates for the spring areas as a whole of
average concentrations show:

$$\text{total dissolved N} \quad 0.38 \quad \text{mg l}^{-1}$$
$$\text{total dissolved P} \quad 0.06 \quad \text{mg l}^{-1}$$

Table 2.1 Composition (mg l^{-1}) of water in spring areas (see
Warncke, 1980).

variable	Tinnet (beginning of upper Gudenå)	Addit (in catchment of Salten Å)	Hald (beginning of River Nørreå)
Na$^+$	13.1	18.2	16.0
K$^+$	0.85	1.4	1.0
Mg^{++}	3.8	3.3	2.9
Ca^{++}	44.1	21.0	36.2
NH$_4$-N	0.08	0.03	0.08
NO$_3$-N	0.01	0.01	0.006
total-N	0.27	0.08	0.37
reactive PO$_4$-P	0.05	0.014	0.07
total filtrable P	-	0.027	0.08
total-P	-	0.054	0.335
Cl$^-$	16.8	21.0	22.5
alkalinity (meq l^{-1})	2.10	0.75	1.81

2.32 POLLUTION

Effluents from domestic and industrial sewage are discharged into waters
of the Gudenå system. In addition to the 190000 people living in the
area, discharges from industrial outfalls calculated on demand of water

amount to the equivalent of about 170000 persons. Pollution from sewage and industrial wastewater is widely reduced through biological treatment and phosphorus removal. A special group of 'water consumers' has been formed by 60 fish farms, which together produce a pollution load corresponding to 60000 persons (Table 2.2).

Table 2.2 Wastewater sources in Gudenå catchment. (PE = person equivalent)

Towns, urbanized areas > 150 PE	300000 PE
Industries with separate outlets	15000 PE
Small villages and farms	45000 PE
Fish farms	60000 PE
Total	420000 PE

The water supply in the Gudenå catchment comes from groundwater wells and approximately 1 m^3 s^{-1} of drinking water is discharged as sewage into the river; this value may be compared with the value of 31 m^3 s^{-1} for the average flow where the river runs into the Randers Fjord. The sewage is of course not discharged uniformly, but mainly as point outlets and often into small receiving waters. In 1980 the sewage treatment plants for about 100 towns with a pollution load above 150 PE used the following methods:

no treatment	2%
mechanical treatment	6%
biological treatment	60%
biological-chemical treatment	32%

The pollution load reaching the river can be described by variables such as biochemical oxygen demand over five days (BOD), total N, total P, etc. (Table 2.3).

The range of concentrations shown for summer reflects the dilution and decomposition of sewage effluents and in the primary production of lakes and streams. The high yearly average value for total nitrogen results from the very high levels which occur in winter, with individual values sometimes as high as 12 mg l^{-1} N. The river is usually relatively well oxygenated, with wintertime values mostly above 8 mg l^{-1} O_2 and summertime values seldom dropping below 5 mg l^{-1} and much higher by day due to photosynthesis by algae and other plants (Fig. 2.6). The concentrations of heavy metals are low, with values at or below the detection limits for all except zinc (as mg l^{-1}: Cr, <0.01; Cu, 0.005; Zn, <0.09; Cd, 0.001; Hg, 0.00025; Pb, 0.025).

Table 2.3 Composition (mg l^{-1}) of water in the Gudenå showing values for annual average and range in summer.

variable		Upper	Middle	Lower
BOD	annual	2.5	3.5	2.8
	summer	1 - 7.5	2 - 7.5	3-4
total N	annual	6	3.0	3.6
	summer	1.5 - 2.3	1.5 - 1.9	0.8 - 1.0
NH$_3$-N	annual	~0.0 - 0.4	0.1 - 1.9	0.1 - 0.7
	summer	0.1 - 0.4	0.1 - 1.9	0.1 - 0.2
NO$_3$-N	annual	0.2 - 8.5	~0.0 - 4.1	~0.0 - 7.6
	summer	0.2 - 1.7	~0.0 - 0.4	0.1 - 0.25
total P	annual	0.2	0.21	0.20
	summer	0.16 - 0.25	0.10 - 0.29	0.15 - 3.8

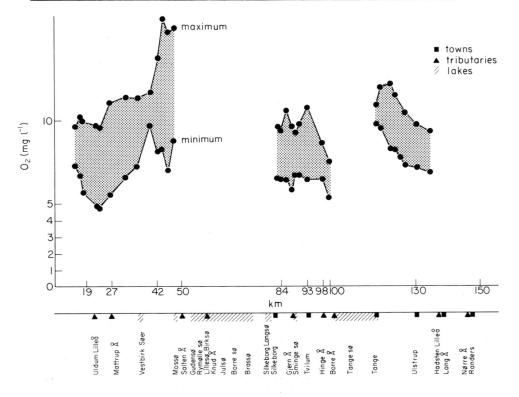

Fig. 2.6 Diel range in dissolved O$_2$ in flowing water stretches of the Gudenå in summer.

2.33   NITROGEN AND PHOSPHORUS LOADINGS

Because of their fertilizing effect on the lakes of the Gudenå system, much emphasis has been placed on the investigation of the sources and importance of nitrogen and phosphorus. Discharge per year within the catchment amounts to approximately 10000 tonnes of BOD, 4100 tonnes of nitrogen and 300 tonnes of phosphorus. Domestic and industrial activities produce approximately 20% of the total nitrogen and 75% of the total phosphorus loadings.

Table 2.4 Nitrogen and phosphorus loadings in run-off from non-point sources in the Gudenå catchment. Voelbaek and Gjelbaek are two minor agricultural areas (each 12 km^2) in the drainage area of the middle Gudenå.

	land-use	nitrogen (kg N ha^{-1} yr^{-1})	phosphorus (kg P ha^{-1} yr^{-1})
UPPER GUDENÅ	agricultural		
reaches with strong slope		28	0.90
reaches with flooded meadows in winter		22.5	0.02 - 0.05
THE LAKE DISTRICT	wood	7 - 10	0.15 - 0.20
MIDDLE GUDENÅ	agricultural	15 - 20	0.30 - 0.35
LOWER GUDENÅ	woods & agricultural	6 - 20	0.15 - 0.20
Voelbaek (sandy)	agricultural	14	0.17
Gjelbaek (clay)		25	0.43
storm sewers	urban		
town of Viborg		14	1.0
town of Skanderborg		8	2.9

The loadings from some non-point sources investigated in 1974-76 are shown in Table 2.4. Detailed mass balances for two minor agricultural areas, Voelbaek and Gjelbaek, showed that about 20% of the nitrogen used as fertilizer was washed out, whereas less than 1% of the phosphorus was so. Analyses of data for the daily runoff from the

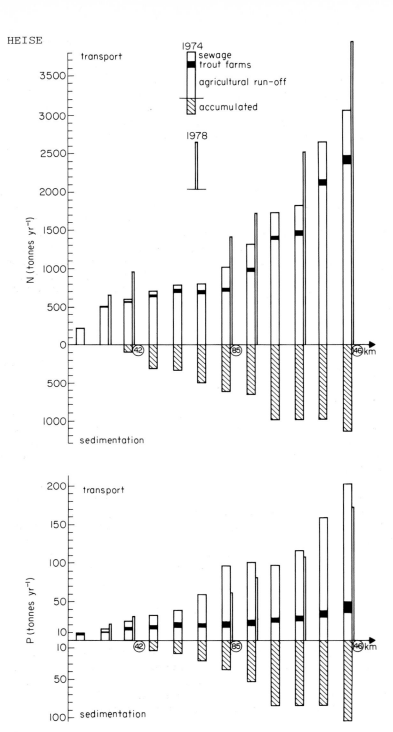

Fig. 2.7 Transport and sedimentation of N and P on selected sites along the Gudenå in 1974, with some comparative data for transport in 1978.

agricultural areas showed a high correlation between water and the nitrogen flux (Strange Nielsen, 1977). The correlation was less evident for phosphorus because of local disturbances caused by septic tank effluents and manure water. The nitrogen loadings per unit area of urban runoff were similar to those of rural areas (Table 2.3), but the phosphorus loadings were higher. As paved areas amount to only 3% of the Gudenå drainage area, urban stormwater runoff causes only local effects.

Annual loadings and transport of nitrogen and phosphorus have been investigated since 1974 at selected sampling sites throughout the whole river system.   About 30% of the nitrogen and 35% of the phosphorus are detained in the lakes and water courses because of sedimentation.   The amount of nitrogen transported in 1978 was higher than in 1974 (Fig. 2.7) due to a higher average water flow. In contrast, the construction of new sewage treatment plant incorporating phosphorus precipitation during 1977-78 led to a reduction in the phosphorus flux downstream of Silkeborg (km 85).

2.4   ECOLOGY

2.41   MACROPHYTES

The combination of being low-lying and having a high content of nitrogen and phosphorus means that the Gudenå and its tributaries have a high potential for producing macrophytes, although the extent of such production depends on local factors such as substratum, flow, water quality (especially turbidity) and shade. Species in the upper Gudenå include the following:

Cladophora fracta	*Butomus umbellatus*	*Batrachium (Ranunculus)*
C. glomerata	*Glyceria maxima*	sp.
Fontinalis	*Potamogeton alpinus*	*Callitriche* sp.
antipyretica	*P. crispus*	*Elodea canadensis*
	P. pectinatus	*Lemna trisulca*
	P. perfoliatus	*Nuphar lutea*
	P. praelongus	*Sium erectum*
	P. pusillus	
	Sparganium erectum	
	S. simplex	

Macrophyte diversity is less in the middle and lower Gudenå. Submerged species include *Butomus umbellatus, Glyceria fluitans, Potamogeton lucens, P. pectinatus, P. perfoliatus, P. praelongus, Sparganium simplex* and *Elodea canadensis*; species at the sides of the river are *Acorus calamus, Glyceria maxima, Phragmites australis, Sparganium erectum* and *Typha latifolia*.

A study of macrophyte production in some tributaries during 1971-75 (Brøndum Jensen & Mathiesen, 1976) failed to establish a general

Fig. 2.8 'The Klosterkanal'. The river Gudenå still runs in the canal dug by the monks at Klostermølle (left-hand photo); there are important spawning beds for trout and grayling just upstream of the Klosterkanal.

picture of the extent to which dissolved nitrogen or phosphorus were limiting factors. However it was clear that massive growths of *Cladophora* can be expected if the average concentration of total phosphorus exceeds 0.05 mg 1^{-1} during the growing season. The authors recorded biomass values for aquatic macrophytes (excluding roots and rhizomes) as follows:

1-40 g m^{-2} dry weight   reaches with shade effects from trees e.g. Spørring Å (Hadsten Lilleå), Mattrup Å; dominants were *Elodea* and *Callitriche*;

50-120 g m^{-2} dry weight   reaches with some shade effects or effects from epiphytes e.g. Mattrup Å downstream of Lake Stigsholm; dominant was *Potamogeton perfoliatus*;

200-300 g m^{-2} dry weight   reaches without shade effects e.g.
                                       upper Gudenå and Hadsten Lilleå.
                                       Dominants were *Potamogeton perfoliatus*,
                                       P. crispus, *P. pectinatus*, *Elodea
                                       canadensis* and *Sparganium simplex*;

500-700 g m^{-2} dry weight   downstream of a fish farm in the lower
                                       part of Mattrup Å, with *Sparganium
                                       simplex* as the dominant (and only)
                                       vegetation.

At their experimental sites in the tributaries, Granslev Å and Gjern Å,
Dawson and Kern-Hansen (1978) recorded maximum biomasses of 200-300 g
m^{-2} and 480 g m^{-2} dry weight, with the dominant species being *Elodea
canadensis* and *Sparganium simplex*, respectively.

2.42 MACROINVERTEBRATES AND USE OF THE SAPROBIEN SYSTEM

Most of the information on macroinvertebrates comes from the use of a
simplified Saprobien system as an aid to controlling the biological
quality of the watercourses. The simplified Saprobien system is based
on the relative occurrence of a number of easily recognisable macroinvertebrates; water quality is distinguished in degrees from I to IV.
The system was issued in 1970 by the Ministry of Agriculture and taken
over by the National Agency of Environmental Protection in 1974. The
following are the most important features:

  I (oligosaprobic) characterized by the occurrence of clean water
     organisms, such as certain caddis fly larvae and stone fly
     nymphs.

 II (β-mesosaprobic) characterized by abundance of species which
     are more or less pollution tolerant; dominance of *Gammarus pulex*,
     Baetis rhodani and various *Caddis* flies, but not *Chironomus* or
     Asellus.

III (α-mesosaprobic) characterized by occurrence of *Chironomus* and
     Tubifex together with some other pollution tolerant organisms
     such as *Asellus* and *Herpobdella*.

 IV (polysaprobic) characterized by the exclusive occurrence of
     saprophilic organisms as *Eristalis*, *Tubifex* and *Chironomus*, the
     last two predominating under the most polluted conditions.

Organisms other than macroinvertebrates may also be valuable as
indicators e.g. *Beggiatoa* for pollution degree IV. The sewage-fungus
community (dominated by *Sphaerotilus*) occurs most frequently in
pollution degree III.

The results of a detailed study of the Gudenå (J. Jensen *et al.*,

1976) are shown in Fig. 2.9. The upper Gudenå is not polluted, the
pollution degree of the river being II or better. The clean water
species include:

Plecoptera. *Perlodes microcephala, Chloroperla burmeisterii.*
Ephemeroptera. Heptagenia sulphurea, Heptagenia fuscogrisea,
Ephemera danica.

The more upstream part of the middle Gudenå and the lower Gudenå are
influenced by the loadings of organic matter (phytoplankton) produced
in the lakes, so evaluation using the Saprobien system has to be done
with care. Comparison of the results obtained with the Danish system
and with the Sládeček (1973) system (modified to take account of the
Danish fauna) shows several differences. For instance, the former did
not reflect the untreated sewage (corresponding to 30000 PE) which was
discharged into the river from Bjerringbro at the time of the investigations, whereas the Sládeček system both indicated this and the self-
cleaning of the river on passing downstream. The pollution index in
this stretch is now about II as a result of biological treatment plants
constructed at Bjerringbro in 1976 and Hadsten in 1978.

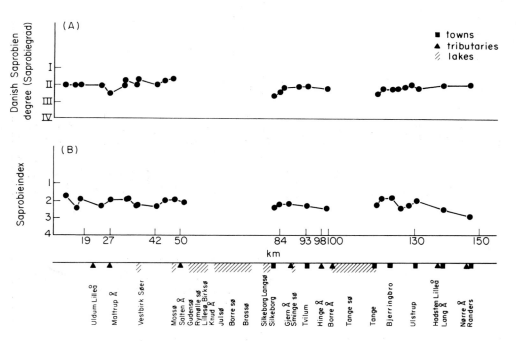

Fig. 2.9 Use of macroinvertebrates to evaluate water quality of
Gudenå according to two different Saprobien systems: A, Danish
simplified system; B, Sládeček (1973) modified for Danish river
fauna. (From J. Jensen *et al.*, 1976; the quality of lower Gudenå has
since improved - see text)

2.43   FISH

2.431   Fauna

The Gudenå, its tributaries and the lakes have the reputation of being good waters for fishing. Many foreign sports fishermen visit the area to compete in 'coarse fishing', while many individual anglers can be seen along the river banks. The trout (*Salmo trutta*) is the dominant of the rivers. Following at nearly 300 sites and involving about 11000 fish, Mortensen (1976) reported that trout was found at 68% of the sites and constituted 60% of the total catch (Fig. 2.10).

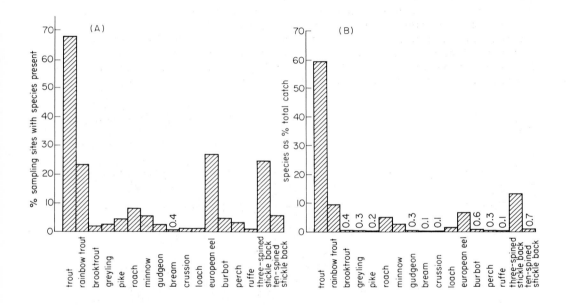

Fig. 2.10 Occurrence of fish at 300 sites in Gudenå catchment during surveys in 1969, 1971 and 1974-75 involving 11000 individual fish: A) Percentage of sample sites with species recorded; B) Percentage of total catch. From Mortensen, 1976)

The diversity of the fish fauna was generally low. About 30% of the sampling sites had only one species and another 50% only two species. Most of these were small streams. In contrast five sites had six species and one site had seven species. Investigations of fish productivity in small tributaries (Mortensen, 1977) showed a maximum value for trout of 22-27 g m^{-2} yr^{-1} in Brandstrup Baek, a stream entering the lower Gudenå at Ulstrup.

2.432 Fish-passes and the Gudena salmon

While many stretches of the Gudenå and its tributaries are suitable for
nursing and growth of sea trout (*Salmo trutta*), obstructions in the form
of dams in many cases prevent the migration of this and other salmonid
fish.   The salmon (*S. salar*) was also common in the middle and lower
Gudenå before the hydropower plant was built at Tange, but impoundments
for the plant flooded key nursery streams by the middle Gudenå.   The
fish-pass constructed in 1920 at the dam was never brought into function.

The evaluations which followed the Gudenå survey carried out in
1973-75 led to the Counties of Vejle, Viborg and Århus deciding to
build a new fish-pass at Tange.   The type of pass chosen was based on
principles developed in 1908 by a Belgian engineer, G. Denil;
Lonnebjerg (1980) described the results of tests for the new fish-pass
at Tange, while the pass was constructed in 1980.

Fig. 2.11 Fish-pass at
Tange. This consists of
7 resting pools and 8
flights, each 6.55 m long
and with a gradient of
20%; the total rise is
about 10 m; water flow
through the pass is
about 150 l s^{-1}.

In order to guide downstream migrants to the exit of the pass, a
barrier of lights has been placed on the upstream side of the intake
for the power station. Downstream, an electric screen has been placed
across the river to prevent upstream migrants from entering the tail-
race tunnels. The total cost of the pass amounted to U.S. $150000 (at
1980 prices).

The migration of fish in the pass at Tange is monitored intensively
and, for example, in November-December 1980 one salmon, 23 sea trout
and 15 brown trout were registered. Eel, roach and perch have also
been found to migrate through the pass. In spite of the occasional
salmon, however, salmon fishing in the Gudenå catchment is not expected
to expand, because of the lost nursery streams and also the heavy netting
which occurs in Randers Fjord. In contrast, there are great possibilities for expansion of sea trout fishing, particularly if other passes
are established where there are obstructions at present e.g. Ry Mølle,
Klostermølle, Vilholt Mølle and Vestbirk hydro-power station.

2.5   RIVER MANAGEMENT

2.51   ADMINISTRATIVE RELATIONSHIPS

The legal provisions for the use and protection of Danish watercourses,
lakes, and other freshwater localities are mainly contained in the
following four acts:

1. Watercourse Act (Vandløbsloven)
2. Environmental Protection Act (Miljøbeskyttelsesloven)
3. Freshwater Fisheries Act (Ferskvandsfiskeriloven)
4. Conservation of Nature Act (Naturfredningsloven)

According to these acts the Counties play a central administrative part
to ensure that the protection of the environment has a high priority
today and will have so in the future too.

The County Councils of Vejle, Viborg and Århus formed 'The Gudenå
Committee' to co-ordinate and collaborate on the management of the
Gudenå. Thus, the Gudenå Committee in 1978 agreed to set up guidelines
to improve the water quality of the Gudenå and the associated lakes,
based on the results of "The Gudenå investigation 1973-75" (Water
Quality Institute, 1976) and the daily work of the Water Authorities
i.e. the County Water Inspections. The County Councils of Vejle,
Viborg and Århus agreed to the guidelines in 1979.

2.52   IMPROVEMENT OF WATER QUALITY IN LAKES

One example of the guidelines is the general emission limit for
phosphorus in sewage discharged from towns upstream of Tange; this is
needed in order to reduce the growth of phytoplankton in the lakes.

The emission limit is 1 mg P l^{-1} and the treatment plants have to fulfil this condition, as follows:

    treatment plants > 5000 PE must establish phosphorus precipitation
        in 1981 or before;
    treatment plants > 2000 PE must establish phosphorus precipitation
        before 1985;
    treatment plants > 500 PE must establish phosphorus precipitation
        before 1990.

The investigation in 1973-75 showed that light and phosphorus limit the growth of phytoplankton, and for some selected lakes a mathematical model has been calibrated to simulate the effect on the water quality of altered nutrient loadings (Heise et al., 1977; Nyholm, 1977). If the phosphorus from wastewater from towns greater than 500 persons was treated with chemical precipitation, growths of phytoplankton were predicted to decrease by 30-40%, at least up to the year 2000. The transparency (Secchi disc depth) was not influenced much more by inserting a 90% phosphorus reduction. The reason for this is the short detention time in the lakes (approximately 30 days) and the relatively high contribution of nutrients from non-point discharges and fish farm outlets, which maintain a high phytoplankton growth rate.

2.53   RIVER MODELLING

For the Gudenå and a number of the tributaries a mathematical model (e.g. Dahl-Madsen and Simonsen, 1974), simulating oxygen fluctuations was applied. The model used is based on the work of Streeter and Phelps (1925) describing 'The Oxygen Sag Curve for Flowing Waters'. In the case of the river Gudenå, however, not only have the decomposition of organic matter and reaeration been taken into account, but also other processes such as oxygen fluctuations due to respiration and photosynthesis by aquatic plants. The oxygen consumption due to nitrification is of minor importance, and so far it has not been taken into consideration.

   Table 2.5 gives the values calculated for certain parameters for the Gudenå and some tributaries (Simonsen et al., 1976). Oxygen concentrations today are in fact satisfactory in the Gudenå system as a whole. The most severe pollution in 1973-75 was found in the tributary Hadsten Lilleå, where untreated sewage from 30000 PE (from the town of Hadsten) was discharged into the stream having a rate of flow of 300 l s^{-1}. Figs. 2.12 and 2.13 show the fluctuations of oxygen and BOD in the Hadsten Lilleå. Calibration of the model was made taking into consideration both the growth of plants and sediment respiration in the Hadsten Lilleå upstream of the sewage discharge. In order to decide the degree of biological treatment, the model was used to simulate the oxygen fluctuations in the stream as a function of the discharge from the town of Hadsten. Using the results of the model, it was decided to establish a biological treatment plant whose

effluent BOD would not exceed 20 mg l^{-1}. The calculation predicted
that the oxygen concentration under those circumstances would not be
less than 2 mg l^{-1} at any location along the river. The treatment plant
now works well under dry weather conditions, and normally the oxygen
concentration in the river in Hadsten Lilleå downstream Hadsten is
higher than 4 mg l^{-1}, However, during storms the plant is hydraulically
overloaded, and the pollution load (as BOD) from the plant rises by a
factor of 5-10 times for some hours.

In summer 1980 some fish kills were recorded in a downstream fish
farm at the same time as some heavy rain storms. As a consequence the
treatment plant was accused of causing these kills. In order to see if
this was the probable explanation, investigations were carried out during
storm events in the summers of 1981 and 1982. The results showed that
the oxygen content in the water body dropped only by about 1 mg l^{-1} O_2
due to disposal of storm water and sewage from the treatment plant. No
fish kills have been reported since the events of 1980. At present the
most likely explanation for the kills would seem to have been the
illegal disposal of manure effluents or other non-point pollution sources.

Table 2.5 Range of values calculated for 20°C for the river parameters, reaeration coefficient, rate of oxygen consumption, total
daily production and total respiration (e.g. Simonsen, 1976)

water-course	reaeration coefficient for oxygen (K_2) d^{-1}	rate of oxy-gen consump-tion due to BOD-decompo-sition (K_1) d^{-1}	total daily oxygen pro-duction (P) g O_2 $m^{-2}d^{-1}$	total daily respiration (R) O_2 $m^{-2}d^{-1}$
UPPER GUDENÅ				
upper reaches	2.0 - 6.1	0.5 - 6.0	3.9 - 17.6	1.7 - 29.9
lower reaches	2.4 - 2.8	0.5 - 0.75	11.8 - 27.7	7.9 - 19.2
MIDDLE GUDENÅ	0.4 - 3.4	0.25 - 2.0	3.8 - 22.5	3.9 - 26.4
LOWER GUDENÅ	0.7 - 1.5	0.25 - 1.75	2.0 - 25.9	3.0 - 29.6
Mattrup Å	1.0 - 9.7	0.5 - 2.0	1.0 - 17.8	9.6 - 46.7
Hadsten Lilleå	2.9 - 4.2	1.7 - 6.9	3.9 - 13.0	12.3 - 21.3
Tange Å	5.1 - 15.9	0.25 - 2.0	1.1 - 4.0	0.5 - 8.0
Nørre Å	0.4 - 3.8	0.75 - 1.7	0 - 8.2	4.0 - 17.1

The calibrated oxygen models have also been used as planning tools
in Nørre Å and Tange Å. In Nørre Å the discharges from a fish farm and
the biological treatment plant in Viborg were evaluated. Calculations
showed that the fish farm discharge was the main cause of oxygen
deficit in Nørre Å. In Tange Å the effect on the oxygen content when

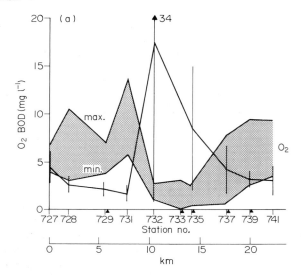

Fig. 2.12 Relationship between dissolved oxygen and BOD in the tributary, Hadsten Lilleå. The shaded limits show the maximum and minimum values for O_2. BOD is plotted as the mean, with the vertical bar indicating maximum and minimum values.

Untreated sewage from the town of Hadsten was discharged between station no. 731-732, causing a rise in BOD concentration of the river water to 34 mg l^{-1}. The 'oxygen sag curve' is obvious.

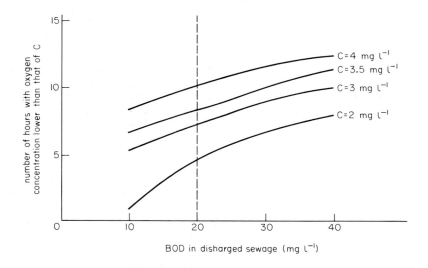

Fig. 2.13 Calculated 'duration curves for oxygen' as a function of different BOD discharges from the town of Hadsten at the 'critical' reach on the Hadsten Lilleå.

pumping river water for irrigation was simulated and evaluated. No
critical oxygen concentrations became apparent; the reason for this is
because organic matter is removed with the irrigation water.

Modelling of processes in river water gives a good understanding of
the likely effects of various types of interference, though results of
the calculations must be interpreted with care. Nevertheless, with
common sense and ecological experience, it is possible to set up guidelines for emission limits for discharges in order to improve the water
quality.

2.54   RIVER MANAGEMENT IN THE FUTURE

In order to fulfil the rules of the Environmental Protection Act the
County Councils have set up objectives for the use of the water systems
(e.g. bathing water, fishing water for migratory or for nursery purposes,
areas for scientific studies in the so-called receiving water plans
(recipientkvalitetsplaner).   The guidelines given by the Gudenå
Committee have now been incorporated into the planning worked out by the
Counties of Vejle, Viborg and Århus, and the Communities of the Gudenå
area.   It is expected that the sewage discharges from towns and industries will be treated biologically and chemically in the next decade to
improve the water quality.

Discharges from fish farms play a greater part today than they did
in 1973-75 and it is evident that the effluents from the fish farms
have to be treated too. New sedimentation techniques, changes in
production and recycling of water are all processes that the fish
farmers have to implement, if the pollution from the fish farms is to
be reduced.

The non-point pollution from agricultural areas and from storm water
pollution has been drawing much attention. Many farmers have been fined
for discharging manure water or silage joice into the receiving waters.
Because of the high price of fertilizers farmers are likely to be more
careful when spreading fertilizers on the fields, so this may reduce
the wash-out of nitrogen and phosphorus from the soil.

Pollution from storm sewers and overflows should be minimized.
This demands retention basins, grids or, in special situations, treatment facilities to handle urban rainwater running to the receiving
waters. As one example, the County of Århus built in 1982 a filtration
plant (sedimentation and sand filtration) for storm water runoff from a
new highway to protect the upper reaches of the tributary Granslev Å.

Many attempts to improve or restore water quality can be expected in
the future. However, the quality of the water alone is not enough to
raise the good quality of the watercourses: banks and substratum have
to be handled with care as well. The 'Klosterkanal' provides an
example where this was not done. The dredging of a 500 m long stretch,

where 500 m^3 sand was removed, caused the reduction of the macroinvertebrates from 60 to 24 species (F. Jensen, 1979). Weed cutting
may also be changed in the future. The Gudenå Committee is aware of the
problems when cutting aquatic macrophytes in the watercourses to improve
the run-off from rural areas. The investigations of Dawson & Kern-Hansen
(1978) have pointed out possible solutions, and in the County of Vejle
trees have been planted on an experimental basis along a stretch of two
kilometers on the banks of Gudenå to shade the aquatic macrophytes.
The Watercourse Act is being revised in 1983, and laws requiring
more gentle weed cutting and restoration of spoiled (canalized) water-
courses can be expected and implemented in the Gudenå area.

Following the Conservation of Nature Act, the Counties of Vejle. Viborg,
Århus have been evaluating boating problems on the river Gudenå system.
Regulations for canoes in the upper Gudenå and many of the tributaries
were given in 1982.   Regulations of the speed of motorboats have
already been made for the middle Gudenå and the lakes and limitations
on the number of motorboats legally permitted are expected in 1987.

Conclusions: Nature around and in the river Gudenå and its tributaries
has many qualities that many people enjoy - thus the Gudenå has to be
used, but not abused.

Acknowledgements

I am most grateful to my colleagues Jens Møller Andersen, Jytte Heslop
Christensen, Jørn Jensen and Lis Stadil for their helpful comments.

REFERENCES

Brøndum Jensen U. & Mathiesen H. (1976) Botanisk-Økologiske under-
    søgelser over vandløbsvegetationen i udvalgte Gudenå vandløb.
    Report to the Gudenå Committee No. 23. (Danish = Botanical-ecolo-
    gical investigations of macrophytes in selected reaches of Gudenå
    and tributaries.
Dahl-Madsen K.I. & Simonsen J. (1974) Vandløbssimulering. *Research
    Report from the Water Quality Institute and the Laboratory of
    Sanitary Engineering, The Technical University of Denmark.*
    (Danish = Water Stream Simulation).
Dawson E.H. & Kern-Hansen U. (1978) Aquatic weed management in natural
    streams. The effect of shade by marginal vegetation. *Verh. int.
    Verein. theor. angew. Limnol.* 20, 1451-1456.
Heise P.B., Nyholm N. & Simonsen J.F. (1978) River Gudenå investi-
    gation 1973-75. *Verh. int. Verein. theor. angew. Limnol.* 20,
    1446-1450.
Hydrological Survey of the Danish Land Development Service (1978)
    10. Beretning, afstrømningsmalinger i Danmark 1917-70. (10. Report,
    Discharge data for Danish Streams 1917-70. Danish, with an English
    summary). Slagelse, Denmark.

Jensen C.F. & Jensen F. (1979) Oprensningens virkning på den specielle rheophile rentvandsfauna ved Klostermøllebro, Gudenå. *Natur-historisk Museum and the County of Vejle*. (Danish = The effect of dredging in the Klostercanal to the Rheophile macroinvertebrates).

Jensen F (1976) Kildeundersøgelser. *Report to the Gudenå Committee* No. 33. (Danish = Investigations of the springs).

Jensen J., Lastein, E. & Hansen I. (1976) Vandløbsbiologi. *Report to the Gudenå Committee* No. 13. (Danish = The macroinvertebrates).

Lonnebjerg N. (1976) Fiskepas af modstrømstypen. *Meddelelser fra Ferskvandsfiskerilaboratoriet, Danmarks Fiskeri- og Havundersø-gelser* 1/80. (Danish = Fishways of the Denil type).

Mortensen E. (1976) Fiskeundersøgelser. *Report to the Gudenå Committee* No. 18. (Danish = Investigations of fish-fauna).

Mortensen E. (1977) Fish production in small Danish streams. In: *Folia limnol. Scand.* **17**. (Danish = Limnology, Reviews and Perspectives).

Nyholm N. (1977) A simulation model for phytoplankton growth and nutrient cycling in eutrophic, shallow lakes. *Ecol. Modelling*.

Simonsen J., Strange Nielsen K. & Heise P. (1976) Intensive vandløbs-undersøgelser. *Report to the Gudenå Committee* No. 11. (Danish = Intensive investigations according to river modelling).

Sládeček V. (1973) System of water quality from biological point of view. Ergebnisse der Limnologie, *Arch. Hydrobiol.* Beiheft 7, Stuttgart.

Strange Nielsen K. & Nyholm N. (1977) The contribution of nutrient from diffuse sources. *Prog. wat. Tech.* **8**(4/5), 111-117.

Streeter H. W. & Phelps E.B. (1925) A study of the pollution and natural purification of the Ohio River. *Public Health Bulletin* No. 146, Treasury Department, U.S. Public Health Service.

Warncke E. (1980) Spring areas: ecology, vegetation and comments on similarity coefficients applied to plant communities. (Diss. Arhus). *Holarctic Ecology* vol. 3, no. 4.

Water Quality Institute (1976) Kilder, søer, vandløb - samlerapport, Gudenåundersøgelsen 1973-75. *Composite report to the Gudenå Committee*. (Danish = Spring areas - lakes - watercourses, The Gudenå investigation 1973-75).

3 · Wye

R. W. EDWARDS & M. BROOKER

3.1   INTRODUCTION

3.11   GENERAL

The scenic beauty of the Wye has been captured by writers, most notably Wordsworth and Shaw, by painters, particularly Turner, Cotman and Fielding, and by such addictive travellers as Borrow and the Reverend Gilpin. During the late 18th century its attractions lured so many to its banks for the 'Wye Tour' that a holiday industry developed to accommodate visitors and carry them by boat from Ross-on-Wye to Chepstow - a journey which took about three days. For many centuries before the arrival of the railways, around 1855, the lower reaches provided an important trade route in the valley, which contributed to the development of settlements, the largest being the City of Hereford with a current population of about 47000.

The river sustained large salmon stocks until the 19th century, when they were decimated both as juveniles (samlets) and as adults to titillate the palates of visitors and to sell elsewhere in the expanding markets made accessible by improved road and rail transport. Strict control of the net fishery in the estuary, and its total prohibition in the river, led to a recovery of stocks in the early 20th century which facilitated an expansion of recreational rod fishing until it became the largest in England and Wales.

The waters of the Wye catchment, like its salmon, have been exported. At the end of the 19th century three reservoirs were constructed in the Elan catchment to supply water for Birmingham (see Fig. 3.8); since that period a further reservoir has been built (Claerwen in 1952) to match the increasing demand of that industrial conurbation. The further development of the water resources of the catchment has been proposed (Water Resources Board, 1973) involving the enlargement of Craig Goch reservoir (see 3.6).

In addition the Wye is of considerable conservational significance. In *A Nature Conservation Review* (Ratcliffe, 1977) the River Wye was classified as a Grade 1 Site of Special Scientific Interest, the safeguarding of such sites being regarded as essential to the success of nature conservation in Britain. Its importance stems largely from its virtual uniqueness as a large, relatively unpolluted river in southern Britain and from its catchment diversity, which gives rise to a biological richness.

3.12 GEOLOGY AND LAND-USE

The Wye, about 250 km long, rises at 677 m on Plynlimon in the northwest of the catchment and flows south-easterly to join the Severn Estuary at Chepstow (Fig. 3.1). It has four major tributaries, the Ithon and Irfon in its upper reaches and the Lugg joining the main river downstream at Hereford and the Monnow at Monmouth (Fig. 3.1).

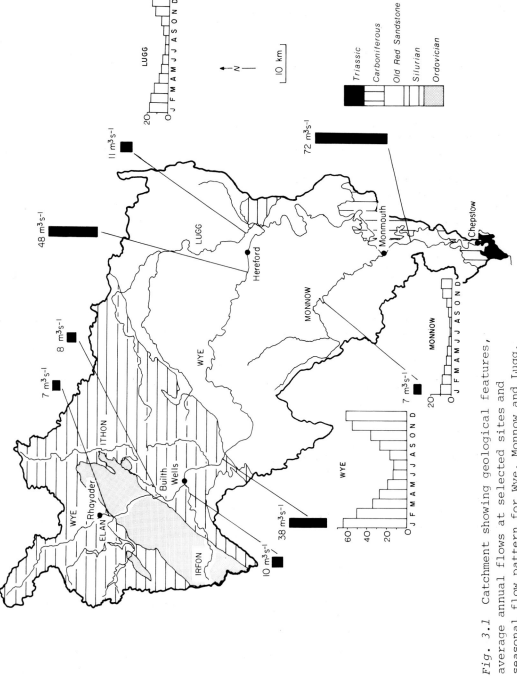

Fig. 3.1 Catchment showing geological features, average annual flows at selected sites and seasonal flow pattern for Wye, Monnow and Lugg.

The Elan, before reservoir development in its catchment, was also a major contributor to the upper Wye but it now exports over 50% of its yield to the West Midlands.

The major geological division of the catchment is between the relatively impermeable lower Palaeozoic rocks (Ordovician and Silurian) in the north-west (41% of catchment), elevated above 200 m, and the Old Red Sandstone (55% of catchment), generally below 120 m. There are outcrops of Carboniferous rocks, principally limestone, downstream of Ross-on-Wye and a small outcrop of Triassic strata between Chepstow and the estuary (Fig. 3.1).

Land-use broadly reflects the topographical and geological features. The elevated rough pastures in the north-west are grazed principally by sheep. On somewhat lower land, intensive sheep rearing predominates with the inclusion of limited dairy farming. On the richer lowlands there is a mixture of dairying, horticulture and arable crops. In 1970 the principle agricultural uses in the catchment were permanent grass (37%), tillage (20%), rough grazing (15%), temporary grass (14%) and forestry (9%).

The population of the catchment, about 200000, represents an average density of 0.47 ha^{-1} and in the upper 42% of the catchment, with no town having a population greater than 4000, the density is only 0.15 ha^{-1}. Although the sewage from isolated dwellings is only treated in septic tanks, over 65% of the population now receives main drainage, sewage being given full biological treatment before discharge to streams and rivers within the catchment.

3.13 HYDROLOGY

The precipitation broadly reflects topography and ranges between 1800 and 2500 mm yr^{-1} in the headwaters to less than 1000 mm yr^{-1} in the lower elevations associated with Old Red Sandstone. In the wetter areas there is a marked seasonal precipitation pattern, the period October-January being wettest and March-June being driest.

Except in the wetter elevated part of the catchment, about half the precipitation is lost in evapotranspiration, a process which also modifies the seasonal distribution of river flows. Fig. 3.1 shows the average flow (discharge) at several recording stations on rivers within the catchment, seasonal rhythms being shown at three of these stations. Except during periods when the Elan valley reservoirs are full and overspill, flow in the Elan is maintained by 'compensation water', currently about 1.3 m^3 s^{-1}. But average flows belie the wide variation which exists. In Fig. 3.2 this variation is shown at Monmouth for only three years (1975-77), the summer of 1976 being the driest on record. Winter flows exceeding 200 m^3 s^{-1} are not exceptional and, during the summer, flows are frequently around 10 m^3 s^{-1}, or 5% of this winter value.

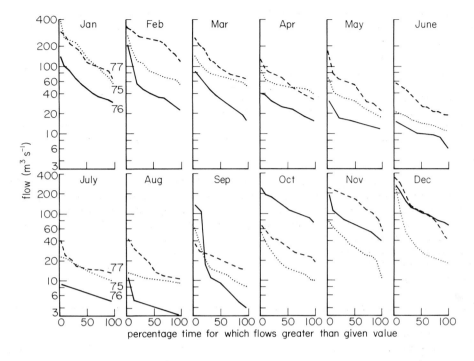

Fig. 3.2 Distribution of daily flows for each month at Monmouth for 1975, 1976 and 1977.

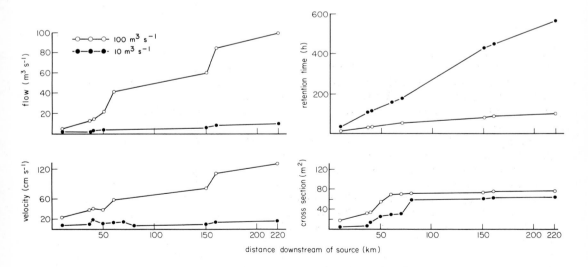

Fig. 3.3 Flows, velocities, retention times and cross-sectional areas of the Wye when flows at Monmouth are 10 and 100 m³ s⁻¹, equivalent to dry summer and wet winter conditions.

Although important to water resource engineers, the flow *per se* has
little ecological relevance except with respect to aspects of water
quality (3.6). Changes in flow are reflected, however, in changes in
velocity, which is reciprocally related to residence time, and in cross-
sectional area, the latter finding tangible expression in depth and
width: all these characteristics have considerable ecological relevance.
Fig. 3.3 shows the effects of changing the equilibrium flows in the
river, from 10 to 100 m^3 s^{-1} at Monmouth, on velocity, residence time
and cross-sectional area. It is based on calculations derived from
iodine tracer studies at predominantly low flows and, whilst it may not
be very accurate at high flows, it demonstrates the comparative response
of parameters to flow - in the middle and lower catchment flow principally affects velocity whereas in the upper catchment both velocity and
cross-sectional area are affected.

At low flows, velocities are generally insufficient to erode sediments but at flows more characteristic of the winter period particles
of a wide size range may regularly be transported downstream. From
turbidity data and studies of labelled bed material (Hey, 1980) it
seems that at flows above 10 m^3 s^{-1} at the Elan Valley confluence
(equivalent to 100 m^3 s^{-1} at Monmouth: Fig. 3.3) the movement of bed
material (>10 cm diameter) will occur.

At low flows (10 m^3 s^{-1}, Monmouth), the total residence time of water
in the main river is about three weeks, which provides ample opportunity
for the development of planktonic algal blooms: in contrast, at high
winter flows the residence time is three or four days. Whilst in the
reaches downstream of the Lugg confluence, flow changes have comparatively little effect on cross-sectional area (Fig. 3.3), above that
confluence, and particularly in the headwaters, substantial effects are
apparent. In the lower reaches, with well-defined banks, such flow
changes will result predominantly in depth changes, whereas in the upper
reaches, with shallower bank profiles, fluctuations in width will also
occur.

3.2   CHEMISTRY

3.21   INTRODUCTION

The waters of this predominantly rural catchment are free from major
sources of sewage and industrial pollution and until relatively recently,
when proposals to regulate the flow of the Wye and abstract considerable
amounts from the lower reaches were considered, there seemed little need
to undertake comprehensive monitoring programmes of water quality.
Consequently, apart from a few simply-measured parameters of water
quality such as ammonia, nitrate, orthophosphate, alkalinity and BOD,
few measurements were made regularly and extensively.

The water quality is related largely to soil structure and the under-

lying geology of the catchment, which determines the potential source of
most materials (other than sulphate and sodium, much of which are
atmospheric in origin and phosphate which is derived from sewage) and to
the pattern of rain-fall, infiltration and run-off which determines the
leaching and dilution of such materials. Man has influenced both the
soils of the catchment by cultivation and the run-off processes by land-
drainage (Jones, 1975) and the nature of land-use: the upland pastures
yield about 15% more run-off, with similar precipitation, than the
coniferous plantations of the neighbouring Severn catchment (Newson,
1979). Inputs are further modified by discharges of sewage effluent,
although in the Wye catchment with a total population of only 200000,
and 65% being connected to sewers, the effect of such effluent is small
except with respect to phosphate concentrations.

Chemical quality may be modified within the river by chemical
processes, such as the precipitation of calcium at high pH, and biological processes, such as silica utilization in diatom growth and denitrification: such processes often prevent the construction of simple
input-output models (Edwards *et al.*, 1978; Houston & Brooker, 1981).

3.22   SPATIAL DISTRIBUTION

As in many rivers, concentrations of most determinands generally increase
steadily downstream in the Wye (Fig. 3.4). The waters draining the
mineral-poor impermeable mudstones in the upper catchment have low concentrations of minerals and nutrients and the more southerly part of
the catchment, particularly that draining the arable lands of the Lugg
basin, is characterized by much richer waters. The major upland east-
bank tributary, the Ithon, also carries higher concentrations than
those in the upper Wye, reflecting drainage from the calcareous Wenlock
beds. In contrast, the upland west-bank tributaries, Elan and Irfon,
are generally less rich than the Wye and the quality of the main river,
below its confluence with the Elan, particularly during periods of
natural low flow, is locally influenced by the acid, nutrient-poor water
from the impoundments of this tributary.

One aspect of particular ecological significance relates to the
relatively high concentrations of iron and manganese in the Elan, the
mean concentrations of 500 μg l^{-1} Fe and 100 μg l^{-1} Mn being about ten
times greater than those in the upper Wye. About 75% of the total iron
and 50% of the total manganese in the Elan are associated with particulate matter and almost half the load of these metals is derived from
the effluent of a water-treatment works and not directly from the
reservoirs. Despite the generally high concentrations of iron and
manganese in the deeper water of the Caban Coch reservoir, oxygen
concentrations remain relatively high (>50% saturation) throughout the
summer stratification period.

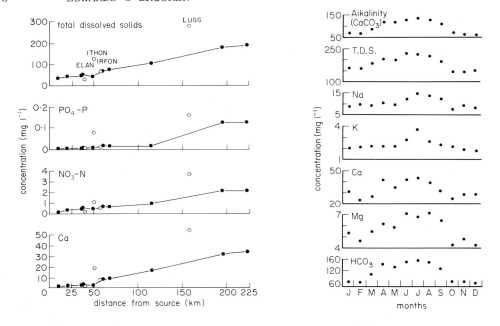

Fig. 3.4 Average concentrations of total dissolved solids, PO₄-P, NO₃-N and Ca at sites down the Wye and in major tributaries.

Fig. 3.5 Seasonal variation in monthly average concentrations of Ca, Mg, Na, K, alkalinity, total dissolved solids and HCO₃ for the period 1975-1977 at Monmouth.

3.23   TEMPORAL VARIATION

The water chemistry of the Wye is characterized by marked seasonal changes, particularly in the lower catchment, where concentrations are substantially higher than in the upper catchment (Fig. 3.5). For example, alkalinity and the concentrations of total dissolved solids and the major ions, such as calcium, magnesium, sodium, potassium and bicarbonate, are generally lowest during the winter and highest during the summer: these differences are predominantly related to the seasonal pattern of flows (Fig. 3.1) and simple concentration-flow regressions of these major ions, after logarithmic transformation, generally account for more than 40% of the variation although they rarely account for more than 70% in the lower reaches (Oborne *et al.*, 1980).

Throughout the catchment highest concentrations of phosphorus also occur principally during summer low-flow periods (Fig. 3.6) and, as expected, concentrations were particularly high during the drought of 1976. Since the major source of phosphorus is relatively constant, being derived from sewage effluents, changes in concentration within the river principally reflect the dilution which such effluents receive.

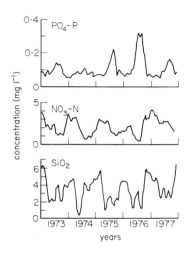

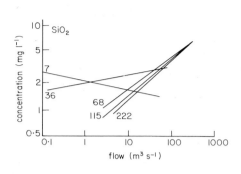

Fig. 3.6 Seasonal variation in concentrations of PO$_4$-P, NO$_3$-N and SiO$_2$ for the period 1973-1977 at Monmouth.

Fig. 3.7 Flow-concentration relationships for dissolved silica (SiO$_2$) at sites down the Wye. The distance (km) from source of each site is shown.

Such behaviour contrasts with that of nitrate (the predominant form of combined nitrogen) which generally has highest concentrations during winter high-flow periods. The increase in concentration was particularly marked during the autumn of 1976 when the drought was terminated by heavy rainfall (Fig. 3.6). Substantial increases (0.1 to 5.0 mg l^{-1} NO$_3$-N) were recorded in some lowland reaches; these probably resulted from flushing processes following the prolonged mineralisation of organic material in the soil during the long dry summer. In certain small lowland tributaries of the Lugg, such as the Frome, high concentrations of nitrate (7 mg l^{-1} NO$_3$-N) occur during summer and are probably derived from nitrate-rich ground waters, which at this period of the year receive little dilution from other more superficial sources.

The concentration-flow relationships of dissolved silica illustrate the complex distribution of substances which are utilized and transformed biologically (Fig. 3.7). At the uppermost site (7 km from source), concentration and water flow were negatively related, behaviour being similar to that of several other major ions, but progressively down the catchment the relationship between concentration and flow became increasingly positive. This rotation seems to be caused by the utilization of silica by diatoms which are abundant for part of the year: at low flows, the appreciably longer residence times (see Fig. 3.3) provide an opportunity for substantial silica removal. During the winter, when diatoms are not growing appreciably, there is no downstream rotation of the concentration-flow relationship, but merely a progressive increase in downstream concentration which reflects the changing catchment topography and soil characteristics. The importance of diatoms in silica depletion is supported by the negative correlation between concentrations of chlorophyll *a*, extracted from suspended material, and dissolved silicate in the lower catchment (Edwards *et al.*, 1978). It has been suggested by Edwards *et al.* (1978) that diatom growth within rivers of the catchment annually utilizes about 3000 tonnes (as SiO_2) and, even on tributaries of short retention time (<4 d), such as the Frome and Trothy, about 10% of the annual load is removed (Houston & Brooker, 1981).

3.24   CATCHMENT LOSSES

Oborne *et al.* (1980) estimated catchment losses of several determinands during two 12-month periods of contrasting hydrology - a dry period from September 1975 to August 1976 and a much wetter period from September 1976 to August 1977 (Fig. 3.8). There are substantial differences in yield between different parts of the catchment related to geology and land-use and between the wet and dry period. Annual yields per unit catchment of several substances, e.g. potassium, calcium, magnesium, nitrate, phosphate, were greater in the lower Wye and Lugg than elsewhere and this difference was generally most marked during the wet period. In contrast, highest yields of sodium, probably derived from rainfall, were in the upper reaches of the river which received the highest annual precipitation. The yield of phosphate could be accounted for solely from sewage effluents: in contrast, less than 20% of nitrate was derived from these sources, even during the dry period. By relating sub-catchment yields to sewered-population densities it is possible to calculate daily per capita outputs of phosphate and nitrate - these were between 1.8 and 3.1 g PO_4-P and 6.8 g NO_3-N, generally similar to values estimated from other studies (Oborne *et al.*, 1980).

3.25   WATER QUALITY MODEL OF THE WYE

Although flow-concentration regression equations may be used as simple empirical models to predict the effects of changing flow in the whole or parts of catchments, in the Wye these regressions rarely account for more than 70% of the variation. Water quality data for the lower reaches have been used to test a more complex model based upon the hydrological separation of river flow into surface run-off, base-flow and effluent components, the quality of each hydrological component being separately described (Oborne, 1981). Models of the behaviour of chloride, phosphorus, total hardness and alkalinity explained between 59 and 88% of test data after calibration (Fig. 3.9) and the performance of the model was considerably better than that of flow-concentration regressions derived from the same data (Sambrook, 1976).

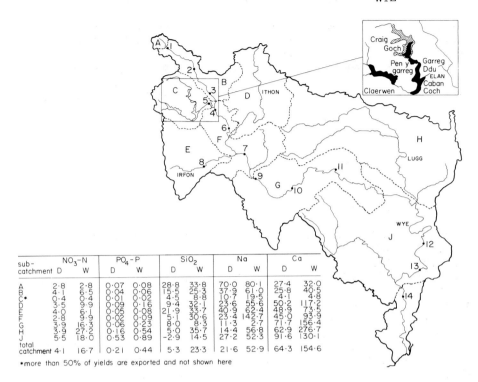

sub-catchment	NO$_3$-N D	W	PO$_4$-P D	W	SiO$_2$ D	W	Na D	W	Ca D	W
A	2·8	2·8	0·07	0·08	28·8	33·8	70·0	80·1	27·4	32·0
B	4·1	6·5	0·04	0·06	15·5	25·3	37·9	61·0	25·8	40·5
C*	0·4	0·4	0·01	0·02	4·5	8·8	10·7	19·5	4·1	4·8
D	3·5	9·9	0·09	0·16	9·4	32·1	23·6	55·6	50·2	117·2
E	4·0	6·1	0·05	0·08	21·9	33·7	40·9	62·4	48·9	73·5
F	2·8	9·9	0·02	0·09	5·1	30·6	23	142·7	45·0	93·9
G	3·9	16·3	0·06	0·23	8·0	8·3	11·3	2·7	71·7	156·4
H	3·9	27·2	0·16	0·54	5·0	35·7	14·4	56·8	62·9	276·7
J	5·5	18·0	0·53	0·89	-2·9	14·5	27·2	52·3	91·6	130·1
total catchment	4·1	16·7	0·21	0·44	5·3	23·3	21·6	52·9	64·3	154·6

*more than 50% of yields are exported and not shown here

Fig. 3.8 Yields (kg ha^{-1} yr^{-1}) of several substances in sub-catchments of the Wye (boundaries denoted by dotted lines) for dry (D) and wet (W) years. Numbers on map indicate sites used for invertebrate sampling (see Fig. 3.11). Insert shows reservoirs of Elan catchment with hatching giving the extent of proposed enlargement of Craig Goch.

3.3 PLANTS

3.31 ALGAE

The algae of the Wye have been described by Furet (1979) from water samples taken at several sites on the main river and at one site on each of the major tributaries. Two groups of species could be distinguished, those which grow planktonically, the 'bloom species' such as the coccoids (*Scenedesmus, Ankistrodesmus*) and discoid diatoms (*Cyclotella, Stephanodiscus, Thalassiosira*), and others which are adventitious, having been derived from benthic growth (*Synedra, Diatoma, Nitzschia*). On algal distribution, according to Furet, the river divides naturally into three zones:

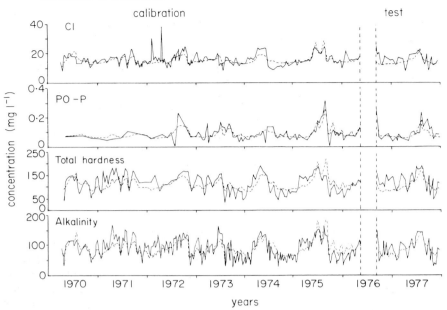

Fig. 3.9 Comparison of actual (-) and calculated (----) concentrations of chloride, PO_4-P, total hardness and alkalinity during calibration and test periods using a flow separation model (Oborne, 1981).

    i) upper zone, low in calcium (<10 mg l^{-1}), extends to km 60 from source and has less than 8 mg m^{-3} chlorophyll a derived from algae suspended in the water, with no obvious seasonal peak. Despite overall lack of seasonality, different algae are prevalent at different times: Meridion circulare occurs during the winter whereas Scenedesmus spp., Ankistrodesmus falcatus, Rhoicosphenia curvata and Navicula cryptocephala are dominant only during the summer. Several benthic diatoms are abundant (Synedra ulna, Diatoma vulgare, Nitzschia palea, N. amphibia)

    ii) middle zone, with calcium concentrations between 10 and 25 mg l^{-1}, extends from km 60 to 75, generally has chlorophyll a concentrations below 20 mg m^{-3} with peaks in June-July and November. As in the upper reach, Meridion circulare occurs during the winter and 'bloom species' (Scenedesmus spp., Cyclotella meneghiniana, Thalassiosira fluviatilis, Carteria spp.) during the summer. Several benthic diatoms are abundant during the spring (Melosira varians, Surirella ovata, Navicula viridula, Ceratoneis arcus, Diatoma vulgare, Pinnularia viridis) and their growth coincides with a decrease in dissolved silica concentrations.

    iii) lower zone, from km 75, with chlorophyll a peak concentrations between 40 (Ithon) and 100 (Lugg) mg m^{-3}, generally in June-July. Species composition is broadly similar to that upstream but Monorhaphidium gracilis, M. tortile and Oocystis crassa become abundant 'bloom species' during the summer.

Of the tributaries, the Elan is surprisingly poor in algae in view of its impounded source; nevertheless *Tabellaria flocculosa* and *T. fenestrata* are more abundant than elsewhere in the catchment.   In contrast, the Ithon, which drains an area of relatively low relief, is very rich in algae which are also abundant in the lower Wye, e.g. *Diatoma vulgare, Cocconeis placentula*.   The Irfon, although generally similar to the Upper Wye, is rich in the Chlorococcales, particularly species of *Scenedesmus*.   The Lugg has the highest algal concentrations in the Wye catchment and, like the Ithon, it drains an area of low relief; it is, however, much richer in plant nutrients, draining fertile farmland, much of it arable, in its lower course.

3.32 BRYOPHYTES

D. G. Merry (pers. comm.) surveyed the bryophytes at 49 riparian sites down the length of the main river and, despite the difficulty of separating aquatic from terrestrial species, of the 114 species he recorded 44 might be regarded as generally associated with wet stream margins or submerged habitats and only seven species as truly aquatic (Table 3.1).   Of the aquatic species which were widely distributed, *Nardia compressa, Scapania undulata* and *Fontinalis squamosa* were generally restricted to upland sites and *F. antipyretica* to the lower catchment.   Sites richest in bryophytes (>20 species per km) were all within 60 km of the source and above an altitude of 130 m.

3.33 PTERIDOPHYTES

These are neither common nor diverse in the riparian habitat; Merry (pers. comm.) recorded only nine species and, of these, five were only local - including *Lycopodium selago* at the highest site (600 m) and *Phyllitis scolopendrium* near the Carboniferous Limestone gorge at Symond's Yat.   The remaining four widely distributed species (*Equisetum arvense, Pteridium aquilinum, Dryopteris filix-mas, Athyrium filix-femina*) are common in Wales and not associated particularly with river banks.

3.34 ANGIOSPERMS

Although several 19th and early 20th century botanists described the aquatic and bank-side flora of limited sections of the Wye and, in more recent times, Perring (1967) and Haslam (1978) have focused on particular sites or tributaries, the first extensive vegetation survey of the main river was undertaken by D. G. Merry in 1976 and 1977.   During Merry's survey, 380 species of angiosperms were recorded, but of these only about 28 were truly aquatic and 88 riparian - the rest being adventitious to the river corridor.   This summary of his study will be confined principally to the aquatic species (Table 3.2).

Table 3.1   Bryophyte species associated with stream margins or truly aquatic*. (Binomials for mosses according to Smith (1978))

mosses	
Sphagnum auriculatum	*Brachythecium plumosus*
Atrichum crispum	*B. rivulare*
Polytrichum commune	*B. rutabulum*
Fissidens crassipes	*Drepanocladus exannulatus*
Campylopus atrovirens	*D. fluitans*
Dichodontium pellucidum	*D. uncinatus*
Dicranella palustris	*Rhynchostegium riparioides**
Cinclidotus fontinaloides	*Amblystegium fluviatile**
Racomitrium aciculare	*A. tenax**
R. aquaticum	
Physocomitrium pyriforme	liverworts
Bryum pseudotriquetrum	
Rhizomnium punctatum	*Conocephalum conicum*
Philonotis fontana	*Lunularia cruciata*
Pohlia proligera	*Marchantia polymorpha*
Hygrohypnum ochraceum	*Pellia epiphylla*
Hyocomium armoricum	*Riccardia pinguis*
Pohlia carnea	*Barbilophozia floerkei*
Climacium dendroides	*Chiloscyphus polyanthus*
*Fontinalis antipyretica**	*Diplophyllum albicans*
*F. squamosa**	*Marsupella emarginata*
Thamnobryum alopecurum	*Nardia compressa**
Calliergon stramineum	*Scapania undulata**
	Solenostoma triste

Table 3.2   Species of aquatic angiosperms found on the main river by Merry (1976-1977)

Ranunculus flammula	*Elodea canadensis*
R. omiophyllus	*Iris pseudacorus*
R. fluitans	*Sparganium erectum*
R. penicillatus	*Lemna minor*
R. peltatus	*Potamogeton crispus*
Rorippa nasturtium-aquaticum	*P. lucens*
Myriophyllum spicatum	*P. pectinatus*
M. verticillatum	*P. perfoliatus*
Callitriche hamulata	*P. polygonifolius*
Myosotis scorpioides	*P. salicifolius*
Veronica beccabunga	*Glyceria fluitans*
Mentha aquatica	*G. plicata*
Littorella uniflora	*Phalaris arundinacea*
Polygonum amphibium	
P. hydropiper	

Except for *Elodea canadensis* and most of the *Potamogeton* spp., the aquatic species showed some broad zonation (Fig. 3.10). At upland sites only *P. polygonifolius* and *Ranunculus omiophyllus* were found. Although not present at the uppermost sites (>350 m), *Callitriche hamulata*, *Glyceria fluitans* and *Myriophyllum verticillatum* occurred only in the upper 100 km of the river: the presence of *M. verticillatum*, normally found in eutrophic lentic habitats, is somewhat unusual. Only two species, *Veronica beccabbunga* and *Polygonum hydropiper*, were confined predominantly to the mid-reaches, the remainder being most abundant in lowland reaches. *Lemna minor* occurs only below the confluence of the Lugg and its occurrence in the main river depends upon production in back-waters of this tributary. Haslam (1978) found several aquatic and riparian species in the Lugg, including *Nuphar lutea*, *Schoenoplectus lacustris* and *Zannichellia palustris*, which have not been recorded in the main river.

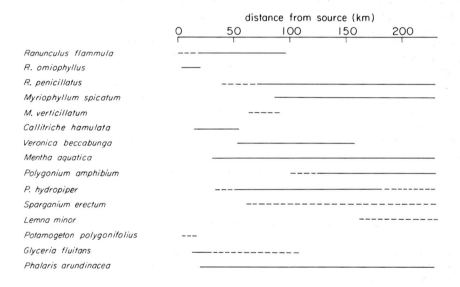

Fig. 3.10 Distribution of aquatic plants down the Wye. Dotted line indicates regions where species are very patchily distributed.

In the lowland reaches and in the Lugg *Ranunculus* spp.* grow profusely during the early summer, with a peak biomass around 2 kg fresh wt m^{-2} being recorded (Brooker *et al.*, 1978). There is considerable inter-year variation in growth which appears to be related to river flows during the growth period (April-June). The dense stands and their subsequent decay cause substantial changes in the oxygen economy

*In the lower Wye there is a mixture of *R. fluitans*, *R. penicillatus* and *R. peltatus* together with hybrids.

of the river, with pronounced daily oscillations in oxygen concentration
during the growth period. During the drought of 1976 the low oxygen
concentrations at night (<1 mg 1^{-1}), caused principally by plant respiration, coupled with high temperatures (>26 °C) caused the mortality
of substantial numbers[*] of adult salmon migrating upstream at the end
of June (Brooker *et al.*, 1977).

Such dense stands of river-plants can also affect dramatically river
hydraulics by increasing frictional resistance to water-flow, with
consequent increases in the danger of flooding, as well as the need to
measure, with increasing frequency, velocity-depth relationships, where
these are used as a basis of flow-gauging. Although plant stands
increased Manning's frictional coefficient roughly ten-fold at low
flows, from 0.02 to 0.25 on the Lugg and from 0.02 to 0.10 on the Wye
(Brooker *et al.*, 1978), the river channels are sufficiently large to
accommodate all normal summer flows, particularly as frictional
effects of plants are reduced at high flows.

Of the 88 or so riparian species, two deserve particular mention:
Allium schoenopraesum has become more widespread over the past century
around Builth Wells, but it occurs at relatively few sites elsewhere
in Britain and *Potentilla rupestre*, found at only one site in the middle
reaches has only been recorded in Britain at two other sites within
Central Wales.

3.4   MACROINVERTEBRATES

3.41   INTRODUCTION

Most studies of riverine macroinvertebrates in the United Kingdom,
other than those in small streams, have concentrated upon polluted
river systems and rivers, like the Wye, which are generally of high
water quality, have been neglected. Of course, many collectors have
visited the Wye over a considerable period and rare species, particularly mayflies, have been recorded from the river (Ratcliffe, 1977).
However, it was not until 1975 that a comprehensive systematic study
of the invertebrate fauna was undertaken. This work had several aims,
including a qualitative assessment of the spatial distribution of
macroinvertebrates in the catchment, particularly in the main river,
using survey methods (Brooker & Morris, 1980a, b; Morris & Brooker,
1979; Lilley *et al.*, 1979); a description of the temporal changes in
macroinvertebrate numbers, chiefly in the upper Wye (Brooker & Morris,
1978; Morris & Brooker, 1980) and an assessment of the drift of macroinvertebrates, particularly in relation to river flow (Brooker &
Hemsworth, 1978; Hemsworth & Brooker, 1979).

[*]About 1000 fish were collected from the river.

3.42   SPATIAL DISTRIBUTION

More than 230 different macroinvertebrates have been recorded from the Wye; these include, in the case of the midge (Diptera) larvae, species groups.   In general the upper reaches are characterised by stonefly (Plecoptera) nymphs in spring, mayfly (Ephemeroptera) nymphs, caddisfly (Trichoptera) larvae and midge larvae being relatively more abundant during the summer months.   In the lower reaches of the catchment the composition of the fauna is more variable, with oligochaetes, mayfly nymphs, caddisfly larvae, beetles and midge larvae all being relatively abundant at different locations and times.

Certain invertebrates (caddisfly larvae, beetles, mayfly nymphs) were found during the surveys to be sparse in the impounded tributary, the Elan, in comparison with the nearby naturally-flowing reaches of the Wye.   Such differences probably resulted from the relatively constant compensation flow that the river received from the impoundment and the iron- and manganese-rich deposits which coat the bed of the river (Brooker & Morris, 1980b; Inverarity et al., 1983).

Total densities of invertebrates varied throughout the catchment from less than 1000 m^{-2} to over 22000 m^{-2}, with highest densities in the lower reaches.   A simple analysis of the number of types of macro-invertebrates and total density indicated that both showed a marked increase downstream of the confluence of the Ithon with the Wye, the contribution of minerals from the Ithon substantially increasing the total cation concentration in the Wye from about 330 to 440 µeq. l^{-1}; Egglishaw & Morton (1965) found similar effects of chemical changes on invertebrates in rivers in the Scottish Highlands.

A more complex assessment of the distribution of invertebrates using rank correlation coefficients and clustering techniques followed by nodal analysis (Brooker & Morris,1980b) revealed that six groups of species or discrete taxa could be used to characterise five site groupings (Fig. 3.11), with the exception of one site each on the Elan and Irfon, these sites being on the main river.   However, a full inter-pretation of those environmental factors responsible for such distri-butions is not yet possible, partly because of the close association between changes in physical and chemical factors throughout the catch-ment and partly because of limited knowledge of the ecology of many of the 'key' species.

However, on the basis of studies in other rivers (Boon, 1979; Hildrew & Edington, 1979), on trichopterans belonging to the Hydro-psychidae, which are an important element of the stream fauna of the Wye, the characteristic grouping of species (Fig. 3.11) seems to be related to current velocity and water temperature.   Although the general pattern of distribution of species was similar in the Wye and the nearby Usk (Hildrew & Edington, 1979) some minor differences were apparent.   *Hydropsyche pellucidula* was found at all sites in the Wye but was not recorded from the upper reaches of the Usk whereas *H.*

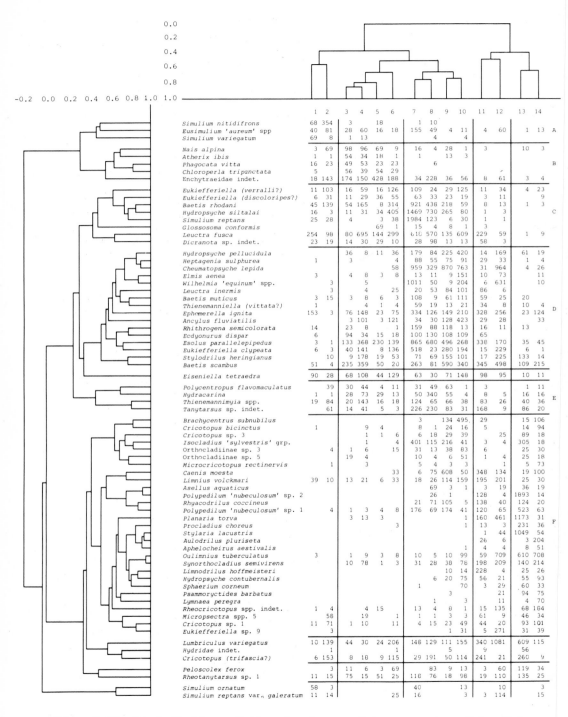

	1	2	3	4	5	6	7	8	9	10	11	12	13	14	
Simulium nitidifrons	68	354	3		18		1	10							A
Eusimulium 'aureum' spp	40	81	28	60	16	18	155	49	4	11	4	60	1	13	
Simulium variegatum	69	8	1	13					4	4					
Nais alpina	3	69	98	96	69	9	16	4	28	1	3		10	3	B
Atherix ibis	1	1	54	34	18	1	1		13	3					
Phagocata vitta	16	23	49	53	23	23	6								
Chloroperla tripunctata	5		56	39	54	29									
Enchytraeidae indet.	18	143	174	150	428	188	34	228	36	56	8	61	3	4	
Eukiefferiella (verralli?)	11	103	16	59	16	126	109	24	29	125	11	34	4	23	C
Eukiefferiella (discoloripes?)	6	31	11	29	36	55	63	33	23	19	3	11		9	
Baetis rhodani	45	139	54	165	8	314	921	438	218	59	8	13	1	3	
Hydropsyche siltalai	16	3	11	31	34	405	1469	730	265	80	1	3			
Simulium reptans	25	28	4		3	38	1984	123	6	30	1	1			
Glossosoma conformis				69	1		15	4	8	1	3				
Leuctra fusca	254	98	80	695	144	299	618	570	135	609	229	59	1	9	
Dicranota sp. indet.	23	19	14	30	29	10	28	98	13	13	58	3			
Hydropsyche pellucidula			36	8	11	36	179	84	225	420	14	169	61	19	D
Heptagenia sulphurea	1		3			4	88	55	75	91	29	33	1	4	
Cheumatopsyche lepida						58	959	329	870	763	31	964	4	26	
Elmis aenea	3		4	8	3	8	13	11	9	151	10	73		11	
Wilhelmia 'equinum' spp.		3		5			1011	50	9	204	6	631		10	
Leuctra inermis		3		4		25	20	53	84	101	86	6			
Baetis muticus	3	15	3	8	6	3	108	9	61	111	59	25	20		
Thienemanniella (vittata?)	1			4	1	4	59	19	13	21	34	8	10	4	
Ephemerella ignita	153	3	76	148	23	75	334	126	149	210	328	256	23	124	
Ancylus fluviatilis			3	101	3	121	34	30	128	423	29	28		33	
Rhithrogena semicolorata	14		23	8		1	159	88	118	13	16	11	13		
Ecdyonurus dispar	6		94	34	15	18	100	130	108	109	65				
Esolus parallelepipedus	3	1	133	368	230	139	865	680	496	268	338	170	35	45	
Eukiefferiella clypeata	6	3	40	141	8	136	518	23	280	194	15	229	6	1	
Stylodrilus heringianus		10	9	178	19	53	71	69	155	101	17	225	133	14	
Baetis scambus	51	4	235	359	50	20	263	81	590	340	345	498	109	215	
Eiseniella tetraedra	90	28	68	108	44	129	63	30	71	148	98	95	10	11	
Polycentropus flavomaculatus		39	30	44	4	11	31	49	63	1	3		1	11	E
Hydracarina	1	1	28	73	29	13	50	340	55	4	8	5	16	16	
Thienemannimyia spp.	19	84	20	143	16	18	124	65	66	38	83	26	40	36	
Tanytarsus sp. indet.		61	14	41	5	3	226	230	83	31	168	9	86	20	
Brachycentrus subnubilus							3		134	495	29		15	106	F
Cricotopus bicinctus	1			9	4		8	1	24	16	5		14	94	
Cricotopus sp. 3			1	1		6	6	18	29	39		25	89	18	
Isocladius 'sylvestris' grp.				1		4	401	115	216	41	3	4	305	18	
Orthocladiinae sp. 3		4	1	6		15	31	13	38	83	6		25	30	
Orthocladiinae sp. 5			19	4			10	4	6	51	1	4	25	18	
Microcricotopus rectinervis		1		3			5	4	3	3	1		5	73	
Caenis moesta						33	6	75	608	50	348	134	19	100	
Limnius volckmari	39	10	13	21	6	33	18	26	114	159	195	201	25	30	
Asellus aquaticus							69	3	1		3	19	36	19	
Polypedilum 'nubeculosum' sp. 2							26	1			128	4	1893	14	
Rhyacodrilus coccineus							21	71	105	5	138	40	124	20	
Polypedilum 'nubeculosum' sp. 1		4	1	3	4	8	176	69	174	41	120	65	523	63	
Planaria torva			3	13	3					1	160	461	1173	31	
Procladius choreus						3				1	13	3	231	36	
Stylaria lacustris											1	44	1049	54	
Aulodrilus pluriseta											26	6	3	204	
Aphelocheirus aestivalis										1	4	4	8	51	
Oulimnius tuberculatus	3		1	9	3	8	10	5	10	99	59	709	610	708	
Synorthocladius semivirens			10	78	1	3	31	28	38	78	198	209	140	214	
Limnodrilus hoffmeisteri									10	14	228	4	25	26	
Hydropsyche contubernalis								6	20	75	56	21	55	93	
Sphaerium corneum							1			70	3	29	60	33	
Psammoryctides barbatus									3			21	94	75	
Lymnaea peregra								1		3		11	4	70	
Rheocricotopus spp. indet.	1	4		4	15		13	4	8	1	15	135	68	164	
Micropsectra spp. 5		58		19			1	1	3	1	61	9	46	34	
Cricotopus sp. 1	11	71	1	10		11	4	15	23	49	44	20	93	101	
Eukiefferiella sp. 9	3								1	31	5	271	31	39	
Lumbriculus variegatus	10	139	44	30	24	206	148	129	111	155	340	1081	609	115	
Hydridae indet.		1				1			5		9		56		
Cricotopus (trifascia?)	6	153	8	18	9	115	29	191	50	114	241	21	260	9	
Peloscolex ferox		3	11	6	3	69		83	9	13	3	60	119	34	
Rheotanytarsus sp. 1	11	15	75	15	51	25	118	76	18	98	19	110	135	25	
Simulium ornatum	58	3					40		13			10		3	
Simulium reptans var. galeratum	11	14				25	16			3	3	114		15	

Fig. 3.11 Distribution of the abundance of taxa groups of benthic invertebrates in relation to site groups in the Wye catchment 1976–1977 (position of sites is shown in Fig. 3.8). Groups are derived from average linkage clustering of Spearman rank correlations.

siltalai, a species found throughout the Usk catchment, was not found in the most downstream 50 km of the Wye. In both rivers *H. contubernalis* and *Cheumatopsyche lepida* were restricted to downstream reaches although *C. lepida* was found further upstream in the Wye. *Diplectrona felix* and *Hydropsyche instabilis*, both species characteristic of headstreams, were not recorded from the Wye - probably because of the sampling locations - but were found in the Usk. Only *H. siltalai*, *H. pellucidula* and *H. contubernalis* were recorded from the N. Tyne in Northern England (Boon, 1978).

The distribution of other invertebrates in the Wye catchment may be related to changes in chemical composition of the water. For example, the snails *Sphaerium* and *Pisidium* were not found at mean concentrations of calcium below *c*. 8 mg l^{-1} and the crustaceans *Gammarus* and *Asellus* have similar distributions. In contrast the mollusc *Ancylus* was more widespread in the Wye catchment, being absent only from reaches with calcium concentrations less than *c*. 3 mg l^{-1}. Maitland (1966) found similar distributions of these organisms in the R. Endrick, Scotland.

Another characteristic of the Wye catchment resulting from river chemistry was the restricted fauna of the Elan, which receives water from Caban Coch reservoir. The stable flow and the iron- and manganese-rich (both soluble and particulate) water from the reservoir and from a water treatment works results in substantial deposits on the bed of the river and probably these have eliminated, or considerably reduced, the numbers of several species which might have been expected to inhabit these reaches. These include *Hydropsyche siltalai*, *H. pellucidula*, *Rhithrogena semicolorata*, *Ephemerella ignita* and the beetles *Esolus parallelepipedus* and *Oulimnius tuberculatus*. Such restrictions, particularly of beetles, are characteristic of pollution with ferric hydroxide (Scullion & Edwards, 1980) but have not been reported from other impounded rivers, such as the Tees in the U.K. (Armitage, 1978).

3.43 QUANTITATIVE ESTIMATES

Any assessment of changes in total numbers of benthic macroinverte-brates requires some knowledge of their distribution in the substrate with depth, since a substantial proportion of the population may not be collected by conventional surface sampling methods. Experiments in the Wye indicated that, overall, 59%, 23% and 18% of benthic invertebrates were recorded in the top (0 - 11 cm), middle (12 - 22 cm) and bottom (23 - 33 cm) levels of special basket samplers buried in the river sub-stratum (Morris & Brooker, 1979). Some organisms were generally con-fined to the top level, e.g. mayfly nymphs, blackfly (Simuliidae) larvae, snails, and could be satisfactorily characterised by surface samples (e.g. Brooker & Morris, 1978) but others were either more uniformly distributed (worms, midge larvae) or increased with depth at certain times (e.g. *Sericostoma* - a trichopteran larva).

Quantitative estimates of invertebrate density derived from intensive

sampling, were made at three sites in the upper catchment. Highest
densities (>12000 m^{-2}) were found at the site furthest downstream (km
48), but at all sites peak estimates were recorded during the summer,
particularly during the drought of 1976.

3.44 DRIFT

The drifting of macroinvertebrates is commonly characterised by seasonal
and daily periodicity. In the upper Wye, daily variations of drift
densities resulted principally from the activities of nymphs of mayflies and stoneflies and blackfly larvae and most often maximum drift
density was recorded during the period of darkness. However, on some
occasions maxima were recorded immediately before or after sunrise (Fig.
3.12) and during the long nights of winter two peaks in density were
recorded. There were some organisms, e.g. *Hydra* and water mites, which
had highest drift densities during daylight and, overall, most midge
larvae drifted during daylight.

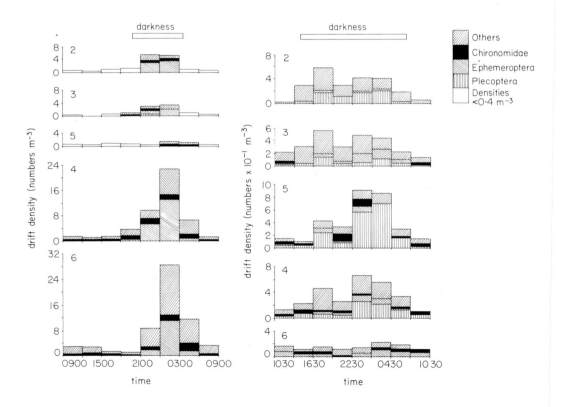

Fig. 3.12 Densities of drifting invertebrates at sites down the Wye
(position of sites shown in Fig. 3.8) over 24 h period in July and
December.

From estimates of mean daily density, which ranged from about 6×10^{-2} to 783×10^{-2} m^{-3} and with maxima generally recorded during July and August, it was estimated that downstream population displacement of selected insect species over the aquatic phase of the life cycle was not likely to be greater than 10 km (Hemsworth & Brooker, 1979) and the normal dispersive flight of adults provided adequate compensation for downstream losses.

The importance of changes in river flow to drifting organisms was illustrated when the flow was experimentally increased on the Wye from about 2 to 5 m^3 s^{-1} by a release of water from the Elan Valley reservoirs during May-June. On the first day of the release the number of macroinvertebrates drifting at a site 20 km below the reservoirs was about seven times that of the preceding day: it declined on the second day but was still three times greater (Fig. 3.13). Some organisms responded immediately to the increased flow, such as larvae of the midge *Rheotanytarsus* whilst others, such as nymphs of the mayfly *Ephemerella,* increased some eight hours later during the night - their normal time of maximum abundance in the drift.

3.45 AUTECOLOGICAL STUDIES

Certain species in the Wye catchment have been studied because of their rarity or intrinsic interest:

a) *Potamanthus luteus*: This mayfly is widely distributed in Europe, but confirmed records in the United Kingdom are restricted to the Thames, Usk and Wye. In the Wye the nymphs were relatively widespread - km 70 and downstream - but estimated densities (<50 m^{-2}) were very low and no specimens were found in tributaries to the main river. Water quality conditions at sites supporting this species varied considerably (Brooker & Morris, 1980a).

Little is known of the life history of this mayfly in the U.K., but imagos and subimagos have been collected at Hereford during July and it seems likely from changes in the size of nymphs that the species is univoltine.

b) *Austropotamobius pallipes*: The oldest records of the crayfish in the Wye catchment date from 1805 when it was reported that stocking of the Irfon took place: the species introduced was referred to as *Astacus fluviatilis* but was probably *Austropotamobius pallipes*. Although not found in routine surveys, a special investigation at 103 sites on 57 streams in the catchment (Lilley *et al.*, 1979) indicated its widespread distribution. It was not found, however, where the average calcium concentration was less than 6 mg l^{-1}. Although crayfish are known to have a high physiological requirement for calcium, other factors are clearly important, e.g. substrate, shading, current velocity, for it was rarely found at sites on the main river.

As in other studies, there were fewer males than females (0.6 : 1 in the Wye) and porcelain disease (*Thelohania contejeani*) was widespread, with at least 3.4% of the population affected.

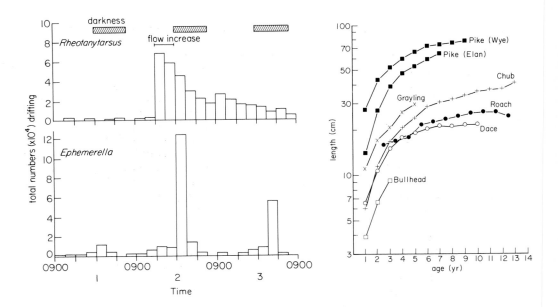

Fig. 3.13 Total numbers per 3 h of larvae of the midge *Rheotany-tarsus* and nymphs of the mayfly *Ephemerella ignita* before and during an increase in water flow resulting from a release from impoundments in the Elan Valley.

Fig. 3.14 Growth rates of several species of coarse fish in the Wye catchment. Pike growth both in the lower Wye and Elan is shown.

3.5 FISH AND FISHERIES

3.51 GENERAL

The Wye catchment contains many fish species (Table 3.3), some, such as the bleak and barbel being of very recent origin and probably introduced by anglers from neighbouring catchments. Of the non-salmonids, the chub, dace and pike are most profuse, judged by anglers' catches, particularly in the lower reaches of the river and its major tributaries. Several coarse fish species (chub, dace, roach, perch, pike) are widely distributed, but none extends its range above Rhayader on the main river where water-falls limit access to the uppermost reaches which are probably too shallow and turbulent to sustain such species. Other species are comparatively local in distribution, the ruffe being limited to the R. Monnow, the rudd to the middle reaches and the carp to the lower reaches of the Wye and the tench to the middle reaches of the Wye and to parts of the Monnow and Lugg catchments.

Table 3.3 Fish species recently recorded in rivers of the Wye catchment

Lampetra planeri	(brook lamprey)
L. fluviatilis	(river lamprey)
Petromyzon marinus	(sea lamprey)
Alosa fallax	(twaite shad)
A. alosa	(allis shad)
Salmo salar	(Atlantic salmon)
S. trutta trutta	(sea trout)
S. trutta fario	(brown trout)
S. gairdneri	(rainbow trout)
Thymallus thymallus	(grayling)
Esox lucius	(pike)
Cyprinus carpio	(carp)
Barbus barbus	(barbel)
Gobio gobio	(gudgeon)
Tinca tinca	(tench)
Abramis brama	(bream)
Alburnus alburnus	(bleak)
Phoxinus phoxinus	(minnow)
Scardinius erythrophthalmus	(rudd)
Rutilus rutilus	(roach)
Leuciscus cephalus	(chub)
L. leuciscus	(dace)
Noemacheilus barbatulus	(stoneloach)
Anguilla anguilla	(eel)
Gasterosteus aculeatus	(3-spined stickleback)
Perca fluviatilis	(perch)
Gymnocephalus cernua	(ruffe)
Cottus gobio	(bullhead)
Platichthys flesus	(flounder)

The growth rates of some species of coarse fish (Fig. 3.14), including the salmonid *Thymallus thymallus* (grayling), have been determined in various tributaries and lengths of main river (A. S. Gee & N. J. Milner, pers. comm.; Hellawell, 1969, 1971, 1972) and although the growth rate of most species compares favourably with that in other catchments in Southern Britain, that of grayling - which is widely distributed in the Wye and its larger tributaries - appears slower than that in some other European waters (Hellawell, 1969). The growth of pike in the main river near Monmouth has been compared with that in the Elan and whilst growth rates in the main river are similar to those in the Stour (Mann, 1976), Vistula (Backiel, 1971) and (lake) Windermere (Frost & Kipling, 1959),

growth rates in the Elan are appreciably lower. This is probably
associated with the lower spring and summer temperatures of this tributary, which receives water only from the Caban Coch reservoir.

Of the anadromous non-salmonids, the Allis and Twaite shads, members
of the herring family and locally distributed in Britain, are of particular interest. Both species migrate upstream as far as Builth Wells
to spawn during the early summer, after which survivors return to sea,
sometimes remaining in the Severn Estuary and Bristol Channel for
several weeks en route. The recently hatched fry follow in the autumn,
but it seems from the work of Claridge & Gardner (1978), who caught shad
on the screens of power stations in the Severn Estuary, that some fry
overwinter in the river, migrating to sea in late-April when river
temperatures rise rapidly.

The sea- and river-lampreys, although not frequently caught, move
upstream as far as Rhayader, over 200 km from the sea, to spawn when
freshwater flows are favourable, the river-lamprey also spawning very
extensively in the Lugg catchment.

Of the catadromous species, the eel, although abundant and commonly
weighing up to 1 kg, is no longer caught commercially within the catchment as it is on the neighbouring Severn where there is also a thriving
elver industry. Until recently an eel-trap was operated on the
Llynfi at the outlet to Llangorse Lake, a site where eel trapping had
been practised for many centuries. The flounder is common in the lower
Wye, its distribution extending upstream to Ross-on-Wye, about 60 km
from the estuary.

Brown trout are extensively distributed in the catchment and their
productivity and mobility have been studied at several sites in the
headwaters (Milner et al., 1978 and 1979). In those tributaries where
2+ and 3+ fish remain and contribute substantially (66 - 88%) to
production, productivity is generally very high, exceeding 15 g m^{-2}
yr^{-1}. In smaller tributaries from which larger trout have migrated
to deeper reaches, productivity is generally lower (<5 g m^{-2} yr^{-1}).
Rainbow trout, which have been stocked in the catchment, are known to
have spawned successfully in localised areas, fry having been found
upstream of Builth Wells in routine surveys during the late 1970s.
Sea trout have never been common the Wye or its neighbours the Usk and
Severn: this is in marked contrast to the generally shorter rivers
further west. Nevertheless, small numbers regularly ascend to the
upper Wye as far as Rhayader.

3.52 SALMON STOCKS

It is Atlantic salmon for which the Wye is justly famous although at
the end of the 19th century stocks were severely depleted by decades
of overfishing. Not only were adults netted throughout the lower
reaches of the rivers in the estuary, but juveniles were caught through-

out the catchment and poaching was rampant in the headwaters. The
stocks became so depleted that the almost valueless netting rights were
purchased cheaply by the Wye Fisheries Association which recovered its
expenses by leasing rod fishing. With the prohibition of river netting
in 1901 and severe restrictions on estuary netting - which was eventually controlled by the River Board - stock recovery was rapid. Gee &
Milner (1980) have analysed the catch statistics since 1905 and demonstrated certain important changes:

    i) although there has been a decline in net catch this has been
compensated by an increase in rod catch, sustaining the mean annual
total catch at about 6800.

    ii) growth rate at sea has declined over the past 70 years for
fish of more than one sea-winter (1 SW)

    iii) the average weight of salmon in both net and rod fisheries has
declined appreciably since 1930, average net caught weights being consistently lower than rod caught weights over this period.

The decrease in average weight of salmon caught is not solely caused
by the declining growth-rate of salmon at sea (ii above), but is principally associated with a change in age composition of the catch.
Before 1945 3 SW was the dominant age class of the rod fishery and
formed 60 - 65% of the catch but since that time 1 SW and 2SW salmon
have increased to comprise 20 and 60% of the catch respectively. 4 SW
fish and previous spawners have also declined, the latter from 7.6 to
1.6% of the catch (Fig. 3.15).

Gee & Milner (1980) attribute the change in age composition principally to the markedly increased catch effort by the rods over the past
few decades, and the preferential exploitation of the older fish, these
being the first to enter in the early spring in contrast to most
grilse (1 SW) which do not enter until late August - towards the end of
the fishing season. If the number of seasonal licences may be used as
an index of rod exploitation, the latter has increased by 650% since
1945. It has been suggested that at current exploitation rates,
virtually all the 3 SW and 4 SW fish are caught, few remaining to spawn.

This explanation of the change in age composition of Wye salmon
stocks, which has considerable significance in fishery management,
hinges on there being a genetic basis to the duration of sea-absence as
claimed by Elson (1973). Although there is now evidence for this
genetic basis (Gardner, 1976) the relative importance of parent age-
composition and environmental factors is not yet established.

A computer simulation model has been developed for salmon stocks of
the Wye, which can explore the effects of fishery exploitation and
other mortality factors on stock abundance and composition (Gee &
Radford, in prep. ; Gee & Edwards, 1980). If reasonable assumptions
about exploitation rates of adult salmon by the rod fishery are made

(5, 50 and 90% for 1, 2 and 3 SW fish) and it is assumed that there is
a genetic basis to sea-absence but with spawners of one sea-age producing some progeny of different sea-age (Gardner, 1976) then the composition of the stock responds in the model similarly to that in the
river (Fig. 3.16) with a substantial increase in the proportion of
grilse. The model also indicates that response of the stocks to
changing exploitation is rapid, a conclusion supported by evidence of
the quick recovery of stocks at the beginning of the 20th century when
netting was virtually banned throughout the Wye system.

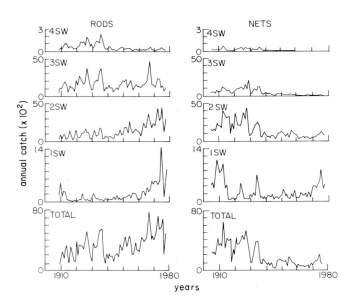

Fig.3.15 Annual rod and net catches of salmon (over a 70-year period)
classified by years at sea.

3.53 ECONOMIC VALUE OF FISHERY

It was estimated from a survey carried out in the mid-1970s and based
on postal questionnaires (Randerson *et al.*, 1978) that there were about
750000 angler visits per year to the river catchment, each angler making
about 20 visits. Of the total angler expenditure of £12.5 x 10^6
(adjusted to prices at September 1979) 88% may be attributed specifically to visits (entrance fees to sites, food, bait, travel), 11% to
basic angler expenditure (club membership, specialist literature,
replacement of tackle etc.)* and only 1% or £0.12 x 10^6 to licence

*Although each licence-holder who fished within the catchment spent on
 average £88.3 p.a. in this category, only 44% of angler visits were
 <u>within</u> the catchment so only £39 of this expenditure may be attributed
 to fishing within the area (total £1.38 x 10^6).

fees. It was suggested (Gee & Edwards, 1980) that because of the high
proportion of expenditure attributable directly to visits, the large
number of coarse fish anglers could contribute up to 50% of this expenditure. Expenditure of less than 30% on salmon fishing was considered
most unlikely and this would give each salmon caught a 'recreational
value' of £736 - contrasting with a 'commercial value' of less than £40
for fish caught in estuary nets.

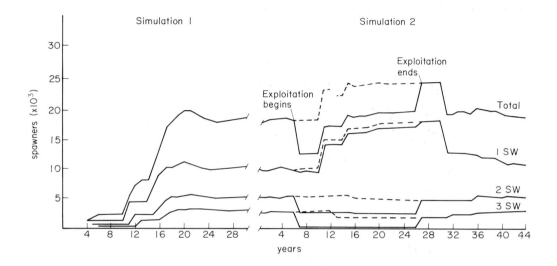

Fig. 3.16 Simulated effect of differential exploitation of large
fish (see text) on population structure. In Simulation 1 the equilibrium population of adult salmon, after recovery from a low number,
but with no exploitation in freshwater, is shown. In Simulation 2
differential exploitation of large fish causes their reduced spawning
escapement and an increase in that of grilse. The total salmon run
(dotted line) increases with exploitation at the particular cropping
level applied: the total escapement, after a brief initial period,
is similar to that without exploitation.

3.6   FUTURE DEVELOPMENTS

In 1973 the Water Resources Board proposed the enlargement of one of
the existing reservoirs in the Elan Valley, Craig Goch, to a size
(volume, 534 x 10^6 m^3; area, 1400 ha) that would make it one of the
largest in Europe, with sources, taken at times of high flow, from the
upper Wye and by transfers from the Severn, Ystwyth and Rheidol. At
that time it was envisaged that it would provide water for the West
Midlands, South Wales, Avon and Thames Valley. Recent reductions in
anticipated demand, coupled with alternative schemes for some demand
areas, have reduced the likely storage needs of Craig Goch reservoir

and it now seems probable that, with demand areas being limited to South Wales and the West Midlands, enough water can be derived from the immediate catchment of the reservoir and the adjacent upper Ystwyth catchment.   South Wales would be supplied by regulation of the Wye, with abstraction at Monmouth, and the West Midlands by regulation of the Severn, with abstraction above Worcester.

   The original scheme, in which water was being transferred to the Wye from several catchments, posed problems of the transfer of both macro-organisms such as fish species, not already present in the catchment, and micro-organisms, particularly fish pathogens.   There was also con-cern over salmon stocks in view of the homing behaviour of salmon and particularly the suggested role of race-specific (i.e. catchment-specific) pheromones in homing behaviour (Solomon, 1973):   even with recent modifications to the scheme, some water is transferred between catchments, including an export from the Wye to the Severn.   Neverthe-less, most effects of the Craig Goch scheme on the Wye will be associa-ted with the supplementation of low summer flows (<14 m^3 s^{-1} at Monmouth) to match abstractions in the lower river.   These possible effects derive principally from changes in velocity, depth and width of the river associated with increased flow and from changes in water quality consequent upon storage (Brooker, 1981; Edwards, 1984).

Fig. 3.17   Aerial view of Craig Goch Reservoir.

Effects in the lower catchment are also likely to be less discernible than those immediately downstream of the regulating discharge because of local disequilibrium effects produced by storage on factors such as oxygen and temperature, and because of the downstream dilution of regulating releases with water from natural sub-catchments.

Edwards (1984) attempts to predict the effects of regulation from Craig Goch reservoir on the Wye. Figs. 3.18 and 3.19 summarize local effects on the biological resources of the river and extensive effects on these resources and on water quality for abstraction. With respect to local effects, riparian and aquatic aspects of conservation are distinguished because in the upper catchment flow changes have a sub-stantial effect on river height and width (see Fig. 3.3). In view of the limited duration and degree of regulation in the summer months, extensive ecological effects are not likely to be pronounced but improvements in the salmon fishery, particularly resulting from an improved oxygen status, are foreseen. Local effects will probably be more diverse, but those resulting from reservoir stratification and hypolimnetic abstraction, such as low oxygen status and temperature together with high concentrations of iron and manganese, could be reduced or avoided by artificial destratification, now regarded as economically feasible for reservoirs of the size proposed.

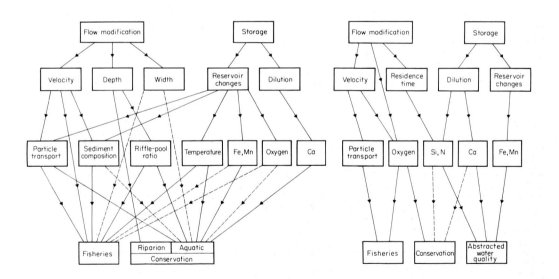

Fig. 3.18 Pathways of possible local effects of river regulation in the upper Wye on conservation and fisheries. Dotted lines indicate where effects are likely to be least significant.

Fig. 3.19 Pathways of possible effects of river regulation on con-servation and fisheries in the lower catchment and on the quality of abstracted water.

REFERENCES

Armitage P.D. (1978) Downstream changes in the composition, numbers and biomass of bottom fauna in the Tees below Cow Green Reservoir and in an unregulated tribuary Maize Beck in the first five years after impoundment. *Hydrobiologia* 58, 145-56

Backiel T. (1971) Production and food consumption of predatory fish in the Vistula River. *J. Fish. Biol.* 3, 369-405

Boon P.J. (1978) The pre-impoundment distribution of certain Trichoptera larvae in the North Tyne river system (Northern England) with particularly reference to current speed. *Hydrobiologia* 57, 167-74

Brooker M.P. (1981) The impact of impoundments on the downstream fisheries and general ecology of rivers. *Advances in Applied Biology*, Vol. VI. Academic Press, pp. 91-152

Brooker M.P. & Hemsworth R.J. (1978) The effect on the release of an artificial discharge of water on invertebrate drift in the River Wye, Wales. *Hydrobiologia* 59, 155-63

Brooker M.P. & Morris D.L. (1978) Production of two species of Ephemeroptera (*Ephemerella ignita* and *Rhithrogena semicolorata*) in the upper reaches of the R. Wye, Wales. *Verh. int. Verein. theor. angew. Limnol.* 20, 2600-604

Brooker M.P. & Morris D.L. (1980a) *Potamanthus luteus* (L.) (Ephemeroptera, Potamanthidae) in the R. Wye. *Entomol. Gaz.* 31, 247-51

Brooker M.P. & Morris D.L. (1980b) A survey of the macroinvertebrate riffle fauna of the River Wye. *Freshwat. Biol.* 10, 437-58

Brooker M.P., Morris D.L. & Hemsworth R.J. (1977) Mass mortalities of adult salmon (*Salmo salar*) in the R. Wye, 1976. *J. appl. Ecol.* 14, 409-17

Brooker M.P., Morris D.L. & Wilson C.J. (1978) Plant-flow relationships in the R. Wye catchment. In: *Proceedings 5th Symposium on Aquatic Weeds, 1978*, 63-70. European Weed Research Council

Claridge P.N. & Gardner D.C. (1978) Growth and movements of the twaite shad, *Alosa fallax* (Lacepede) in the Severn Estuary. *J. Fish. Biol.* 12, 203-11

Edwards R.W. (1984) Predicting the environmental impact of a major reservoir development. In: Roberts R.D. & Roberts R.M. (Eds) *Planning and Ecology.* Chapman & Hall

Edwards R.W., Oborne A.C., Brooker M.P. & Sambrook H.T. (1978) The behaviour and budgets of selected ions in the Wye catchment. *Verh. int. Verein. theor. angew. Limnol.* 20, 1418-422

Egglishaw H.J. & Morgan N.V. (1965) A survey of the bottom fauna of streams in the Scottish Highlands. Part II. The relationship of the fauna to the chemical and geological conditions. *Hydrobiologia* 30, 305-34

Elson P.F. (1973) Genetic polymorphism in Northwest Miramachi salmon, in relation to season of river ascent and age at maturation and its implications for management of the stocks. *International Commission for Northwest Atlantic Fisheries.* Research Document 73/76, 1-6

Frost W.E. & Kipling C. (1967) The determination of the age and growth of pike (*Esox lucius* L.) from scales and opercular bones. *J. Cons. perm. int. Explor. Mer.* 24, 314-41

Furet J.E. (1978) Algal studies on the River Wye system. Ph.D. Thesis.
    University of Wales (Cardiff)

Gardner M.L.G. (1976) A review of factors which may influence the sea-
    age and maturation of Atlantic salmon *Salmo salar* L. *J. Fish.*
    Biol. 9, 289-327

Gee A.S. & Edwards R.W. (1980) The recreational exploitation of the
    Atlantic salmon in the River Wye. *International Symposium on*
    Fisheries Resources Allocation (FAO), Vichy, France. 28 pp.

Gee A.S. & Milner N.J. (1980) Analysis of rod and net catch statistics
    for Atlantic salmon, *Salmo salar*, in the R. Wye, 1905-1977, and
    implications for fishery management. *J. appl. Ecol.* 17, 41-57

Haslam S.M. (1978) River plants: the macrophytic vegetation of water-
    courses. Cambridge Univ. Press

Hellawell J.M. (1969) Age determination and growth of the grayling,
    Thymallus thymallus L. of the River Lugg, Herefordshire, tributary
    of the River Wye. *J. Fish. Biol.* 1, 373-82

Hellawell J.M. (1971) The autecology of the chub, *Squalius cephalus*
    (L) of the R. Lugg and Afon Lynfi. 1. Age determination, population,
    structure and growth. *Freshwat. Biol.* 1, 29-60

Hellawell J.M. (1972) The growth, reproduction and food of the roach
    of the R. Lugg, Herefordshire. *J. Fish. Biol.* 4, 469-86

Hemsworth R.J. & Brooker M.P. (1979) The rate of downstream displace-
    ment of macroinvertebrates in the upper Wye, Wales. *Holarctic*
    Ecology 2, 130-36

Hey R.D. (1980) Regime equations for gravel-bed rivers. In:
    Engineering Problems in the Management of Gravel-bed Rivers.
    International Workshop: University of East Anglia

Hildrew A.G. & Edington J.M. (1979) Factors facilitating the co-
    existence of hydropsychid caddis larvae (Trichoptera) in the same
    river system. *J. Anim. Ecol.* 48, 557-76

Houston J. & Brooker M.P. (1981) A comparison of nutrient sources and
    behaviour in two lowland subcatchments of the R. Wye. *Wat. Res.* 15,
    49-57

Inverarity R.J., Rosehill G.C. & Brooker M.P. (1983) The effects of
    impoundment on the downstream macroinvertebrate riffle fauna of the
    River Elan, Mid-Wales. *Environmental Pollution* (A) 32, 245-267

Jones A. (1975) Rainfall, run-off and erosion in the upper Tywi catch-
    ment. Unpublished Ph.D. Thesis. University of Wales

Lilley A.J., Brooker M.P. & Edwards R.W. (1979) The distribution of
    the crayfish, *Austropotamobius pallipes* (Lereboullet) in the upper
    Wye catchment, Wales. *Nature in Wales* 16, 195-200

Maitland P.S. (1966) *Studies on Loch Lomond. 2. The Fauna of the*
    River Endrick. Univ. Glasgow, Scotland

Mann R.H.K. (1976) Observations on the age, growth, reproduction and
    food of the chub *Squalius cephalus* (L.) in the River Stour, Dorset.
    J. Fish. Biol. 8, 265-88

Milner N.J., Gee A.S. & Hemsworth R.J. (1978) The production of brown
    trout (*Salmo trutta* L.) in the tributaries of the upper Wye, Wales.
    J. Fish. Biol. 13, 599-612

Milner N.J., Gee A.S. & Hemsworth R.J. (1979) Recruitment and turnover
    of populations of brown trout, *Salmo trutta*, in the upper River Wye,
    Wales. *J. Fish. Biol.* 15, 211-22

Morris D.L. & Brooker M.P. (1979) The vertical distribution of macro-
    invertebrates in the substratum of the upper reaches of the River
    Wye, Wales. *Freshwat. Biol.* 9, 573-83
Morris D.L. & Brooker M.P. (1980) An assessment of the importance of
    the Chironomidae (Diptera) in Biological Surveillance, 195-202.
    In: Murray D.A. (Ed.) *Chironomidae, Ecology, Systematics, Cytology
    and Physiology.* Pergamon Press
Newson M.D. (1979) The results of ten years' experimental study on
    Plynlimon, mid-Wales, and their importance for the water industry.
    J. Inst. Eng. Sci. 33, 321-33
Oborne A.C. (1981) The application of a water-quality model to the R.
    Wye, Wales. *J. Hydrol.* 52, 59-70
Oborne A.C., Brooker M.P. & Edwards R.W. (1980) The chemistry of the
    River Wye. *J. Hydrol.* 45, 233-52
Perring F. (1957) Llandrindod Wells. August 4th-11th, 1956. Field
    Meetings. *Proc. bot. Soc. British Isles* 11, 416-18
Randerson P.F., Edwards R.W. & Shaw K. (1978) An economic evaluation
    of recreational fishing in areas of Wales, 203-218. In: *Recreational
    Freshwater Fisheries: their Conservation, Management and Development.*
    Water Research Centre Conference, Oxford 1977
Ratcliffe D.A. (1977) *A Nature Conservation Review*, 2 vols. Cambridge
    U. P., Cambridge
Sambrook H.T. (1976) Water quality: Wye. Unpublished M.Sc. Thesis.
    University of Wales (UWIST)
Scullion J. & Edwards R.W. (1980) The effects of coal industry pollu-
    tants on the macroinvertebrate fauna of a small river in the South
    Wales coalfield. *Freshwat. Biol.* 10, 141-62
Smith A.J.E. (1978) *The Moss Flora of Britain and Ireland.* 706 pp.
    Cambridge U. P., Cambridge
Solomon D.J. (1973) Evidence for pheromone-influenced homing by
    migrating Atlantic salmon, *Salmo salar* (L.). *Nature, Lond.* 244,
    231-32
Water Resources Board (1973) *Water Resources in England and Wales.*
    H.M.S.O., London

4 · Ebbw

R. W. EDWARDS, P. F. WILLIAMS & R. WILLIAMS

4.1   INTRODUCTION

The River Ebbw Fawr, which discharges into the Usk Estuary at Newport just before its confluence with the Severn Estuary (Fig. 4.2) is fed by two major tributaries, the Ebbw Fach and the Sirhowy, and several smaller ones. The river catchment, with a population of about 160000, is very small (226 km^2) near the eastern perimeter of the South Wales Coalfield. The justification for its inclusion in this book is based on its dramatic biological response, recorded intensively during the 1970s, to the abatement of industrial pollution.

The decline of the river began two centuries earlier, in 1764, with the establishment of the first major iron-works in the catchment at Tredegar and the rapid development of other works such as those at Sirhowy, Nantyglo and Ebbw Vale (Fig. 4.1). By the end of the 18th century there was a concentration of major iron-works along the north-ern rim of the coalfield at the heads of the valleys where there were supplies of iron-ore and of both limestone, used as a flux in the iron-making process, and coal, suitable for coking. The general location of these early iron-works at the headwaters of the rivers of the coalfield was important in establishing the patterns of later development and in its consequences with respect to river quality.

Fig. 4.1 An early 19th century engraving of industrialization on the banks of the R. Ebbw Fach at Nantyglo.

Iron-making had occurred within the catchment before the mid-18th century but these earlier furnaces were dependent on charcoal as a fuel and water power to operate bellows and this 'tyranny of wood and water' had produced only modest quantities of iron. It seems that the technological opportunities resulting from Abraham Darby's use of coke as a fuel (1756) and Smeaton's improved blast engine (1765), coupled with the demands of the Seven Years War (1756-63) and American War of Independence (1775-83), were responsible for the growth of the iron-making industry in the late 18th century.

At Ebbw Vale the development of steel-making started in 1854 although the first system was soon replaced by a process invented by Henry Bessemer in 1856. The other major iron-works within the catchment, not adapting to the new technology, declined and closed towards the end of the 19th century. Ebbw Vale flourished, however, to become one of the major integrated producers of iron, sheet-steel and tin-plate in the U.K., making up to 1 million tonnes of steel annually until in 1977 it too reduced its activities, to tin-plating and galvanising.

Early coal mining in the catchment exploited outcrops, and soil and debris were frequently separated from the coal by scouring with water from local rivers. During the period 1820-1840 the first shafts were introduced to exploit deeper seams, which extended over much of the catchment. This early development of the coalfield was dependent on the needs of iron-making but with improved transport, firstly by canals and later by railways, the production of coal increased and by the beginning of the 20th century the industry in South Wales employed over 200000 miners and raised 50 million tonnes per year. In the Ebbw catchment, the many small mines of the 19th century amalgamated, and the few resulting collieries continued to produce large quantities of coal. During the 1970s, although only 7 deep mines remained (Fig. 4.2), there was an extensive legacy of spoil heaps or tips above ground and interconnecting shafts and galleries below ground which continued to influence the water quality and hydrology of the river.

Urbanization in the deeply incised valleys of the catchment was squeezed along the valley floor (Fig. 4.2) with consequences for river management. Early industrial settlements had no sewerage systems or supplies of piped water. Edmund Head in 1841 (Gray-Jones, 1970) wrote:

> "Very few or none of the houses have privies: sewerage is
>   entirely unknown and the surface drainage is imperfect,
>   there is a great accumulation of filth around the houses .."

With such poor sanitary conditions it is not surprising that cholera epidemics occurred (1832, 1849, 1866), that of 1849 killing almost 2% of the inhabitants of the catchment. In 1878 the first reservoir, Llangynidr, was built to provide piped water and within a decade a sewerage system was constructed which conveyed wastes to small ponds before their discharge into the river. This system was not improved until 1912 when a trunk sewer was built to convey wastes to the coast.

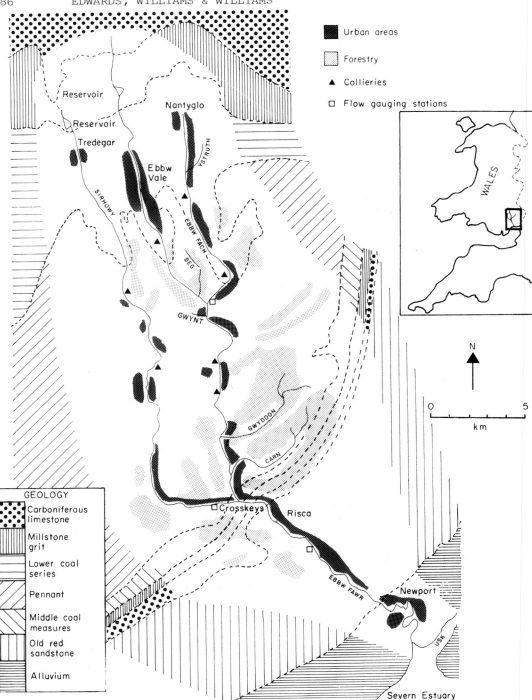

Fig. 4.2 Map of catchment (showing main land-use, geology, urban areas, tributaries, collieries).

This sewer, as in other valleys of the coalfield, was sited generally
adjacent to the river, with some lengths being along the river bed.
The capacity of such sewers was frequently exceeded by the intrusion of
flood-waters, inadequately separated from foul sewage, and the overflow
of sewage into rivers during storms was frequent. Roads were similarly
located adjacent to rivers and were raised appreciably to avoid flooding
in an area of capricious and plentiful rain. Tributaries, culverted
under such roads, were frequently inaccessible to fish migrating upstream from the main river, due to inadequacies in culvert design.

The major decline in water quality of the rivers of the Ebbw catchment took place in the early 19th century soon after the development of
iron-making. Certainly at the end of the 18th century both the rural
character of much of the valley and the good quality of the fishery
seemed beyond dispute for according to Edmund Jones writing in 1778
(Gray-Jones, 1970) the river abounded with eels, trout and sewin (sea-
trout) and Coxe in his 'Tour throughout Monmouthshire, 1781', wrote:

> "I spent many a happy day
> On Sirhowy's banks so gay
> Gathering nuts that grew about
> And deceiving the old trout"

Nevertheless there were local pollution problems even then, for Edmund
Jones noted that fish were becoming less plentiful, the river water
'being often troubled with the pond waters scouring the coal works'.
Some decades later there is evidence that other rivers of the coalfield
had become polluted and their fish stocks decimated by the coal- and
iron-workers. Of the Taff, some twenty kilometres to the west, Hansard
(1834) wrote:

> "The vicinity of Merthyr Tydfil has greatly contributed to
> render certain proportions of the Taf unworthy of anglers'
> attention. The poisonous matter discharged into it from
> the iron-works, and the lawless practices of the forge-men,
> continually diminish the stock of fish. In dry seasons these
> depredators assemble in bands, and, wading into the streams
> armed with sledge hammers, contrive, by violently striking
> the stones under which the trout are concealed, to destroy
> an incredible quantity of fish of all sizes".

By 1861, from the Report of the Commissioners appointed to enquire
into Salmon Fisheries (England and Wales), it was clear that the Ebbw
was grossly polluted and, although there were a few trout remaining,
these could not be eaten because of the taste imparted by tars and acids:

> "There are works up the river for making what is called
> pyroligneous acid. The Ebbw receives all the sewage and
> all the filth from Bryn Mawr, Nant-y-Glo, Coalbrook Vale,
> Blaina, Cwm Celyn, Ebbw Vale, Beaufort and Victoria Works.
> Then at Abercarn there are the tar-works and at Risca there
> is another pyroligneous acid manufactory".

It was about a century later that the Usk River Authority proposed pollution control measures to improve river quality for the purpose of industrial abstraction and to restore life to the river by:

   i)   commissioning a full-scale treatment plant at the
         Ebbw Vale steel-works ;

  ii)   constructing a large capacity trunk sewer to carry
         sewage to a coastal outfall;

 iii)   improving the standard of waste treatment at five coal
         washeries remaining in the catchment.

The full treatment of the steel-works effluent, the most significant aspect of pollution abatement, was not accomplished until the autumn of 1973 although there were periods during the previous four years when partial treatment was achieved. The replacement of the trunk sewer was started in 1971, construction being progressively upstream and reaching Cwm in 1980. Only sections within the valley of the Ebbw Fach remain to be completed.

4.2   GEOLOGY, HYDROLOGY AND WATER QUALITY

4.21   GEOLOGY

The South Wales Coalfield is a complex syncline bounded by an escarpment of Old Red Sandstone (Fig. 4.2). Below the Coal Measures - deposited in a brackish or freshwater environment and classified by George (1970) into Upper, Middle and Lower Coal Measures - are strata of Carboniferous Limestone and Millstone Grit. The Middle and Lower Coal Measures, which over most of the coalfield can only be reached by sinking deep shafts, have the thickest and most important coal seams. Associated with these measures, in some areas, is iron ore, in the form of chalybite, occurring as nodules or bands. The exposure of these measures, and the associated limestone and iron-ore, at the northern rim of the coalfield, were major factors in the early industrialisation of the area. The Upper Coal Measures contain Pennant Sandstone and form the undulating plateau of the Ebbw catchment.

The Carboniferous rocks have been extensively mineralised, the sulphides of lead, zinc, copper-iron and nickel being common with those of cobalt and iron-nickel also occurring. Although not commercially exploitable, extensive mining operations for iron ore and coal have exposed these sulphides to oxidation, allowing heavy metals to leach into local waters.

4.22   HYDROLOGY

Annual precipitation in the catchment is high, varying from 1650 mm in the northern uplands to about 1200 mm near the mouth of the river. Fig. 4.3 shows the seasonal pattern of precipitation at Tredegar, 3 km from

Ebbw Vale, during the period 1969-1978. At this site, representative
of the headwaters, precipitation during the winter (November-January)
is approximately twice that of the period April-July.

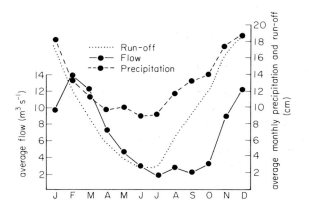

Fig. 4.3 Monthly averages of daily flows for 1977-1979 near Risca
and average monthly precipitation and run-off for the northern catchment during the period 1969-1978.

Of this precipitation about 70% flows down the river, evapotranspiration accounting for most of the difference. River flows have a pronounced seasonal rhythm (Fig. 4.3), mean river flows in the winter being
about seven times those of July. Low summer flows reflect the enhanced
evapotranspiration rates during this period as well as reduced precipitation. A reduced January flow probably reflects the storage of snow,
particularly at the high altitudes (up to 500 m) in the northern part
of the catchment. Whilst net precipitation increases in August, river
flows do not increase appreciably until November, the delay being caused
by soil and underground storage.

Of the average flow of the river at Rhiwderin, near Risca (Fig. 4.2),
about 7 m^3 s^{-1}, minewaters contribute 0.5 m^3 s^{-1} or 7%. During drought
periods the river flow may decrease to values below 1 m^3 s^{-1} and during
such periods water pumped from underground mines is extremely important
in maintaining flows, representing more than 50% of the river flow.
Although there are outcrops of carboniferous limestone on the northern
rim of the catchment (Fig. 4.2) these do not appear to provide a major
component of storage (Parsons & Hunter, pers. comm.).

4.23   EFFLUENT TREATMENT

The major improvements in water quality during the 1970s resulted from
the introduction of improved methods of waste treatment at Ebbw Vale
steel-works, the treatment plant operating intermittently in 1971 and
becoming fully effective late in 1973.

  Effluents from steel-making and associated activities derive principally from pickling (sulphuric and hydrochloric acids, ferrous sulphate,
heavy metals), electro-plating (phenols derived from phenol-sulphonic
acid), tin-plating (chromium) and coke-making (ammonia, cyanide, phenols):
tars and oils are generally present and temperatures are high.

  Before 1970 treatment consisted of acid neutralization, initially
with powdered lime and subsequently with milk of lime, followed by a
brief period of settlement. Neutralization procedures were inadequately
controlled and violent fluctuations in pH occurred. Although some
metals were partially precipitated, other pollutants were discharged to
the river. The ferrous sulphate became oxidized and hydrated leaving
the upper reaches of the river deficient in oxygen and rich in particulate and colloidal ferric compounds. Improved treatment involved the
separation of strong and rinse pickling-liquors, followed by rapid
oxygenation, pH regulation with calcium carbonate, and then solids
removal. Additionally, closure of the coke-ovens in 1972 reduced the
inputs of ammonia, cyanide and phenol. The replacement of sulphuric
acid by hydrochloric acid for rinsing in 1975 led to further improvements, for calcium chloride formed in treatment, being more soluble
than calcium sulphate, does not coat lime particles and impede reaction.

  The coal washeries within the area have over the past twenty years
also introduced more effective techniques to remove particulate wastes
from discharges to the river. There remains, however, the problem of
disposing of wastes and although at some sites some fine particles are
disposed of underground, most waste material is tipped and, unless the
tips are stabilized and re-vegetated, surface run-off carries such
particles to water-courses.

  During the 1970s, the trunk sewer, which carried wastes to Newport,
but which, being overloaded frequently, discharged sewage to the river
during rainy periods, was gradually replaced by one of larger capacity.

4.24   WATER QUALITY

Before the treatment plant of the Ebbw Vale steel-works became fully
operational in 1973 several aspects of water quality prevented the
establishment of a diverse flora and fauna (Fig. 4.4). Dissolved
oxygen concentrations were occasionally negligible throughout the length
of the river despite the high reaeration capacity of the river which
was associated with its shallow and generally-fast current. Although
intermittent discharges of sewage exacerbated the problem, the major

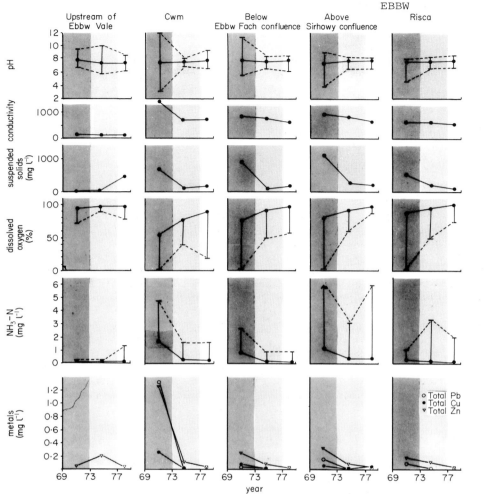

Fig. 4.4 Aspects of water quality at sites down the Ebbw Fawr during the period of improvement. Three phases are shown:

1) up to 1973 the steel-works effluent was not fully operational;
2) 1973-1976 effluent quality improved dramatically;
3) 1976-1980 improved quality sustained.

Average values are shown by continuous lines and, where relevant, range limits are shown by hatched lines.

cause of deoxygenation was the considerable discharge of ferrous salts. Values of pH oscillated widely, the range at Cwm, 3 km downstream of Ebbw Vale, being from 3 to 12. Even 13 km downstream the pH range was between 4 and 9. Several soluble poisons were present in high concentrations, particularly ammonia, phenols and heavy metals (Cr, Ni, Cu, Zn, Cd, Pb) and the presence of these would have prevented the establishment of most organisms. The average suspended solids concentration, except in the headwaters, exceeded 400 mg l^{-1}, well above the range

regarded as acceptable for a viable fishery (E.I.F.A.C., 1964). Whilst
a high proportion of these solids was derived from coal wastes (pumped
mine-waters, washery discharges, tip run-off) there were high concentrations of total iron, averaging about 300 mg l^{-1} below the steel-
works. These subsequently settled and coated the bed with heavy
deposits of ferric compounds.

After 1973 there were widespread improvements in water quality.
Dissolved oxygen concentrations, although occasionally low, were generally above 50% air saturation (Fig. 4.4) and pH fluctuations were
reduced to acceptable limits (pH 6 - 9) for many organisms (Scullion &
Edwards, 1980a). Total iron concentrations downstream of the steel-
works were reduced one hundredfold, contributing to the dramatic
decrease in total suspended solids concentrations but these, which are
now derived principally from the activities of the coal industry, have
since remained at about 200 mg l^{-1}, a value too high for the maintenance
of a successful fishery (Scullion & Edwards, 1980b) and for the establishment of a normal benthic fauna (Scullion & Edwards, 1980a). The
concentrations of soluble poisons also decreased, particularly of heavy
metals and phenols. Nevertheless occasionally high concentrations of
ammonia still occur in the lower catchment.

The toxicity of soluble poisons to fish may be summed when their
concentrations are expressed as fractions of their toxic threshold
concentrations (Brown, 1968). Fig. 4.5 shows that median and 95 percentile summed values for soluble poisons at sites downstream of the
steel-works during the period 1974-76 were less than 0.1 and 0.4
respectively and generally within a range considered tolerable to a
wide range of organisms (Alabaster et al., 1972).

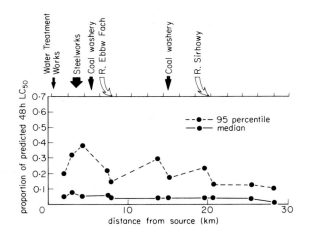

Fig. 4.5 Median and 95-percentile values of calculated toxicity to
rainbow trout of water at sites down the Ebbw Fawr during the period
1974-1976.

In most rivers the concentrations of many water quality determinands
are related to water flow, a log. concentration - log. flow regression
accounting for much of the variation in quality, but in the Ebbw, comparatively few, e.g. total hardness, are so related (Fig. 4.6a). The
concentration of suspended solids is particularly flow-dependent in
many rivers, concentrations being considerably increased at high flows.
This is not so in the Ebbw (Fig. 4.6b) where coal washeries exert their
major effect on river quality at low flows and catchment run-off, particularly from spoil heaps, is most pronounced at high flows during periods
of rainfall. Similarly, chloride concentrations are not markedly flow-
dependent (Fig. 4.6c). In many catchments sewage effluent provides a
major constant source of chloride and consequently river concentrations
decrease at high flows. In contrast, in the Ebbw catchment, with no
major treatment works, sewage enters principally at high flows as storm
sewage. However, the annual contribution of chloride from sewage, even
if all were discharged to the river, is only about 1000 tonnes or 20%
of the total chloride yield of the catchment. The steel-works effluent,
influencing river concentrations at low flows and mine drainage, making
its principal contribution after periods of heavy rainfall, are also
major sources of chloride.

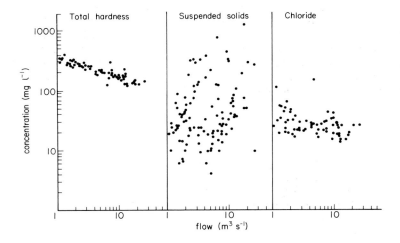

Fig. 4.6 Relationship between flow and aspects of water quality
(water hardness, suspended solids and chloride) at Rhiwderin (gauging
station on Ebbw Fawr in Fig. 4.2) during the period 1976-1979.

4.3 PLANTS AND MACROINVERTEBRATES

4.31 INTRODUCTION

Sampling of the benthic flora of the Ebbw was carried out intermittently between 1958 and 1970, demonstrating that there was no change in the gross impoverishment of the river during this period (Benson-Evans & Williams, 1970), before improvements in water quality occurred following treatment of the steel-works effluent. Between 1970 and 1975 regular, more-detailed surveys of both plants and macroinvertebrates were carried out at sites on the main river and its major tributaries, the principal objectives being to record the sequence of recolonization stages downstream of the steel-works effluent as water quality improved and to assess the influence of tributaries on such recolonization.

During each survey an estimate of cover of the major benthic plant communities was made at each site and samples taken for subsequent microscopic examination. Macroinvertebrates were collected by kick-sampling, a procedure previously used in surveys of neighbouring catchments (Edwards *et al.*, 1972).

Until 1972 there was no visible plant growth in the main river downstream of Ebbw Vale. Even sewage fungus was restricted to the middle and lower reaches where it covered the bed of the river for most of the year (Fig. 4.7). In the spring of 1972 visible areas of green algal growth were noticed in downstream sites and a year later these had extended upstream to Cwm. Nevertheless in those reaches immediately below the Ebbw Vale steel-works sewage fungus continued to dominate until 1975, the end of the period of detailed surveys: in contrast its importance at downstream sites had declined by 1974.

Macroinvertebrates, particularly naidid and tubificid worms, chironomid midges and mayflies (principally *Baetis rhodani*) were present in small numbers at downstream sites in 1970, the first year of systematic observations (Fig. 4.8). Some of these, particularly the mayflies, could well have drifted downstream from tributaries. Recolonization at Cwm, 1 km below Ebbw Vale, was delayed until 1973 despite the high density of macroinvertebrates in the unpolluted headwaters. Although there was a general increase in invertebrate abundance throughout the investigation at downstream sites, nowhere did abundance approach that of sites in the headwaters and major tributaries. Nevertheless during 1974 and 1975 the density of macroinvertebrates in the lower reaches was similar to that in other rivers of the coalfield, such as the River Taff (Edwards *et al.*, 1972).

4.32 COMMUNITY DIVERSITY

In order to describe aspects of the change in structure of both plant and animal communities, several indices were computed. Fig. 4.9 shows values for one of these, the Shannon Diversity Index, which reflects

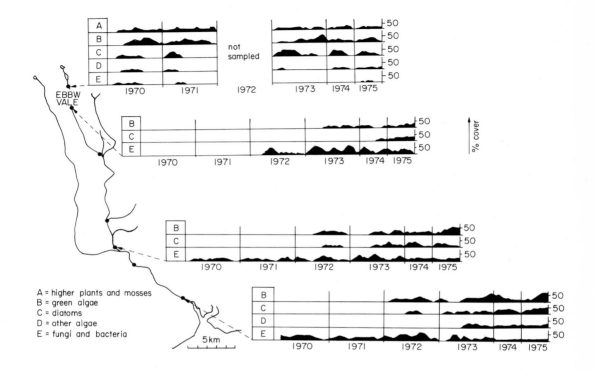

Fig. 4.7 Seasonal changes in percentage cover of groups of plants and bacteria at sites above and below Ebbw Vale during 1970-1975. In the period 1970-1973 surveys were carried out in spring, summer, autumn and winter; in 1974 and 1975 only spring and autumn surveys were conducted (see also Figs. 4.8 - 4.11).

not only the number of taxa but also the equitability of their abundance and has been widely used in surveys of rivers in South Wales (Edwards *et al.*, 1972; Williams *et al.*, 1974; Williams *et al.*, 1976).

Shannon diversity values for communities of plants and bacteria increased dramatically during the summer of 1972 with algae dominating except immediately below Ebbw Vale where fungi and bacteria contributed most to the diversity of the benthic communities. Plant diversity continued to increase during 1974 and 1975, with particularly high values during the summer, and was very similar to that recorded for polluted rivers elsewhere in the coalfield (Edwards *et al.*, 1972; Williams *et al.*, 1974). Indeed, diversity values in 1975 were very similar to those in unpolluted reaches upstream of Ebbw Vale and higher than those in certain tributaries, e.g. Ebbw Fach.

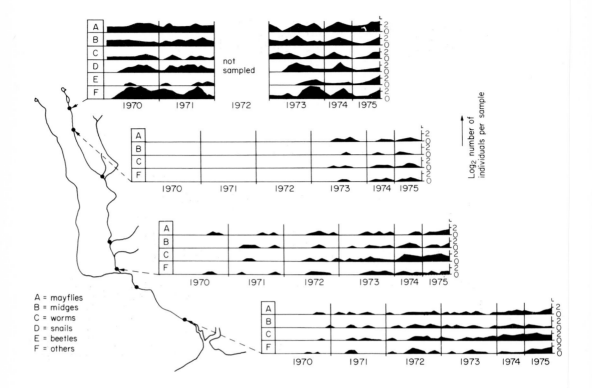

Fig. 4.8 Seasonal changes in the abundance of macroinvertebrates at sites above and below Ebbw Vale during 1970-1975.

Values of macroinvertebrate diversity shown in Fig. 4.9 were lower than those of plants but this reflects, to some extent, the level of identification rather than a fundamental difference in 'community' structure. In the early years when only small numbers of animals were found, those drifting downstream could have been very significant. As with plants, there was a marked increase in diversity during the summer of 1972 in the middle and lower reaches to values comparable with un-polluted sites which was sustained until the end of the study period. Immediately downstream of Ebbw Vale comparable diversity values were not achieved until 1975.

The Trent Biotic Index, based on the variation in response of macro-invertebrates to predominantly-organic pollution and on a scale from 0 (grossly polluted) to 10 (clean), has been widely used to classify river quality (Woodiwiss, 1964). In Fig. 4.10 values based on the total sample are shown for the period 1970-1975 but it is possible that certain sample components have drifted downstream, particularly from clean tributaries, and influenced values of the index. The downstream

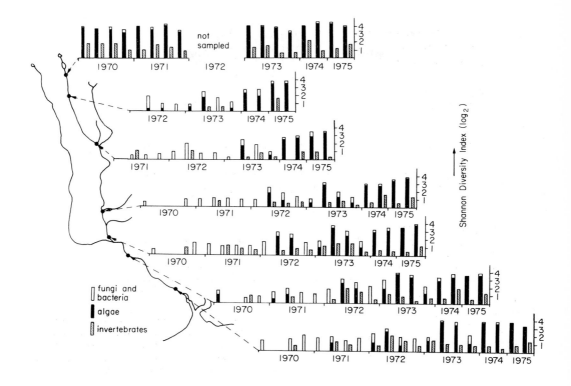

Fig. 4.9 Seasonal changes in the Shannon Diversity Index for bacteria, plants and macroinvertebrates at sites above and below Ebbw Vale during 1970-1975.

improvement, seen first in 1972, occurred progressively later with nearness to Ebbw Vale. By 1975 values of the Trent Biotic Index were still generally below 5 as compared with 6 or 7 upstream of the steel-works effluent.

4.33 QUALITATIVE ASPECTS OF COMMUNITY STRUCTURE

Whilst there were no benthic plant communities in the 5 km reach below Ebbw Vale in 1971 and 1972, water quality was such that a sewage fungus association dominated by *Fusarium aqueductum* was able to grow in the middle and lower reaches. With increasing distance downstream *Fusarium* was mixed with bacterial zoogloea. The upstream dominance of *F. aqueductum* is interesting for it is a major component in only about 0.6% of sewage-fungus outbreaks in rivers of the U.K. (Curtis & Curds, 1971). Several workers describe it as favouring acid waters (Paulsen, 1962) although Painter (1954) demonstrates its ability to grow well in

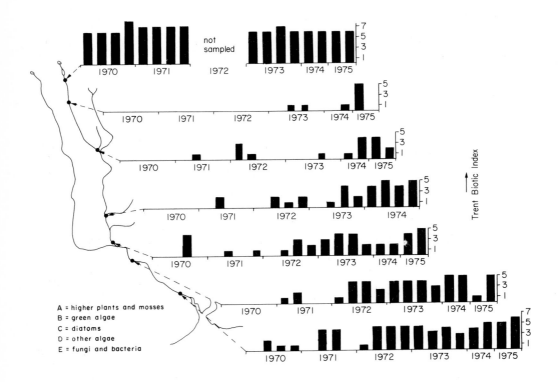

Fig. 4.10 Seasonal changes in the Trent Biotic Index for macro-invertebrates at sites above and below Ebbw Vale during 1970-1975.

culture within the pH range 4 to 9 - a characteristic which might account for its dominance in the River Ebbw where pH ranged between 3 and 12 at upstream sites (Fig. 4.4) during the period 1970 to 1973. In 1972 *Sphaerotilus natans*, the usual dominant of the sewage-fungus association (Curtis & Curds, 1971), became more abundant at downstream sites as did *Stigeoclonium tenue*, a green alga often associated with polluted areas (Curtis, 1969), which, in the absence of *Cladophora glomerata*, is often an indication of combined high concentrations of heavy metals and nutrients (Whitton & Say, 1975; Whitton, 1979). Another alga which became widespread during these early recovery stages in the lower reaches was *Navicula avenacea*, where it was found in mucilage tubes rather than the normal free-living condition. Similar tubes were found in a neighbouring river, the Afon Lwyd, in 1962 when that river was recovering from pollution after the closure of a steel-works (Benson-Evans & Williams, 1970).

These initial stages of recovery were followed by an increase in algal diversity, particularly with the establishment of diatoms of the

genera *Navicula, Achnanthes, Nitzschia, Surirella, Gomphonema* and
Diatoma. In the summer of 1973 *Cladophora* and two diatoms, *Cocconeis
placentula* and *Diatoma vulgare*, became widespread in the lower stretch.
Cladophora did not recolonize the stretch immediately below Ebbw Vale
during the study period.

Field observations in the initial recovery phase of the early 1970s
were supplemented by *in situ* exposure of *Cladophora* and the mosses
Rhynchostegium riparioides and *Fontinalis antipyretica* in perspex and
nylon cages at sites in the main river for periods up to 20 days during
the winter and at a control site of a clean tributary from which 'experimental' plant material was obtained. A scale of damage from 0 (healthy)
to 4 (dead and decomposing) was used and Fig. 4.11 summarises the
results for *Cladophora*; mosses behaved very similarly. At Cwm, 3 km
downstream of Ebbw Vale, no observations were undertaken until 1973 when
death of *Cladophora* was rapid; in 1974 there was a slow breakdown of the
plant material - behaviour consistent with the absence of *Cladophora*
from the site during this period. Immediately above the Ebbw-Sirhowy
confluence, no observations were made after 1973, at which time, although
water quality was much improved, plant damage was still occurring. At
the lowest site studied, Newport, little damage occurred in 1973 and
1974; this is consistent with the natural recolonization of the lower
reaches with *Cladophora* during this period.

After 1975 intermittent surveys have shown that diverse algal communities containing *C. glomerata* have extended upstream to within 5 km of
Ebbw Vale, with a mixed community of *Stigeoclonium tenue*, diatoms
(including *Navicula mutica*) and sewage-fungus occurring immediately
below Ebbw Vale.

Were the Ebbw Fawr unpolluted, bryophytes would almost certainly be
abundant, with liverworts such as *Scapania undulata, Solenostoma triste*
and *Nardia compressa* upstream and the mosses *Fontinalis squamosa, F.
antipyretica* and *Rhynchostegium riparioides* more downstream. Aquatic
encrusting lichens (e.g. *Verrucaria* spp., *Dermatocarpon fluviatile*),
algae (e.g. *Phormidium, Lemanea, Hildenbrandia*) and several filamentous
and encrusting members of the Chlorophyta would probably be widespread
(N.T.H. Holmes, pers. comm.).

Most of the later surveys have, in contrast with earlier observations,
focussed on macroinvertebrates and have shown continued faunal recovery.
The Trent Biotic Index score had by 1977 improved to between 5 and 7
at all sites, except immediately downstream of Ebbw Vale where it
remained at two. The fauna (Table 4.1) was regarded as typical of
moderately polluted rivers of the coalfield, containing molluscs, mayflies, caddis-flies, midge larvae, leeches and worms (Enchytraeidae,
Tubificidae and Naididae). Although *Baetis rhodani*, the most tolerant
of the mayflies to pollution, was exceptionally abundant, others
(*Ephemerella ignita* and *Ecdyonurus dispar*) had begun to recolonize
the lower reaches. Despite this marked improvement in species richness
the biotic score based on that of Chandler (1970) was, in 1977, only

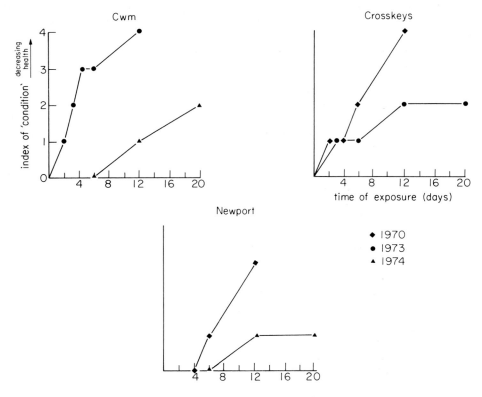

Fig. 4.11 Condition of *Cladophora* placed in cages for up to 20 days at sites in the Ebbw Fawr. Condition is represented on scale of 0 (healthy) to 4 (dead and decomposing).

between 15 and 30% of that which might be expected for an unpolluted river of the region.

4.4 FISH

4.41 DISTRIBUTION IN CATCHMENT

Before the improved waste-treatment at Ebbw Vale steel-works in 1973, fish were confined to the headwaters of the Ebbw Fawr and to the tributaries. The two major tributaries, the Ebbw Fach and Sirhowy, were both intermittently polluted by washery effluents and material from colliery waste tips adjacent to the river. Despite these problems several species were present in both the Sirhowy (minnow, stickleback, bullhead, stoneloach, brown trout, eel) and Ebbw Fach (stickleback, stoneloach, brown trout) and fishing clubs, particularly on the lower

Table 4.1 Macroinvertebrates found at sites in the Ebbw Fawr downstream of Ebbw Vale in September 1979.   (Data from Welsh Water Authority)

Oligochaeta
    Enchytraeidae
    Tubificidae
      Tubifex tubifex
      Limnodrilus hoffmeisteri
      Potamothrix hammoniensis
      Aulodrilus pluriseta
    Naididae
      Nais elinguis
      N. barbata
      N. bretscheri
      Stylaria lacustris
    Lumbriculidae
      Lumbriculus variegatus

Hirudinea
      Erpobdella octoculata
      Helobdella stagnalis
      Glossiphonia complanata

Mollusca
      Potamopyrgus jenkinsi
      Lymnaea peregra
      Ancylus fluviatilus
      Sphaerium sp.
      Planorbis sp.
      Physa fontinalis

Crustacea
      Gammarus pulex
      Asellus aquaticus

Insecta
  Ephemeroptera
      Baetis rhodani
      Ephemerella ignita
  Trichoptera
      Rhyacophila dorsalis
      Hydropsyche pellucidula
      Polycentropus flavomaculatus
      Hydroptila sp.
  Diptera
    Chironomidae
    Tipulidae
    Empididae
    Tabanidae
    Muscidae

Hydracarina

reaches of the Sirhowy, had stocking programmes for brown trout. A few
of these stocked fish entered the grossly-polluted Ebbw Fawr and were
subsequently found in the vicinity of small streams where there was
localised improvement in water quality. The other cleaner tributaries,
draining upland moors and conifer plantations, were small and their
flow unreliable during dry summers; of these the Carn and Gwyddon (Fig.
4.2) were most significant. Generally, the small tributaries contained
bullheads and brown trout and a few held sticklebacks and stoneloach
(Fig. 4.12). Roach and chub were caught in the lower Sirhowy in 1972
but it seems likely that these had escaped from the canal near the
Carn (Fig. 4.2) through a defective sluice. At this time there were
also trout and stoneloach in the Ebbw Fawr at the Sirhowy confluence,
suggesting that water quality was suitable for fish in this local area,
at least for short periods.

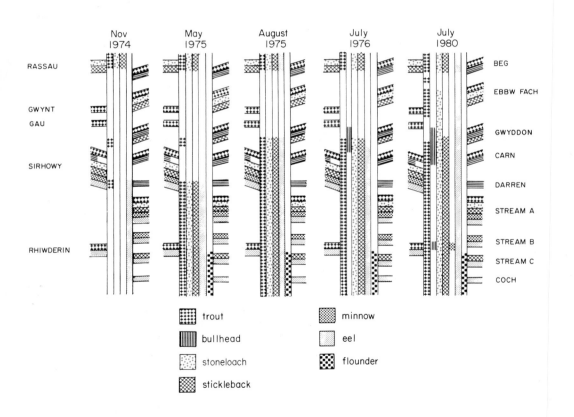

Fig. 4.12 Pattern of fish recolonization in the Ebbw Fawr since 1970.

In the years after 1973 there was a rapid recolonization of the main river by several species (Fig. 4.12). Late in 1974 eels extended upstream to the Sirhowy confluence and brown trout occurred at the mouths of both the Sirhowy and Gwyddon. In 1975 there were dramatic changes, leading to the recolonization of the Ebbw Fawr by brown trout, eels, stoneloach, stickleback and flounder, all but the flounder extending their range upstream to the Gwyddon, only about 16 km from Ebbw Vale. In 1976 the bullhead was recorded in a reach at the mouth of the Gwyddon.

By 1980 trout, stoneloach and eels had extended upstream from the sea to sites within a few kilometres of the steel-works effluent. Although the recolonization of stoneloach and eels was natural, that of trout was aided by a stocking programme carried out by the Welsh Water Authority in 1979 and 1980. Bullhead remained generally confined to areas adjacent to clean tributaries, such as the Carn and Gwyddon, but an isolated occurrence was recorded in the lower reaches at a site near Risca where minnows were also found. The slow recolonization of the river by minnow may be partly explained by its very limited distribution within the catchment, the only substantial population being in the upper reaches of the Sirhowy, and its apparent restricted mobility (Kennedy & Pitcher, 1975). A list of all fish species recorded recently within the catchment is given in Table 4.2.

4.42 DENSITY, GROWTH AND CONDITION

The densities of trout in tributaries (Table 4.3) are generally similar to those found elsewhere in the coalfield but most fish are small, few achieving a length greater than 20 cm, and consequently both biomass, densities and potential recruitment are low. Bullheads, although less widespread, reach high densities in the Beg, Gwyddon and Carn. Stoneloach and minnow are particularly abundant in the Sirhowy (the latter species being concentrated particularly in the headwaters) where they achieve densities of 144 and 107 per 100 m^2 respectively.

The largest populations of native trout in the Ebbw Fawr were found close to the mouths of the Carn, Gwyddon and Sirhowy. In 1976 at the mouth of the Sirhowy maximum densities ranged from 7 - 8 per 100 m^2 with biomasses between 810 and 1150 g per 100 m^2. In 1980, at the mouths of the Carn and Gwyddon similar maximum densities were recorded; these compare favourably with those found in other rivers of the coalfield with established but vulnerable trout populations (Learner et al., 1971; Scullion & Edwards, 1980b). Outside these very limited areas influenced by the dilution water of clean tributaries, trout densities were low.

Upstream migration of the flounder was probably prevented by Bassaleg Weir, 2 km above the tidal limit: in parts of this downstream reach, which were sampled, it was present at densities up to 13 per 100 m^2. Stickleback were found in backwaters and river margins downstream of tributaries with resident populations.

Table 4.2 Fish species recorded recently in the Ebbw system.

Petromyzonidae	*Lampetra planeri*	brook lamprey
Salmonidae	*Salmo trutta*	brown trout
Cyprinidae	*Gobio gobio*	gudgeon
	Phoxinus phoxinus	minnow
	Rutilus rutilus	roach
	Leuciscus cephalus	chub
	L. leuciscus	dace
Cobitidae	*Noemacheilus barbatulus*	stoneloach
Anguillidae	*Anguilla anguilla*	eel
Gasterosteidae	*Gasterosteus aculeatus*	three-spined stickle-back
Percidae	*Perca fluviatilis*	perch
Cottidae	*Cottus gobio*	bullhead
Pleuronectidae	*Platichthys flesus*	flounder

Table 4.3 Density and biomass (per 100 m^2) of trout and bullhead in tributaries of the Ebbw Fawr 1970-72.

River	trout		bullhead	
	density	biomass (g)	density	biomass (g)
Beg	24 - 113	411 - 1282	6 - 117	29 - 267
Gwynt	31 - 71	866 - 1359	-	-
Gwyddon	61 - 167	828 - 2429	19 - 247	38 - 545
Carn	49 - 82	1122 - 1688	45 - 113	92 - 244
Ystruth (off Ebbw Fach)	9 - 91	173 - 648	-	-
Sirhowy	1 - 9	402 - 571	-	-
Sirhowy tribu-taries	3 - 125	220 - 1390	-	-

During 1975 and 1976 brown trout caught in the Ebbw Fawr, near the Sirhowy confluence, were tagged, and from subsequent recaptures in this reach their growth rate could be assessed. The growth of many fish was remarkably fast, the average specific annual growth rate of four and five year-old fish being 36 - similar to those in unpolluted rivers of similar temperature regime in southern Britain (Williams & Harcup, 1974). There is evidence from scale examination that slow-growing fish

from small tributaries, such as the Gwyddon, grow much faster in the
Ebbw Fawr but that the growth rate of those descending from the much
larger Sirhowy does not change appreciably.

Average condition factors of trout captured in the Ebbw Fawr during
1975-76, derived from weight and length measurements, were about 1.35
and well above the value of unity often regarded as adequate for trout
in reasonable condition (Frost & Brown, 1967).

4.43   TROUT MOVEMENTS AND RECRUITMENT WITHIN THE CATCHMENT

An extensive marking programme of native brown trout in the lower
reaches of several small tributaries of the Ebbw Fawr and in the Sirhowy,
together with the marking of brown trout stocked into the Sirhowy during
the period 1971-78 (Williams & Harcup, 1974; Cresswell & Williams, 1979;
Turnpenny, 1979), has facilitated an appreciation of the broad features
of movement.

Native brown trout in the small tributaries, although relatively
numerous, are not large and therefore egg production, or potential recruitment, is low. It has been estimated that in the Gwyddon, Carn and
Beg, the three more important unpolluted tributaries, the average
density of eggs is only about 900 per 100 m^2. The total egg production
of these streams, principally because of their small size, is very low
(355×10^3 Gwyddon; 121×10^3 Carn; 15×10^3 Beg). With the resulting
low fry production there is little of the downstream emigration which
occurs in many river systems soon after emergence from the gravel (Le
Cren, 1973) and such emigration is further hindered by physical barriers
(Williams, 1980).

In most river systems which support trout some fish move from main
rivers into small tributaries to spawn. In the Ebbw system, the two
larger rivers containing trout, Sirhowy and Ebbw Fach, are intermittently polluted by coal wastes and sewage and contain low trout
densities (Table 4.3 for Sirhowy). These trout are prevented from
moving into several clean tributaries for spawning by impassable culverts and other barriers and the gravels in the larger tributaries
(Sirhowy and Ebbw Fach) are mostly unsuitable for spawning because of
the low permeabilities principally associated with fine particles
derived from colliery spoil heaps and coal washeries (Turnpenny &
Williams, 1980).

As with fry, it seems that the downstream movement of older fish
from the small, clean tributaries is very restricted for they move very
little, most having a home range of less than 15 m over periods of
several months (Harcup, 1978).

In consequence of this immobility of native trout caused by the
topography, physical development and pollution of the catchment, the
general recolonization of the Ebbw Fawr would be expected to be slow

following an improvement in water quality. It seems that the relatively
fast but very localized recolonization during the period 1973-76 has
been assisted both directly and indirectly from stocking of brown trout
within the lower reaches of the Sirhowy where in the period 1974-76
about 4000 were introduced. Turnpenny (1979) estimated that about 13%
of all trout caught in the Ebbw Fawr during these years were fish
stocked in the Sirhowy. This downstream emigration is supported by
studies of Williams & Harcup (1974) who found that extensive downstream
movements occurred and that on one occasion up to 80% of stocked fish
were lost, presumably to the Ebbw Fawr, within a week of their introduction into the lower Sirhowy. As stocking was with relatively large
fish (20 - 30 cm) it seems likely that stocking also led to the downstream displacement of native fish and after large fish were introduced
to the lower reaches of the Sirhowy substantial increases in the density
of native trout occurred in the Ebbw Fawr (Turnpenny, 1979).

   The recolonization of the Ebbw Fawr with trout, by immigration, will
not necessarily lead to substantial improvements in recruitment within
the catchment for gravels in the river seem generally unsuitable for
spawning. Turnpenny & Williams (1980) showed that most ova planted in
substrates at sites in the Ebbw Fawr died, whereas at sites on clean
tributaries most survived: in the Sirhowy, polluted by coal wastes,
intermediate results were obtained. Both embryo survival and alevin
length were related to substrate permeability which influenced oxygen
supply (Fig. 4.13).

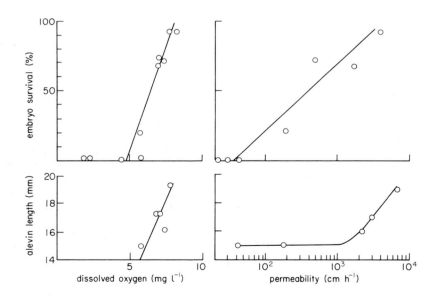

Fig. 4.13 Relation between success of embryo development (embryo
survival and alevin length) and substrate quality (permeability and
dissolved oxygen content of interstitial water).

The concentrations of some metals in sediments of the Ebbw Fahr, when compared with other rivers of the catchment, were very high - particularly Cr, Fe, Zn and Pb (Table 4.4). In the case of certain metals, e.g. Co, Cd, Mn, the most contaminated site was in the headwaters of the Ebbw Fawr, upstream of steel-works discharges. Contamination probably results from the leaching of metalliferous wastes adjacent to the river derived from earlier mining activities. Despite these high metal concentrations, their effect on the survival of fish embryos seems minimal because of their generally low solubilities at normal pH values.

Table 4.4 Mean concentrations (mg kg^{-1}) of heavy metals in sediments of the Ebbw Fawr, Sirhowy and the Gwyddon in 1975.

site	Cr	Mn	Fe (x 10^3)	Co	Cu	Zn	Cd	Pb
Gwyddon (baseline)	36.8	495	15.9	25.5	39.0	117	<0.2	27.0
Sirhowy	11.5	987	18.6	31.1	53.1	114	<0.2	30.3
Ebbw Fawr (headwaters)	6.4	5229		52	65.4	20665	78.2	565
Ebbw Fawr (Cwm)	136	1347		65	146	384	3.6	224
Ebbw Fawr (Crosskeys)	39.7	784		48	80.3	225	<0.2	143

4.44 FISH SURVIVAL AND WATER QUALITY

That the improvement in water quality in the Ebbw since 1973 has been adequate to support fish life, has been demonstrated by their capture within the river. But the capture of fish does not indicate the length of time they might survive in the system and so independent assessments of the adequacy of water quality to support them have been made. In other industrial river systems it has been demonstrated that if other water quality characteristics are satisfactory, natural fish populations will be sustained where the median concentrations of soluble poisons, expressed as fractions of their lethal threshold concentration to rainbow trout, do not exceed about 0.25 (Alabaster *et al.*, 1972). Turnpenny & Williams (1980) have shown that during the period 1974-76 at all sites down the river, this median fraction of the 48 h LC_{50} did not exceed 0.1 (Fig. 4.5). Curiously, the calculated toxicity of sites in the headwaters was similar to that downstream of Ebbw Vale: this toxicity is derived largely from metals which seem to emanate from ground and tip drainage.

Further evidence of the adequacy of water quality has been acquired from experiments in which trout (Fig. 4.14) and several other fish species have been retained in cages at sites within the river. Such fish are inevitably stressed and the adequacy of feeding, important in long-term tests, presents a problem. Nevertheless, in the Ebbw the median period of survival of trout at sites only 5 km downstream of the Ebbw Vale steel-works increased from about 0.5 h in 1970 to several days after 1974. More importantly, at downstream sites fish generally survived the test period, about two months for rainbow trout and up to three weeks for other species (bullhead, minnow, stickleback, stone-loach, brown trout).

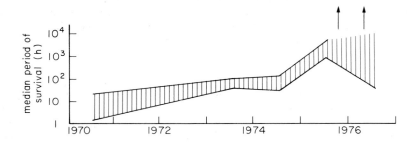

Fig. 4.14 Improvement in survival times of caged trout in the Ebbw Fawr, 1970-76.

4.5 CONCLUSION

The efforts made by the Usk River Authority during the 1960s to improve the water quality of the grossly-polluted Ebbw Fawr, which led to major pollution abatement programmes during the early 1970s, has provided an excellent opportunity to study the recovery of a river, formerly almost devoid of living organisms over most of its length. Furthermore the fact that the major effluent discharge and the one which has improved most dramatically, the effluent from the Ebbw Vale steel-works, was located in the headwaters, has meant that the recovery sequence could be followed down the whole river, a distance greater than 30 km. Chemical and biological surveillance has demonstrated an improvement in water quality leading to the establishment of associations of organisms of increasing diversity. Surveillance has been supported by *in situ* survival tests which have demonstrated the increasing ability of the river to support both plants and animals. In the case of fish, and particularly brown trout, it has been possible to compare survival of captive fish in the river with that expected from prevailing water quality and laboratory toxicity data.

It is perhaps the work on the establishment of fish populations in the river that demonstrates most clearly the dramatic change which has taken place in the past decade. Seven species are now found in the main river, some distributions extending downstream from the steel-works to the estuary. Although siltation of the river bed, in part by coal wastes prevents natural spawning, a 'put and take' fishery of brown trout has been established with the aid of regular stocking.

Despite these recent major improvements, the Ebbw Fawr, along with other rivers in the industrial and highly-urbanised valleys of the South Wales Coalfield, will continue to be affected by pollution problems. Such factors as the linear urban settlements mostly near the river, mining subsidence leading to the periodic fracture of the sewers, which are often routed along river-banks, high rainfall and shallow soils causing unsteady river-flows and the intermittent discharge of storm-sewage, historical accumulation of mining spoil which is easily washed into water-courses, are all significant constraints on further improvement.

The capital costs of improving the effluent of the steel-works (£8 million*), of constructing a new trunk sewer (£13 million) and of modifications to coal washeries and colliery spoil disposal arrangements (£1 million), must be coupled with the considerable on-going costs of effluent treatment plant at the Ebbw Vale steel-works. Apart from the benefits to amenity and recreation, particularly fishing, the river now provides water of acceptable quality for abstraction for a variety of industrial uses in the lower reaches, thus saving 'imported' water of higher quality for potable and agricultural supply.

Acknowledgements

The authors wish to thank Dr K. Benson-Evans (University College, Cardiff), Mr J. Bartlett (UWIST) and the many staff of the Welsh Water Authority, particularly Mr M. Henderson, Mr M. Hunter, Mr F. Jones and Mr R. Streeter, who provided information and commented upon the text, and Mrs M. Thomas and Mrs S. Fish for the preparation of typescript and figures.

REFERENCES

Alabaster J.S., Garland J.H.N., Hart I.C. & Solbé J.F. de L.G. (1972) An approach to the problem of pollution and fisheries. *Symp. Zool. Soc. Lond.* <u>29</u>, 77-114

Benson-Evans K. & Williams P.F. (1970) Biological review of 1960-70 position in some Usk River Authority streams. *Usk River Authority Annual Report* <u>5</u>, 75-85

*all costs adjusted to 1980 values

Brown V.M. (1968) The calculation of the acute toxicity of mixtures of poisons to rainbow trout. *Wat. Res.* 2, 723-33

Chandler C.M. (1970) A biological approach to water quality management. *Wat. Pollut. Control* 66, 415-21

Cresswell R.C. & Williams R. (1979) Studies on trout stocking in an industrial river and their management implications. *Proc. 1st British Freshwater Fisheries Conference*, Univ. Liverpool, pp. 285-95

Curtis E.J.C. (1969) Sewage fungus: its nature and effects. *Wat. Res.* 3, 289-311

Curtis E.J.C. & Curds C.R. (1971) Sewage fungus in rivers in the United Kingdom: the slime community and its constituent organisms. *Wat. Res.* 5, 1147-59

Edwards R.W., Benson-Evans K., Learner M.A., Williams P.F. & Williams, R. (1972) A biological survey of the River Taff. *J. Wat. Pollut. Control* 71, 144-60

E.I.F.A.C. (European Inland Fisheries Advisory Commission) (1964) Water quality criteria for European freshwater fish. *Report on Finely Divided Solids and Inland Fisheries EIFAC/T1.* FAO Rome

Frost W.E. & Brown M.E. (1967) *The Trout*, 316 pp. Collins, London

Gray-Jones A. (1970) *A History of Ebbw Vale*, 271 pp. Gray-Jones

Harcup M.F. (1978) *Fish Movement in some Streams of the Ebbw Catchment.* M.Sc. Thesis, University of Wales (UWIST)

Kennedy G.J.A. & Pitcher T.J. (1975) Experiments on homing in shoals of the European minnow, *Phoxinus phoxinus* L. *Trans. Am. Fish. Soc.* 104, 454-57

Le Cren E.D. (1973) The population dynamics of young trout (*Salmo trutta*) in relation to density and territorial behaviour. *Rapports et Procès-verbaux des Réunions*, Conseil International pour l' Exploration de la Mer 164, 241-47

Learner M.A., Williams R., Harcup M.F. & Hughes B.D. (1971) A survey of the macro-fauna of the River Cynon, a polluted tributary of the River Taff (South Wales). *Freshwat. Biol.* 1, 339-67

Painter H.A. (1954) Factors affecting the growth of some fungi associated with sewage purification. *J. gen. Microbiol.* 10, 177-90

Paulsen B.B. (1962) Pollution of the lower reaches of the Otteraa. *Vattenhygien* 18, 115-27

Scullion J. & Edwards R.W. (1980a) The effects of coal industry pollutants on the macro-invertebrate fauna of a small river in the South Wales coalfield. *Freshwat. Biol.* 10, 141-62

Scullion J. & Edwards R.W. (1980b) The effects of pollutants from the coal industry on the fish fauna of a small river in the South Wales coalfield. *Environ. Pollut. (Series A)* 21, 141-53

Turnpenny A.W.H. (1979) *Aspects of the Recovery of Fish Populations in the River Ebbw Fawr.* Ph.D. Thesis, University of Wales (UWIST)

Turnpenny A.W.H. & Williams R. (1980) Effects of sedimentation on the gravels of an industrial river system. *J. Fish. Biol.* 17, 681-93

Whitton B.A. (1979) Plants as indicators of river water quality. In: James A. & Evieson L. (Eds.) *Biological Indicators of Water Quality*, pp. 5.1-5.34. Wiley, Chichester

Whitton B.A. & Say P.J. (1975) Heavy Metals. In: Whitton B.A. (Ed.) *River Ecology*, 725 pp., pp. 286-311. Blackwell, Oxford

Williams P.F., Benson-Evans K. & Jones A.K. (1974) Further analysis of
    data from biological surveys of the River Taff. *J. Wat. Pollut.
    Control* 75, 576-83
Williams R. (1980) *The Fish of the Ebbw System.* Ph.D. Thesis,
    University of Wales (UWIST)
Williams R. & Harcup M.F. (1974) The fish populations of an industrial
    river in South Wales. *J. Fish Biol.* 6, 395-414
Williams R., Williams P.F., Benson-Evans K., Hunter M.D. & Harcup M.F.
    (1976) Chemical and biological studies associated with the recovery
    of the River Ebbw Fawr (1970-1974). *J. Wat. Pollut. Control* 75,
    428-44
Woodiwiss F.S. (1964) The biological system of stream classification
    used by the Trent River Board. *Chemy Ind.,* 443-47

5 · Mersey

D. G. HOLLAND & J. P. C. HARDING

5.1   INTRODUCTION

Ever since the start of the Industrial Revolution in the 18th century, many rivers in Britain have been subjected to a wide range and excessive volume of man's wastes. The river described here, the Mersey, has a long history of particularly severe pollution and the local author, Charles Kingsley, who knew the river well, probably had it in mind when he wrote:

> "Dank and foul, dank and foul.
>
> By the smoky town with its murky cowl;
>
> Foul and dank, foul and dank,
>
> By wharf and river and slimy bank;
>
> Darker and darker the farther I go,
>
> Baser and baser the richer I grow."

Today, despite many years of pollution control measures, that inheritance is still evident for all to see. An indication of this severe pollution is the fact that the Mersey, with several of its main tributaries, forms a major component of the region's dirty rivers, which themselves make up 26% of the lowest water quality rivers of England and Wales (National Water Council, 1981). It naturally follows that this poor water quality has an adverse effect on the biology of the river system, which is typified largely by pollution-tolerant flora and fauna and limited species diversity.

The description of the biology of the Mersey catchment presented in this chapter has been compiled by two Water Authority biologists who are involved in practical day-to-day studies. Although the catchment has received little detailed study in the past, the authors are fortunate in having access to a large body of data collected during routine work. In order to describe the present situation, it is necessary to give first an historical account of the area, which suggests how the biology of the river may have been affected by the changing conditions induced by man. The past history is then linked to the relationship between present-day water quality and plant and animal communities. Since the theme of change will be evident throughout the account, some attempt will also be made to forecast future trends.

5.2 GENERAL DATA

DIMENSIONS
Area of catchments

Etherow	155.7 km^2	
Goyt (excl. Etherow)	216.4 km^2	Sources of Etherow, Goyt and
Tame	145.3 km^2	Tame all in range 460-520 m
Mersey (excl. E/G/T	161.2 km^2	

total Mersey   678.6 km^2

Highest point in catchment 631 m (Kinder Scout)

CLIMATE
Precipitation

top of Etherow	2400 mm yr^{-1}
Tame-Etherow-Goyt	900-2400 mm yr^{-1}
Mersey	800-900 mm yr^{-1}

Temperature
(Data for whole area of the former: Mersey & Weaver River Authority, First Periodical Survey, 1969)

Annual data:

		Bright sunshine:	
mean daily minimum	mostly 6-7 ^{O}C	mean daily near coast	4-4.5 h
mean daily maximum	12-13 ^{O}C	mean daily on high ground	
mean	c. 9.5 ^{O}C		3-3.5 h

GEOLOGY
(i) Carboniferous outcrop of both Namurian (Millstone grit) and West-phalian (coal measure) ages on the high ground of the upper catchment.

(ii) Glacial sand and gravel, and alluvium in narrow bands following the courses of the Etherow, Goyt and Tame and then broadening out into the Mersey flood plain.

(iii) Boulder and laminated clays on the low ground between rivers.

LAND USE

hill farming	46%	urban	34%
heathland and poor pasture	27%	arable	3%

POPULATION
Total in catchment (1971 Census)   828000 people

PHYSICAL FEATURES OF RIVER

On-stream reservoirs and major weirs

total number of reservoirs including 39 (totalling 22700 m^3 capacity
4 service

total number of major weirs   23 (as per Reservoir Act limit)

Discharge at selected points
data given for modal flows, i.e. most frequentiy occurring

	natural	artificial	total
Etherow above Goyt	0.3	0.54	0.93 m^3 s^{-1}
Goyt above Tame	1.45	0.93	2.37
Tame above Goyt	0.66	0.81	1.47
Mersey (Irlam Weir)	2.78	2.66	5.44

Artificial = compensation from reservoirs + sewage discharge + industrial discharge

Seasonal changes in flow (snow melt etc)
    Effect of limited snow melt mostly absorbed by reservoirs

Influence of regulation
    Artificial flow = Reservoir compensation + industrial and sewage
    effluents

Etherow	0.54	=	>83%	+	<17%
Goyt	0.93	=	>27%	+	<73%
Tame	0.81	=	>27%	+	<73%

Important abstraction points
    Etherow consented) $0.12 \ m^3 \ s^{-1}$ sewer
    Goyt (consented) $0.49 \ m^3 \ s^{-1}$ river and sewer

Substratum
    Mostly stable stone and gravel beds in Etherow, Goyt and Tame—giving
way to silt in the slower Mersey.

Temperature

	mean annual values (^{O}C)
Etherow above Goyt	10-11
Goyt above Tame	10-12
Tame above Goyt	11-13
Mersey (Irlam Weir)	11-13

MANAGEMENT
    Water quality management within the North West Water Authority is
undertaken principally by the Scientist's Department of the Rivers
Division, whose duties cover the entire region. Close cooperation with
Head Office staff enables this function to be related to the work of the
three dual purpose Divisions on the management of water supply and waste
water treatment.   The Rivers Division Scientist is supported by:
 (i) a team of River Inspectors who liaise with individuals and
     organizations instrumental in causing river pollution;
(ii) a technical support team providing expertise in the biology and
     chemistry of water quality control.

5.3   THE MERSEY CATCHMENT

The Mersey catchment above the tidal limit covers 203000 hectares (Fig. 5.1), an area only half that of some of the major British rivers, such as the Thames, the Severn and the Trent. It is characterised by extensive urbanisation. Most of the upper river valleys are well developed and the Mersey and its tributaries pass from town to town almost without break.

The upper Mersey catchment consists of three similar sized tributaries, the Tame, Etherow and Goyt, but the defined point at which the Mersey actually begins has been the subject of debate for many years. Although the river named on present-day maps as the Mersey starts at Stockport, where the Goyt and Tame join, Palmer (1950) claims that 13th century documents give the Etherow, rising on the Pennine Hills 40 km to the east, as the source of the Mersey. For ease of description, the Mersey will be taken from this source in the Pennines as the Mersey-Etherow. Below the confluence with the Goyt, the river will be referred to as the Mersey-Goyt, which then joins with the Tame to become the Mersey proper at Stockport.

The Mersey-Etherow and these two major tributaries rise at similar altitudes (460 - 520 m) on the high peaty plateaus of a section of the Pennine range from Saddleworth Moor in the north to High Peak in the south (Fig. 5.2). Similar gradients (10 m km^{-1} or more) result in fast currents and high aeration rates, so that even in stretches subject to heavy organic pollution, concentrations of dissolved oxygen are generally much higher than they would be in lowland rivers of comparable size. Downstream from the confluence with the Tame the gradient of the Mersey decreases markedly as it passes through a narrow gorge and onto a broad flood plain to the south of Manchester. 32 km from Stockport the Mersey drops over Irlam Weir to unite with the Manchester Ship Canal.

The Mersey catchment above Irlam Weir has a high proportion of urban development (24%) and there is a rapid run-off to the river from all the impermeable surfaces. Pearson (1939) quotes water level rises after heavy rain of up to 5 m in 3 to 4 hours and, because of the flat gradient of 1 m km^{-1} between Stockport and Irlam, flooding has been a severe problem for several centuries. The earliest attempts to control flooding can be traced back to the end of the 18th century when local landowners constructed a variety of floodbanks, many of which were ill-designed and subject to frequent breaching. The present-day flood control measures of high channel embankments and flood relief basins have largely overcome the threat of widescale inundation.

In medieval times the lower Mersey meandered across a broad boggy floodplain before broadening out into the Mersey Estuary west of Warrington. Today, however, the situation is more complex as a result of extensive alterations made to the river to enable merchant ships to pass between the Irish Sea and docks in the heart of Manchester. Downstream of Irlam Weir, the Mersey joins the Manchester Ship Canal, which

Fig. 5.1 Mersey system, showing sampling sites.   (u/s, upstream; d/s, downstream)

was constructed in the late 19th century on the course of the Irwell,
a major tributary flowing south-west from Manchester. The combined
natural flow of the Mersey and Irwell continues in the Ship Canal for
14 km, and picks up two smaller tributaries before being diverted away
from the canal at Rixton. After a further 10 km the Mersey discharges
over Howley Weir into the tidal reach of the estuary at Warrington.
Downstream of Rixton the Ship Canal skirts the course of the Mersey to
connect with the estuary at Eastham Locks.

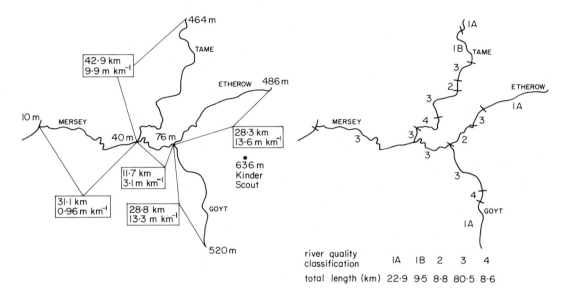

Fig. 5.2 Examples of altitude, river distance and gradient in the
Mersey catchment.

Fig. 5.3 River quality classification of Mersey catchment in 1980.

5.4 HISTORICAL CHANGES

5.41 THE INDUSTRIAL REVOLUTION

Many small settlements were present on the Mersey catchment in Roman
times, but intensive urban development of the area did not begin until
the start of the Industrial Revolution. Cotton shipped into Liverpool
formed the foundation of a massive industry which took advantage of the
abundant supplies of clean soft river water and established many spin-
ning and weaving mills in the valleys over a wide area of South
Lancashire, North Cheshire and the area now termed Greater Manchester.
The availability of water was improved further by the construction of
a large number of reservoirs between 1840 and 1870 to supply the valley

factory towns and a large part of Manchester.

The development of communications, first of canals, then later of railways and the Manchester Ship Canal, made the import of raw materials and the dispersal of finished goods much easier. At the same time the South Lancashire coal industry supplied the area's fuel needs. During the 1800s industrial interests diversified into paper and machine-making, dyeing and bleaching and textile printing. In the present century the manufacture of man-made fibres has gradually taken the place of the cotton industry, but this industry in turn is now going into a period of recession. In more recent years industry has diversified further to include engineering, chemicals and many other interests.

With the expansion of Manchester and with additional townships springing up in the surrounding valleys, refuse, sewage and industrial wastes found their way into the river system and water quality deterio-rated. However, Bracegirdle (1973) records that even as late as 1830 drinking water was still being taken from the Irwell in Manchester – undoubtedly in ignorance of health risks. By the mid 1800s it was becoming clear that sewage-contaminated drinking water was an excellent vehicle for the transmission of diseases such as cholera and typhoid, which by then had become endemic in the area.

Apart from the adverse effect on the quality of human life, declining water quality in the 19th century had a severe impact on fish stocks in the river. The only available detailed documentary evidence is that by Corbett (1907), who described the demise of salmon and coarse fisheries in the Irwell through to the Mersey in the period between 1820 and 1850. As industry continued to expand, water quality deteriorated further so that by 1870 even the commercial fishery downstream in the Mersey Estuary had become affected and was eventually destroyed completely.

Sadly there is an apparent shortage of documentary evidence about the 19th century conditions in the catchment upstream of the confluence with the Irwell, but because of the similarities of industrial develop-ment throughout the Manchester area in this period, Corbett's recollec-tions of the general condition of the Irwell can be taken as represen-tative of all the rivers. In particular, Corbett recorded a visible deterioration in the Irwell from about 1855 to the mid 1880s, by which time water-weeds had disappeared from the river, and the banks to a height of nearly 1 m above summer level were also devoid of plant growth.

During this period local concern was being expressed about what was happening to the river. The first of a series of Acts of Parliament aimed at controlling river pollution (The Mersey Etc. Protection Act, 1862) was passed and the first river inspector ever appointed commenced duties in 1867.

5.42 THE TWENTIETH CENTURY

Documentary evidence becomes easier to find after 1892 when the Mersey
and Irwell Joint Committee was formed as the first statutory body
specifically charged with the responsibility of controlling pollution
to the river system. In 1906 the Committee reported that 444 manufacturers discharged wastes to the Mersey and Irwell catchments, and,
with the addition of no less than 419 sewage works effluents, there was
clearly considerable scope for improvement.

Despite the introduction of statutory pollution control powers in
the latter half of the 19th century, the appalling river water quality
conditions at the beginning of the 20th century showed very little
improvement over the next fifty years. Indeed it was not until the
Rivers (Prevention of Pollution) Act 1951 that effective powers of
persuasion were made available to the controlling authority of the time
(Mersey River Board). Further tightening of legislation has enabled
considerable progress to be made over the last thirty years and this
latter period has seen an impressive reduction in the number of sewage
and industrial waste effluents discharged to river. Chemical data from
1953 onwards demonstrate clearly the changes which have taken place
(Fig. 5.4). The 5-day Biochemical Oxygen Demand (B.O.D.) values can be
interpreted as representing the amount of organic material in the river,
whilst percentage saturation values of dissolved oxygen express the
effect of the B.O.D. on the river. During 1962-64 there was a particularly marked improvement all over the catchment demonstrated most
strikingly by the increase in dissolved oxygen. Several important
changes in the polluting discharges account for this.

One clear-cut example of the kinds of pollution and improvement that
have occurred throughout the catchment is found in the Mersey-Etherow.
Ever since the beginning of the 20th century one of two paper manufacturing companies in Britain using the sulphite process for digesting.
wood had been discharging highly polluting acidic organic wastes to
Glossop Brook, a tributary of the Mersey-Etherow. The company ceased
to operate in 1963 after having failed throughout its entire period of
operation to provide adequate treatment facilities for its wastes. At
the same time extensions to Glossop Sewage Works brought to an end
twenty years of exceptionally poor sewage treatment. Prior to these
two events the pollution produced by the two effluents, in combination
with that of a textile printing works, had made the water quality of
the Mersey-Etherow amongst the worst in the Manchester area, equalled
only by the Irwell and Tame.

Immediately after 1963 the improvement in the Mersey-Etherow could
be traced downstream chemically through the Mersey-Goyt to below
Stockport. In the 20 years since then, water quality conditions
throughout the upper Mersey catchment have, on the whole, continued to
improve at a more gradual pace.

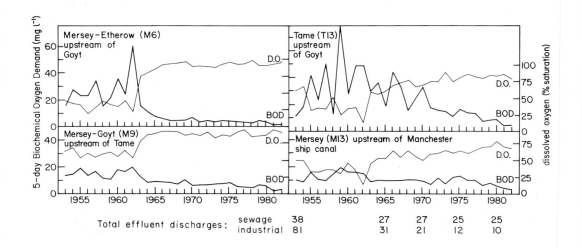

Total effluent discharges: sewage   38   27   27   25   25
                              industrial 81   31   21   12   10

Fig. 5.4 Changes in 5-day B.O.D. and dissolved oxygen (as % satura-
ted) at key sites in Mersey catchment over period 1953-1982, together
with number of effluent discharges in whole catchment. Chemical data
based on mean of monthly readings.

Fig. 5.5 The Tame at Stockport upstream of the Mersey.

In addition to organic pollution from treated and untreated sewage and industrial waste, the Mersey and its two tributaries have been subject to other forms of pollution, such as heavy metals, cyanide, detergents (Fig. 5.6), phenol and chlorine. Industrial development of the river valleys included the establishment of several factories that made use of these toxic substances in a variety of processes. In some cases elevated concentrations provided significant contributions to the industrial wastes being discharged to the rivers. However, it is only in the last few years that the importance of the heavy metals has become apparent.

Fig. 5.6 Mersey at Woolston Weir, Warrington, showing flocs of detergent foam.

A well documented and clear-cut example of the kind of problem which can occur is again provided by the Mersey-Etherow. Factory closures and improvements to sewerage and sewage treatment had brought about such a marked decrease in organic loads by the mid 1970s, that hopes were high for the establishment of a mixed coarse/trout fishery. However, attempts to stock the river met with no success and it soon became apparent that most of the Mersey-Etherow was toxic to fish. The use of caged fish and laboratory fish toxicity tests established the existence of a previously unsuspected source of toxic waste to the river and chemical analysis of water and the moss *Fontinalis anti-pyretica* enabled the poison to be identified as zinc from a specialised paper treatment process (Fig. 5.7). Following this discovery the firm responsible for the input began to reduce the amount of zinc entering

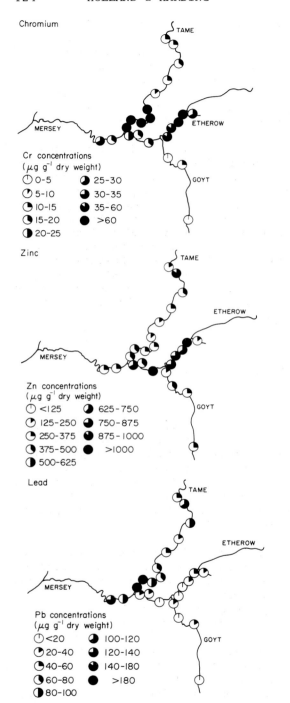

Chromium

Cr concentrations
(μg g^{-1} dry weight)
0-5   25-30
5-10   30-35
10-15   35-60
15-20   >60
20-25

Zinc

Zn concentrations
(μg g^{-1} dry weight)
<125   625-750
125-250 750-875
250-375 875-1000
375-500 >1000
500-625

Lead

Pb concentrations
(μg g^{-1} dry weight)
<20   100-120
20-40   120-140
40-60   140-180
60-80   >180
80-100

A. <u>Chromium</u>. Concentrations increase markedly downstream of the effluent from Dukinfield S.T.W., which treats tannery waste. High levels were found also in the Etherow, providing evidence for short-term releases from industrial premises.

B. <u>Zinc</u>. The high concentrations in the Etherow were due to continuous input of metal from an industrial source reaching the river via old culverts.

C. <u>Lead</u>. A marked increase occurred on the Tame downstream of the effluent from Denton S.T.W. The high concentrations probably result from battery manufacture.

Fig. 5.7 Cr, Zn and Pb in samples of shoot tips (2-cm long) of *Fontinalis antipyretica* collected between October 1979 and January 1980.

the river, but progress was slow because the metal reached the river via
an extensive underground system of derelict culverts and contaminated
aquifers. However, by late 1980 zinc leaching from the culverts had
decreased markedly (Say *et al.*, 1981), and following the closure of the
works in 1982 it seems likely that trout and coarse fish will be able
to spread up the river in the near future. The history of the zinc
and organic pollution and its impact on plant and animal life in the
Mersey-Etherow has been summarised by Harding *et al.* (1981).

In seeking to improve river water quality, one of the approaches of
the controlling bodies has been to arrange for industrial wastes to be
diverted into the sewers, so that they may be treated more effectively
in admixture with sewage. The second approach has been to improve
industrial waste treatment plants and sewage works. Increased capital
investment and closure of small inefficient sewage works, leading to
new regional treatment facilities, have all contributed to improved
river quality, and the factory closures that have occurred have also
helped to lessen the impact of industrial discharges. Despite the
reduction in numbers of discharges, the total volume of all effluent
to river has changed little over the years, as the remaining sewage and
industrial works have increased their output and new discharges have
come into being.

5.5   THE MERSEY SYSTEM TODAY

5.51   INTRODUCTION

It is clear from the previous section that from the beginning of the
Industrial Revolution the Mersey has been subjected to a long history
of severe pollution, with marked and sometimes dramatic changes taking
place from time to time. Whilst the trend in recent years has been
towards substantial improvement, it is unlikely that a stable situation
has been reached. A description of present conditions can apply only
to a single point in time, but is needed in order to relate the plant
(5.62) and animal communities (5.63) to their environment.

Most industrial concerns pass their waste effluents to sewage works
for treatment and only the Goyt still receives the impact of a significant number of direct industrial discharges. As a result, the largest
contribution today to the still relatively poor water quality of the
Mersey catchment comes from the many sewage works and also from untreated sewage overflowing from inadequately sewered areas.

An important feature of the Manchester conurbation, in common with
most long-established urban areas, is the existence of a large number
of overflows on the sewerage system (at least 220 upstream of Irlam
Weir). The frequency of operation of these overflows is influenced by
the condition and capacity of the sewers they serve and since many of
these sewers are decrepit and overloaded they tend to surcharge to

river at irregular and frequent intervals with a minimum of stimulus
from blockage or rainfall. Because of their irregular behaviour their
effect on the river is difficult to document and assess.

The water quality of the catchment can be summarised conveniently in
the terms of the classification scheme in current use by water authorities (National Water Council, 1981). The main chemical criteria
employed are dissolved oxygen, B.O.D. and ammonia, with modifications
where necessary, to allow for biological data, toxic substances and
general water usage. The classification is used by water authorities
for setting water quality management policies and five classes of river
are recognised:

Class 1A	Good quality	Water of high quality suitable for potable supply abstractions ; game or other high class fisheries ; high amenity value.
Class 1B	Good quality	Water of less high quality than Class 1A but usable for substantially the same purposes.
Class 2	Fair quality	Waters suitable for potable supply after advanced treatment ; supporting reasonably good coarse fisheries ; moderate amenity value.
Class 3	Poor quality	Waters which are polluted to an extent that fish are absent or only sporadically present ; may be used for low grade indus- trial abstraction purposes; considerable potential for further use if cleaned up.
Class 4	Bad quality	Waters which are grossly polluted and are likely to cause nuisance.

The classification of the Mersey catchment using this scheme (Fig.
5.3: see p. 119) can be compared with the chemical data presented for
five key river sampling points (Table 5.1). A description of the more
important influences on the various parts of the river system will help
to explain why so much is of poor or bad water quality.

5.52 MERSEY

In the Mersey-Etherow all that remains of the severe pollution described
in 5.4 is the input from Glossop Sewage Works, which is now operated to
a reasonable standard, and the steadily declining concentrations of
zinc. The river is also subject to intermittent contamination by other
toxic metals, notably chromium, copper and cadmium, but these probably
rarely reach concentrations high enough to endanger the biota. Unlike
the situation with zinc, inputs of these metals usually take the form
of short-term contributions from spillages and contaminated surface
run-off to the Glossop Brook. These incidents are very difficult to

Table 5.1 Summary of data on water chemistry at five sites during 1979-1980. (Total alkalinity and elements other than O_2 as mg l^{-1})

		Tame u/s of Goyt	Etherow u/s of Goyt	Goyt u/s of Etherow	Goyt u/s of Tame	Mersey u/s of Manchester Ship Canal
discharge	n	70	16	2	86	78
(m^3 s^{-1})	$\bar{x}$	4.55	3.20	2.48	5.48	13.9
	s.d.	4.35	2.31		4.35	12.4
suspended		93	15	2	104	91
solids		34.2	16.6	25.5	18.6	46.7
		38.04	9.15		11.0	55.8
conductivity		93	15	2	104	91
(μS cm^{-1})		617	347	415	395	544
		186	118		92	152
O_2 (%)		84	15	1	96	77
		84.0	91.3	94.3	87.1	74.8
		9.1	7.1		8.9	13.4
B.O.D.		93	15	2	104	90
		9.26	3.95	4.3	5.10	7.3
		4.56	2.22		3.01	4.42
C.O.D.		93	15	2	104	88
		60.4	25.7	42.5	32.6	48.5
		23.1	9.41	-	10.8	17.3
pH		93	15	2	104	91
		7.24	7.16	7.45	7.23	7.18
		0.21	0.21		0.22	0.16
total alk.		92	15	2	104	90
($CaCO_3$)		91.4	47.3	72.5	64.7	93.4
		29.2	17.7		24.4	32.9
Cr		59	5		48	70
		0.015	0.013		0.017	0.015
		0.005	-		0.012	0.004
Zn		81	9		53	80
		0.10	0.16		0.07	0.09
		0.08	0.08		0.04	0.05
Cd		59	1		38	59
		0.006	0.007		0.006	0.006
		0.003	-		0.002	0.002
Pb		61	3		43	62
		0.06	0.03		0.04	0.05
		0.05	-		0.02	0.05
Cl		93	15	2	104	90
		65.6	37.6	35.5	39.0	57.8
		30.0	17.9		20.7	24.6
NH_3-N		93	15	2	104	91
		1.88	0.54	0.85	0.78	2.21
		1.29	0.37		0.61	1.37
NO_2-N		93	15	2	104	91
		0.31	0.11	0.15	0.16	0.22
		0.24	0.11	-	0.10	0.16
NO_3-N		93	15	2	104	91
		4.22	2.48	2.12	2.82	3.45
		1.83	1.16		0.78	1.15
PO_4-P		91	15	2	103	99
		1.19	0.34	0.27	0.48	0.88
		0.99	0.32		0.33	0.65

detect by analysis of water samples alone, but evidence of their occurrence may be gained by the chemical analysis of submerged river plants, which accumulate metals from the water passing over them (Say *et al.*, 1981).

The next section of the river, the Mersey-Goyt, receives additional organic loading from the Goyt, one large sewage works and a textile works. Unfortunately, the benefit of natural river self-purification is counter-balanced by each polluting input so that water quality through this reach shows little downstream change.

The same comments apply to the first lowland section of the Mersey from Stockport to the Manchester Ship Canal. Contributions of heavy organic pollution from the Tame, the large sewage works serving Stockport, two smaller sewage works and untreated sewage from a poorly sewered part of Manchester all combine to depress water quality. The situation deteriorates even further where the Mersey joins with the Manchester Ship Canal, which is itself affected by several badly polluted Irwell tributaries and imposes an additional organic burden. A slight recovery takes place before the Mersey finally discharges to the Estuary at Warrington.

5.53   GOYT

The Goyt is characterised by having more direct industrial discharges than any other river on the Mersey catchment. Four textile processing firms, a paper mill, a manufacturer of brake linings and two sewage works together provide a heavy organic load to the river. The largest industrial contributor discharges kier wash liquor and a significant amount of chlorine from bleaching processes to the head of the Goyt. Fortunately, the steep gradient immediately downstream provides excellent aeration in the river and a high rate of self-purification reduces the effect of the total load of organic material, whilst the chlorine is rapidly lost by volatilization. Dilution from a clean tributary (Sett) helps self-purification and the overall result is a downstream improvement. However, the final output of the Goyt to the Mersey is a poor quality river in need of further improvement; hopefully, present work on extending the sewage treatment facilities in the valley will bring this about.

5.54   TAME

In spite of dramatic improvements that have taken place in the last twenty years (Fig. 5.4), the Tame gives the best impression of what most of the Mersey catchment would have been like at the end of the 19th century. From source to the Mersey at Stockport the Tame Valley is completely urbanised and the river is bounded on both banks by housing and industrial development. The situation today is dominated by the effluents from six sewage works, all treating a wide variety of

industrial wastes and by the frequent operation of several sewer overflows. Two of the works are grossly overloaded and contribute substantial sub-standard effluents to the river.

The upper stretch, which supports a fishery, is relatively unpolluted,
but a paper works and then the six sewage works discharge to the river
and the water quality deteriorates progressively downstream. In the
lower stretch untreated sewage and continuous growths of sewage-fungus
dominate the scene.

Although less well documented than the Mersey-Etherow, the Tame also
provides examples of pollution by metals from industrial sources. The
treated effluents from Dukinfield and Denton Sewage Works contain
relatively high concentrations of chromium and lead, respectively, due
to the wastes from tanning processes and battery manufacture. The
effects of these two inputs on the metal composition of *Fontinalis* can
be seen clearly in Fig. 5.7, and their impact can be traced downstream
into the Mersey below Stockport despite dilution from the Mersey-Goyt.

5.6   LIFE IN THE RIVER

5.61   INTRODUCTION

Most of the data used for the following descriptions are taken from
surveys conducted since 1970 by the North West Water Authority and
previously by the Mersey and Weaver River Authority. Analysis of invertebrate communities has proved an invaluable tool in the day-to-day
management of water quality, but it is only with the passage of time
that the value of a long-term review has become apparent. The authors,
as two members of a large team of scientists, have found that the
exercise of describing the current situation has greatly enhanced their
understanding of the biology of the Mersey - a benefit which has been
passed straight back into the operation of routine river surveillance.

In the most recent years of this short history of biological investigation further techniques have been developed to expand the Water
Authority's ability to monitor and understand the changes taking place.
Analysis of heavy metals in river plants and surveys of plant distribution have now become part of the routine monitoring programme and
together with fish surveys give a comprehensive picture.

5.62   PLANTS

5.621 General comments

As is the case with many British rivers the Mersey has until recently
received very little attention from local botanists, and historical
documentation of plants in the catchment is non-existent. In pre-

historic times the whole Mersey valley was densely wooded and growths
of photosynthetic plants in the river itself must have been severely
restricted by dense shading. However, as man began to clear the
forests for farming and timber, increased illumination, accompanied by
increased nutrient run-off from the cleared land, must have led to a
marked increase in submerged plant growth throughout the Mersey system.
A few clues to the nature of these early plant communities can be seen
in the present vegetation, especially in the upper reaches of the
Mersey-Etherow and Goyt, which have never been subject to the kind of
pollution affecting the rivers further downstream. In both cases the
bed is rocky and the vegetation is dominated by mosses and liverworts,
with four species, *Hygrohypnum ochraceum, Fontinalis squamosa,
Racomitrium aciculare* and *Scapania undulata*, being especially characteristic. Since occasional specimens of these species have been
recorded recently in fast-flowing riffles further downstream in both
the Mersey-Etherow and Goyt, it does seem probable that these four
bryophytes were abundant in both rivers prior to the Industrial Revolution.

   It is much more difficult to obtain evidence concerning the possible
early distribution of rooted aquatic angiosperms in the catchment.
Although the middle and lower Mersey has probably always contained areas
of gravel, sand and mud suitable for the establishment of rooted plants,
none have remained unaffected by pollution or artificial channel management. Only one angiosperm, *Ranunculus fluitans*, has a present-day
distribution that suggests it may have been present in the catchment
before the onset of severe pollution. Surveys of the distribution of
R. fluitans throughout the N.W.W.A. area (Harding, 1981) indicate that
although the species is unable to form large beds in rivers affected by
severe organic pollution, occasional plants may survive for long periods
in such situations even in competition with dense growths of pollution-
tolerant plants. Isolated plants of *R. fluitans* occur on gravel substrata in the middle stretches of the Goyt, Mersey-Goyt and Mersey, and
it is tempting to speculate that these plants represent the last
survivors from beds that may have flourished in the Mersey system from
the time the forests were first cleared to the beginning of the Industrial Revolution.

   As pollution intensified during the 19th century there seems little
doubt that many submerged plants were eliminated rapidly from the
system, or became severely restricted in their occurrence. Corbett's
(1907) brief description of the plantless state of the Irwell in
Manchester in the mid 1880's can be used to infer that similar conditions prevailed in the Mersey at that time, and his obvious delight at
the appearance of the pollution-tolerant alga *Vaucheria* in the Irwell
in 1905 can be taken as evidence that even the most resilient photosynthetic plants were absent during the latter half of the 19th century.
In the face of continuous organic pollution heterotrophs must have
played an important part in the ecology of the river at that time, and
even today visually obvious growths of sewage-fungus occur regularly
in the lower Tame and intermittently elsewhere. Stretches of river

affected by such growths often also support the tolerant filamentous alga *Stigeoclonium*, but there is no documentary evidence to suggest that this alga was present when pollution was more severe. Similarly, the pollution-tolerant angiosperm *Potamogeton pectinatus*, which has dominated the vegetation of the lower Tame, Mersey-Goyt and Mersey for at least ten years, has probably increased in abundance as organic pollution has decreased.

Apart from pockets of woodland on the lower Mersey-Etherow and Goyt, the catchment today is well illuminated, and with the exception of a small area of the Goyt affected by chlorine pollution, photosynthetic plants occur throughout the system. However, chemical conditions are still changing quite rapidly, and it seems likely that only the most upstream parts of the Mersey-Etherow, Tame and Goyt support plant communities that have reached stable equilibria with their environment. Despite this the system today contains a wide variety of plant communities influenced to varying degrees by the different types of pollution that still occur. Several workers (e.g. Haslam, 1978; Westlake, 1975) have stressed the importance of physical factors such as bedrock geology and channel topography on plant communities and the following brief description of the vegetation of the Mersey attempts to distinguish between these various influences.

5.622 Mersey

Thirty-three species of macrophyte were recorded from the Mersey-Etherow during 1978 (Harding, 1979). Submerged or emergent angiosperms were present at all sites, but the vegetation was in all cases dominated by attached filamentous algae (*Cladophora, Vaucheria*) or by submerged bryophytes. This is probably due to the largely rocky nature of the river bed and the relatively high current speeds at most sites, but it is also possible that rooted aquatic macrophytes have not yet realised their full potential range of distribution within the river. Improvements in the water quality of the Mersey-Etherow have taken place much more quickly in recent years than in the Mersey-Goyt or Mersey downstream, and whilst communities of algae and bryophytes probably change quickly in response to these improvements, some of the angiosperms recorded from the lower part of the Mersey-Etherow have been established for only a relatively short period (Harding *et al.*, 1981). Conditions at several sites appeared suitable for the growth of certain angiosperms (e.g. *Potamogeton crispus, P. pectinatus, P. natans, Ranunculus fluitans*) that were recorded elsewhere in the Tame and Goyt, and plants such as these might be expected to colonise the river in the future.

A downstream zonation of macrophytes was apparent in the Mersey-Etherow in 1978, and several species showed clearly defined ranges of occurrence. This zonation is illustrated most clearly by the seven submerged bryophytes. Two, *Scapania undulata* and *Racomitrium aciculare*, were confined almost exclusively to the top site, upstream of any inputs of treated sewage. The relative abundances of both *Fontinalis squamosa* and *Hygrohypnum ochraceum* were also greater at this site than further downstream. In contrast, *Rhynchostegium riparioides* was absent here, but appeared as abundant growths immediately down-stream of a minor sewage effluent and continued as the dominant bryo-phyte at all other sites in the Mersey-Etherow. *Fontinalis antipyretica* was recorded only at sites downstream of Glossop Brook, the tributary receiving intermittent pollution from the Glossop area. Another moss, *Amblystegium riparium* is particularly tolerant of organic pollution in rivers and appears to be a reliable indicator of marked nutrient enrich-ment within the area of the North West Water Authority. Its presence in the Mersey-Etherow was restricted to sites downstream from Glossop Sewage Works.

This zonation of bryophyte species appeared to be largely a response to downstream increases in nutrient concentrations, and none of the observed changes occurred in the first part of the Mersey-Etherow affected by high concentrations of zinc. In fact, the only obvious sign of the toxic effect of zinc on plants was the absence of the alga *Cladophora* immediately downstream of the zinc input, in a section which carries levels of nutrients that should be high enough to ensure its success. The marked sensitivity of *Cladophora* to zinc (Whitton, 1979) seems the only plausible explanation for its absence from the shallow, well-illuminated riffles that occur in this part of the river. It seems very likely, therefore, that *Cladophora* will spread upstream as further reductions in zinc pollution take place. *Cladophora* was also absent from the lower reaches of Glossop Brook, which although not carrying high concentrations of zinc, does receive intermittent chromium pollution (Say *et al.*, 1981).

Despite the numerous polluting discharges and sewer overflows affec-ting the Mersey and the catchment upstream of Stockport, the eight sites surveyed on the Mersey-Goyt and Mersey downstream of Stockport supported a fairly wide variety of macrophytes in 1978 (26 species recorded). A high proportion of the species found in the lower Goyt and lower Tame were recorded also in the Mersey-Goyt and the Mersey, and it seems probable that small populations of even quite sensitive species are maintained in the main river by continuous downstream inoculation (Fig. 5.6). *Potamogeton pectinatus* (inoculated from the Tame) and *Cladophora* dominated the vegetation at all sites on the Mersey and the other species were mostly present at a few sites only.

5.623 Goyt

Forty species of macrophyte were recorded from the Goyt. The uppermost site surveyed, upstream of Whaley Bridge, supported a very diverse and dense vegetation dominated by mosses (especially *Rhynchostegium riparioides* and *Fontinalis squamosa*), reflecting the excellent water quality in this part of the river and the stable nature of the rocky bed.

At Whaley Bridge, where the river receives the severely polluting effluent from the textile works containing organic waste and free chlorine from bleaching processes, true aquatic plants were absent and only bankside macrophytes (e.g. *Agrostis stolonifera* and *Phalaris arundinacea*) were recorded in the water. Even these had an unhealthy bleached appearance. Progressively downstream the vegetation showed a steady increase in diversity and cover, with first the pollution-tolerant *Amblystegium* appearing and then the slightly less-tolerant *Cladophora* and *Rhynchostegium*. Further improvement by dilution from a small clean tributary, the Sett, was indicated by an increase in the abundance of *Rhynchostegium* relative to that of *Amblystegium*, but high nutrient concentrations throughout the Goyt resulted in an extensive cover of *Cladophora* from the Goyt to the Mersey at Stockport. Plants of *Ranunculus fluitans* were recorded at sites downstream of the Sett and into the Mersey-Goyt, although nowhere did this species reach high cover values.

5.624 Tame

Upstream of any sources of pollution the Tame had a vegetation that was fairly typical of a small upland stream flowing over millstone grit; that is, it was dominated by bryophytes (e.g. *Fontinalis antipyretica, Rhynchostegium*) and with relatively few rooted angiosperms. There was some indication, however, that even these upper sites were subject to nutrient enrichment, with quite large growths of *Cladophora* in places. Further enrichment was indicated downstream as, successively, the abundance of the *Cladophora* increased, *Scapania undulata* disappeared, *Amblystegium* appeared and *Rhynchostegium* started to decrease in abundance. Downstream of the entry of Chew Brook, which at the time of the survey was heavily polluted by organic waste from a paper mill, there was a particularly marked decrease in the abundance of *Rhynchostegium*, and a corresponding increase in the cover of the more tolerant *Amblystegium*.

Several rooted macrophytes (e.g. *Potamogeton crispus, P. natans, Sagittaria sagittifolia, Sparganium emersum, S. erectum*) were recorded in the mid-section of the Tame in artificially created slow-flowing stretches with soft muddy beds. However, turbulent riffles alternated with these much slower stretches downstream to the confluence with the Mersey-Goyt, and *Amblystegium, Fontinalis antipyretica* and *Cladophora* continued to occur. In the lower Tame, beds of *Potamogeton pectinatus* were common and most sites were dominated by either this or *Cladophora*. As in the Mersey-Etherow, there was a stretch in the lower Tame below Dukinfield Sewage Works in which *Cladophora* was absent from an entire 0.5 km length (Fig. 5.8). Suitable substrata for *Cladophora* were abundant at the site, and the alga reappeared at the next site downstream. It seems probable that the high chromium concentrations revealed by analysis of *Fontinalis antypyretica* collected from the site in 1979 may have been responsible for the discontinuous distribution of *Cladophora* in 1978. Surveys of other rivers in N-W England (Harding,

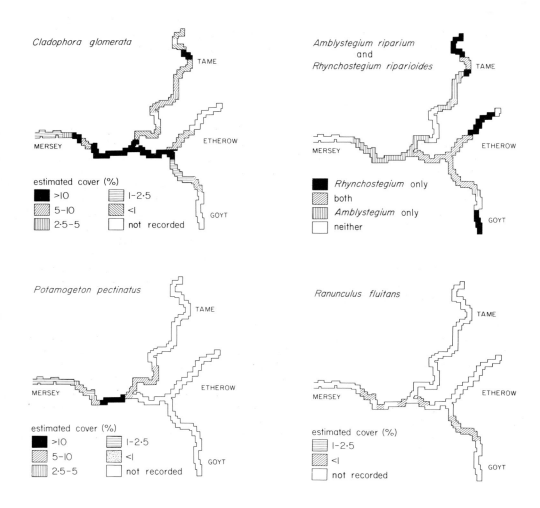

Fig. 5.8 Distribution and estimated cover of selected macrophytes in the Mersey catchment, as shown by a survey of 0.5 km lengths of river in summer 1978.

1979) have shown that *Cladophora* may be absent from rivers carrying
very high loads of organic material with low levels of metals, but it
seems unlikely that the absence of the alga from this particular site
is due to organic pollution, since it reappeared further downstream
despite inputs from several other large sewage works.

5.625 *Changing distribution*

It will be clear from the above descriptions of vegetation in the
catchment that plant communities have undergone rapid change in the
system and that it is unlikely that any species has attained a pattern
of distribution within the system that will remain stable in the near
future. This picture of instability is reinforced by examination of the
distribution of individual species throughout the catchment. For
example, it can be seen in Fig. 5.8 that five species (*Cladophora
glomerata, Amblystegium riparium, Rhynchostegium riparioides, Potamogeton pectinatus, Ranunculus fluitans*) showed highly discontinuous patterns of distribution and abundance. These are likely to change
markedly in response to pollution control measures in the near future,
as well as by inoculation with other rooted species from outside the
catchment or between the tributaries of the Mersey.

In addition to the effects of changing water quality, plant growth
in the Mersey catchment as a whole has obviously been influenced by the
introduction of alien species in the past. For example, the banks of
the Mersey through Stockport are coloured pink in the summer by *Impa-
tiens glandulifera* which first spread in England in the 19th century.
During 1980 dense beds of the exotic species *Elodea nuttallii* were
recorded for the first time in the Mersey at Warrington. This macrophyte is at present invading rivers and standing waters in N-W. England,
and it will be interesting to observe its future success at colonising
rivers of the Mersey catchment.

5.63   INVERTEBRATES

5.631 *General comments*

Of all the facets of plant and animal life in the Mersey the distribution of invertebrates perhaps gives the best demonstration of the
effect of man's activities on the catchment. The Upper Goyt Valley
is the most unspoiled part of the system and can be used to suggest
the conditions of former times. The wide range of invertebrate
species present today is usually dominated by *Gammarus pulex* and
representatives of the Plecoptera, Ephemeroptera, Trichoptera and
Chironomidae and is characteristic of many clean upland rivers in other
parts of Britain. Prior to the Industrial Revolution it is likely that
such a fauna would have been common throughout much of the upper catchment, at least downstream to the head of the Mersey lowland plain where
Stockport now stands. Further downstream, slower-water communities

would probably have been more prominent, but there are no clues to their possible composition because of the absence today of comparable lowland unaffected by pollution.

From the beginning of biological survey work in 1970, it was known that a wide range of water quality conditions prevailed and that many stretches were also subject to short-term fluctuations. The Trent Biotic Index (T.B.I.) (Woodiwiss, 1964), which depends on a sequential loss of key invertebrate groups with increasing organic pollution, was quickly established as a very useful means of transmitting the results of biological surveys to non-biological water quality control staff. Throughout the period all biological samples have been examined in the field in order to provide a quick appraisal of river conditions. Changes in T.B.I. values since 1970 (Fig. 5.9) suggest that water quality changes have taken place.

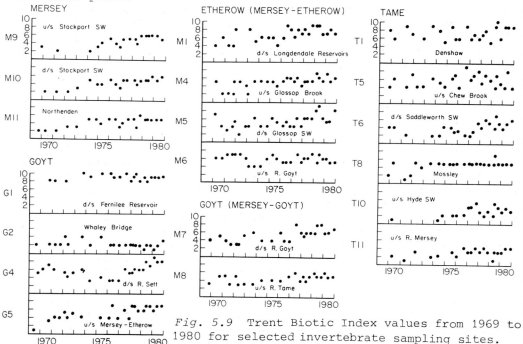

Fig. 5.9 Trent Biotic Index values from 1969 to 1980 for selected invertebrate sampling sites.

5.632 Mersey

The headwaters of the Mersey-Etherow flow from peaty moorland and are generally somewhat unproductive because of the soft, slightly acidic water (pH 5.5 – 7.0). The invertebrate community present upstream of all known sources of pollution usually includes Nemouridae, *Leuctra* spp., Polycentropidae, *Rhyacophila* spp., *Baetis* spp., Dytiscidae and Chironomidae. An increase of faunal variety, from the seven taxa in 1971 to 25 taxa in 1980 and a gradual rise of T.B.I. values to 7-9

(Fig. 5.9) suggest that conditions for invertebrates have improved, but without any readily identifiable cause.

During the period of review the river upstream of Glossop Brook has shown marked improvement from a community once dominated by tubificid worms and chironomid larvae and often supporting dense growths of sewage-fungus. In 1971 there was a reduction of organic inputs upstream, but continuing minor pollution still produces evidence in the fauna of organic enrichment. Several of the upstream groups, including Plecoptera and Dytiscidae, are suppressed, leaving *Baetis, Rhyacophila,* Polycentropidae and Chironomidae dominant, together with the Tubificidae which appear as a response to the organic inputs. The effects of zinc superimposed on organic pollution are described on p.139. Whilst accidental industrial pollution and the operation of sewer overflows on the Glossop Brook have an adverse effect on the Mersey-Etherow, most organic pollution comes from Glossop Sewage Works further downstream. The fauna below the works is usually dominated by *Baetis, Asellus,* Chironomidae and Tubificidae. There has been a positive benefit to the river from extensions completed in 1973 and various cleaner-water organisms have been recorded since then, with T.B.I. values showing an upward trend throughout the period to the present range of 5 to 8.

Steep gradients downstream to the Goyt confluence produce high self-purification rates with improved water quality, resulting in Trichoptera and Simuliidae being part of the dominant fauna. Up to the early part of the 1970s the downstream improvement had been reversed by the addition of the poorer quality Goyt, but a gradual restoration of water quality of this tributary has had a beneficial effect on the Mersey-Goyt downstream, pushing T.B.I. values up to 6-8. Whilst the invertebrate community continues to be dominated by the more pollution-tolerant taxa such as *Baetis* spp., *Asellus*, Chironomidae and Tubificidae, clean-water indicators such as Trichoptera and Ephemeroptera are usually present also. These latter organisms are eliminated lower downstream by the effluent from Hazel Grove Sewage Works, but even here a long-term improvement is still evident.

The next influence comes from the badly polluted Tame and below Stockport the Mersey fails to make any recovery as the beneficial effects of self-purification are cancelled out by additional organic loadings from three sewage works and a number of frequently or continuously operating sewer overflows. The biological picture downstream to the Manchester Ship Canal is thus fairly uniform with T.B.I. values at all survey sites varying between 4 and 6 (Fig. 5.9). The community of pollution-tolerant organisms consists of *Baetis* spp., *Asellus,* *Erpobdella* spp., *Lymnaea,* Chironomidae and Tubificidae.

Although the present condition of the Mersey is unsatisfactory it was in a much worse state in the early 1970s. Downstream from Stockport Sewage Works the fauna was limited to Chironomidae and Tubificidae, with *Asellus* appearing only occasionally. Improvements to the works were completed in 1973 and produced an immediate response in the river.

Baetis spp. appeared in 1974 for the first time and marked the start of
the upward trend evident in Fig. 5.9; the total number of invertebrate
species recorded has been steadily increasing ever since.

Little is known about the biology of the Mersey from the confluence
with the Manchester Ship Canal, but occasional samples from just above
the tidal limit at Warrington have only ever produced chironomids and
tubificid worms, thus confirming the poor water quality indicated by
the chemical data.

5.633 Goyt

Downstream from the pollution-free stretch of the upper Goyt the
effluent from the textile works at Whaley Bridge has a particularly
severe impact on the fauna. The deterioration to a Tubificidae/
Chironomidae community makes the biggest contrast of any on the Mersey
catchment with T.B.I. values dropping from 8-10 upstream to 1-3 downstream. Its severity may be due in part to the free chlorine discharged.

Despite a succession of textile, papermill and sewage effluents
adding to the organic load to the river, changes in the faunal community demonstrate a downstream recovery which is due to the excellent
aeration provided by the steep gradient. The first stages are seen in
a limited appearance of *Baetis* spp. and *Asellus*, but conditions improve
with the diluting effects of the Sett. *Baetis, Asellus* and Simuliidae
become firmly established, although high numbers of Chironomidae and
Tubificidae continue as a direct response to the polluted river upstream. In recent years the faunal variety below the Sett has been
steadily increasing with the tributary providing an inoculation of
cleaner-water invertebrates, such as Plecoptera, Heptageniidae,
Ephemerella ignita and *Rhyacophila* spp. so that T.B.I. values have risen
to 7-9. Although a pollution-tolerant fauna still dominates this stretch,
the increasing T.B.I. values over the last few years (Fig. 5.9) provide
a convincing picture of steady improvement. This is due probably to
the gradually decreasing volumes of industrial effluent combined with
tighter quality control.

5.634 Tame

In the absence of any major discharges the type of invertebrate community of the headwaters of the Tame down to the Chew Brook has changed
little over the last ten years. The water quality indicated by T.B.I.
values of 5-10 and a dominant fauna of *Baetis, Ephemerella, Rhyacophila*
and Chironomidae falls well short of that of both the Mersey-Etherow
and the Goyt headwaters. *Asellus* appears in the lower part of the
stretch confirming the slight downstream deterioration shown by the
sequence of plant species and probably results from contaminated surface water run-off. The immediate response to the effluents from

Saddleworth Sewage Works and the paper mill on Chew Brook is the loss of *Ephemerella* and *Rhyacophila*. *Baetis* spp. and *Asellus* have become established only since the completion of extensions to the sewage works in 1977.

From this point downstream a succession of sewage effluents, sewer overflows and industrially polluted tributaries causes a gradual deterioration.   T.B.I. values generally remain within the range 2 - 5, due to a pollution-tolerant fauna of *Baetis* spp., *Asellus, Erpobdella* spp., *Limnaea peregra,* Chironomidae and Tubificidae. The worst conditions prevail near the bottom end of the Tame where the effluent from Hyde Sewage Works eliminates all but Chironomidae and Tubificidae, and extensive growths of sewage-fungus are recorded throughout the year. The reappearance of *Asellus* upstream of the confluence with the Goyt marks a final slight recovery, though its distribution throughout the lower Tame serves to emphasise the polluted state of this tributary.

5.635 Changing distribution

It will be clear that the water quality of the Mersey has undergone many changes over a long period of time and that both plant and invertebrate communities reflect these changes. The distribution of individual species throughout the catchment provides additional information for interpreting survey data and forecasting future trends.

As an example, Harding *et al.* (1981), looking at the Mersey-Etherow in isolation, failed to detect any abnormalities of downstream distribution associated with the zinc pollution problem, but Fig. 5.10 highlights some quite striking anomalies in distribution through the catchment as a whole. In view of the relatively high organic content of the river the absence or scarcity of *Ancylus fluviatilis, Limnaea peregra, Erpobdella* spp., *Asellus* and Simulidae is somewhat surprising. The absence of *Asellus* downstream from Glossop Brook is particularly notable, considering it has been present in the tributary on at least 50% of all sampling occasions since 1970. Studies of zinc and lead polluted rivers in West Wales have shown molluscs, leeches, worms, crustaceans and fish to be particularly susceptible to these heavy metals (Hynes, 1960). The range of zinc concentrations (0.25 - 0.5 mg l^{-1}) determined in the Mersey-Etherow in 1979 and 1980 (Say *et al.*, 1981) has also been found to limit the distribution of molluscs in West Wales (Jones, 1940 ; Newton, 1944). It seems highly probable, therefore, that the limited distributions mentioned for the Mersey-Etherow are caused by zinc. If this is so, then as the forecasted decrease in concentration takes place these animals should extend their distribution upstream. *Gammarus pulex*, which is remarkably scarce throughout the rivers of the Mersey catchment (Holland, 1976), appeared in the Mersey-Etherow for the first time in 1979 and may be included in this expected upstream invasion.

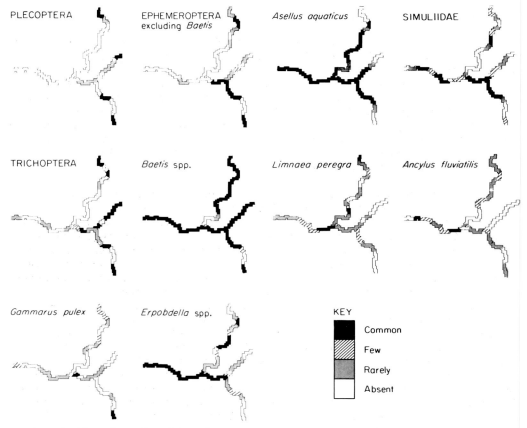

Fig. 5.10 Distribution of representative invertebrate taxa in the Mersey catchment, 1978-1980. (For details, see text)

Of the various invertebrates in the river upstream of the zinc input, *Baetis vernus, Rhyacophila* and Polycentropidae appear to be resistant to the zinc pollution downstream. Since *Baetis* spp. and the Trichoptera have been recorded frequently since 1971, it is obvious they have a resistance to the even higher zinc concentrations which must have occurred in earlier years before control measures started to take effect. The decreased abundance of Plecoptera is probably solely a response to the presence of organic material, as Jones (1940) has shown them to have a high tolerance to zinc.

The lower Tame appears to provide a parallel to the Mersey-Etherow in that *Erpodella, Lymnaea* and *Ancylus* are unexpectedly absent or scarce (Fig. 5.10). It seems possible that a similar explanation, the influence of heavy metals, applies here too; the metals in this case are chromium and lead from tannery and battery manufacturing processes.

5.64 FISH

Apart from Corbett's (1907) description of fish life in the Irwell and
the Mersey below Salford in the early years of the Industrial Revolution, there is no information on the distribution of fish on the Mersey
before pollution took its toll. The physical types of river suggest
that the entire catchment upstream of Stockport would have supported a
salmonid fishery, with the probable inclusion of fast-water coarse
fish such as grayling, chub and dace. In the slower waters of the lowland Mersey downstream from Stockport conditions would have been more
suitable for coarse fish, but would still have been available for
migrating salmon and sea trout.

Fish were, of course, the first noticeable casualties of the
declining water quality in the Industrial Revolution and over a short
time interval in the mid 1800s fish populations were destroyed in all
but the top reaches of the upper catchment. Ever since those years
poor water quality has continued to prohibit the re-establishment of
viable fisheries and it is only in the last ten years or so that any
recovery has taken place. Part of the biological investigations which
have been in progress on the Mersey since 1970 have been directed
towards the fishless state of the catchment. Between 1973 and 1978
trout in cages placed at strategic sites were found in some instances
to survive only a few days or even only hours (Fig. 5.11). These
short survival times emphasised how wide the gap was between the
existing water quality and that required for supporting fish life.
As part of the continuing investigations, samples of industrial and
sewage wastes were subjected to static laboratory fish toxicity tests.
The results showed that few sewage effluents were even slightly toxic
and those that were had no toxic significance after dilution in the
river. However, several industrial effluents were found to contain
highly toxic constituents and these provided explanations for the
fishless state of the upper Goyt (textile works) and the upper Tame
(paper mill). The toxic concentrations of zinc in the Mersey-Etherow
became known only as a result of the moss analyses. Elsewhere, it
was concluded from the chemical data that low dissolved oxygen concentrations caused by sewer overflows and poor quality sewage effluent
were responsible for the lethal conditions.

The extent of present pollution damage to the fisheries is shown in
Fig. 5.12. River sections regularly fished by anglers are limited to
the headwaters of the Mersey-Etherow, Goyt and Tame, with short
stretches lower downstream in the Mersey-Etherow and Goyt offering
newly established fisheries. Upstream of the zinc pollution on the
Mersey-Etherow natural perch populations and artificially maintained
stocks of brown and rainbow trout form the basis of a restricted
fishery. Another limited but natural brown trout fishery with bullheads and stone loaches occupies a short reach of the Goyt upstream of
Whaley Bridge. A similar fishery exists in the top stretch of the Tame
upstream of the Chew Brook with several coarse fish (roach, perch,
grayling, chub, carp) being recorded in small numbers from time to time,

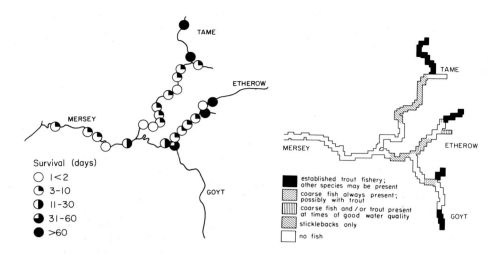

Fig. 5.11 Results of fish survival tests in cages during surveys from 1974-1978 (Mersey-Etherow, 11 surveys; Mersey, 3 surveys; Tame, 17 surveys).

Fig. 5.12 Distribution of fish in 1980, based both on electro-fishing surveys and information provided by anglers.

probably as a short-lived result of the stocking activities of local angling clubs.

The few changes which have taken place over the last ten years were preceded by several unsuccessful attempts to establish fisheries in the Mersey-Etherow at the Goyt confluence and the Mersey below Stockport under unsuitable water quality conditions. Since 1975, however, the declining zinc concentrations in the Mersey-Etherow have permitted the establishment of a fishery dominated by gudgeon and accompanied by small numbers of other species (perch, brown trout, roach, tench) near the Goyt confluence.

After the closure of a paper mill on the Sett in 1970, trout and grayling became established in the tributary through to the confluence with the Goyt. Over the last five years, as conditions in the Goyt have gradually improved, the fish population of the Sett has extended into the main river for a short distance downstream of the confluence, and has been permanently resident there since 1980.

In the near future it is anticipated that fish in the Mersey-Etherow will move upstream as zinc concentrations decrease further. It is also hoped that the water quality of the Goyt will continue to improve and thus allow the fish to extend their distribution. The presence of sticklebacks in the Mersey-Goyt holds out good prospects for this part of the river system. Whilst sticklebacks in the mid reaches of the Tame point to the near suitability of the river for coarse fish, there are, in fact, no immediate prospects for improvement in water quality.

5.7   DISCUSSION

A 200-year history of development of industry and housing on a large part of the Mersey catchment has left a legacy of pollution problems which makes the Mersey one of the worst systems in Britain today. Man's activities have produced an example of a river system heavily used in both upland and lowland stretches. The various rivers are subject to wide and rapid fluctuations in flow and the pollution created by the large volume of sewage and industrial wastes has a particularly severe impact on flora and fauna at times of low flow. The present account (5.6) of plant and animal life illustrates both the marked differences in water quality across the catchment and also the many changes which have taken place with time. Indeed, from the onset of the Industrial Revolution stable conditions have probably never been maintained anywhere for long, and it is also unlikely that a state of equilibrium will be achieved in the near future in view of further expected improvements.

The main thrust of biological investigation has from the outset in 1970 centred on examination of invertebrate communities. The response of animal life to different kinds of pollution has provided an invaluable tool for monitoring water quality. As other lines of investigation have been introduced, understanding of the life processes in the river has increased. As an example, the macrophyte survey work initiated in the area in 1978 has shown that the interplay of distribution of the two bryophytes, *Amblystegium riparium* and *Rhynchostegium riparioides*, can be used in a similar fashion to describe water quality. Increasing organic pollution is suggested by the sequence: neither present $\longrightarrow$ *Rhynchostegium* only $\longrightarrow$ both present $\longrightarrow$ *Amblystegium* only $\longrightarrow$ neither present (Fig. 5.8). The high tolerance of these bryophytes to heavy metals removes what would otherwise be a complicating factor. The alga *Cladophora* on the other hand, is sensitive to heavy metal contamination and gaps in its distribution provide a useful pointer for further investigation.

Although each biological survey technique has been developed singly, the results can be brought together to produce an integrated picture of the biology of the river system. The invertebrates have been seen to respond most quickly to changes in water quality, whilst the macrophyte communities change more slowly and present a long-term summary of conditions. Fish populations are the last to recover in response to improving water quality. Secure fisheries will be a most worthy end-product of several decades of intensive effort to restore the river system to an acceptable quality.

Acknowledgements

The authors are grateful to the North West Water Authority for permission to publish the data used in this paper. They also acknowledge their debt to the many colleagues who have contributed physical help and constructive discussion to what is essentially a team effort.

REFERENCES

Bracegirdle C. (1973) *The Dark River*, 175 pp. John Sherratt & Son, Altrincham.

Corbett J. (1907) *The River Irwell*, 168 pp. Abel Heywood & Son, Manchester.

Haslam S.M. (1978) *River Plants*, 396 pp. Cambridge U.P., Cambridge.

Harding J.P.C. (1979) *River Macrophytes of the Mersey and Ribble Basins Summer 1978*. North West Water Authority Rivers Division, Ref. No. TS-BS-79-1. British Lending Library loan collection.

Harding J.P.C. (1981) *Macrophytes as Monitors of River Quality in the Southern N.W.W.A. Area*. North West Water Authority Rivers Division, Ref. No. TS-BS-81-2. British Lending Library loan collection.

Harding J.P.C., Say P.J. & Whitton B.A. (1981) River Etherow: plants and animals of a river recovering from pollution. *Naturalist, Hull* 106, 15-31.

Holland D.G. (1976) The distribution of the freshwater Malacostraca in the area of the Mersey and Weaver River Authority. *Freshwat. Biol.* 6, 265-76.

Hynes H.B.N. (1960) *The Biology of Polluted Waters*, 202 pp. Liverpool U.P., Liverpool.

Jones J.R.E. (1940) A study of the zinc-polluted River Ystwyth in north Cardiganshire, Wales. *Ann. appl. Biol.* 27, 368-78.

National Water Council (1981) *River Quality: the 1980 Survey and Future Outlook*, 39 pp. National Water Council.

Newton L. (1944) Pollution of the rivers of West Wales by lead and zinc mine effluent. *Ann. appl. Biol.* 31, 1-11.

Palmer W. (1950) *The River Mersey*, revised edition, 250 pp. Robert Hale, London.

Pearson S. (1939) The Mersey and Irwell Basin. *The Royal Engineers Journal*. March, 1-12.

Say P.J., Harding J.P.C. & Whitton B.A. (1981) Aquatic mosses as monitors of heavy metal contamination in the River Etherow, Great Britain. *Environ. Pollut. Ser.B.* 2, 295-307.

Westlake D.F. (1975) Macrophytes. In: Whitton B.A. (Ed.) *River Ecology*, 725 pp., pp. 106-28. Blackwell, Oxford.

Whitton B.A. (1979) Plants as indicators of river water quality. In: James A. & Evison L. (Eds) *Biological Indicators of Water Quality*, pp. 5/1-5/34. Wiley, Chichester.

Woodiwiss F.S. (1964) The biological system of stream classification used by the Trent River Board. *Chemy Ind.* 11, 443-47.

6 · Tees

B. A. WHITTON & D. T. CRISP

6.1 INTRODUCTION

The Tees is one of a number of rivers draining the eastern slopes of the Pennines in northern England. Three in particular, the Tyne, the Wear and the Tees, have been the subject of a wide range of studies (Horne, 1977). All three are similar in that the rivers are predominantly fast-flowing and relatively unpolluted for much of their length; they do however become highly polluted towards the estuaries, where much of the population and industry is located.

Just over half a century ago, the Tees, the southern-most of these three rivers, received the first detailed survey of any British river. The authors (Butcher *et al*, 1937) showed that the inflow of a downstream tributary, the Skerne (Fig. 6.1), with very hard and grossly polluted water, led to major changes in the main river. This study has been widely quoted in the literature since then, although the condition of the Skerne has long been improved.

The upper part of the Tees catchment became a focus for research in the late 1960s and early 1970s. The construction of a new reservoir on the main axis of the river at Cow Green threatened an area of arctic-alpine vegetation famous for its flowering plant 'rarities'. Largely as the result of a gift of £100000 from Imperial Chemical Industries Ltd., many studies were made in Upper Teesdale, with the aims of establishing just what was so special about the area and advising on conservation of the remaining parts of high biological importance. Some studies on streams were included. In spite of the many interesting papers (see Clapham, 1978), no clear explanation for the rarities emerged, though the post-glacial history of the area, severe winter climate, exposed surfaces of highly metamorphosed limestone and perhaps also the presence of old mine workings are all important factors. The changes brought about by the Cow Green reservoir on the downstream Tees are one of the topics discussed in the present chapter.

A third burst of research took place in the mid-1970s, this time on the main river. A major water transfer scheme was planned to bring water from the Tyne to both the Wear and the Tees (Burston & Coats, 1975) and completed in 1983. This involved the construction of the largest reservoir in Britain (Kielder Water) on the N. Tyne, abstraction of water from the Tyne and its transfer southwards by means of a tunnel. Apart from initial tests, water has so far been transferred to the Tees only once.

In addition to research stimulated by special problems, fundamental studies on hydrology, chemistry and biology have been made by the Freshwater Biological Association, the University of Durham and other organizations. A number of these studies have dealt with streams in the Moor House National Nature Reserve, located in the upper part of the catchment. Hydrological records and routine chemical surveys have been made by the responsible water management body for many years.

Useful summaries are given in the Annual Reports of the Wear and Tees
River Board (1951-1965) and of the Northumbrian River Authority (1967-
1972).

In spite of the fact that there have probably been more scientific
publications on the Tees and its tributaries than on any other British
river, there has never been an integrated study of the ecology of the
river since the survey by Butcher *et al.* (carried out in 1929-1932).
The available information is patchy and the reader of this chapter will
no doubt be aware of many obvious questions which it is impossible to
answer.

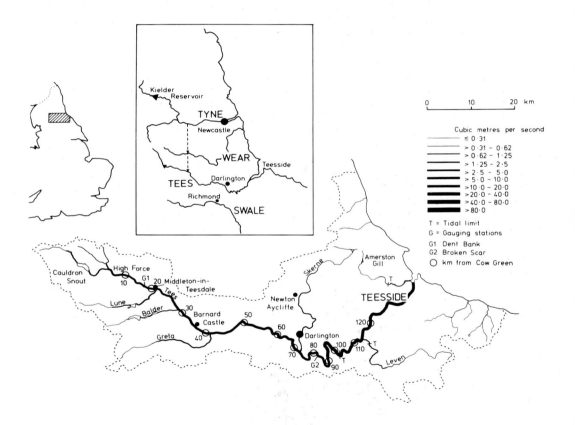

Fig. 6.1 Catchment of Tees and adjacent coastal streams, showing
tidal limits, mean flows and distances downstream from Cow Green
dam. Insert shows route of Tyne - Tees water transfer scheme. (Map
is redrawn from River Pollution Survey, England and Wales 1975 -
published by Department of the Environment and Welsh Office, 1978-
by permission of HMSO)

6.2 GENERAL DATA

The Tees was separated into upper and lower stretches in 1970 by the completion of the Cow Green Reservoir. Km points used here refer to distance downstream of the dam, whether or not reference is to an event subsequent to construction of the dam. (Km values are those used by Durham University.)

PHYSICAL FEATURES OF CATCHMENT
area (including streams entering estuary, but not coastal streams) 1805.9 km^2
   highest point   762 m
   height below Cow Green dam   463 m
   length of river:
      dam - tidal limits (High Worsall)   103.0 km
      dam - North Sea   141 km

CLIMATE
Precipitation for period 1916-50 ranged from 2285 mm yr^{-1} just south of highest point to 635 mm yr^{-1} near mouth of river.

ABSTRACTION
On 1 September 1976, towards end of period of dry weather. (Data from J. Brady of Northumbrian Water Authority)

	distance from Cow Green dam (km)	abstraction (m^3 s^{-1})	estimated natural discharge (m^3 s^{-1})
Broken Scar abstraction	64.4	2.31	
Blackwell abstraction	69.6	0.30	
Low Worsall abstraction	106.5	1.00	
discharge to sea		1.39	1.78

WATER TEMPERATURE
(Data of Crisp, 1977; see also Smith, 1968 and Crisp et al. 1975)
monthly water temperatures during 1971-75 at:

	mean	min.	max.
Maize Beck	7.2 °C	1.8 °C (Feb.)	14.4 °C (July)
regulated Tees (75 m downstream of dam)	6.8 °C	1.6 °C (Feb.)	13.7 °C (Aug.)

IMPOUNDMENTS

reservoir	stream	area (ha)	capacity (m^3 x 10^6)	year filled	compensation discharge (m^3 s^{-1})
Cow Green	Tees	312	40.9	1972	0.45
Selset	Lune	111	15.3	1960	
Grassholme	Lune	57	6.1	1915	0.34
Balderhead	Balder	117	19.7	1965	
Blackton	Balder)				
	)	78	6.0	1894/6	0.18
Hury	Balder)				
Hurworth	Skerne	23	0.73	c1870	
Crookfoot	Amerston Gill (to Greatham Creek)				

MANAGEMENT
Water management in catchment is the concern of the Northumbrian Water Authority, except for Hurworth and Crookfoot reservoirs owned by the Hartlepools Water Co.

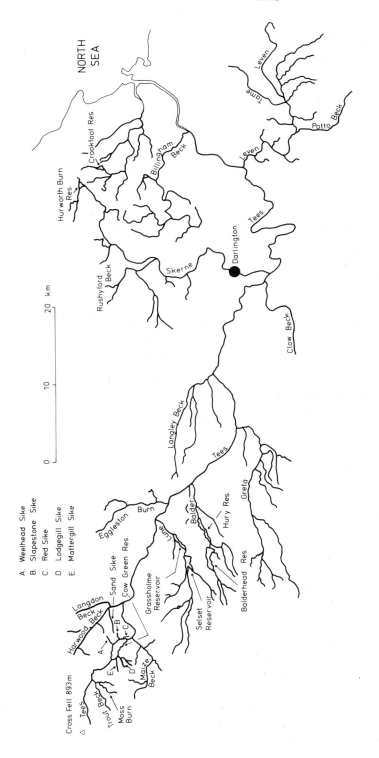

Fig. 6.2   Tees, showing tributaries and reservoirs.

6.3 GEOGRAPHY AND GEOLOGY

In spite of its relatively small size, there are marked variations in
climate and land use within the catchment (Smith, 1970). Annual precipitation varies from over 2000 mm in the headstream area to 640 mm or
less in the lowlands. The upland area has in general a greater amount
of cloud cover and temperatures are lower. Minima can fall below
freezing point in all months at Moor House (556 m). Snow can persist
at higher altitudes for some weeks.

   The highest parts of the catchment are mostly covered in peat or are
heather moors. Most of the middle and much of the lower altitudes are
agricultural, with pasture tending to give way to crops the nearer the
location is to the coast. There are many villages and a number of small
towns, but Darlington (population 97200) is the only major town whose
effluent reaches the Tees system above the tidal limit. The upland
parts of the Tees and its tributaries mostly lack a tree cover on their
banks, but some stretches of the middle river and extensive stretches
of the lower river have a dense cover on one or both banks (Fig. 6.3).

   The geology of the region is described by Dunham (1948). A number
of streams in the Cow Green region receive springs draining the metamorphosed limestone and flow over alternating bands of limestone and
other types of rock. The coarsely crystalline marble ('sugar limestone') here has a low magnesium content, in marked contrast to the
Magnesian Limestone of parts of the Skerne catchment. There are a
number of old lead or barytes mines in the upper part of the catchment,
but at present only one barytes mine is still operational. The upstream part of the Skerne lies within the Durham coalfield.

   The possibility cannot be ruled out that upland lead and barytes
mining had an obvious impact on the river in the 18th and 19th centuries,
but supporting evidence, which is available for the Wear, appears to be
lacking for the Tees. Present-day ecological effects of the old mine
workings are probably only very local, though the possibility of more
widespread effects at times of sudden flood needs critical study.

   Until the closure of the Cow Green dam (in 1970), the Tees was
unusual in having an upland, slow-flowing stretch (the Weel) with
well-developed stands of macrophytes (6.62). Downstream of the dam
there are two natural barriers to upstream fish migration, Cauldron
Snout (km 0.5) and High Force (km 10). The latter, at least, has
almost certainly been a complete barrier to migratory fish, since
there is a vertical drop of 21 m.

6.4 HYDROLOGY

6.41 DISCHARGE

The Tees, which rises at an altitude of 762 m, has the steepest average channel slope (2.7 m km^{-1}) and steepest headwater slope (4.3 m km^{-1}) of any of ten U.K. rivers listed by Newson (1981). By the time the river has been joined by Trout Beck (550 m), its mean discharge is about 1.3 m^3 s^{-1}. Between there and Cow Green (490 m), the mean discharge is augmented to about 2.5 m^3 s^{-1}. The Tees is unusual among British rivers in being of comparatively large size at a relatively high altitude.

The mean discharge at Broken Scar, the main gauging station (catchment 817 km^2) is 16.6 m^3 s^{-1}. The minimum natural discharge at Broken Scar is about 0.8 m^3 s^{-1}, but the present regulation of the river raises the minimum summer discharge to about 5 m^3 s^{-1}.

The Tees is a notoriously 'flashy' river and Butcher *et al.* noted that at times the rise in water level is so rapid that the flood water passes down the river 'like a tidal wave' and the first waters of a spate are very turbid. The Base Flow Index (Institute of Hydrology, 1978) can be regarded as an indication of flashyness. Streams with a B.F.I. approaching 1.0 have relatively even flow regimes, while streams with low values are relatively flashy. All except one of the upland tributaries (down to and including the Greta) have B.F.I. values (Table 6.1) less than 0.25, while all the lowland ones have values of 0.25 or more. These values may be compared with 0.31 and 0.29 for the upper reaches of the Wye and Severn respectively (M. D. Newson, pers. comm.).

Because snow can last for some weeks at higher altitudes, the eventual onset of mild weather can lead to spates, as documented for a small tributary by Crisp & Gledhill (1970). Snow-melt sometimes leads to floods in the main river (Smith, 1971), but there is no regular period for such floods each spring, as there is for some European rivers. Study of old records for the region suggests that major floods have occurred at all times of year.

During recent decades the construction of reservoirs and programmes for regulation and abstraction have favoured a reduction of flow extremes. Most other changes brought about by man's activities will probably have had the opposite influence, though their quantitative importance is not clear. 95% or more of the catchment was probably once covered by forest. Subsequently much of this was destroyed and the upland area became covered by peat or peaty soils. Within historical times much of the peat was drained, the extent of drainage depending on the economics of upland farming and other types of land use. Within the past five years there has been another surge of upland drainage, largely as a result of government subsidies.

One quantitative study is that of Conway & Miller (1960), who examined four small (4 - 8 ha) catchments on blanket peat in the Upper Tees catch-

Table 6.1 Values of Base Flow Index, calculated from hydrographs, for a selection of tributaries of the Tees and for two stations on the main river. (Information on Eggleshope Burn and Carl Beck from P. A. Carling and for other stations from Institute of Hydrology)

	National Grid Reference	Base Flow Index
Tributaries		
Trout Beck	NY 759336	0.15
Langdon Beck	NY 852309	0.21
Harwood Beck	NY 850309	0.20
Eggleshope Burn	NY 984288	0.34
Carl Beck	NY 946229	0.23
R. Balder	NY 928183	0.18
R. Greta	NZ 034100	0.21
Dale Beck	NZ 260156	0.25
R. Skerne (Preston-Le-Skerne)	NZ 292238	0.53
R. Skerne (Bradbury)	NZ 318285	0.64
Rushyford Beck	NZ 280290	0.78
Mardon Stell	NZ 323274	0.46
Clow Beck	NZ 282101	0.54
R. Leven (Leven Bridge)	NZ 445122	0.45
R. Leven (Easby)	NZ 585087	0.63
Billingham Beck	NZ 408227	0.29
Main River		
R. Tees (Dent Bank)	NY 932260	0.24
R. Tees (Low Moor)	NZ 364105	0.39

ment. The least managed of these had an extensive *Sphagnum* cover and only one drainage channel. At the other extreme was a catchment which had been severely burnt and had an extensive pattern of drainage channels. Relative to the natural catchment, the burnt and drained catchment had a far more sensitive response to rainfall, with earlier and higher peak discharges, but gave very little discharge in dry weather. Of the other two catchments, the hydrological performance of the burnt one was not so different from that of the natural catchment, whereas performance of the drained one was quite similar to that of the one which had been both drained and burned. Although this particular study suggests that drainage has a much greater effect than simply the removal of the peat, discussion on the management of the Tees catchment has often assumed that peat plays an important buffering role on stream discharge. Jansson (1982) concluded from a statistical survey of 45 catchments in northern Sweden that bogs do not act as buffering reservoirs, so there is clearly a need for further study within the Tees catchment.

6.42   TRANSPORT OF ERODED PEAT

Several studies have quantified the transport of peat by streams in the catchment. Crisp (1966) found that in a small headstream catchment (83 ha), about 77% of the input rainfall flowed down the stream, while the rest was lost as evaporation. About 30% of the stream discharge occurred during about 3% of the time and during these same spates (>0.2 m^3 s^{-1}) the stream transported about 75 t dry weight of eroded peat, some 80% of the total transported each year. Crisp & Robson (1979) found that there was a quantitative relationship between the concentration of peat in suspension and the quantity transported per unit time; the peat concentration reached a peak before the pea, discharge. Peat concentrations of up to 500 g m^{-3} water occurred during spates of up to 1 m^3 s^{-1} and, during one exceptionally large spate (c 1.5 m^3 s^{-1}), reached 2000 g m^{-3}. These quantities and patterns are probably typical of headwater tributaries of the Tees arising on blanket peat, where peat is eroding at a relatively steady rate.

A less frequent and more violent form of erosion also occurs from time to time.   During a thunderstorm on the evening of 6 July 1963, about 6000 m^3 of peat slid from the face of Meldon Hill and passed down Lodgegill Sike into the Tees (Crisp et al., 1964). Some of the peat blocks were deposited on the banks of the river and subsequently broke down slowly. However, sufficient of this material remained in the river and became finely divided to give a peat concentration of 2200 g m^{-3} at the Broken Scar gauging station the next morning. A similar phenomenon happened in the catchment of Langdon Beck on 17 July 1983.

It is evident that considerable quantities of peat enter the Tees from its upper tributaries. Much of this peat may be deposited in the lower stretches of the river or washed out to sea, but a certain amount settles within the tributaries and the upper river during the recession of each spate.

6.5   PHYSICO-CHEMICAL FEATURES

Small 3rd and 4th-order streams encounter the widest temperature range. The most upstream ones sometimes freeze almost entirely in severe winter weather, although there is probably always some moving water if part of the flow is spring-fed. In contrast, summer stream temperatures sometimes exceed 20° for short periods. Crisp (1977) reported that the mean annual temperature of Maize Beck was 7.2^o and of the regulated Tees just downstream of the dam, 6.8^o. The mean temperature of the river increases by about 2° by the time it has reached the tidal limit. Crisp & Howson (1982) showed in a study of six streams that there is an approximately linear relationship between mean air temperature and mean water temperature (over 5- or 7-day periods), except for periods when the mean air temperature is below 0°. Thermal stratification in the Cow Green reservoir is rare and generally of short duration

(Crisp, 1977), so has little influence on downstream temperature. The reservoir has, however, modified the temperature immediately downstream of the dam in several ways. The annual temperature fluctuation is reduced by 1 - 2° and there is a marked reduction in diel fluctuations. The spring rise in water temperature is delayed by 20 - 50 days and the autumn fall by 0 - 20 days.

Some indication of changes in water chemistry on passing down the river is given in Table 6.3. The composition of the water at most sites does, however, vary markedly according to flow conditions, with water which is much softer, rich in humic materials and with lower conductivity and pH values when discharge is high.

The chemistry of the Skerne is sufficiently different from that of the main river to bring marked changes in the concentrations of several elements between Tees km 67.5 and km 78.8. Increases in Mg and Ca reflect the influence of the Magnesian Limestone and pumped mine water; increases in Na, P and Cl are due to effluents entering the Skerne from Darlington and other sewage treatment plants. The Skerne is also polluted by phenols and cyanide from a coke-works.

Table 6.3 Mean values (mg l^{-1}) for selected variables in 0.2 μm filtrable water sampled at four sites in the Tees system during a one-year study (Holmes & Whitton, 1981a). Data of Carling (1983) for rainfall in the Great Eggleshope Beck basin are included for comparison.

	Sand Sike	Tees (km from dam)			rainfall
		0.1	67.5	78.8	
Na	3.6	2.6	7.8	30.4	3.16
K	0.71	1.31	2.16	3.71	1.30
Mg	1.42	0.86	5.27	12.7	0.51
Ca	40.0	8.47	32.4	51.3	0.47
Fe	0.14	0.32	0.21	0.27	
Zn	0.069	0.044	0.029	0.038	
NH_4-N	0.017	0.044	0.026	0.280	
NO_3-N	0.166	0.195	0.889	1.05	
PO_4-P_i	0.025	0.028	0.126	1.27	
Cl	6.6	5.0	15.7	39.1	

6.6   PLANTS

6.61   SOURCES OF INFORMATION

The 1929-32 survey (Butcher *et al.*, 1937) paid particular attention to
the photosynthetic organisms, especially the microalgae of the main river
and the macrophytes below the entry of the Skerne. The survey led to
generalizations on observations at 24 sites and also descriptions of
several new algal species, which have since been shown to be widespread
in rivers of many countries. The (now drowned) Weel was studied in some
detail during 1967-68 by Proctor (1971, 1978), his account also including
casual observations as far back as 1937. Selected sites in the lower
stretches of the Tees and Skerne were surveyed during 1964-65 (Whitton
& Dalpra, 1968) and the Tees from the entry of the Skerne to km 100 in
June-July 1970; the emphasis in both studies was on comparison with the
earlier survey. A much more detailed survey was made in 1975 (Holmes &
Whitton, 1977), six years after the closure of Cow Green dam. Records
were made for both 'river' and 'bank' in each 0.5 km length downstream
of the dam. The records included subjective estimates (expressed on
1 - 5 scale) for both the relative biomass of each species and the
percentage of the area covered by the species. The occurrence of
Grimmia agassizii on rocks subject to very high current speeds (Holmes,
1976) provided the second record for this species in the British Isles.

The phytobenthos of six sites down the river system was studied at
monthly intervals over one year during 1976/77 (Holmes & Whitton,
1981a). Besides floristic records, estimates were made of the relative
proportions both of various life-forms and individual species. Phytoplankton was sampled at the same time as the phytobenthos at the two
most downstream sites and also on three different occasions at other
sites (Holmes & Whitton, 1981b); comparisons were made with the phytoplankton of adjacent rivers at the same time.

The only upland streams studied are those subject to the influence of
the metamorphosed limestone (Carter, 1972; Hughes & Whitton, 1972;
Livingstone & Whitton, 1984). The blue-green alga *Calothrix parietina*,
an important species in Sand Sike, has been the subject of a number of
experimental studies and its behaviour in the laboratory compared with
that in the stream (see Livingstone & Whitton, 1984a, for references).
The fixation of CO_2 and N_2 by *Rivularia* in Red Sike is described by
Livingstone & Whitton (1984b).

6.62   PHYTOBENTHOS AND MACROPHYTES

The vegetation of the upland streams is usually dominated by algae,
though bryophytes (e.g. *Acrocladium cuspidatum*, *Philonotis fontana*) are
often conspicuous immediately below the source of calcareous springs,
and several rooted macrophytes can occur in slow-flowing stretches.
Rooted macrophytes were well represented in the Weel (see below). Nowadays submerged and floating-leaved species occur only in the mid- and

A

B

C

D

E

Fig. 6.3 Views of Tees system. A, Harwood Beck just above gauge; B, Cow Green dam; C, view downstream from B; D, Eggleston (km 25.6), showing extensive area of cobbles exposed under low flow; E, Hurworth (km 72.8) downstream from Croft Bridge, showing stretch with dense growths in summer of *Ranunculus fluitans* (in photo), *Cladophora* & *Fontinalis*.

lower stretches. Bryophyte crops are highest immediately below the Cow
Green dam and in general become less important on passing downstream.
Below the entry of the Skerne, only one bryophyte (*Fontinalis antipyretica*) remains abundant, in contrast to flowering plants which show
a marked increase in abundance. Other abundant or frequent species
below the Skerne are *Ranunculus fluitans, Potamogeton pectinatus,
Cladophora glomerata, Vaucheria sessilis* and, slightly further downstream, *Myriophyllum spicatum*. Only non-vascular species occur in 'river'
habitats downstream of km 113.0.

   Rivularia is abundant in streams on the metamorphosed limestone,
though it appears to be restricted to those parts which receive both
peat and limestone drainage, possibly due to the importance of organic
phosphate in such waters (Livingstone & Whitton, 1984a). These streams
usually show conspicuous growths of diatoms in spring, mostly species
of *Cymbella*, several of which form small hemispherical colonies. The
development of obvious growths of other species (especially filamentous
Conjugales) appears to be determined at least as much by recent flow
history as season. Filamentous green algae are often abundant in the
larger, soft-water streams in late winter, these usually being species
of *Ulothrix*. Such growths develop during periods of low flow and
bright sunshine and, if there is no scouring flood, may be succeeded by
a mass development of the large hemispherical colonies of *Didymosphenia
geminata*.

   In spite of the presence of old mine workings in the area, it is doubtful if the slight elevation in the concentrations of heavy metals in the
water is sufficient to have much influence on phytobenthos biology. An
isolate of *Calothrix parietina* from Sand Sike did however prove to be
much more resistant to zinc than might be expected from ambient zinc
levels (Shehata & Whitton, 1981). This alga grows in microhabitats
alternatively wetted and dried and is perhaps at times subject to zinc
enrichment in this zone.

   Proctor (1971) suggested that differences between his own observations
in the Weel and those of R. W. Butcher might have been due in part to
the complicated history and effects of the Cow Green barytes mine.
Wastes from the mine appear to have favoured the growth of aquatic
vegetation. One way this might have been brought about was due to the
destructive effects of geese. Some 60 - 70 geese were summered on the
Weel until 1937 when many died, apparently due to ingesting barytes
sludge. The species found by Proctor in 1967-68 were, in decreasing
order of abundance: *Potamogeton alpinus, Myriophyllum alterniflorum,
Equisetum fluviatile, E. palustre, Callitriche platycarpa, C. intermedia,
Potamogeton natans, Ranunculus aquatilis*.

   During the study of six sites down the river system (Holmes & Whitton,
1981a), a spring 'burst' of diatoms occurred at four sites, the two most
upstream and the two most downstream. The latter occurred a month later
than the former, in spite of higher downstream temperatures. It seems
likely that the higher irradiation received at the surface of the sub-

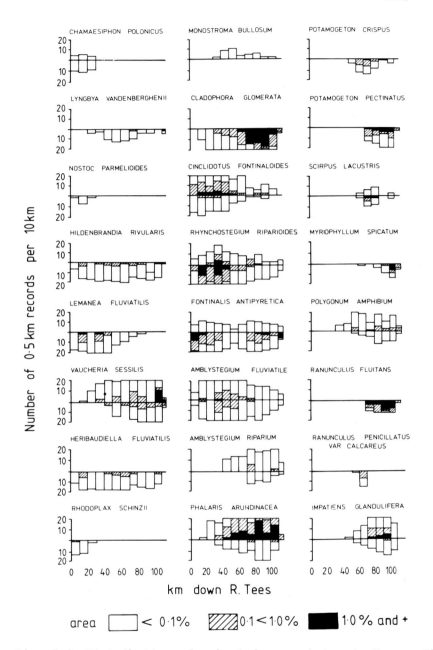

Fig. 6.4 Distribution of selected macrophytes in Tees. The histograms show the number of 0.5 km lengths from which each species was recorded in the 20 lengths of each successive 10 km of river; plotted above the line are records from bank habitats, below the line, those from river habitats. In addition to giving presence/absence data, the histograms also show the percentage cover of a species in each 0.5 km length. (Data of Holmes & Whitton, 1977)

stratum in the shallow upstream waters was an important factor. The
lack of a diatom burst below the Cow Green dam and at km 25.4 may have
been due to the lower silica concentrations in the water there, this
almost certainly resulting from the removal of silica in the reservoir.
Apart from the lack of a spring diatom burst and the greater cover of
bryophytes, the site below the Cow Green dam differed in other ways.
Filamentous green algae of many species were abundant in late summer
and early autumn and there was a conspicuous autumn cover of *Phormidium*.
General observations (the authors and P. D. Armitage, pers. comm.) have
confirmed that bryophyte crops are consistently higher here than at the
same site prior to regulation or elsewhere in the river.

Another apparent effect of regulation is evident much further downstream. Apart from the populations in the Weel, submerged angiosperms
were restricted in 1929-32 to the river below the entry of the Skerne.
This was again the situation in 1965. However in 1975 four species
were established upstream of the Skerne, although none were abundant.
Potamogeton crispus, *Zannichellia palustris* and *Myriophyllum spicatum*
had spread 25, 9.5 and 4.5 km, respectively, upstream from their
previous known most upstream localities. *Ranunculus penicillatus* var.
calcareus (Fig. 6.4) was an apparent invader to the river and crops of
this species above the Skerne have increased still further since 1975.
It seems probable that the four submerged angiosperms which became
established above the Skerne did so during the period after the full
regulatory effects of Cow Green Reservoir started in 1971. Although
Butcher *et al.* described the Tees above the Skerne as 'virtually
unhabitable by higher plants', this does not apply now.

Three macrophyte species appear to have been lost between 1929-32
and 1975. Of these, *Nitella opaca* disappeared before 1964-65, while
Potamogeton perfoliatus and *P. pusillus* have disappeared since then.
None have been found in tributaries. *Elodea canadensis*, present in
1965, was absent in 1970, but present again in 1975. It seems quite
likely that the main river population was washed out by an exceptional
flood in March 1968 and that the species was re-inoculated from tributaries where it is known to be present.

The macrophytic green alga *Enteromorpha flexuosa* was absent from the
main river in 1929-32 and 1975, yet was widespread during 1964-65 and
1970 in the Skerne and below its inflow. Its absence from the main
river in 1975 is especially striking in view of the fact that it was
present in Clow Beck, which presumably provided an inoculum to the
Tees at km 71.6; it was also abundant elsewhere in north-east England
in 1975.

Holmes & Whitton (1981a) reported that surveys of photosynthetic
plants of the Tees and tributaries have shown about 600 species. In
view of the large number of streams to be studied and their marked
differences in water chemistry, it seems probable that the complete
flora of flowing waters in the Tees catchment is not far short of
1000 species.

6.63 PHYTOPLANKTON

The survey of phytoplankton in 1975 showed that the most important
groups were, in decreasing order of abundance, diatoms, green algae and
cryptomonads. Most taxa were ones which have been found also in the
benthos. Of the sixteen taxa found only in the plankton, only two
(*Rhodomonas lacustris, Actinastrum hantzschii*) were ever frequent and
many could easily have been overlooked during surveys of the benthos.
Nine of the species found only in the plankton were motile: *Lepocinclis
ovum, Lobomonas* sp., *Gymnodinium* sp., *Ochromonas* sp., *Gonium pectorale,
G. sociale, Lobomonas* sp., *Pteromonas angulosa, Pyramimonas* sp.

Among the diatoms, the density of empty frustules was usually of a
similar order to that of live cells, so counts made only on cells
cleared in acid would have given a false impression. The density of
all types of live cell ranged from 100 to 5000 cells (or units) ml^{-1}.
These values are quite similar to those in the Tyne, Wear and Swale,
but about one order of magnitude less than in the Neckar (p. 333). The
three studies made on the same day of the sites on the Tyne and Tees
associated with the present water transfer scheme showed that plankton
densities were only slightly higher in the Tyne. So far no studies
have been made of the fate of plankton on passing through the tunnel.

The plankton flora immediately downstream of the Cow Green dam
probably reflected the composition of plankton in the reservoir, the
most frequent species being *Melosira italica, Achnanthes* spp.,
Asterionella formosa, Diatoma elongatum and *Synedra* spp. Live cells
of all these species were found further downstream, but overall their
densities tended to decrease on passing downstream. It is uncertain
what effect passage over High Force has on the viability of cells, as
the downstream sample site at km 25.7 may receive inocula from the
Lunedale reservoirs as well as Cow Green. While *Melosira italica*
became infrequent by km 67.5, another centric diatom, *Cyclotella
meneghiniana,* which was absent below the dam, became frequent in the
downstream river, expecially in August when it was the most abundant
species at km 78.8.

6.7   INVERTEBRATES

6.71   SOURCES OF INFORMATION

The most detailed accounts are for the Upper Tees and the tributaries
within and around the Moor House National Nature Reserve. These include
the Simuliidae (Davies & Smith, 1958; Phillipson, 1957; Wotton, 1976,
1977, 1978), Plecoptera (Brown *et al.*, 1964), Ephemeroptera (Crisp &
Nelson, 1965) and Chironomidae (Birkett, 1976).   In addition, papers by
Coulson (1959) and Nelson (1971) have records of aquatic Tipulidae and
other groups.   There are also accounts of the benthic invertebrates of
four headstream sites within the Moor House Reserve (Armitage *et al.*,
1975) and in tributaries within the Cow Green area (Armitage *et al.*,
1974).   Whitehead (1929) and Brown & Whitehead (1938) list several species,
but all ones included by Butcher *et al.*

The 1929-33 study (Butcher *et al.*, 1937) included a survey at sites
down the whole of the non-tidal part of the main river. Macan (1957)
added two species of Ephemeroptera to the faunistic list. A general
survey of the fauna was made also in April 1976 (by P. D. Armitage;
data held by Freshwater Biological Association).   The effects of impoundment and regulation upon the invertebrate fauna just downstream of Cow
Green have received considerable attention (see 6.74).

Table 6.3 Percentage composition of fauna from two collections of
samples taken at different altitudes. Collection 1 is a composite of
kick samples taken in May, August and October over several years
from four stations in headstreams on Moor House N.N.R. (Armitage *et
al.*, 1975); collection 2 is a similar composite for the main river,
one major tributary and several minor tributaries within Cow Green
area (Armitage *et al.*, 1974). (Representatives of some minor groups
were excluded from this analysis)

	composition (%)	
	collection 1	collection 2
Plecoptera	34.7	21.0
Ephemeroptera	26.2	44.4
Diptera	11.3	8.2
Trichoptera	10.3	5.0
Coleoptera	8.3	4.1
Crustacea	4.5	12.1
Annelida	2.8	1.6
Hydrachnellae	1.6	0.6
Mollusca	0.3	3.0
	100.0	100.0

6.72   THE FAUNA AND ITS DISTRIBUTION

A total of 135 taxa have been recorded for the Tees system in the various papers listed above. (Species limits and nomenclature used here follow those of Maitland, 1977.) The records include 76% of known British Plecoptera, but only 19% of gastropod molluscs.

Some indication of the relative proportions of various taxa among upstream invertebrate populations is given in Table 6.3. The table also illustrates the decrease in importance of Plecoptera and increase of Ephemeroptera with decreasing altitude, though the Ephemeroptera decrease again in the downstream stretch of the main river. The differences shown in the table for Crustacea and Mollusca are largely a reflection of local conditions.

The 1976 survey of the main river showed that overall the most common invertebrates were *Caenis rivulorum, *Naididae, Tubificidae, *Baetis rhodani, Hydropsyche siltalai,* Orthocladiinae, *Limnius volckmari, *Elmis aenea* and *Rhithrogena semicolorata.* Some of these (indicated by *) were apparently more abundant in 1976 than in the 1929-32 survey. The most important changes are probably immediately below the Cow Green dam and in the stretch of the river below the entry of the Skerne, though the records are insufficient for a detailed comparison to be made for the latter.

6.73   DRIFT AT UPSTREAM SITES

Drift of benthic organisms has been followed at two upstream sites (Table 6.4). In the small headstream, about 4×10^6 individual animals (c 1.6 kg dry weight) passed a point in the stream annually. Within individual spates the drift concentration has been found (Crisp & Robson, 1979) to reach its peak before that of the hydrograph, a result similar to that found for peat fragments (6.42). For most organisms concentrations were low during the spate peak and increased again as

Table 6.4 Comparison of estimates of mean discharge, total annual drift past a point and of mean concentration in terms of numbers and weight for a small headstream (Rough Sike: Crisp & Gledhill, 1970) and an unregulated major tributary (Maize Beck: Armitage, 1977). Data from a southern chalk stream are included for comparison (East Stoke Mill stream: Crisp & Gledhill, 1970).

	Rough Sike	Maize Beck	East Stoke Mill Stream (Crisp & Gledhill, 1970)
mean discharge ($m^3 s^{-1}$)	0.04	1.48	0.71
total drift (kg yr^{-1})	1.4	15.6	28.0
mean conc. (number m^{-3})	2.92	4.09	0.912
mean conc. (mg m^{-3})	1.0	0.3	1.2

the spate receded. In contrast, terrestrial casualties peaked with the
spate. As terrestrial casualties formed 10 - 15% of annual drift in
terms of numbers and had a greater mean weight per individual than the
aquatic component, the terrestrial contribution was appreciable. Despite
these within-spate patterns, on a round-the-year basis the concentration
(number or weight per unit volume of water) and transport (drift per unit
time in terms of weight) showed a highly significant positive correlation
with stream discharge. The relationships between discharge and numbers
were fitted best by the equation:

$$C_n = 1.5 \ Q^{0.72} \quad \text{and} \quad T_n = 40744 \ Q^{1.71}$$

where C_n is individuals m^{-3}, T_n is individuals h^{-1} and Q is discharge
(expressed as $m^3 \ s^{-1}$).

   The observation that drift concentration increased with discharge
agrees with that found for the Wye (see pp. 70-71) by Brooker & Hemsworth
(1978) during regulation releases.

6.74   INFLUENCE OF IMPOUNDMENT AND REGULATION DOWSTREAM OF COW GREEN
       RESERVOIR

Information on the pre-regulation benthos of the main river some 500 m
downstream of the present dam and in the lower stretch of the unregulated
Maize Beck is given by Armitage *et al.* (1974) and accounts of both after
regulation by Armitage (1976, 1978a, b). The following is a simplified
summary of the main findings.

   Both population density and biomass increased substantially in the
Tees following regulation; Maize Beck, whose faunal composition and
abundance was similar to that of the Tees prior to regulation, showed
no obvious change over the same period. Species changes on the Tees
included both losses (mainly Plecoptera) and gains, though all were
species which occurred elsewhere in the area. Probably of more importance were changes in the relative abundance of groups. The abundance
of *Gammarus*, Chironomidae and Mollusca has increased in the Tees and
the abundance of Ecdyonuridae nymphs has decreased. In addition, there
has been the appearance of large numbers (but small biomass) of *Hydra*
and *Nais*.

   Following closure of the Cow Green dam in June 1970 there was a
rapid build-up of zooplankton in the reservoir and of *Hydra* in the
reservoir and the river downstream of the dam. Armitage & Capper
(1976) estimated that 150 kg (dry weight) of this material was discharged from the reservoir annually, comprising about 136 kg of Cladocera,
13 kg of Copepoda and 1 kg of *Hydra*. About 98% of this material was
discharged during July-October and for much of this period the concentrations in the water column were about 1300 Cladocera and 200
Copepoda m^{-3}. Densities had however dropped to 2% or less of these
values by km 6.5.

In contrast to this drift of material from the reservoir, drift of benthos from the river bed was changed only slightly as a result of regulation. There was an increase in *Nais* in the drift, reflecting the greatly increased importance in the benthos. Comparison with Maize Beck also suggested that nocturnal peaks of *Baetis* and *Simulium* were less pronounced in the regulated river.

The various observations on seston, benthos and drift suggest that impoundment has also had a marked effect on trophic relationships within at least the first 0.5 km. Organic sediment, whose importance as a primary or secondary food source for detritivores and as case-building material for some Chironomidae is becoming apparent (McLachlan *et al.*, 1979; Walentowicz & McLachlan, 1980), has almost certainly decreased markedly downstream of the dam. Extrapolation from results for Rough Sike (6.42) to the Tees at Cow Green suggests that about 4×10^6 km of peat enters Cow Green reservoir annually, while less than 0.2×10^6 kg of organic sediment is discharged downstream. On the other hand, the 150 kg of small animals from the reservoir is a rich source of direct or indirect food for some species and is taken up rapidly by the benthic system. During August and September the rate of uptake averages about 0.15 g m^{-2} d^{-1}, though it may reach four times as much (Armitage & Capper, 1976).

It is not yet possible to assess the importance for the ecosystem of the input of reservoir animal material and loss of detrital material. However, the former is only 4% of the latter and is concentrated within a four-month period. Although no studies have yet been made, the qualitative composition of the peat fraction discharged from the reservoir is also likely to be changed by its passage through the reservoir, with a smaller average particle size but perhaps an increased bacterial content.

The post-regulation density and biomass of benthos is substantially higher in the Tees than in the Maize Beck (Armitage, 1978b). This may represent an increase in available food for fish. Trout, but apparently not bullheads, consume the Cladocera and Copepoda from the reservoir (Crisp *et al.*, 1978). Trout also feed avidly on terrestrial casualties such as earthworms, slugs and larval Tipuliidae, which one would expect to be reduced considerably downstream of the impoundment.

6.75 COMMENTS ON INDIVIDUAL SPECIES

The large mayfly *Ephemera danica* was described by Butcher *et al.* as uncommon and nymphs were recorded only at Eryholme (km 83.9). As the nymphs burrow in sand and fine gravel, it would not be expected to be widespread or abundant in the Tees, but it appears more widespread than indicated. Nymphs have been recorded at Dubby Sike (Cow Green) during electrofishing (Armitage *et al.*, 1974) and large numbers of imagines were observed in the late 1970s in the Cotherstone area

(A. Clarke, pers. comm.) and a single nymph at Red Sike in 1983. Either the earlier workers overlooked the species at some sites or its status has changed.

The lack of records for the crayfish *Austropotamobius pallipes* by Butcher *et al*. is almost certainly due to its being overlooked. The species is in fact widespread both in the main river in the Barnard Castle area and in the lower parts of some tributaries. It is also found at High Coniscliffe (c km 58) and the R. Leven.

6.8 FISH

6.81 SOURCES OF INFORMATION

Published information on fish in the main rivers includes: Longstaffe (1954), Jackson (1930), Pentelow (1932), Pentelow *et al*. (1933a, b), Southgate *et al*. (1932), Brown *et al*. (1970), Butcher *et al*. (1937), Crisp *et al*. (1974, 1975), Hancock (1977). The Tees fishery for migratory salmonids is summarized in the annual reports of the Northumbrian Water Authority and its predecessors. Butcher *et al*. dealt mainly with the distribution of species in the main river. In contrast, Crisp *et al*. covered distribution, population density, growth, production and reproduction, but only in the river system above the confluence with Maize Beck. Hancock surveyed the main river from Cauldron Snout to Hurworth and described species distribution, but gave growth and population data only for the brown trout.

6.82 SPECIES DISTRIBUTION

At least twenty-two species are present in the Tees system (Table 6.5).

Table 6.5 Fish species present in Tees system. B, listed by Butcher *et al*. (1937); H, additions by Hancock (1977); N, further records of Northumbrian Water Authority.

N	brook lamprey	N	bream	N	10-spined
B	salmon	B	minnow		stickleback
B	sea and brown trout	H	rudd	B	perch
N	rainbow trout	B	roach	B	bullhead
B	grayling	B	dace	B	flounder
B	pike	B	chub		
N	carp	B	stone loach		
N	barbel	B	eel		
B	gudgeon	B	3-spined stickleback		

Two others mentioned by Longstaffe, the sturgeon and the burbot, have
long since disappeared, the former already being a rarity in 1854.
Butcher *et al.* listed only fifteen species, but several were probably
overlooked, while others have been introduced since then.

The rainbow trout is present as a stocked fish in some of the Lunedale and Baldersdale reservoirs and, not only can downstream escapes
occur, but there is evidence of breeding in Carl Beck. The bream was
introduced by the Water Authority at Croft in 1974/75. A carp of about
230 g was taken at Over Dinsdale in 1979 and small barbel were found at
Croft in 1978. Butcher *et al.* recorded the three-spined stickleback
as far up the estuary as Yarm and also in the Skerne. Nowadays, at
least, the species is certainly more widespread.

Butcher *et al.* found that dace, chub and minnows were the most frequent fish in the lower river, with gudgeon and roach less frequent
and pike and perch rare. Of these, all except the minnow decreased in
numbers above Croft (km 72.8) and the most upstream site for dace,
roach and chub was at Piercebridge (km 56.3). Hancock's (1977) observations are consistent with these earlier ones, but they do extend the
ranges of some species, with bullhead and minnow at Cauldron Snout
(km 0.5) and stone loach at Middleton-in-Teesdale (km 19). If the
Tees system as a whole is considered, there are records for minnows
at 503 m altitude and of bullhead at 628 m (Crisp *et al.*, 1974, 1975).

Eels were described by Butcher *et al.* as common throughout the Tees,
but despite intensive sampling for over a decade, they have not been
recorded at or above Cauldron Snout (Crisp *et al.*, 1974, 1975) and it
is doubtful whether they occur upstream of High Force.

Among Salmonidae, the brown trout occurs throughout almost the
length of the river. According to Butcher *et al.* the grayling was
common in the middle stretch of the river, with its upper limit at
about km 48. It now occurs at least as far upstream as Dent Bank (A.
Clarke, pers. comm.). Butcher *et al.* mentioned a local tradition that
the grayling was introduced about 1890. It would appear, therefore,
that since its original introduction it has slowly spread upstream.
The changing status of the salmon and sea trout in the Tees is covered
in 6.84.

6.83 POPULATION DENSITY, GROWTH AND PRODUCTION

The larger, older brown trout tend to occur at low density and relatively low biomass in the main river (Table 6.6). These fish spawn in
the tributaries, which act as nursery streams and contain mostly
smaller, younger fish at higher density and biomass. Recruitment in
the main river is largely by downstream dispersal of fish from the
tributaries and headwaters. There is considerable variation in the
annual production of different tributaries (Table 6.7). The annual
survival rates of both brown trout and bullhead are high, as can be

Table 6.6 Population density and biomass of brown trout and bull-head based on data of Crisp *et al.* (1974, 1975, unpublished) and Hancock (1977).   O, species absent; -, not estimated.

	brown trout		bullhead	
	numbers m^{-2}	$g\ m^{-2}$	numbers m^{-2}	$g\ m^{-2}$
Tributaries				
Trout Beck	0.1 -0.2	2.0 - 4.3	0.04	-
Weelhead Sike	0.25-1.37	1.6 -16.6	0	0
Dubby Sike	0.21-0.92	1.7 - 6.9	0	0
Lodgegill Sike	0.03-0.13	0.3 - 5.3	0	0
Mattergill Sike	0.04-0.19	1.0 -11.8	0	0
Gt. Dodgen Pot Sike	0.14-0.30	3.0 - 6.8	0	0
Moss Burn	0.02-0.15	0.8 - 1.6	0	0
Tees				
Tees Bridge	-	-	0.18-0.25	0.48-0.87
Tees above Weel	0 -0.2	0.01- 0.37	0.33-0.96	0.41-0.74
Tees below Cauldron Snout	0.02-0.06	1.1 - 3.7	0.06-0.62	0.38-1.38
Middleton-in-Teesdale (km 19)	0.07	-	-	-
Egglestone Abbey (km 25.5)	0.03	-	-	-

Table 6.7 Estimates of fresh weight production (g m^{-2} year^{-1}).   O, species absent; -, not estimated; *, based on tray samples.

	production (g m^{-2} yr^{-1})	
	brown trout	bullhead
Tributaries		
Trout Beck	2.2	-
Weelhead Sike	8.6	0
Dubby Sike	6.0	0
Mattergill Sike	5.9	0
Gt. Dodgen Pot Sike	3.5	0
Moss Burn	2.2	0
Cow Green tributaries	4.4	0
Tees		
Tees Bridge	-	*7.4
Tees in reservoir basin (Cow Green)	0.7	1.1
Tees below Cauldron Snout (c̲ km 0.6)	-	1.1

seen by comparing results summarized in Table 6.8 with the values of
0.095 and 0.047, respectively, found for chalk streams in southern England by Mann (1971).

Table 6.9 gives details of the age of fish going to sea. About 95%
of Tees salmon were found to go to sea at age 2 or 2+ by Pentelow *et al.*
(1933a) and this percentage is similar to those for the Tyne, Tweed,
Forth, Wye and Findhorn, but different from those for the Spey and Don.
Size is an important factor governing the time of seaward migration of
both salmon and sea trout. Thus those salmon which went to sea at age
2 were the larger specimens of their year class.

Table 6.8 Annual survival rate (fraction of fish present at time t
which survive to time t + 1 years)

site(s)	annual survival rate (but confounded by movements)	
	brown trout	bullhead
Trout Beck	0.52	–
Tees Bridge	–	0.30
Tees & tributaries at Cow Green	0.57	–
Cow Green tributaries	0.40	–
Tees at Cow Green	0.75	0.33
Tees below Cauldron Snout (c km 0.6)	0.49	0.52

Table 6.9 Percentages of salmon and sea trout smolts going to sea
at each age based on combined data for 1930 and 1931 (Pentelow *et
al.*, 1933a)

age (years) at migration	salmon	sea trout
2	56.6	44.6
2+	38.4	29.6
3	5.0	24.5
3+	–	1.3

6.84   THE TEES FISHERY AND POLLUTION

Coarse fishing occurs chiefly in the lower stretches of the river,
where some stocking is done. The brown trout, which occurs throughout
the river, is however the main quarry of anglers. Stocking with hat-
chery trout is a procedure of long-standing in the Tees. 64750 trout
were distributed in the Tees between 1904 and 1915 and further stocking

took place in seven different years up to the time of the survey by
Butcher *et al*. Hancock (1977) referred to the stocking of 10420 trout
in the Tees below Middleton-in-Teesdale in 1974, 8460 in 1975 and 5900
in 1976. It is instructive to examine this in more detail. If the
trout population density between Middleton-in-Teesdale and Hurworth is
taken as 0.03 fish m^{-2} (probably a low estimate), then there are about
65000 trout in this stretch and Hancock's data suggest that 30 - 40%
of these are of takeable size. At the time of stocking, the stock fish
of takeable size are equivalent to 3.7% of the indigenous fish of similar size. The remaining stock fish are undersized at the time of
stocking and, given growth and survival rates similar to those of the
indigenous fish, will reach takeable size two years after stocking and
be equivalent to some 10 - 30% of the number of indigenous fish reaching takeable size at the same time. This suggests that stocking may
have an appreciable impact on the Tees fishery, though this rests on
the assumption that survival and growth of stocked fish are similar to
that of indigenous fish.

During the late 19th and 20th centuries the Tees had a substantial
fishery for migratory Salmonidae. The Skerne apparently also had at
one time a reputation as an important salmon river, but lost this at an
early stage of the Industrial Revolution. The Tees fishery consisted
of two elements, a rod and line fishery within the river and a net
fishery within the estuary, the latter being the more important. There
has been a drastic decline of both salmon and sea trout in the main
river within the past century (Table 6.10), with a major change taking
place shortly after the First World War. Butcher *et al*. state that the
first official recognition that there was a serious pollution problem
in the Tees was in 1920 when the Tees Fishery Board reported substantial pollution mortality of smolts in the estuary. If 1920-24 is
accepted as the start of serious pollution problems in the Tees, then
the virtual destruction of this fishery took only a decade.

Table 6.10 Summary of Tees rod catches of salmon and sea trout
between 1886 and 1978, based on reports by Northumbrian Water Authority and its predecessors.

years	mean catch (no. of fish yr^{-1}) (± 95% confidence limits)		% sea trout
	salmon	sea trout	
1886–1912	502.0 ± 11.07	28.5 ± 30.3	5.4
1914–1930	144.5 ± 57.3	15.6 ± 9.54	9.8
1931–1939	0.7 ± 0.94	0.8 ± 0.92	53.8
1951–1978	0	1.4 ± 1.79	100.0

Recently there have been signs that rehabilitation of the fishery is possible, with an adult sea trout found in a small tributary of the Lune in 1976 and a salmon near Barnard Castle in 1980. By 1983 it was clear that salmon were spawning successfully within the Tees system. The most likely explanation for this change is the depressed state of the economy of Teesside during the late 1970s and early 1980s, which led to a reduced pollutant load in the estuary.

In the autumn of 1983 there was a severe pollution of the Tees by "flux oil". The pollutant entered the river near High Force and there was a substantial fish kill. The casualties included several hundred adult salmon and sea trout. This gives some indication of the recovery of stocks up to that time. The effects of this pollution upon the recovery will not be clear for some years.

6.85   EFFECTS OF REGULATION

An account of the effects of the Cow Green regulation scheme upon the populations of fish within 0.6 km of the dam was given by Crisp *et al*. (1983) (Table 6.11), of regulation upon stomach contents by Crisp *et al*. (1978) and of the effects of changed temperature regime upon the growth and metabolism of brown trout (Crisp, 1977). The main changes observed among trout were an increase in mean population density (by c 40%) and a small, but statistically significant, increase in growth rate. There was a threefold increase in the mean population density and biomass of the bullhead and this was accompanied by changes in reproductive performance. This improvement in conditions for trout and bullhead may reflect an improved food supply. However, as growth of individual fish is limited by temperature rather than food supply, this improvement has been exploited through an increase in population density, rather than by improved growth of individuals.

6.9   CONCLUDING COMMENTS

In spite of the many changes to the Tees brought about by man's activities, most of the river upstream of the estuary is in good condition. The changes which have taken place can be grouped broadly as either chemical or hydrological.

The main sites for pollution in the Tees system are the R. Skerne and the estuary. The Skerne probably underwent a progressive deterioration during the 19th century and was at its worst in the middle of this century. The predicted cessation of a coke-works effluent should lead to a marked improvement in the condition of this tributary, making it suitable for an increased range of organisms, including a number of fish species. Provided this river is subject to careful management, it could develop into an important amenity.

Table 6.11 Simplified summary of changes in the population characteristics of brown trout and bullhead in the Tees below Cauldron
Snout as a result of regulation of the Tees. Values observed before
regulation are followed by values observed after regulation. (Derived from Crisp *et al.*, 1983)
-, inadequate data for analysis; +, no significant change;
    *, P <0.05; **, P <0.01.

	brown trout	bullhead
population density (no. 100 m^{-2})	3.5 to 4.9*	9.9 to 33.4*
biomass (g m^{-2})	-	0.5 to 1.4**
fecundity (eggs per female	+	148 to 116 (for 7 cm female)*
fecundity (eggs m^{-2} yr^{-1})	+	2.3 to 8.4
growth rate	small increase**	+ (small increase in age group 0 only)
age at first sexual maturity	-	50% to 84% (% mature in age group II)
instantaneous rate of "loss"	+	0.7 to 1.3*
production (g m^{-2} yr^{-1})	0.3 to 0.4	0.5 to 3.6

In contrast to the Skerne, pollution in the estuary became a serious
problem only in this century, with a sharp drop in the condition of the
river in the 1920s continuing until the 1970s. The recent passage of
migratory salmon and trout appears to be due in part to improvements
in the quality of industrial effluents, but probably even more to an
overall reduction in pollution resulting from economic decline in the
region. It is to be hoped that any future resurgence of industrial
activity is matched by still further improvements in effluent quality.

Superficially the hydrological changes in the Tees system appear
more profound than those due to pollution, since they are unlikely
ever to be reversed. Construction of the Cow Green reservoir destroyed
the Weel and has led to some changes in chemistry and in plant and
animal populations in the downstream river. However, with the exception of the stretch immediately below the reservoir, these changes
appear relatively slight. The fact that the dam acts as an upstream
barrier to migration is of minor importance, because High Force, the
waterfall 10 km downstream of the dam, has always acted as another
such barrier. The Tyne-Tees water transfer scheme, whose ecological
significance at one time caused considerable apprehension, also seems

likely to have little impact on the Tees, both because of the infrequency with which the system is used and because the chemistry and ecology of the donor river are quite similar. Overall, therefore, the hydrological management of the river has probably had much less influence on its ecology than a superficial glance at the map would suggest. In view of this and the decreasing influence of pollution, the Tees should certainly be considered among those British rivers where nature conservation is an important criterion in making future decisions on management.

Acknowledgements

Many people have supplied data or helpful conversation. These include: P.D. Armitage, P.A. Carling, A. Clarke and E.M. Ottaway, from the Freshwater Biological Association; M.D. Newson from the Institute of Hydrology; N.T.H. Holmes from the Nature Conservancy Council; J.A. Brady, J. Cave, A.S. Champion, J.W. Hargreaves, P. Johnson, W.O. Ord, J.R. Pomfret and W.J. Walker, from the Northumbrian Water Authority.

REFERENCES

Armitage P.D. & Capper M.H. (1976) The numbers, biomass and transport downstream of microcrustaceans and *Hydra* from the Cow Green reservoir (Upper Teesdale). *Freshwat. Biol.* 6, 425-32.

Armitage P.D., MacHale A.M. & Crisp D.T. (1974) A survey of invertebrates in the Cow Green basin (Upper Teesdale) before inundation. *Freshwat. Biol.* 4, 369-98.

Armitage P.D., MacHale A.M. & Crisp D.T. (1975) A survey of the invertebrates of four streams in the Moor House National Nature Reserve in Northern England. *Freshwat. Biol.* 5, 479-95.

Armitage P.D. (1976) A quantitative study of the invertebrate fauna of the River Tees below Cow Green reservoir. *Freshwat. Biol.* 6, 229-40.

Armitage P.D. (1977a) Invertebrate drift in the regulated River Tees, and an unregulated tributary Maize Beck, below Cow Green dam. *Freshwat. Biol.* 7, 167-83.

Armitage P.D. (1977b) Development of the macroinvertebrate fauna of Cow Green Reservoir (Upper Teesdale) in the first five years of its existence. *Freshwat. Biol.* 7, 441-54.

Armitage P.D. (1978a) The impact of Cow Green reservoir on invertebrate populations in the River Tees. *Freshwat. Biol. Ass. U.K., Annual Reports* 46, 47-56.

Armitage P.D. (1978b) Downstream changes in the composition, numbers and biomass of bottom fauna in the Tees below Cow Green reservoir and in an unregulated tributary Maize Beck, in the first five years after impoundment. *Hydrobiologia* 58, 145-56.

Birkett N.L. (1976) Chironomidae (Diptera) trapped in a Pennine stream, including two species new to Britain. *Ent. Gaz.* 27, 161-70.

Brooker M.P. & Hemsworth R.J. (1978) The effect of the release of an artificial discharge of water on invertebrate drift in the R. Wye, Wales. *Hydrobiologia* 59, 155-63.

Brown J.M. & Whitehead H. (1938) The Trichoptera or caddisflies of Yorkshire. *Naturalist, Hull,* 260-65.

Brown V.M., Cragg J.B. & Crisp D.T. (1964) The Plecoptera of the Moor House National Nature Reserve, Westmorland. *Trans. Soc. Br. Ent.* 16, 123-34.

Brown V.M., Shurben D.G. & Shaw D. (1970) Studies on water quality and the absence of fish from some polluted English rivers. *Water Research* 4, 363-82.

Burston U.T. & Coats D.J. (1975) Water resources in Northumbria with particular reference to the Kielder Water Scheme. *J. Instn Wat. Engrs* 29, 226-51.

Bussell R.B. (1979) Changes of river regime resulting from regulation which may affect ecology : a preliminary approach to the problem. *Central Water Planning Unit, U.K. 1979.* pp. 48.

Butcher R.W., Longwell J. & Pentelow F.T.K. (1937) Survey of the River Tees. Part III. The non-tidal reaches - chemical and biological. *Wat. Pollut. Res. Techn. Pap.* 6, 1-189 (HMSO London).

Butcher R.W. (1933) Studies on the ecology of rivers 1. On the distribution of macrophytic vegetation in the rivers of Britain. *J. Ecol.* 21, 58-91.

Carling P.A. & Reader N.A. (1982) Structure, composition and bulk properties of upland stream gravels. *Earth Surface Processes and Landforms* 7, 349-65.

Carling P.A. (1983) Particulate dynamics, dissolved and total load, in two small basins, northern Pennines, U.K. *Hydrological Sciences - Journal* 28, 355-75.

Carter J.R. (1972) The diatoms of Slapestone Sike, Upper Teesdale. *Vasculum* 57, 35-41.

Clapham A.R. (Ed.) (1978) *Upper Teesdale. The Area and its Natural History.* 238 pp. Collins, London.

Conway V.M. & Millar A. (1960) The hydrology of some small peat-covered catchments in the northern Pennines. *J. Inst. Wat. Engrs* 14, 415-24.

Coulson J.C. (1959) Observations on the Tipuliidae (Diptera) of the Moor House National Nature Reserve, Westmorland. *Trans. R. Ent. Soc. Lond.* 3, 157-74.

Crisp D.T. & Gledhill T. (1970) A quantitative description of the recovery of the bottom fauna in a muddy reach of a mill stream in Southern England after draining and dredging. *Arch. Hydrobiol.* 67, 502-41.

Crisp D.T. & Howson G. (1982) Effect of air temperature upon mean water temperature in streams in the north Pennines and English Lake District. *Freshwat. Biol.* 12, 359-67.

Crisp D.T., Mann R.H.K. & Cubby P.R. (1983) Effects of regulation of the River Tees upon fish populations below Cow Green reservoir. *J. appl. Ecol.* 20, 371-86.

Crisp D.T., Mann R.H.K. & McCormack J.C. (1974) The populations of fish
    at Cow Green, upper Teesdale before impoundment. *J. appl. Ecol.* 11,
    969-96.

Crisp D.T., Mann R.H.K. & McCormack J.E. (1975) The populations of fish
    in the River Tees system on the Moor House Nature Reserve, Westmor-
    land. *J. Fish. Biol.* 7, 573,93.

Crisp D.T., Mann R.H.K. & McCormack Jean C. (1978) The effects of im-
    poundment and regulation upon the stomach contents of fish at Cow
    Green, Upper Teesdale. *J. Fish. Biol.* 12, 287-301.

Crisp D.T. & Nelson J.M. (1965) The Ephemeroptera of the Moor House
    National Nature Reserve, Westmorland. *Trans. Soc. Brit. Ent.* 16,
    181-87.

Crisp D.T., Rawes M. & Welch D. (1964) A Pennine peat slide. *Geog. J.*
    130, 519-24.

Crisp D.T. & Robson S. (1979) Some effects of discharge upon the trans-
    port of animals and peat in a north Pennine headstream. *J. appl.
    Ecol.* 16, 721-36.

Crisp D.T. & Robson S. (1982) Analysis of fishery records from Cow
    Green Reservoir, Upper Teesdale, 1971-80. *Fish Mangmt* 13, 65-78.

Crisp D.T. (1966) Input and output of minerals for an area of Pennine
    moorland : the importance of precipitation, drainage, peat erosion
    and animals. *J. appl. Ecol.* 3, 327-48.

Crisp D.T. (1977) Some physical and chemical effects of the Cow Green
    (Upper Teesdale) impoundment. *Freshwat. Biol.* 7, 109-20.

Davies L. & Smith C.D. (1958) The distribution and growth of *Prosi-
    mulium* larvae (Diptera: Simuliidae) in hill streams in northern
    England. *J. Anim. Ecol.* 271, 335-48.

Dunham K.C. (1948) *Geology of the Northern Pennine Orefield. Vol. 1.*
    282 pp. H.M.S.O., London.

Edwards R.W., Densen H.W. & Russell P.A. (1979) An assessment of the
    importance of temperature as a factor controlling the growth rate of
    brown trout in streams. *J. Anim. Ecol.* 48, 501-7.

Elliott J.M. (1975a) The growth of brown trout, *Salmo trutta* L., fed
    on maximum rations. *J. Anim. Ecol.* 44, 805-21.

Elliott J.M. (1975b) The growth rate of brown trout (*Salmo trutta* L.)
    fed on reduced rations. *J. Anim. Ecol.* 44, 823-42.

Elliott J.M. (1975c) Weight of food and time required to satiate brown
    trout (*Salmo trutta*). *Freshwat. Biol.* 5, 51-64.

Gorham E. (1956) On the chemical composition of some waters from the
    Moor House National Nature Reserve. *J. Ecol.* 44, 375-82.

Hancock R.S. (1977) The impact of water transfers and associated regu-
    lation reservoirs on the fish populations of the Tyne, Tees and
    Swale. *Proc. Eighth British Coarse Fish Conference*, 137-58.

Holmes N.T.H., Lloyd E.J.H., Potts M. & Whitton B.A. (1972) Plants of
    the River Tyne and future water transfer scheme. *Vasculum* 57,
    56-78.

Holmes N.T.H. & Whitton B.A. (1977) The macrophytic vegetation of the
    River Tees in 1975 : observed and predicted changes. *Freshwat.
    Biol.* 7, 43-60.

Holmes N.T.H. & Whitton B.A. (1981) Phytoplankton of four rivers, the
    Tyne, Wear, Tees and Swale. *Hydrobiologia* 80, 111-27.

Holmes N.T.H. & Whitton B.A. (1981) Phytobenthos of the River Tees
    and its tributaries. *Freshwat. Biol.* 11, 139-63.

Holmes N.T.H. (1976) The occurrence and ecology of *Grimmia agassizii*
    (Sull. & Lesq.) Jaeg. in Teesdale. *J. Bryol.* 9, 275-78.

Horne J.E.M. (1977) *A Bibliography of Rivers Tyne, Wear, Tees and
    Swale*. 54 pp. Occas. Publs. No. 3, Freshwater Biological Assoc,
    U.K.

Hughes M.K. & Whitton B.A. (1972) Algae of Slapestone Sike, Upper
    Teesdale. *Vasculum* 57, 30-35.

Institute of Hydrology (1978) *Low Flow Study*. *Rep. No. 1*. 61 pp.
    Institute of Hydrology (U.K.), Wallingford.

Jackson W.H. (1930) Pre-pollution days in the Tees. *Salmon Trout
    Mag.* 58, 87-89.

Jansson B.-O. (1982) Myrat, svampar och statistik. Bogs, sponges and
    statistics. *Vatten* 38, 50-58. (In Swedish, English summary)

Lavis M.E. & Smith K. (1972) Reservoir storage and the thermal regime
    of rivers, with special reference to the River Lune, Yorkshire.
    Sci. total Environ. 1, 81-90.

Livingstone D. & Whitton B.A. (1984a) Water chemistry and phosphatase
    activity of the blue-green alga *Rivularia* in Upper Teesdale streams.
    J. Ecol. (in press).

Livingstone D. & Whitton B.A. (1984b) Diel variations in nitrogen and
    carbon dioxide fixation by the blue-green alga *Rivularia* in an up-
    land stream. *Phycologia* (in press).

Longstaffe W.H.D. (1954) *The History and Antiquities of the Parish of
    Darlington, in the Bishoprick*. 374 pp. + appendices. Proprietors
    of Darlington & Stockton Times, Darlington.

Macan T.T. (1957) The Ephemeroptera of a stoney stream. *J. Anim.
    Ecol.* 26, 317-42.

Macan T.T. (1979) Key to British freshwater Crustacea - Malacostraca.
    Sci. Publs Freshwat. Biol. Ass. U.K. 32, 1-72.

Maitland P.S. (1977) *A Coded Checklist of Animals occurring in Fresh
    Water in the British Isles*. 76 pp. Inst. of Terrestrial Ecology,
    Edinburgh.

McLachlan A.J., Pearce L.J. & Smith J.A. (1979) Feeding interactions
    and cycling of peat in a bog lake. *J. Anim. Ecol.* 48, 851-61.

Mann R.H.K. (1971) The populations, growth and production of fish in
    four small streams in southern England. *J. Anim. Ecol.* 40,
    155-90.

National Water Council (1981) *The 1980 Survey and the Future Outlook*.
    National Water Council, 1 Queen Anne's Gate, London SW1H 9BT.
    (Name now changed to Water Authorities Association).

Nelson J.M. (1971) The invertebrates of an area of Pennine moorland
    within the Moor House National Nature Reserve in northern England.
    Trans. Soc. Br. Ent. 19, 173-235.

Newson M.D. (1981) Mountain streams. In: Lewin J. (Ed.) *British
    Rivers*. Allen & Unwin.

Northumbrian River Authority (1967-1972) *Ann. Reps. for Years Ending
    31st March 1966-31st March 1971*. Northumbrian River Authority,
    Newcastle-on-Tyne.

Northumbrian River Authority (1973) *Report on Survey of Water Re-
    sources.* 191 pp. Northumbrian River Authority, Newcastle-on-Tyne.
Northumbrian Water Authority (1975-   ) *Annual Water Reports 1974/75-*
        Northumbrian Water Authority, Gosforth. (Published
    annually; analytical data not included in this series of reports.)
Northumbrian Water Authority (1976) *Report on Experimental Releases
    from Cow Green Reservoir on the R. Tees (23rd-28th June 1976).*
    Directorate of Planning and Scientific Services, Northumbrian Water
    Authority, Gosforth.
Pentelow F.T.K., Southgate B.A. & Bassindale R. (1933a) I. The rela-
    tion between the size, age and time of migration of salmon and sea
    trout smolts in the River Tees. *Fishery Invest., London.* Ser. 1,
    3, No. 4, 1-10.
Pentelow F.T.K., Southgate B.A. & Bassindale R. (1933b) II. The pro-
    portion of the sexes and the food of smolts of salmon (*Salmo salar*
    L.) and sea trout (*Salmo trutta* L.) in the Tees Estuary. *Fishery
    Investigations, London.* Ser. 1, 3, No. 4, 11-14.
Pentelow F.T.K. (1932) The food of brown trout (*Salmo trutta* L.).
    J. Anim. Ecol. 1, 101-7.
Phillipson J. (1957) The effect of current speed on the distribution
    of the larvae of the black-flies: *Simulium variegatum* (Mg.) and
    S. monticola Fried. (Diptera). *Bull. Ent. Res.* 48, 811-19.
Proctor H.G. (1971) Aquatic macrophytes in the Wheel of Tees.
    Vasculum 56, 59-66.
Proctor H.G. (1978) Changes in the large aquatic flora of the Cow
    Green basin. In: Clapham A.R. (Ed.) *Upper Teesdale. The Area and
    and its Natural History.* 238 pp., pp. 191-94. Collins, London.
Shehata F.H.A. & Whitton B.A. (1981) Field and laboratory studies on
    blue-green algae from aquatic sites with high levels of zinc.
    Verh. int. Verein. theor. angew. Limnol. 21, 1466-71.
Smith K. (1968) Some thermal characteristics of two rivers in the
    Pennine area of northern England. *J. Hydrol.* 6, 405-16.
Smith K. (1969) The baseflow contribution to runoff in two upland
    catchments. *Water & Water Engineering,* Jan. 1969, 18-20.
Smith K. (1970) Weather and climate. In: Dewdney J.C. (Ed.) *Durham
    County and City with Teesside.* 522 pp., pp. 58-74. British
    Association, Durham.
Southgate B.A., Pentelow F.T.K. & Bassindale R. (1932) An investi-
    gation into the causes of death of salmon and sea trout smolts in
    the estuary of the River Tees. *Biochem. J.* 26, 273-84.
Walentowicz A.T. & McLachlan A.J. (1980) Chironomids and Particles :
    a field experiment with peat in an upland stream. In: Murray D.A.
    (Ed.) *Chironomidae Ecology, Systematics, Cytology and Physiology.*
    Pergamon.
Wear & Tees River Board (1951-1965) *Annual Reports for Year Ending
    31st March 1951-31st March 1965.* Wear and Tees River Board,
    Darlington.
Whitehead H. (1929) Plecoptera (stone-flies) of Yorkshire. *Natura-
    list, Hull* 875, 403-7.
Whitton B.A. & Dalpra M. (1968) Floristic changes in the River Tees.
    Hydrobiologia 32, 545-50.

Wotton R.S. (1976) The distribution of blackfly larvae (Diptera: Simuliidae) in Upper Teesdale, Northern England. *Hydrobiologia* 51, 259-63.

Wotton R.S. (1977) Sampling moorland stream blackfly larvae (Diptera: Simuliidae). *Arch. Hydrobiol.* 79, 404-12.

Wotton R.S. (1978) Life-histories and production of blackflies (Diptera: Simuliidae) in moorland streams in Upper Teesdale, Northern England. *Arch. Hydrobiol.* 83, 232-50.

7 · The Winterbourne Stream

A. D. BERRIE & J. F. WRIGHT

7.1   INTRODUCTION

The Winterbourne Stream is the only tributary of the River Lambourn, a chalk stream which drains the Berkshire downs in southern England. With a total length of 8 km, of which only the lower 2 km have permanent flow, it might be regarded as an unusual choice for detailed investigation. The reason for the attention now being given to a number of similar streams in England will become apparent later, but first it is important to describe the features which characterize all winterbourne streams.

Chalk outcrops over much of southern and eastern England, and its porous structure gives rise to streams with a high groundwater component to their discharge. The headwaters and tributaries of these chalk streams are of an intermittent nature, their flow largely being determined by the level of water in underground storage. In winter, when the aquifer is replenished, water is discharged into the stream through springs and the stream bed itself. The water-table drops as the aquifer is depleted through the summer due to lower rainfall and consequently the upper limit of the running water moves down the valley. The stream thus has an intermittent section where flow is seasonal and a perennial section downstream, where flow is permanent. This intermittent section is commonly known as a winterbourne, i.e. a winter stream, and this name is given to the stream which forms the subject of this chapter, despite the fact that it is perennial in its lower reaches.

Our understanding of the structure and functioning of chalk streams has increased considerably in the last 20 years (Westlake *et al.*, 1972; Le Cren & Lowe-McConnell, 1980), but until recently winterbournes received very little attention. In an attempt to rectify this, Casey & Newton (1972) and Casey & Ladle (1976) published accounts of the water chemistry, discharge regime, macrophytes and invertebrate communities at a series of sites along the South Winterbourne stream, a tributary of the River Frome in Dorset. During this survey, Ladle & Bass (1975) discovered a blackfly, *Metacnephia amphora*, new to science. They also examined a small chalk stream in Dorset, the Waterston stream (Ladle & Bass, 1981), which dried up for a period of four months during a minor drought in 1973.

Perhaps the greatest stimulus to the investigation of winterbournes resulted from the apparent need to develop groundwater pumping schemes to meet the predicted increase in consumption of water (Water Resources Board, 1973). One of the largest demands was for London and a scheme involving the Lambourn and Winterbourne valleys was planned and commissioned to provide an additional supply in dry periods (Brettell, 1971). The water is pumped from boreholes located some distance from the rivers and piped to outfalls in the perennial sections from where it flows downstream to be abstracted again from the R. Thames near London. This avoids the need for expensive pipelines and provides additional water in the rivers at times when the flows would otherwise be low. Suitable facilities already exist for treating the river water

before passing it to the supply system. During the development of the
scheme a team from the University of Reading carried out ecological
studies on the R. Lambourn and the Winterbourne Stream to obtain base-
line information prior to any ecological effects of the scheme. The
research programme for the Winterbourne Stream involved an extensive
survey of the macrophytes and macroinvertebrates in 1972. This was
followed by more limited observations in 1976/77 when extreme drought
in summer 1976 led to operation of the groundwater scheme between
August and November. High rainfall during the autumn and winter of
1976 and the return of water to many of the study sites on the Winterbourne Stream provided the opportunity to make some observations on
recolonization by the macroinvertebrate fauna in the aftermath of the
drought. Observations on the fish were made from 1971 to 1976.

Ecological studies have also been undertaken in connection with at
least four other groundwater schemes in the chalk regions of southern
and eastern England. These involve pumping groundwater into the Candover
Brook in Hampshire (Southern Water Authority, 1979) and into the Rivers
Thet, Rhee and Little Ouse in eastern England (Klee, 1974; 1976);
lining the bed of the Gussage stream in Dorset and altering its flow
regime (Furse, 1977a & b); and pumping groundwater direct to supply
from the upper catchment of the R. Wylye in Wiltshire with effects on
flow in the headwaters and in tributaries (Avon & Dorset River Authority, 1973).

In reviewing the conservation value of sites in Great Britain,
Ratcliffe (1977) highlighted the special character of chalk streams and
listed the R. Lambourn as a good example of an unspoilt chalk stream.
Winterbournes appear to support some scarce species, they may provide
nursery areas for the trout populations of chalk streams and they are
under threat from a variety of sources. These are some of the practical reasons why winterbournes have received attention in the past few
years. They also have much to tell us about the response of organisms
to drought conditions, methods and rates of recolonization and the
importance of physical factors in determining the composition of
biological communities. Whilst these topics have received considerable
emphasis in relation to standing waters (Wiggin et al., 1980) only now
is attention being focused on temporary streams (Williams & Hynes,
1976; 1977).

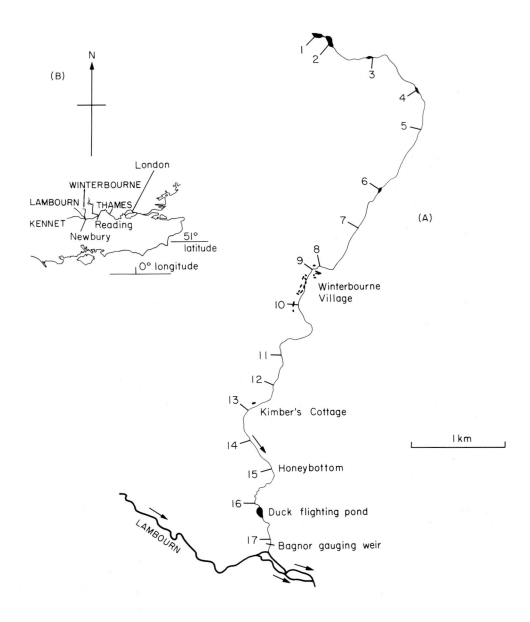

Fig. 7.1 A) Winterbourne Stream, showing the 17 macroinvertebrate
sampling sites and additional locations mentioned in the text.
B) Location of the Winterbourne Stream in S-E. England.

7.2   GENERAL DATA

DIMENSIONS
Area   49 km^2
Length   8 km (total length, including inter-
                                    mittent section)
Highest point in catchment 230 m
Drop in height   32 m (from 113 m to 81 m)

LOCATION
51.5° N   1.5° W

CLIMATE
Precipitation 705 mm yr^{-1} for 1941-1970
    driest month = April (44 mm)
    wettest month = November (73 mm)
    120-140 wet days per year (>1 mm rain)

Temperature
    mean January air   3.5-4 $^\circ$C
    mean July air   16.5-17 $^\circ$C

GEOLOGY
Upper chalk is exposed over much of the catchment, but along the valley
floor there are river and valley gravels with some alluvium. On higher
ground in the northern part of the catchment the upper chalk is covered
by clay-with-flints, whilst further south the higher ground is capped
by Eocene sands and clays overlain by Plateau gravel.

LAND USE
Arable farming and pasture land in chalk upland areas; trees are restricted to occasional copses. Woods and heathland occur on the higher
ground in the southern part of the catchment, where the chalk is overlain by sands, clays and finally gravelly drift.

POPULATION
Approximately 500 people in small villages and isolated farms.

PHYSICAL FEATURES OF RIVER

On-stream reservoirs and major weirs
No reservoirs or major weirs. A gauging weir and a duck-flighting pond
(Fig. 7.1) may influence the movement of some fish and invertebrates.

Flow at selected points
See Fig. 7.2 for mean monthly discharge (1972-1977) at gauging weir.

Seasonal changes in flow
In a typical year peak discharge occurs in March (~0.25 m^3 s^{-1}) followed
by a steadily decreasing hydrograph to early winter (<0.1 m^3 s^{-1}) when
discharge starts to rise again (see Fig. 7.2 and text for further details)

Influence of regulation
Between August and November 1976, water from boreholes high in the
catchment was used to augment the Winterbourne stream above the gauging
weir (Fig. 7.1). Some water was also added at the head of the perennial
section to maintain the amenity value of the stream. Pumping delayed

the return of water to the intermittent section of the stream through a lowering of the water table.

Important abstraction points

None on the Winterbourne itself, but Chieveley pumping station (SU 476751) takes on average 148.5 m^3 day^{-1} from a borehole located near the edge of the catchment.

Substratum

Terrestrial vegetation or emergent macrophytes normally dominate the intermittent section of the stream, but submerged and emergent macrophytes, gravel and silt are all important components of the substratum in the perennial section (see Table 7.2).

Temperature

A thermograph located upstream of the gauging weir indicates a very regular water temperature regime under normal meteorological conditions in which the winter minimum is c 4 °C and the summer maximum is c 17.5 °C. (See Fig. 7.2 and text.)

7.3 PHYSICO-CHEMICAL BACKGROUND

The course of the stream is shown in Fig. 7.1 together with 17 sites
chosen for a survey in 1972. The perennial head is normally taken to
be between sites 12 and 13 just above Kimber's Cottage. However, the
Winterbourne Stream does not dry out to the perennial head every year
and in fact parts of the intermittent section remained wet in 1972.

Fig. 7.2 presents mean monthly discharge at the gauging weir on the
Winterbourne Stream before its confluence with the R. Lambourn. Since
the major outfall of the pumping scheme is just upstream of the weir,
it includes both natural and pumped discharge for July 1976 (test pumping) and for August to November 1976 (operational pumping). In a
typical year, the peak discharge occurs in March followed by a steady
decline through the summer and autumn when at least part of the intermittent section of the stream becomes dry. Winter rains then lead to
an increase in the height of the water-table in the catchment, breaking
of the springs and inundation of the entire intermittent section. This
is reflected by the rising hydrograph at the weir. The very stable
flow regime observed at the weir is a consequence of the high proportion
of the discharge derived from groundwater and the low proportion of
surface runoff. In 1973 (Fig. 7.2) a minor drought was experienced
whilst in 1975 a peak spring discharge in excess of 400 l s^{-1} occurred.
This was followed by a prolonged period of low rainfall which resulted
in a failure of the springs in 1976 and a major drought.

Fig. 7.2 also includes the monthly maximum and minimum stream
temperatures recorded at a site just upstream of the gauging weir and
groundwater outfall. The natural groundwater enters the stream through
the river bed and springs at a temperature of approximately 11 $^{\circ}$C
before being influenced by the seasonally varying air temperature.
Superimposed upon the very regular annual temperature regime is evidence of higher summer maxima in years of drought (1973, 1976) and
warmer winter minima when discharge is exceptionally high (1975).

The chemical composition of the water is similar to a number of
other chalk streams previously investigated in the region (Westlake *et
al.*, 1972). Monthly water samples (taken by the Thames Water Authority)
near the gauging weir show that many chemical determinands maintain
relatively stable values with time, despite seasonal changes in discharge. This feature has also been observed in a chalk spring (Casey,
1969) and in the South Winterbourne stream in Dorset (Casey & Newton,
1972; Casey & Ladle, 1976), but is less pronounced in larger chalk
streams and rivers (Casey, 1969; Casey & Newton, 1973).

Between 1972 and 1977, pH was recorded over the range 7.2 - 8.3
(n = 90) and levels of dissolved oxygen were invariably high (8.0 -
14.3 mg l^{-1}, n = 89) at the time of sampling. Suspended solids were
almost always low (range 0.5 - 21.7 mg l^{-1}, n = 75) resulting in very
clear waters for most of the year.

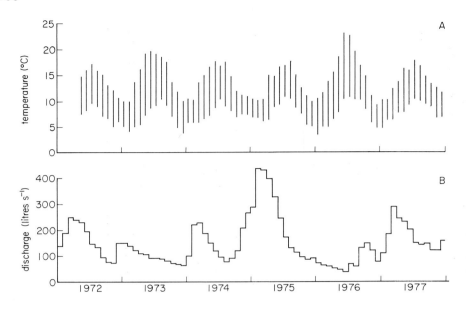

Fig. 7.2 Records of temperature and discharge. A. Monthly maximum and minimum temperatures from a thermograph just upstream of both site 17 and groundwater outfall. B. Mean monthly discharge at Bagnor gauging weir. (Discharge between July and November 1976 includes pumped discharge.)

Table 7.1 presents further water quality data in the form of annual means and ranges. Total oxidized nitrogen which is due almost entirely to nitrate-N, failed to demonstrate a significant positive correlation with discharge as observed for nitrate by Casey & Newton (1972) in the South Winterbourne stream; the drought years of 1973 and 1976 did, however, have the lowest yearly means. Particularly low values were noted during both test pumping (1.4 mg l^{-1}) and operational pumping (4.0 mg l^{-1}) of the groundwater scheme in 1976.

Orthophosphate never exceeded 100 µg l^{-1} P over the period, with the exception of the first four months of 1972 when values up to 261 µg l^{-1} P were recorded. Alkalinity normally maintained relatively stable values and there was no inverse relationship between discharge and alkalinity, as observed by Casey (1969) in the R. Frome. Results for sodium, potassium, magnesium and calcium also showed remarkable stability with time and Casey (1969) and Casey & Newton (1972) present similar findings for a chalk spring and the South Winterbourne stream respectively. A high value for sodium of 23.4 mg l^{-1} in the Winterbourne Stream in November 1975 may have been a result of road salting.

Table 7.1 Water quality data for the Winterbourne Stream at the Bagnor gauging weir. Yearly mean values and ranges expressed as mg l^{-1} and based on 10 or more analyses. (Data provided by Thames Water Authority)

determinand	Year					
	1972	1973	1974	1975	1976	1977
alkalinity as CaCO$_3$	252 (230–260)	252 (244–264)	251 (229–257)	241 (135–259)	248 (238–257)	243 (194–254)
sodium	7.6 (6.3–8.6)	8.1 (6.1–10.8)	9.6 (7.1–11.8)	12.5 (9.2–23.4)	8.6 (6.5–12.3)	9.5 (8.6–11.9)
potassium	1.5 (1.2–1.8)	1.3 (1.0–2.8)	1.6 (1.1–2.4)	1.5 (0.9–1.9)	1.4 (1.2–1.7)	1.5 (0.9–2.2)
magnesium	1.7 (1.4–2.0)	1.7 (1.5–1.8)	1.7 (1.4–2.4)	1.7 (1.6–1.8)	1.8 (1.5–2.1)	1.7 (1.3–2.0)
calcium	108 (95–116)	107 (87–115)	113 (104–120)	111 (108–114)	110 (98–121)	118 (109–142)
total oxidized nitrogen as N	6.7 (5.5–7.8)	5.5 (4.9–6.3)	6.0 (5.0–8.9)	6.5 (5.5–8.5)	5.3 (1.4–6.1)	6.3 (5.5–7.3)
orthophosphate as P	0.08 (0.01–0.26)	0.02 (0.01–0.06)	0.05 (0.03–0.10)	0.06 (<0.03–0.10)	0.02 (0–0.06)	0.03 (<0.03–0.10)

7.4   STUDY SITES

In 1972 a sampling programme was established to investigate the macro-phyte and invertebrate communities of the stream. Sampling sites for the invertebrate programme were chosen along the entire length of the stream (Fig. 7.1) to give comprehensive coverage of both the intermittent (sites 1-12) and permanent stream (sites 13-17). Sites 1-4, 6 and 8 were ponds in the stream channel but all other sites were 25 m sections of stream.

In February 1972 the invertebrate survey included each of the 17 sites, after which attention was focused on sites 8-17 in a bimonthly programme which continued to the end of the year. Several factors led to the decision to exclude the upper sites from further detailed study. Between sites 1 and 7 the stream has very little amenity value and for much of its length was channelled in a ditch. In contrast, downstream of the pond at site 8, the Winterbourne Stream had considerable amenity value as it flowed first through Winterbourne village and then continued alongside a minor road towards Bagnor. Its lower reaches also supported a trout population. Finally, since sites 8-17 encompassed both intermittent and perennial sections of the stream, the scientific interest of comparing the fauna of the contrasted sections was still retained.

7.5   MACROPHYTES

Each of the 25 m long sites from 9-17 was mapped bimonthly to record the occurrence of major macrophytes and substrata. Information was also collated on mean widths and shading (Table 7.2). Note in particular the increase in mean width downstream and the impact of total and partial shading by trees and bushes on the mean cover of macrophytes during the year.

Rorippa nasturtium-aquaticum (= *Nasturtium officinale*) occurred at all sites except the heavily shaded site 13, but its pattern of growth differed between intermittent and perennial sites. At intermittent sites 9-11 it was unimportant between February and June but showed spectacular growth by August. In contrast, at sites 15-17 the area of *Rorippa* in February was not markedly different from August. Since *Rorippa* is killed by frost, the effect of cold may be greater on the shallow exposed reaches of sites 9-11, thereby reducing the area during winter. At sites 15-17 the high discharge from March to June (Fig. 7.2) probably removed much of the *Rorippa* present in February, causing a reduction in area before the onset of summer growth. The depth and current speed of the water at the downstream sites probably prevented *Rorippa* from extending far from the banks and thus attaining the high percentage cover recorded at sites 9-11.

Another species, *Apium nodiflorum*, was important only in the inter-mittent zone (sites 9-12) and, although it dominated a number of sites

Table 7.2 Features of sites 9–17 in 1972.   O = unshaded; I = partially shaded; II = total shading by trees.   A = *Apium nodiflorum*; C = *Callitriche* sp.; E = emergent vegetation including grasses; M = *Mentha aquatica*; N = *Rorippa nasturtium-aquaticum*; R = *Ranunculus* spp.; V = *Veronica beccabunga*

site number	9	10	11	12	13	14	15	16	17
shading	O	O	O	O	II	I	I	O	O
mean width (m)	1.0	2.6	3.3	3.5	4.5	4.2	4.3	4.3	6.4
mean area in 1972 as %									
macrophytes	61	71	72	66	2	40	28	47	43
silt	9	15	9	8	73	19	56	33	47
gravel	30	14	19	26	25	41	16	20	10
area of macrophytes each month as %, with dominant									
February	54-A	68-C	51-A	46-A	0	63-C	22-N	34-N	55-N
April	33-A	75-E	57-E	42-R	1-M	37-V	18-N	22-N	35-N
June	68-A	83-E	75-E	67-N	0	36-C	21-N	34-R	35-N
August	67-N	73-C/N	93-N	96-N	5-M	31-E	39-N	66-N	45-N
October	DRY	59-N	96-N	84-N	2-M	28-C	43-N	70-N	46-N
December	82-E	68-E	59-E	59-N	3-M	45-M	28-N	55-N	45-N

Fig. 7.3 (above)   The intermittent section (site 11).   In spring, emergent grasses and gravel predominate before discharge decreases and *Rorippa* increases in area during late summer.

Fig. 7.4 (below)   The perennial section (site 15).   A substratum of of gravel and silt supports *Ranunculus* in mid-stream and *Rorippa* along the left-hand margin, whilst further silt has accumulated under trees on the right-hand bank.

in February, it tended to be overgrown by other plants, particularly *Rorippa* in summer. Casey & Ladle (1976) found *Apium* well represented near the source of the South Winterbourne stream and Furse (1977b) also noted that *Apium* dominated winterbourne sites on the Gussage.

   Ranunculus penicillatus var. *calcareus* and *R. peltatus* were both present in the Winterbourne Stream and the maximum area was recorded at perennial sites around August, as in the Lambourn at Bagnor (Ham *et al.*, 1981). *Callitriche* sp. occurred in small quantities in both inter-mittent and perennial sites but only *Mentha aquatica* appeared able to tolerate the low light intensities to be found at the heavily shaded site (13) near the perennial head.

7.6   INVERTEBRATES

7.61   FAUNA IN 1972

The primary aim of the 1972 survey was to obtain comprehensive species lists for study sites 8-17. Each sample was a 5-minute collection with a pond-net, obtained by both kicking and sweeping all available biotopes at the study site in proportion to their occurrence. At sites 1-7, limited sampling took place in February and June 1972 but sites 8-17 were sampled (4 samples per site) every second month from February to December except for site 9 which was dry in October. Hence, the data set for the major study sites (8-17) consists of 4 samples x 6 months = 24 qualitative samples for each site (20 only for site 9).

   Table 7.3 presents the number of taxa in each major invertebrate group for study sites 8-17 and also indicates the number of additional taxa confined to sites 1-7. 217 taxa were recorded at sites 8-17 in 1972 but only 16 additional taxa were found at sites 1-7. Of the 16, just four were regarded as typical of cold or intermittent waters, the remaining species being characteristic of ponds, ditches and small trickles (Wright *et al.*, 1984).

   Within the 10 major study sites (8-17) there was a richer fauna downstream from site 14 to 17 as the Winterbourne Stream approached its confluence with the Lambourn. Attempts to interpret this as being the result of greater habitat variability as the stream increased in size or through colonization from the Lambourn would be speculative. However, there were a number of clearly defined patterns of distribution between sites 8 and 17 for several of the major invertebrate groups. The Gastropoda, Ephemeroptera and Trichoptera all increased in species richness downstream. In contrast, site 8 was a pond situated in the course of the stream and this allowed the Hemiptera and Coleoptera to develop their most diverse fauna at this site. A number of community changes were associated with the perennial head between sites 12 and 13. The Hemiptera, Coleoptera and Chironomidae, which were well represented in the pond and intermittent section, decreased in species richness at the perennial head. But it could be argued that this was a response to

Table 7.3 Number of taxa recorded during 1972 survey confined to sites 1 - 7, and number at each of sites 8 - 17. Groups marked with an asterisk were not necessarily identified to species; in the Oligochaeta, the families Naididae and Enchytraeidae were not identified to species and in the Chironomidae, many specimens were taken to genus only.

	taxa confined to sites 1-7	8	9	10	11	12	13	14	15	16	17	total taxa at sites 8-17
Tricladida	1	2	2	2	–	–	2	3	3	2	4	4
Acanthocephala*	–	1	1	1	1	1	1	1	1	1	1	1
Nematoda*	–	1	1	1	1	1	1	1	1	1	1	1
Gastropoda	–	4	4	5	4	4	4	4	7	7	10	11
Bivalvia	–	–	–	2	3	4	4	4	3	3	2	5
Oligochaeta(*)	–	6	6	6	5	4	8	8	8	8	9	11
Hirudinea	–	–	1	–	2	5	5	5	6	5	5	6
Hydracarina*	–	1	1	1	1	1	1	1	1	1	1	1
Cladocera*	–	1	–	–	–	–	–	1	–	1	–	2
Ostracoda*	–	1	1	1	1	1	1	–	1	1	1	1
Copepoda*	–	1	1	–	–	–	–	1	1	–	1	1
Malacostraca	–	2	2	1	1	3	3	2	3	3	3	4
Ephemeroptera	1	2	3	3	3	3	3	5	6	5	8	10
Plecoptera	–	–	–	1	–	–	–	1	1	1	1	2
Odonata	–	1	1	–	–	–	–	–	–	–	–	1
Hemiptera	–	8	2	2	4	4	–	1	1	3	5	13
Coleoptera	6	28	12	15	13	12	7	12	12	17	18	46
Megaloptera	–	–	–	–	–	–	1	1	1	1	1	1
Trichoptera	2	5	8	10	13	14	15	18	19	22	24	31
Chironomidae*	5	18	16	18	19	18	11	15	23	25	35	48
other Diptera*	1	7	13	11	11	11	12	11	13	13	13	17
number of taxa	16	89	75	80	82	86	79	95	111	120	143	217

the heavy shading at site 13 which, lacking abundant macrophyte growths (Table 7.2), may not offer suitable living conditions. On the other hand, the Tricladida, Oligochaeta and Megaloptera increased in species richness between sites 12 and 13 and are known to be more characteristic of permanent water. The fact that the stream did not dry down to the perennial head in 1972 may be another factor which allowed a number of invertebrates, including *Gammarus pulex*, to colonize the intermittent section, thereby blurring the faunal distinction between the perennial and intermittent sites.

7.62 CLASSIFICATION OF SITES

Species lists have been compiled for the study sites (Wright *et al.*,

1984; Wright, 1984 and, by the use of cluster analysis, this information
has been used to produce a classification of the sites. Since 24
samples were taken at each site (except site 9), each species was given
a frequency of occurrence from 0 - 24.   (The frequencies for site 9
were standardised to the same range by multiplying by 24/20.) Two
dissimilarity coefficients were used on the site x species frequency
matrix followed by group average clustering (UPGMA of Sneath & Sokal,
1973) in each case (Fig. 7.5).

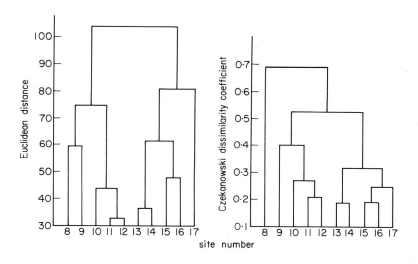

Fig. 7.5 Classification of sites 8-17 based on the invertebrate
communities using dissimilarity coefficients (Euclidean distance and
Czekanowski coefficient) followed by group average clustering.

The dendrogram based on Euclidean distance followed by group average
clustering provided a clear indication of the importance of the discharge regime, since intermittent sites (8-12) formed one group and
perennial sites (13-17) formed a second group. This feature was particularly notable in view of the enhanced opportunities for migration in
1972 when the stream did not dry out to the perennial head. Within the
group of perennial sites, site 17 was the most distinct, probably
reflecting the species richness at this site. The dendrogram based on
Czekanowski dissimilarity coefficient followed by group average
clustering, provided a slightly different perspective. The pond (site
8) was first delimited from the remaining sites before the next major
division successfully differentiated the intermittent (9-12) from the
perennial sites (13-17). The physical environment of the pond
contrasted strongly with the stream sites (9-17) and this was reflected
in the many taxa only recorded from the pond. In addition, a number of
typical stream invertebrates failed to colonize the pond.

Ham *et al.* (in prep.), using similar clustering techniques, also demonstrated a major division between the invertebrate communities of intermittent and perennial sites on the Lambourn, as did Furse (1977a) for the Gussage, using a combination of ordination and clustering procedures.

7.63   FAUNAL GROUPINGS

Table 7.4 lists the taxonomic groups which were normally identified to genus or species. They have been divided into three major groups as follows: species present at all sites; species restricted to the intermittent sites (8-12); and species restricted to the perennial sites (13-17). Within the intermittent sites, pond species restricted to site 8 are distinguished with an asterisk, as are species in the perennial stream which were restricted to site 17, adjacent to the Lambourn.

In autumn 1972, when discharge remained too high for the stream to dry to the perennial head and the distinction between intermittent and perennial sites was not reinforced, some species normally restricted to one zone may have moved downstream or upstream. Species which retained their distinctive distributions are likely to be those most character-istic of either intermittent or perennial sites.

The occurrence of *Gammarus pulex* at all ten sites throughout 1972 implies an ability to survive in damp gravel during periods of drought or to migrate upstream from the perennial zone when water returns. The latter explanation is favoured since only large specimens of *G. pulex* were found in reinundated sections of the Lambourn, which suggested upstream migration of large individuals (Ham *et al.*, in prep.). Clifford (1966) noted that only small amphipods (*Crangonyx*) survived summer and autumnal dry periods by burrowing deep into the damp substratum.

Of the Ephemeroptera, only *Baetis vernus* occurred at all 10 sites, but *B. rhodani* and *Ephemerella ignita* were found at 9 and 8 sites, respectively. These three species were also successful at colonizing the intermittent section of the Lambourn (Ham *et al.*, in prep.).

The ability of Trichoptera of the family Limnephilidae to exploit intermittent sites is already well documented (Clifford, 1966; Williams & Williams, 1975; Wiggins *et al.*, 1980) and three species were recorded at each of the 10 sites. Furse (1977a) also recorded one of these, *Limnephilus lunatus*, as an early colonizer of winter-bourne reaches of the R. Piddle in Dorset.

Since there are approximately 500 species of Chironomidae known to occur in the British Isles, it was not surprising that they also contributed the most taxa to colonize the 10 sites.

Finally, four species of oligochaetes also occurred at 10 sites and all appear to have some resistance to drought. *Eiseniella tetraedra* can aestivate when pools dry out (Reynolds, 1977) and several authors record its ability to survive drought (Hynes, 1958; Legier & Talin, 1973; Casey & Ladle, 1976; Furse, 1977a). In general, tubificids are largely confined to permanent waters, but Williams & Hynes (1976) observed that some tubificids, probably *Tubifex tubifex*, secreted mucous envelopes around themselves in drought conditions and Casey & Ladle (1976) and Furse (1977a) produced evidence of *T. tubifex* colonizing intermittent sites. Of the two lumbriculids, *Lumbriculus variegatus* is known to be capable of resisting drought conditions by encysting (Cook, 1969). Although *L. variegatus* and *Stylodrilus heringianus* were both found in the Winterbourne Stream (Table 7.4) and also the South Winterbourne (Casey & Ladle, 1976), Ham *et al.* (in prep.) did not find either in the intermittent section of the Lambourn following the severe drought of 1976.

There are a number of additional taxonomic groups with species restricted to intermittent sites, compared to the wide-ranging species in the previous column of Table 7.4 although no oligochaetes are confined to the intermittent zone. *Niphargus aquilex* is widely distributed in groundwater in southern England and Wales (Gledhill *et al.*, 1976) and is liable to occur where this reaches the surface. It has also been recorded at intermittent sites on the Lambourn (Ham *et al.*, in prep.) and the Gussage (Furse, 1977a).

The group with the greatest number of species colonizing the intermittent sites was the Coleoptera and, although nine species were restricted to the pond and some were facultative colonizers of streams, a few were typical of intermittent streams. Legier & Talin (1973) reported that Coleoptera were better represented in intermittent than perennial streams and Hynes (1958) found that after an unprecedented drought, additional species of dytiscids colonized the Afon Hirnant in Wales for a limited period of time. Nevertheless, it is clear that the Winterbourne Stream has an unusually rich fauna. Without further investigation we can only speculate on the possible causes of this. The pond at site 8 has been shown to attract Coleoptera, but farms alongside the stream and houses in Winterbourne village may also have small water bodies from which species can colonize the stream. The lower part of the catchment supports heathland and copses which may offer suitable refuges for species not usually associated with the few bodies of open water found on chalk downland. The lack of drying out of this part of the Winterbourne Stream during and immediately prior to 1972 also means that colonization was possible over an unusually long period of time.

During an extensive survey of 98 chalk stream sites including sites 1-17 on the Winterbourne Stream, Cooling (1981) recorded *Hydroporus marginatus, H. pubescens, Agabus bipustulatus* and *Colymbetes fuscus* from intermittent sites only. In contrast, he recorded *Agabus biguttatus* and *Hydrobius fuscipes* from both intermittent and perennial

Table 7.4 A list of taxa recorded at every major study site (8-17), together with those restricted either to intermittent (8-12) or perennial sites (13-17). All taxa with some records from both intermittent and perennial sites are excluded unless they were recorded at every site.

taxon	present at 10 sites	restricted to intermittent sites (8-12); *species restricted to site 8 (pond)	restricted to perennial sites (13-17); *species restricted to site 17, adjacent to R. Lambourn
Malacostraca	Gammarus pulex	Niphargus aquilex	
Ephemeroptera	Baetis vernus	Cloeon dipterum*; Caenis horaria	Centroptilum pennulatum*; Paraleptophlebia submarginata*; Centroptilum luteolum; Ephemera danica; Caenis rivulorum.
Coleoptera	Helophorus spp.	Laccophilus minutus*; Guignotus pusillus*; Coelambus confluens*; Graptodytes pictus*; Hydroporus incognitus*; Agabus sturmi*; A. nebulosus*; Acilius sulcatus*; Laccobius bipunctatus*; L. striatulus*; Coelambus impressopunctatus; Hydroporus marginatus; H. memnonius; H. nigrita; H. pubescens; Agabus biguttatus; A. didymus; Colymbetes fuscus; Dytiscus marginatus; Hydrobius fuscipes; Anacaena limbata	Potomonectes depressus ssp. elegans*; Brychius elevatus; Oreodytes sanmarki; Platambus maculatus; Agabus chalconatus; Hydraena gracilis; Limnebius truncatellus; Oulimnius tuberculatus; Riolus subviolaceus
Trichoptera	Limnephilus lunatus; Halesus sp. Stenophylax sp.	Limnephilus vittatus; Glyphotaelius pellucidus	Chaetopteryx villosa*; Melampophylax mucoreus*; Bereodes minutus*; Lype reducta; Hydroptila sp. Ithytrichia sp.; Oxyethira sp; Beraea maurus; Odontocerum albicorne; Anthripsodes albifrons; Adicella reducta; Brachycentrus subnubilus; Sericostoma personatum

Chironomidae	Macropelopia sp.; Conchapelopia gp.; Brillia modesta; Prodiamesa olivacea; Micropsectra sp. A	Psectrotanypus various*; Chironomus sp. B*; Micropsectra sp. C*; Natarsia sp.; Diamesa insignipes; Limnophyes sp.	Eukiefferiella sp. B*; Orthocladius sp.*; Epoicocladius flavens*; Polypedilum sp. A*; Larsia sp.; Cricotopus sp. D; Heleniella ornaticollis; Phaenopsectra sp.; Polypedilum sp. B; Polypedilum sp. C; Polypedilum sp. D; Rheotanytarsus sp.; Stempellina bausei
Oligochaeta	Tubifex tubifex; Lumbriculus variegatus; Stylodrilus heringianus; Eiseniella tetraedra		Psammoryctides barbatus; Peloscolex ferox; Aulodrilus pluriseta
Odonata	Agrion splendens		
Bivalvia	Pisidium personatum		Pisidium milium
Hemiptera	Sigara nigrolineata*; S. lateralis*; Notonecta maculata; Corixa punctata; Sigara concinna		Gerris sp. nymph* Corixa panzeri*
Plecoptera	Nemoura cinerea		Nemurella picteti
Tricladida			Polucelis felina
Gastropoda			Valvata piscinalis*; Physa fontinalis
Hirudinea			Theromyzon tessulatum
Megaloptera			Sialis lutaria

sites in other streams. These two species were restricted to intermittent sites in the Winterbourne Stream in 1972 but not in 1976
(Cooling *et al.*, in prep.). Foster (1980) has examined the records of
the two 'subterranean' species *Hydroporus marginatus* and *Agabus
biguttatus* in the U.K. and although the former does appear to be
closely associated with hard water, the latter has a much wider distribution and its occurrence in hard water is probably because this
provides suitable subterranean habitats.

The Ephemeroptera, Trichoptera and Odonata which were limited to
the intermittent section (Table 7.4), were still or slow-flowing water
forms which occurred in the pond at site 8 or in nearby sites (9, 10)
downstream.

Chironomidae were also well represented, with both still water
(*Psectrotanypus varius*) and cold water (*Diamesa insignipes*) forms plus
a member of the Metriocnemini (*Limnophyes* sp.) which are mainly terrestrial or semi-aquatic. All the Hemiptera except *Sigara concinna* (site
11 only) occurred in the pond (site 8) but some also penetrated further
downstream.

Pisidium personatum was the only bivalve limited to the intermittent
stream but *P. casertanum* also occurred predominantly in this zone and
both species are known to be capable of tolerating drought conditions
(Boycott, 1936).

The plecopteran *Nemoura cinerea,* confined to the intermittent stream,
has been widely reported in these conditions (Legier & Talin, 1973;
Casey & Ladle, 1976; Ham *et al.*, in prep.). It has also been recorded
in the Afon Hirnant in Wales in the aftermath of a drought (Hynes,
1961); the species had not been recorded previously in this stream and
did not reappear after passing through a single generation. Khoo
(1964) found that *N. cinerea* had a long egg diapause and also showed
that hatching from a single batch of eggs occurred over a long period.
Since the eggs have a gelatinous coat which protects them from drying
out amongst the stones of a dry river-bed, the species has an effective method of coping with the intermittent nature of the flow regime,
provided that nymphs can grow and emerge before the stream dries in
summer.

Taxa restricted to perennial sites need less detailed consideration
than those of the intermittent section. The major feature which
distinguishes the two lists is the greater species richness of the
Ephemeroptera, Trichoptera, Chironomidae and Oligochaeta in the
perennial zone and the occurrence of a number of species in other
taxonomic groups (Tricladida, Gastropoda, Hirudinea, Megaloptera) which
were absent from the intermittent section. In contrast, it was notable
that fewer Coleoptera were restricted to the perennial zone than to the
intermittent zone. The majority of the species recorded exclusively
from the perennial zone also occurred in the Lambourn and relied on
continuous discharge throughout the year for their survival.

7.64   THE 1976 DROUGHT

When a drought appeared imminent early in 1976, a new sampling programme was initiated to record changes during and after the drought (Cooling *et al.*, in prep.). In April 1976 the upstream limit of water was the perennial head at site 13 and by August 1976 the upstream limit was site 15. Only sites 15-17 remained wet until December, when flow returned to 13 and 14 and by February 1977 to all of sites 8-17. No macrophytes were eliminated by the drought but several showed poor growth and sites which remained wet in 1976 became heavily silted.

Although the number of taxa at the lowest two perennial sites (16, 17) remained fairly stable between April 1976 and August 1977, the fauna at site 15, which became the upstream limit of water in 1976, suffered a substantial decrease in the number of taxa, but recovered during the study period. Sites upstream of 15, which dried out, showed a rapid recolonization upon the return of water and this was still continuing in August 1977.

7.7   FISH

The brown trout *Salmo trutta* and the bullhead *Cottus gobio* are the most common fish species in the Winterbourne Stream, as they are in the Lambourn. Three-spined stickleback *Gasterosteus aculeatus*, ten-spined stickleback *Pungitius pungitius* and brook lamprey *Lampetra planeri* occur locally, particularly in beds of *Rorippa* in the lower reaches of the perennial section. The grayling *Thymallus thymallus* is common in the Lambourn but was never caught in the Winterbourne Stream although it has been reported in the lower part. The Winterbourne Stream has been used successfully as a nursery stream for young salmon (*Salmo salar*). These were introduced by Thames Water Authority as part of their programme to re-establish the species in the catchment of the Thames (M. Bulleid, pers. comm.).

7.71   BULLHEADS

The bullhead population of the Winterbourne Stream was studied in detail by Green (1975) who obtained samples by electric fishing short sections of the stream. He found that bullheads survived for up to three years but that fish in their first year were always the majority of the population. Numbers increased during the summer with the recruitment of fry and then fell through the rest of the year. Some fish spawned when one year old but most did not do so until the end of their second year and very few survived beyond that age.

A survey of bullheads at 17 sites was carried out from 27 March to 18 April 1973. Population densities ranged from 0.2 - 1.2 m^{-2} with a mean of 0.69 m^{-2} at the 15 sites in the perennial section. No bullheads

were found at the two sites sampled in the intermittent section but this
may be due to a small weir at Kimber's Cottage which may obstruct upstream movement of bullheads. Green found that the highest population
densities occurred at sites with a high cover of macrophytes and with
coarse gravel or loose stones. Areas of exposed, compact gravel or of
silt were less-well populated.

The uppermost study site in the perennial section, at Kimber's
Cottage, was first sampled in March 1972. At that time no O group bullheads were found and the population consisted of I group fish only.
This suggests that no fry had hatched there in 1971. The stream dried
as far down as Honeybottom for a short period in the autumn of 1969 as
a result of trial pumping operations. The site at Kimber's Cottage
may not have been recolonized by bullheads within the following 15
months, but a further year had allowed recolonization to occur. Fry
were found at the site after the 1972 breeding season. This indicates
that bullheads do move upstream to colonize available areas but that
the rate of movement is probably slow. Recolonization of the Gussage
by non-salmonid species was also slow (Furse, 1977a).

During 1972-73 Green estimated the biomass and production of bullheads at three sites on the Winterbourne Stream (Table 7.5). Biomass
varied greatly at different times and both biomass and production were
highest at the uppermost site; the missing age-class (mentioned above)
at this site is apparent in Table 7.5. The results are similar to
those obtained for sites on the Lambourn (Green, 1975) and on the Bere
Stream and the R. Tarrant (Mann, 1971).

Table 7.5 Estimates of biomass and production of bullheads in
1972-73 (from Green, 1975).

site	biomass ($g\ m^{-2}$)		production ($g\ m^{-2}\ yr^{-1}$)				
	min	max	egg	0	I	II	total
A (near 13 in Fig. 7.1)	0.98	10.84	0.6	15.9	0.0	2.8	19.3
B (between 15 and 16 in Fig. 7.1)	0.11	2.89	0.3	2.2	0.6	0.4	3.5
C (" " " " "	0.09	8.94	0.2	2.9	0.5	0.5	4.1

Green found that bullheads fed entirely on benthic invertebrates.
The fry ate ostracods, cladocerans and chironomid larvae while larger
fish fed mainly on *Gammarus* but also took larval stages of various
aquatic insects. The observations are similar to those from other
chalk streams (Mann & Orr, 1969; Abel, 1973) and the differences that
occur are probably due to the local abundance of particular invertebrates.

7.72   TROUT

The trout population was sampled by electric fishing at least once each
year from 1972 to 1976. This always covered the 1.5 km from the
flighting pond to site 13 and sometimes included the 230 m between the
pond and the gauging weir or a stretch above site 13. On most occasions two fishings were carried out and all fish caught on the first
fishing were marked by injection of a spot of dye so that they could
be identified on the second fishing. A different batch mark was used
on each occasion. The results have not been fully examined and only a
preliminary account of some aspects is presented here.

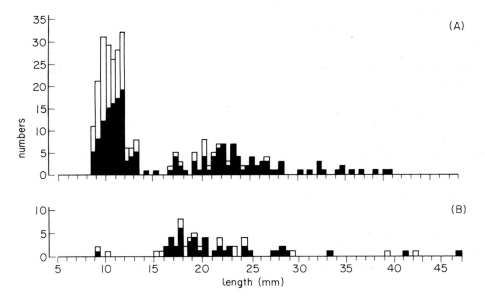

Fig. 7.6 Length frequency distributions of trout from above the
flighting pond in (a) January 1972, and (b) December 1972. The
black areas indicate marked fish that were recaptured.

The size structure of the trout population above the flighting pond
in January and December 1972 is shown in Fig. 7.6. These were second
fishings and recaptured fish are indicated in black. The January
histogram corresponds conveniently with three age groups which can be
deduced from experience of ageing Winterbourne trout from their scales.
The large group of fish under 15 cm in length are young trout which
hatched in the spring of 1971. The group between 15 and 30 cm in
length includes trout that are one and two years old while the fourteen
individuals over 30 cm in length are probably three and four years old.
By comparison the December histogram shows that very few young trout
hatched in spring 1972 and the numbers of older fish are considerably
lower than in January. These impressions are confirmed when the recap-

ture data are used to obtain estimates of the trout population at both times (Table 7.6). The calculation for the young fish must be treated with caution since the efficiency of catching them is lower than for the larger fish and they are more liable to die as a result of handling during the marking operation. Similar changes in the trout population and especially in the recruitment of young fish were noted in 1971 and 1972 in the Chitterne Brook, a winterbourne tributary of the R. Wylye in Wiltshire (Avon & Dorset River Authority, 1973).

Table 7.6 Population estimates for trout above site 16 (in Fig. 7.1) derived from numbers marked and recaptured.

date	age class	number marked	number re-captured with marks	total re-capture	population estimate
January 1972	0	254	105	199	481
	>0	130	82	97	154
December 1972	0	4	1	3	12
	>0	55	43	61	78

The next fishing during the mild drought in August 1973, showed that the number of larger fish had fallen to an estimated 17 and indicated a better hatch of young trout that year. During 1974 and 1975 the numbers of older trout rose to around the December 1972 levels but the high population of January 1972 was not equalled. In three other small chalk streams Mann (1971) has shown that the biomass and production of trout are generally less than those of bullheads and that 0 group trout contribute most to total production.

In August 1972 electric fishing was carried out for about 1 km above Kimber's Cottage and 20 trout were caught. Four of these had been marked below Kimber's Cottage in December 1971 and the group included no 0 group trout. When this section was fished again in December 1972 no trout were caught. Between 1971 and 1973 a substantial number of trout were marked in the perennial section of the Winterbourne Stream and in the part of the Lambourn downstream from the confluence. Good recoveries were obtained for fish in their home streams but there were only a few cases in which trout were found to have moved between the two streams. In contrast, results from a trap at the lower end of the Chitterne Brook showed that in 1972 mature trout migrated into it from the R. Wylye after flow began in the winter and that adult trout and large numbers of young trout left the brook before it dried in the summer (Avon & Dorset River Authority, 1973). However, in other years substantial numbers of fish had been found stranded when the brook dried. A study on the Candover Brook (Solomon & Temple-

ton, 1976) showed that there was an annual pattern of local movements
in the trout population but that less than 1% made extensive movements.
Spawning fish moved as far as 1.5 km above the perennial head when the
winterbourne section was flowing but most failed to return downstream
before the section dried again. It is still not clear whether winterbournes act as nursery streams which supplement recruitment of trout
in the main river.

   Young trout fed mainly on *Gammarus* and the larval stages of aquatic
insects but older trout relied less on these and took substantial
quantities of surface food and fish. One third of all large trout
examined from the Winterbourne Stream had eaten fish, mainly bullheads
and sticklebacks. Thus, young trout utilized the same food as bullheads
but older trout extended their range of food.

7.8 CONCLUSIONS

The Winterbourne Stream has substantial growths of macrophytes, a rich
invertebrate fauna and viable populations of fish. Clear differences
exist between the communities of plants and invertebrates in the
perennial and intermittent parts of the stream despite the fact that
the stream does not always dry to the perennial head each year. The
data on fish are more limited but indicate that neither bullheads nor
trout are successful in exploiting the intermittent zone. Williams &
Hynes (1977) have suggested that the invertebrates found in temporary
streams may be divided into three main groups. First, stream species
capable of surviving drought for limited periods; second, facultative
colonizers capable of living in both lotic and lentic waters; and
third, species highly adapted and often restricted to temporary waters.
These groups can be recognised in the Winterbourne Stream. The intermittent zone contains a number of invertebrates that have not been
recorded at many other sites in Great Britain whereas the perennial
zone is essentially similar to other chalk streams.

   Several small chalk streams in the south of England are already
affected by water engineering schemes and others are likely to be
affected in the future. If these schemes produce substantial changes
in the flow conditions of winterbournes they may have detrimental
effects on species whose survival depends on an intermittent flow
regime. Information is still lacking on the precise requirements of
these species and on their adaptations for surviving and exploiting
the alternating wet and dry conditions.

Acknowledgements

This study was financed by the Thames Conservancy and the Water
Resources Board and by their successors, Thames Water Authority and the
Central Water Planning Unit, as part of an ecological study of chalk
streams. We are grateful to both organizations for their support and

to successive Heads of the Department of Zoology in the University of Reading for making facilities available for the work. Special thanks are due to the landowners in the Winterbourne valley who gave us access to the stream and to colleagues who took part in the field surveys and in processing the samples, particularly Dr P.D. Hiley, Dr G.P. Green, Mrs A.C. Cameron, Mrs P.R. McLeish and Mrs M.E. Wigham. Thames Water Authority provided data on discharge and water chemistry together with some of the information for Section 7.2. We are also grateful to Mr D.A. Cooling who developed the cluster package; Mrs A.M. Matthews who drew the figures; and Mr M.T. Furse who provided helpful comments on the manuscript.

REFERENCES

Abel R. (1973) The trophic ecology of *Cottus gobio* L.. Unpublished Thesis, University of Oxford, England

Avon and Dorset River Authority (1973) Upper Wylye investigation. *A report on the Effects of Abstraction from a Chalk/Upper Greensand Aquifer*, 120 pp.

Boycott A.E. (1936) The habitats of freshwater Mollusca in Britain. *J. Anim. Ecol.* 5, 116-86

Brettell E.J. (1971) *Report on the Lambourn Valley Pilot Scheme, 1967-1969*, 170 pp. Thames Conservancy.

Casey H. (1969) The chemical composition of some southern English chalk streams and its relation to discharge. *River Auth. Ass. Yb. 1969*, 100-13

Casey H. & Newton P.V.R. (1972) The chemical composition and flow of the South Winterbourne in Dorset. *Freshwat. Biol.* 2, 229-34

Casey H. & Newton P.V.R. (1973) The chemical composition and flow of the River Frome and its main tributaries. *Freshwat. Biol.* 3, 317-33

Casey H. & Ladle M. (1976) Chemistry and biology of the South Winterbourne, Dorset, England. *Freshwat. Biol.* 6, 1-12

Clifford H.F. (1966) The ecology of invertebrates in an intermittent stream. *Invest. Indiana Lakes & Streams.* VII, 2, 57-98

Cook D.G. (1969) Observations on the life history and ecology of some Lumbriculidae (Annelida, Oligochaeta). *Hydrobiologia* 34, 561-74

Cooling D.A. (1981) Records of aquatic Coleoptera from rivers in southern England. *Entomologist's Gaz.* 32, 103-13

Cooling D.A., Wright J.F., Ham S.F. & Berrie A.D. (in prep.) The effects of the 1976 drought on the invertebrate fauna of a small chalk stream in Berkshire, England

Foster G.N. (1980) Subterranean statistics II. *The Balfour-Browne Club Newsletter* 15, 3-7

Furse M.T. (1977a) *An ecological study of the Gussage, a lined winterbourne 1973-1977*. A report to the Water Research Centre, 363 pp.

Furse M.T. (1977b) The ecology of the Gussage, a lined winterbourne. *Annual Report, Freshwat. biol. Ass. U.K.* 45, 30-6

Gledhill T., Sutcliffe D.W. & Williams W.D. (1976) A Revised Key to
   the British species of Crustacea: Malacostraca occurring in Fresh
   Water with notes on their Ecology and Distribution, 72 pp. *Sci.*
   Publs Freshwat. biol. Ass., U.K. 32

Green G.P. (1975) The food and production of the bullhead, *Cottus*
   gobio L. in the River Lambourn. Unpublished thesis, University of
   Reading, England

Ham S.F., Cooling D.A. & Berrie A.D. (in prep.) The distribution of
   macroinvertebrates in the River Lambourn, Berkshire, England

Ham S.F., Wright J.F. & Berrie A.D. (1981) Growth and recession of
   aquatic macrophytes on an unshaded section of the River Lambourn,
   England, from 1971 to 1976. *Freshwat. Biol.* 11, 381-90

Hynes H.B.N. (1958) The effect of drought on the fauna of a small
   mountain stream in Wales. *Verh. int. Verein. theor. angew. Limnol.*
   13, 326-33

Hynes H.B.N. (1961) The invertebrate fauna of a Welsh mountain stream.
   Arch. Hydrobiol. 53, 344-88

Khoo S.G. (1964) Studies on the biology of stoneflies. Unpublished
   thesis, University of Liverpool, England

Klee C.A. (1974) *Some aspects of the ecology of the River Thet*, 27 pp.
   Great Ouse River Authority

Klee C.A. (1976) *Report on Fish Ecology and the Groundwater Develop-*
   ment Scheme, 80 pp. Great Ouse River Division, Anglian Water
   Authority

Ladle M. & Bass J.A.B. (1975) A new species of *Metacnephia* Crosskey
   (Diptera: Simuliidae) from the south of England, with notes on its
   habitat and biology. *Hydrobiologia* 47, 193-207

Ladle M. & Bass J.A.B. (1981) The ecology of a small chalk stream and
   its responses to drying during drought conditions. *Arch. Hydrobiol.*
   90, 448-66

Le Cren E.D. & Lowe-McConnell R.H. (1980) *The Functioning of Fresh-*
   water Ecosystems, 588 pp. Cambridge U. P., Cambridge, England

Legier P. & Talin J. (1973) Comparaison de Ruisseaux permanents et
   temporaires de la Provence Calcaire (Comparison of permanent and
   temporary streams of calcareous Provence). *Annls Limnol.* 9, 273-92

Mann R.H.K. (1971) The populations, growth and production of fish in
   four small streams in southern England. *J. Anim. Ecol.* 40, 155-90

Mann R.H.K. & Orr D.R. (1969) A preliminary study of the feeding
   relationships of fish in a hard-water and a soft-water stream in
   southern England. *J. Fish Biol.* 1, 31-44

Ratcliffe D.A. (Ed.) (1977) *A Nature Conservation Review.* Vols. 1 &
   2, 401 & 320 pp. Cambridge U. P., Cambridge, England

Reynolds J.W. (1977) The earthworms (Lumbricidae and Sparganophilidae)
   of Ontario. *Roy. Ont. Mus. Life. Sci. Misc. Publ.* 1-141

Sneath P.H.A. & Sokal R.R. (1973) *Numerical Taxonomy*, 573 pp. W.H.
   Freeman & Company

Solomon D.J. & Templeton R.G. (1976) Movements of brown trout *Salmo*
   trutta L. in a chalk stream. *J. Fish Biol.* 9, 411-23

Southern Water Authority (1979) *The Candover Pilot Scheme, Final*
   Report, 165 pp.

Water Resources Board (1973) *Water Resources in England and Wales.*
    Volume 1: Report, 67 pp. Volume 2: Appendices, 56 pp. H.M.S.O.,
    London

Westlake D.F., Casey H., Dawson H., Ladle M., Mann R.H.K. & Marker,
    A.F.H. (1972) The chalk stream ecosystem. In: (Eds. Z. Kajak &
    A. Hillbricht-Ilkowska) *Proc. I.B.P. UNESCO Symposium on Producti-
    vity Problems of Freshwaters,* Kazimierz Dolny, pp. 615-35. Warszawa-
    Kraków

Wiggins G.B., Mackay R.J. & Smith I.M. (1980) Evolutionary and ecolo-
    gical strategies of animals in annual temporary pools. *Arch.
    Hydrobiol., Suppl.* 58, 97-206

Williams D.D. & Hynes H.B.N. (1976) The ecology of temporary streams.
    I. The faunas of two Canadian streams. *Int. Rev. ges. Hydrobiol.*
    61, 761-87

Williams D.D. & Hynes H.B.N. (1977) The ecology of temporary streams.
    II. General remarks on temporary streams. *Int. Rev. ges. Hydrobiol.*
    62, 53-61

Williams D.D. & Williams N.E. (1975) A contribution to the biology of
    Ironoquia punctatissima (Trichoptera: Limnephilidae). *Can. Ent.*
    107, 829-32

Wright J.F., Hiley P.D., Cooling D.A., Cameron A.C., Wigham M.E. &
    Berrie A.D. (1984) The invertebrate fauna of a small chalk
    stream in Berkshire, England, and the effect of intermittent flow.
    Arch. Hydrobiol. 99, 176-99

Wright J.F. (1984) The chironomid larvae of a small chalk stream
    in Berkshire, England. *Ecol. Ent.* 9, 231-8

8 · Lot

H. DÉCAMPS, J. CAPBLANCQ & J. N. TOURENQ

8.1 ¯INTRODUCTION

The Lot provides a typical example of a river domesticated by man.
Diverse developments have profoundly modified flow in order to make the
river navigable and to produce electricity. Since the 14th century,
the construction of navigation routes has transformed the middle and
lower Lot into a succession of 62 stages. Nowadays the passage of
water through one-third of these routes is used for electricity production and there are nine large hydroelectric dams in the upper part of
the catchment.

All these transformations have been accompanied by wastes originating
from domestic, industrial and agricultural sources. The river has thus
changed rapidly during recent decades. Towards the end of the 1960s,
diverse observations showed the existence of zones of doubtful water
quality with, in numerous places, a turbid or greenish appearance, due
apparently to the proliferation of algae. It was important to understand the causes of these events. Such has been the principle objective of the studies made on the development of materials in suspension,
particularly the phytoplankton, in waters of the Lot (Bordes *et al.*,
1973; Dauta, 1975a, b; Massio, 1976; Denat, 1977).

Fig. 8.1A Entry into the
Lot of the grossly pollu-
ted Riou Mort; note the
sharp transition between
the two waters.

Fig. 8.1B Weir on the
Lot downstream of the
Riou Mort; the boulders
have frequent growths of
the moss *Rhynchostegium
riparioides*.

8.2   GENERAL DATA

DIMENSIONS

Area of catchment	118240 km^2
Length of river dealt with in chapter	313 km below confluence of Truyère and Haut-Lot
Confluence of Truyère and Haut-Lot	
Highest point in catchment	1858 m (Plomb du Cantal)
Drop in height down 313 km stretch	230 m

CLIMATE

Precipitation   mean 986 mm yr^{-1} (1971-1975)
Occurs mainly in winter, with snow at the higher altitudes

GEOLOGY

Crystalline and intrusive rocks in the upstream part, calcareous rocks in the middle and Tertiary rocks covered by alluvial formations in the lower part of the basin.

POPULATION

Mean density   31 people km^2
Principal conglomerations in the catchment
    Decazeville   26000 people
    Cahors   19000 people
    Fumel   14000 people

LAND USE

Pasturage in the upper and middle parts of the catchment
Agriculture (fruit, cereals) in the lower catchment
Industrial zones at Decazeville and Fumel

PHYSICAL FEATURES

Discharge at selected points (for period 1962-1975: all as m^3 s^{-1})

	mean annual mean	minimum mean annual	maximum mean annual
Truyère - Haut-Lot			
confluence (km 167)	110	70	170
Cahors (km 330)	150	110	200

Seasonal changes in discharge (for period 1962-1975)
At km 330 mean monthly rate ranged from 14 m^3 s^{-1} in August 1967
                                   to 515 m^3 s^{-1} in January 1966

Influence of regulation
The reservoirs in the upper part of the catchment are filled by the end of the winter and may be restored again during summer. According to the year, variation can reach as much as 10^8 m^3 s^{-1} depending on the natural flow. In summer peaks of 40 - 60 m^3 s^{-1} are superimposed in a basal flow of 7 m^3 s^{-1} at km 178.

TEMPERATURE

In summer 1976, at the top of one stretch of downstream river (km 330), temperatures reached 29 $^{\circ}$C near the surface and 23 $^{\circ}$C at the bottom (6 m).

8.3 GEOGRAPHY AND HYDROLOGY

The lot rises at 1295 m altitude, flowing for 491 km before its entry to the Garonne at an altitude of 22 m (Fig. 8.2). The basin contains three main geological regions (Fig. 8.3):

Basin of the Truyère and Haut-Lot, with crystalline and intrusive
    rocks; leached brown soils, flanked by calcareous ones on
    the souther border;
Basin of the middle Lot, with calcareous Jurassic rocks from
    Causses of Quercy;
Lower basin with Tertiary rocks covered by alluvial formations;
    soils of calcareous marl or poorly permeable calcareous clay.

The Atlantic, Mediterranean and montane influences combine to give rain, particularly in winter, this being accentuated in the mountains by snow, notably in the north-east part of the basin (Fig. 8.3).

The populations density of the Lot basin (31 inhabitants km^{-2}) is well below the national average. Uncultivated zones, with few people, occur in the mountains and the Causses, limestone plateaus essentially given over to breeding animals. Agriculture follows the alluvial axis and spreads out in the lower basin where fruit and cereal production dominate. As for industry, which is relatively little developed, this is concerned especially with agro-food (milk in upstream areas, fruit and vegetables downstream) and, in association with the coal-mining regions of Decazeville and Fumel, the extraction of materials and metallurgy. These zones are, together with Cahors, the source of the principal pollutants of the Lot (Fig. 8.2).

The valley of the Lot is thus very diverse. The elongated form of the basin and the localization of important centres near the river lead to conflicting interests: the needs for drinking water, irrigation, industries, electricity production, fishing and leisure. The growth of these uses, aggravated by the weakness of low river flows and pollution of all sorts, makes it apparent that improvement in water quality is an essential preliminary for all economic activity in the basin (Anonymous, 1973).

With a mean annual discharge of 150 m^3 s^{-1} at its confluence with the Garonne, the Lot carries 46% of the precipitation falling on the catchment (986 mm yr^{-1}). The natural flow regime corresponds to that of a Mediterranean type river: high from December to April, with a maximum in February, low from July to September. In summer the rarity of rain and intensity of evapotranspiration bring about low flows, to the extent that in one year out of ten the mean discharge for August falls below 9 m^3 s^{-1} at the junction with the Garonne. At Cahors (km 330), the ratio of minimum to maximum discharges, which averaged 1 : 20 for the period 1962-75, can drop to 1 : 40. Another feature is that the regime of the Lot depends closely on the zone situated upstream of the Lot - Truyère confluence, which covers only 46% of the basin, but contributes 62% of the discharge (Fig. 8.4).

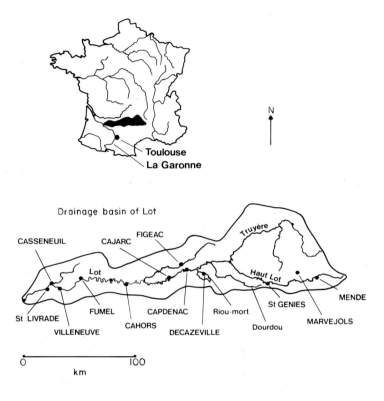

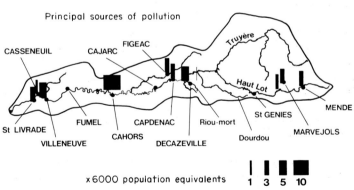

Fig. 8.2 The Lot basin, showing the main towns and principal sources of pollution (after Tourenq *et al.*, 1978).

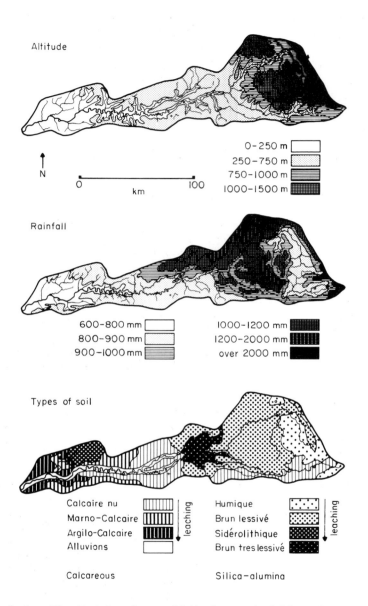

Fig. 8.3   The Lot basin: altitude, rainfall and soil (after Tourenq *et al.*, 1978).

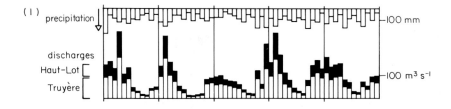

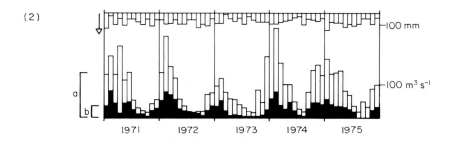

Fig. 8.4 Precipitation and natural discharge:
(1) At the confluence of Haut-Lot and Truyère (km 178); discharges
from the Truyère are always more than that of the Haut-Lot.
(2) At Cahors (km 330), discharges added between km 178 and km 330 (b)
are low in comparison with the total discharge (a).

The natural regime is profoundly modified by the construction of
large reservoirs for hydroelectric purposes: two on the Haut-Lot (22
x 10^6 m^3 capacity) and seven on the Truyère (538 x 10^6 m^3). As a
rule these reservoirs are filled by the end of winter in order to maintain summer-flows, but their management varies and, according to the
year, there is a variation of ± 10^8 m^3 in natural summer discharges.
In fact, with a summer base-flow of 6 m^3 s^{-1} at the confluence of the
Lot and Truyère, from 40 to 60 m^3 s^{-1} is added over a 4 - 6 h period
each day. Further, these fluctuations are shifted in an often anarchic
fashion by small hydroelectric plants. Consequently the times for
renewal of the water are changing constantly in consecutive stretches
of the middle and downstream course of the river; these stretches are
1 to 5 km long and 2 to 10 m deep. To give an example, if there is a
discharge of 20 m^3 s^{-1}, the mean current does not exceed 10 cm s^{-1} in
the middle and lower river and the passage of the upper water takes
one month (Fig. 8.5).

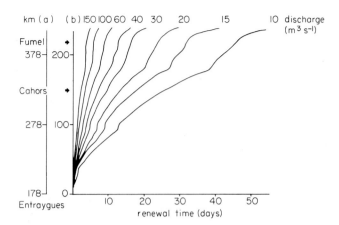

Fig. 8.5 Change in the time for passage of water for different discharges below the Haut-Lot - Truyère confluence (after Tourenq *et al.*,
1978) :
(a) km from the source of the Lot
(b) km from the confluence of the two rivers.

8.4 TRANSPORT OF MATERIALS

8.41 DISSOLVED SUBSTANCES

The waters of the Lot are essentially calcareous bicarbonate (Table 8.1);
Ca^{2+} and HCO_3^- represent on the average 62% of the cations and 82% of
the anions, calcium bicarbonate contributing from 59% (Truyère) to 79%
(Haut-Lot at km 431) of the conductivity of the water.

However, the chemical characteristics of the water differ entirely
between the Truyère and the Haut-Lot upstream of their confluence
(Table 8.1). The Truyère drains a crystalline basin, characterized by
waters which are more acidic and three times less mineralized than those
of the Haut-Lot, whose basin is predominantly calcareous. Besides,
Truyère water has a higher silica content, often exceeding 10 mg l^{-1}
SiO_2. On passing downstream the water of the Truyère increases progressively in Ca^{2+} and HCO_3^-, whilst the contents of Na^+, K^+, Mg^{2+} and
SiO_2 remain more or less constant. The ratio of Haut-Lot as to Truyère,
which is influenced by the hydraulic management of the large reservoirs
upstream, determines the chemical composition of water downstream of
the confluence (Angelier *et al.*, 1978; Capblancq & Tourenq, 1978).
The concentration (S) of dissolved mineral elements depends on the
ratio (X) of Truyère discharge : (Truyère + Haut-Lot) discharge at the
confluence (km 178). The following relationships have been established:

Table 8.1 Chemical composition of Truyère and Lot water at six stations. (Means for concentrations measured from 1971 to 1975, together with standard deviation. (After Capblancq & Tourenq, 1978)

	Truyère	Haut-Lot	middle and lower Lot		
			km 271	km 358	km 431
conductivity (μmhos. cm⁻¹ at 20 °C)	56.25 ± 6.12	167.5 ± 13.1	143.6 ± 10.5	186.5 ± 9.8	207.3 ± 15.4
HCO_3^- (meq l⁻¹)	0.46 ± 0.07	1.59 ± 0.08	1.14 ± 0.09	1.57 ± 0.15	1.87 ± 0.14
SO_4^{--} (meq l⁻¹)	0.08 ± 0.02	0.10 ± 0.03	0.21 ± 0.04	0.32 ± 0.07	0.25 ± 0.05
Cl^- (meq l⁻¹)	0.03 ± 0.01	0.06 ± 0.01	0.14 ± 0.02	0.18 ± 0.02	0.14 ± 0.02
total anions (meq l⁻¹)	0.58 ± 0.09	1.75 ± 0.12	1.50 ± 0.16	2.07 ± 0.23	2.26 ± 0.21
Ca^{++} (meq l⁻¹)	0.27 ± 0.06	1.22 ± 0.23	0.86 ± 0.17	1.14 ± 0.15	1.64 ± 0.26
Mg^{++} (meq l⁻¹)	0.14 ± 0.02	0.41 ± 0.10	0.38 ± 0.04	0.35 ± 0.03	0.40 ± 0.05
Na^+ (meq l⁻¹)	0.14 ± 0.02	0.15 ± 0.02	0.18 ± 0.02	0.17 ± 0.02	0.19 ± 0.02
K^+ (meq l⁻¹)	0.03 ± 0.01	0.04 ± 0.01	0.04 ± 0.004	0.04 ± 0.01	0.04 ± 0.01
total cations (meq l⁻¹)	0.57 ± 0.10	1.82 ± 0.36	1.45 ± 0.23	1.69 ± 0.21	2.27 ± 0.33
SiO_2 (mg l⁻¹)	8.42 ± 1.17	5.40 ± 1.0	6.10 ± 0.61	6.24 ± 0.63	6.40 ± 0.93
NO_3 - N (μg l⁻¹)	316 ± 38	385 ± 55	380 ± 47	402 ± 60	524 ± 86
PO_4 - P (μg l⁻¹)	8.96 ± 2.5	13.5 ± 4.3	10.6 ± 3.32	10.23 ± 2.63	12.94 ± 3.54
total Fe (μg l⁻¹)	310 ± 22	240 ± 34	570 ± 110	480 ± 55	390 ± 80

at km 220

$(S) mg l^{-1}$ = 172.5 - 112.5 X $(r = 0.852, 26 DF)$

at km 271

$(S) mg l^{-1}$ = 223 - 88.2 X - 24.7 Q $(r = 0.776, 42 DF)$

at km 358

$(S) mg l^{-1}$ = 254 - 80.3 X - 21.8 Q $(r = 0.623, 38 DF)$

(where Q = logarithm of discharge at station considered).

During the year the SiO_2 content decreases in spring when the diatom population is growing, whilst the content of filtrable reactive phosphorous is stripped when the Chlorococcales proliferate. In the Lot the lower concentrations of nitrate (20 - 30 μg l^{-1}) and filtrable reactive phosphorus (0.1 μg l^{-1}) have been found with chlorophyll a concentrations of the order of 20 to 30 μg l^{-1}. The contents of filtrable reactive and total (filtrable + particulate) phosphorus, which are the highest in winter when drainage of the basin is intense and biological activity least, return to 5 and 30 - 90 μg l^{-1}, respectively, during summer. The ratio of mean nitrate to mean filtrable reactive phosphorus is of the order of 25 : 1 in winter and autumn, 60 : 1 in summer and, in certain stretches, over 800 : 1 in July and August.

Filtrable organic materials (< 0.5 μm diameter) form 27 - 86% of the total organic material transported by the Lot; their concentration is relatively constant ($\bar{x}$ = 7.73 mg l^{-1}, SD = 3.1). They are weakly degradable; BOD_5 has a mean value of 2.8 mg l^{-1} and rarely exceeds 5 mg l^{-1}. Organic pollution is then reduced and the water of the Lot is generally well oxygenated. However, a thermal and chemical stratification, accompanied by pronounced oxygen deficits at depth, appears in certain stretches when the discharge drops below 20 $m^3 s^{-1}$.

At km 220, the Lot receives with the entry of the Riou Mort, which is the outlet for the Decazeville industrial basin, a heavy load of suspended materials and diverse heavy metals. In 1976-77, samples taken from the Riou Mort showed: Mn, 0.35 - 2.65 mg l^{-1}; Cu, 0.04 - 0.2 mg l^{-1}; Zn, 5.1 - 60 mg l^{-1}; Co, 0.11 - 1.28 mg l^{-1}. The influence of these additions can still be detected 40 kilometers downstream, where concentrations of 0.34 mg l^{-1} Zn and 0.02 mg l^{-1} Cd have been recorded (Say, 1978).

Altogether the Lot transports around 1.04 x 10^6 tonnes yr^{-1} of dissolved materials (Capblancq & Tourenq, 1978), with almost 40% coming from the lower basin which covers a fifth of the total basin and where chemical erosion is 166 tonnes $km^{-2} yr^{-1}$. Nitrate and filtrable reactive phosphorus transported during periods of high flow from the upper to the lower basin are 0.85 - 3 kg N $km^{-2} d^{-1}$ and 23 - 66 g P $km^{-2} d^{-1}$, respectively (means from November to April).

8.42   PARTICULATE SUBSTANCES

The Lot transports mineral and organic particles in variable amounts
(Table 8.2). During the year the relationship 'algal organic materials:
detrital organic materials:mineral materials' changes from 1 : 2 : 3 in
summer to 1 : 20 : 90 in winter. (Algal organic materials are deduced
from counts.) The 1 - 20 μm fraction largely predominates; it includes
the largest part of the phytoplankton and is the main cause of the
turbidity (Table 8.3). In 1974-76, the lower part of the Lot transported about 10000 - 80000 tonnes yr^{-1} of mineral and 3000 - 35000 tonnes
yr^{-1} of organic materials (Décamps & Casanova-Batut, 1978).

Table 8.2 Concentration of suspended solids (mg l^{-1} : mean, SD) at
different times of year calculated from samples collected at 15
stations on the middle and lower Lot. (After Massio, 1975)

	mineral particles	organic particles	
		detritus	algae
20.08.74	1.85 ± 0.75	1.18 ± 0.60	0.69 ± 0.46
03.12.74	1.40 ± 0.65	1.25 ± 0.62	0.15 ± 0.21
10.03.75	16.74 ± 1.59	3.58 ± 1.05	0.19 ± 0.24
22.07.75	4.22 ± 1.14	2.74 ± 0.91	1.06 ± 0.57

Table 8.3 Size distribution of mineral and organic particles in
summer 1975, together with their influence on turbidity. (All data
given as mean and SD)

size range (diameter in mm)	mineral particles (mg l^{-1})	organic particles		turbidity (formazin units)
		detritus (mg l^{-1})	algae (mg l^{-1})	
total	4.42 ± 0.83	4.02 ± 0.65	1.50 ± 0.31	4.37 ± 0.41
32 mm	0.77 ± 0.18	0.58 ± 0.10		0.46 ± 0.12
32 >∅> 20	1.10 ± 0.17	0.73 ± 0.11		0.38 ± 0.07
20 >∅> 1	2.55 ± 0.62	4.21 ± 0.62		2.92 ± 0.33
1 > ∅				0.62 ± 0.22

Mineral particles come especially from the upper part of the basin,
notably the Dourdou basin with its sandstone and red clays of the
Permian. The Riou Mort basin is the source of metallic inputs (Table
8.4). Further, in certain sections of the Lot, gravel extraction has

disturbed the bed of the river, undermining the banks and adding to
the suspended solids. Finally, the upstream dams provoke sharp variations in discharge which lead to resuspension of sediments on the river
bed (Galharague & Giot, 1974, 1975a; Décamps et al., 1976).

Table 8.4 Concentrations of various elements in sediments of the
Lot (μg g^{-1}) above and below the entry of the Riou Mort

	km above entry	km	below entry	
	2	3	47	198
Mn	800	1300	1300	1400
Cu	100	280	200	100
Zn	350	6200	6000	2900
Cd	5	210	180	40
Hg	<0.3	9.5	6	2.1
Pb	120	820	600	600

The organic particles include algae and detritus. Their content
increases in summer, as the decreased discharge permits the development
of phytoplankton (Fig. 8.6). Algae produced in the reservoirs of the
Haut-Lot and Truyère feed the lower part of the river continuously.
The latter offers ideal conditions for proliferation during summer.

As for turbidity (Galharague et al., 1975b, c, d), it becomes
important on the one hand at times of high flow when particles,
especially mineral ones, attain densities of the order of 100×10^6 l^{-1}
and, on the other hand, when flows are low, when particles, especially
organic ones, exceed 200×10^6 l^{-1} (Fig. 8.7).

Overall these suspended particles maintain a close relationship with
dissolved substances and bottom sediments (Fig. 8.8). Erosion and
direct input of wastes putting mineral and non-algal organic materials
into suspension. In contrast, a reduction inflow leads to sedimentation of suspended materials and, in combination with increased temperature and light flux, an increase in algal biomass. Materials in
suspension interact with each other in the water column. Thus, mineral
particles interfere with phytoplankton growth by, for example, reducing
radiant energy and adsorbing nutrients.

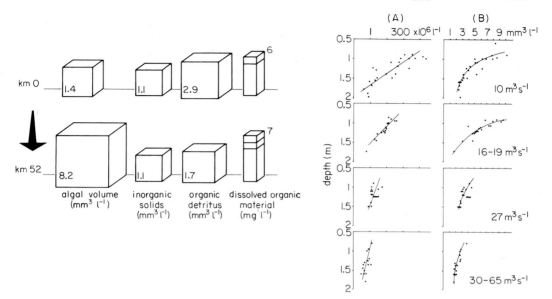

Fig. 8.6 Change in suspended materials and dissolved organic matter in the middle Lot during a passage of water over a distance of 52 km in summer 1975 (after Décamps *et al.*, 1979)

Fig. 8.7 Variation in turbidity (Secchi disc depth) as a function of A) number, and B) volume of 2 - 20 μm diameter particles at different discharges.

8.5   PHYTOPLANKTON

8.51   DEVELOPMENT OF POPULATIONS

The upper course of the Lot is rapid and its microphytic vegetation is composed almost entirely of benthic diatoms which have been brought into suspension (Dauta, 1975a). Planktonic algae appear in the upper reservoirs; they develop further in the stretches of the middle and lower river course where their biomass varies inversely with discharge, ranging from 0.1 to 35 g m^{-3} by weight and from 0.1 to 45 mg m^{-3} as chlorophyll *a* (Dauta, 1975b; Massio, 1976; Denat, 1977; Capblancq & Dauta, 1978).

Over 90% of the plankton consists of diatoms and greens. The former, with *Stephanodiscus* sp., *Melosira varians, M. granulata, Fragilaria crotonensis, Cyclotella* sp. dominate upstream and form essentially the whole crop in spring. The latter, with nanoplanktonic Chlorococcales (*Scenedesmus* sp., *Ankistrodesmus falcatus, Micractinium pusillum, Dictyosphaerium* sp., *Selenastrum* sp., *Coelastrum microporum, C. reticulatum*, unicellular (*Chlamydomonas* sp.) or colonial Volvocales (*Eudorina, Gonium, Pandorina morum*), develop in the middle and lower

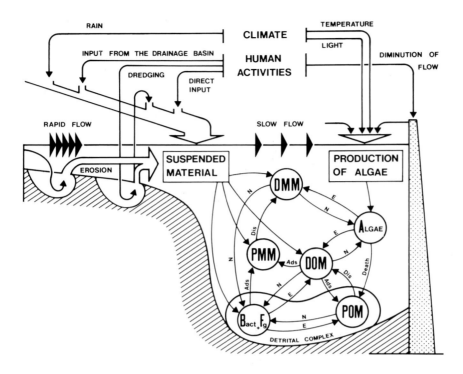

Fig. 8.8 Interactions in a body of water and influence of human activities (after Décamps *et al.*, 1979: by permission of Plenum Press).

stretches, forming over 50% of the biomass in summer. Bursts of *Mougeotia* sp. and blue-green algae (*Anabaena* sp., *Aphanizomenon flos-aquae*) can appear in calmer zones in July and August.

Because of turbulence, the distribution of the phytoplankton is usually homogeneous. However, a vertical gradient of temperature develops with discharges less than 10 m^3 s^{-1}. The algal biomass then increases in the deepest water, as does the proportion of diatoms, the phaeophytin/chlorophyll *a* ratio and the contents of nitrate, ammonium and phosphate. As a result of this sedimentation of dead algae and their breakdown, an under-saturation of oxygen develops near the bottom (Casanova-Batut, 1977; Capblancq & Dauta, 1978). But this stratification is always ephemeral; an increase in flow for a few hours re-establishes the vertical homogeneity of phytoplankton and chemistry.

The seasonal development of phytoplankton in the Lot differs little from that described for numerous meso-eutrophic lakes of the temperate zone. However, the hydraulic management of the river increases the contrasts. Thus, it is when renewal of the water is slow and temperatures are high, both linked with feeble flows, that there occur high concentrations of phytoplankton dominated by greens (Fig. 8.9).

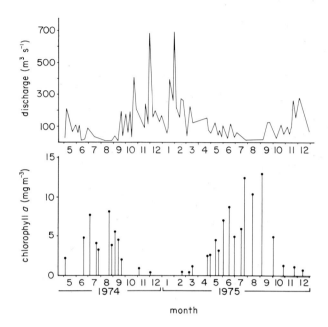

Fig. 8.9 Changes in discharge and chlorophyll *a* content at km 278 from May 1974 to September 1975.

8.52 PRIMARY PRODUCTION

The attenuation of light by phytoplankton in the Lot is relatively weak in comparison with that due to mineral particles and organic detritus. More precisely, it has been shown that the coefficient of vertical attenuation (ε) is maximal in the blue (εblue = 1.64 - 3.36; $\bar{x}$ = 2.65), minimal between the green (εgreen = 0.83 - 1.6; $\bar{x}$ = 1.23) and the red (εred = 0.9 - 1.5; $\bar{x}$ = 1.235). One can deduce from these values the transmission of photosynthetically active radiation (PAR) and the depth at which PAR drops to 1% of that immediately under the surface. In the middle and lower course, depth of this euphotic zone (Z_{eu}) ranges from 0.7 - 5.3 m; it reaches a mean of 3.1 m during the period of phytoplankton development.

   Photosynthetic production by algae at various sites is shown in Fig. 8.10. Neglecting inhibition found in flasks immediately below the surface, the relation between photosynthesis and light can be expressed by the equation (Talling, 1957; Vollenweider, 1965):

$$a = a_{opt} \frac{I/I_k}{\sqrt{1 + (I/I_k)^2}} \qquad (1)$$

where a_{opt} = photosynthesis at light saturation or optimal photo-
       synthetic capacity

and I_k = intensity at the threshold of light saturation

The parameter a_{opt} and I_k vary as a function of the structure of the
phytoplankton population. The optimum photosynthetic capacity (a_{opt})
varies from 0.22 - 25 mg C mg chl a^{-1} (Fig. 8.11). These extremes
correspond, respectively, to high flows and low temperatures on the one
hand and low flows and high summer temperatures on the other hand.
With the former the phytoplankton biomass is feeble, composed of diatoms
partly from the plankton of the upstream reservoirs and partly from the
benthos. With the upper extreme, the population consists of nanoplankton
(greens and cryptomonads); the photosynthetic capacity is thus quite high,
since during 64 measurements made between June and October, a_{opt} reached
a mean value of 8.6 ± 2.3 mg C mg chl a^{-1} h^{-1}.

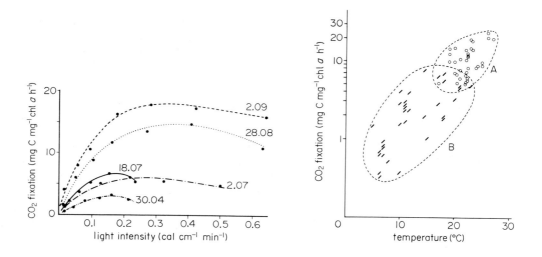

Fig. 8.10 Photosynthetic production by algae in the Lot as a function of light intensity on five different dates.

Fig. 8.11 Optimum photosynthetic capacity of Lot algae as a func-
tion of temperature:
A.   Phytoplankton predominantly of greens;
B.   Phytoplankton predominantly of diatoms.

In the Lot photosynthetic capacity varies with the nature of the phytoplankton, apparently independently of the biomass. Thus, the summer nanoplankton dominated by Chlorococcales is much more active than the spring plankton of diatoms from the upstream reservoirs (Fig. 8.11). Further, the photosynthetic capacity of the phytoplankton varies inversely with the mean volume of the cells (Fig. 8.12).

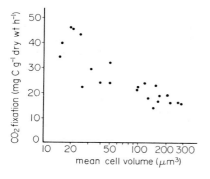

Fig. 8.12 Optimum photosynthetic capacity of Lot plankton as a function of mean cell volume

The light flux at the threshold of saturation (I_k) ranges from 0.03 to 0.12 cal cm^{-2} min^{-1}, with a mean of 0.973 ± 0.02 cal cm^{-2} min^{-1} during the period of phytoplankton development. For this intensity, the ratio, a / a_{opt} is 0.77 ± 0.04. The parameter I_k depends not only on a_{opt}, but also the initial slope K of the experimental curve of the efficiency of conversion of light energy. Now, contrary to observations made in marine or lacustrine environments, K appears very variable in the Lot (Fig. 8.9). There are two possible explanations for these variations: (i) Pigment extraction by acetone which is effective for diatoms is much less for certain Chlorococcales. This would explain the variations of the curves for photosynthetic assimilation as a function of light when the results are expressed relative to chlorophyll *a*. (ii) Photosynthetic activity varies inversely with the volume of the algae (Fig. 8.12) and, therefore, in the Lot the phytoplankton becomes more active the greater the proportion of nanoplanktonic Chlorococcales. In support of this explanation is the fact that algal biomass has been evaluated by two independent methods (chlorophyll *a* and volume) and both give similar results.

Since the phytoplankton biomass is usually distributed homogeneously in the water, vertical differences in photosynthesis depend only on the influence of depth on light flux. Under these conditions, instantaneous production per unit area of water surface (ΣA) obeys the equation (Talling, 1957):

$$\Sigma A = \frac{A_{opt}}{\varepsilon} . f (I'_o/I_k)$$

(where $A_{opt} = B.a_{opt}$).

In effect, in the Lot the relationships are straightforward
-   between photosynthetic production in the water column ΣA and the
    ratio A_{opt}/ε (Fig. 8.12):

$$\Sigma A = A_{opt}/\varepsilon. (2.21 \pm 0.15)$$

-   between the ratio $\Sigma A.\varepsilon/A_{opt}$ and I'_o/I_k (Fig. 8.13), suggesting
    that daily variations in ΣA follow the logarithm of I'_o if I_k,
    A_{opt} and ε remain constant.

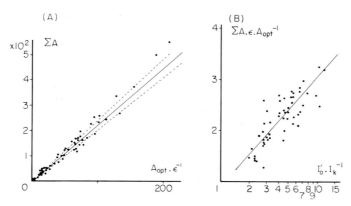

Fig. 8.13 Relations in the Lot between:
A.   Photosynthetic production ΣA and the ratio A_{opt}/ε.
B.   The ratios $\Sigma A.\varepsilon/A_{opt}$ and I'_o/I_k.

Therefore, the daily total of photosynthetic production can be estimated (Talling, 1957) by:

$$\Sigma\Sigma A = \Sigma A. [Ln(2\bar{I}_o/I_k)/Ln(2I_o/I_k)].0.9\Delta t$$

(where $\bar{I}_o$ is mean light flux throughout the day immediately
             below the surface

     I_o is light flux during period of exposure

     Δt is length of day, in hours.)

The results obtained (Capblancq & Dauta, 1978) show that almost 80%
of annual algal production occurs in the three summer months.   Daily
production was:

	min.	max.	mean
October-May	2	245	68 mg C m^{-2} d^{-1}
May-June	237	1377	745
July-September	504	10650	2320

Overall, algal production can be estimated to be 275 g C m^{-2} yr^{-1}, which
indicates a raised trophic level.

8.53 PHYTOPLANKTON GROWTH

The increase in the phytoplankton crop results from the difference
between gains measured by photosynthetic production and losses. Zooplankton are not well developed into the stretches of the Lot, so
predators can be neglected and losses are limited to respiration and
sedimentation of dead algae.

Photosynthetic production and the development of the crop has been
followed in a single body of water during its passage downstream over a
distance of 52 km. This was carried out from 20 to 29 August 1975 at a
time when discharge was 16 - 19 m^3 s^{-1} (Dauta et al., 1975; Capblancq &
Décamps, 1978). During this period phytoplankton growth consisted
almost entirely of the nanoplanktonic Coelastrum reticulatum, whose
density ranged from 500 to 50000 cells ml^{-1}. The corresponding
increase in biomass, expressed as organic carbon calculated from the
relationships between cellular carbon and biovolume (Strathmann, 1967)
and chlorophyll and biovolume, takes the form:

$$C = C_0 . e^{0.135 t} \qquad (3)$$

t being the time (in days) for passage of the water and C_0 the carbon
content at the beginning. The daily rate of photosynthesis being
measured (Table 8.5) and the daily increase in biomass (P) being calculated from equation (3); the difference A - P represents losses due
to respiration and degradation. This has been calculated as about 80%
of the carbon incorporated each day by photosynthesis. In fact the
total loss daily reaches a mean of 1.22 ± 0.46 mg C mg chl a^{-1} h^{-1},
about 15 ± 3.1% of the total for optimum photosynthesis. In order to
oxidize the carbon lost in this way, the quantity of oxygen needed
reaches only 20 - 33% of the quantity actually consumed in the Lot. A
large part - over 50% - is thus used for degrading dissolved and organic
materials.

The scheme in Fig. 8.14 has served as the basis for the study of the
influence of different factors on the growth of the algae (Delclaux,
1980). The assimilation of carbon as a function of temperature and
light has been calculated by integrating results from various depths
according to equation (1). Respiration has been assumed to be constant
and estimated at 10% of the value for optimum photosynthesis. As for
the assimilation of nitrogen U_N and phosphorus U_P, this has been linked
with the concentration of nitrate and filtrable reactive phosphorus
through a Michaelis-Menten equation. Algal mortality has been estimated by the increase in the phaeophytin : chlorophyll a ratio with time.
Lastly, the regeneration of nitrogen and phosphorus to the dissolved
and particulate phases has been estimated as a function of the total
oxygen consumed, using the stoichiometric relationship C : N : P.

The model making the best use of the results obtained for the Lot
resembles those of Droop (1973), Lehman et al. (1975) and Jørgensen
(1976). The rate of phytoplankton growth is considered there as
dependent on the internal cellular concentration of nutritional elements

Table 8.5 Phytoplankton metabolism during passage of a water body over a 52 km distance during summer 1975.   (After Capblancq & Décamps, 1978)

time since the start (in days)	A rate of photosynthesis ($mg\ C\ m^{-3}\ d^{-1}$)	P increase in biomass over 24 h ($mg\ C\ m^{-3}\ d^{-1}$)	P/A	A − P respiration (converted to C equivalent) ($mg\ C\ m^{-3}$)	O nocturnal use of oxygen (converted to C equivalent) ($mg\ C\ m^{-3}\ h^{-1}$)	$\dfrac{A - P}{O} \times 100$ min.
0	296	51	0.17	245	–	–
1	136	74	0.54	62	13.2	20
2	644	89	0.14	555	78.1	30
3	670	100	0.15	570	–	–
4	575	113	0.20	462	81.0	24
5	390	108	0.28	282	36.0	33
6	518	129	0.25	389	83.0	20
7	676	159	0.23	517	65.5	33
8	919	188	0.20	731	–	–

(Q), which itself varies as function of cellular division and assimilation. Cellular division constitutes in effect a dilution factor and
assimilation is linked to the external and internal concentrations of
nutritional elements. The validity of this model has been tested on
eight species in culture (Dauta, 1982a, b) and in one stretch of the
Lot (Capblancq *et al*., 1982).

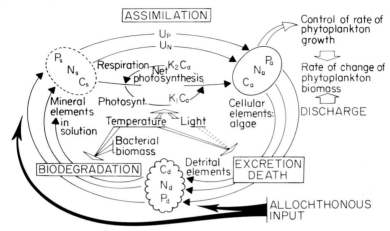

Fig. 8.14 Nutrient cycle in the Lot.   (For guide to symbols see text)

This model (Figs 8.15 - 8.17) provides an explanation of the development of the phytoplankton crop under apparently unfavourable conditions
of phosphate nutrition. By storing phosphate, the algae can multiply
for several days by using their reserves, thus confirming observations
made in batch culture for *Scenedesmus crassus* (Delclaux & Guerri, 1980;
Dauta *et al*., 1982). It can be seen that curves simulating growth of
the crop are very sensitive to the initial values of Q_N and Q_P, the
internal concentrations of nitrogen and phosphorus. In particular,
variations in Q_P are closely linked to those of environmental filtrable
reactive phosphorus.

The phytoplankton thus increases progressively in successive
stretches of the Lot when climatic conditions are favourable. During
its downstream passage the algal crop increases as a function of the
length of exposure to light, to the concentration of nutrient elements
available and to the importance of cellular reserves. The length of
exposure of the algae to light influences the net gain in carbon;
for any particular site, this depends on the relationship between the
depth of the euphotic zone (Z_{eu}) and that of the water column, more
shallow sites ($Z/Z_{eu} \leq 1$) favouring the growth of the algal crop. However,
the importance of cellular reserves must also be considered, notably
those of phosphorus which is apparently a limiting element. A large
proportion of the total phosphorus seems to be associated with complex
organic detritus (dissolved and particulate organic materials); this is

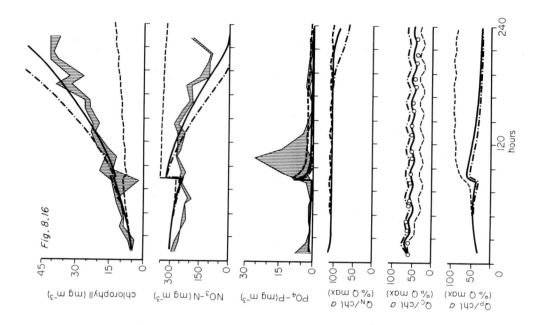

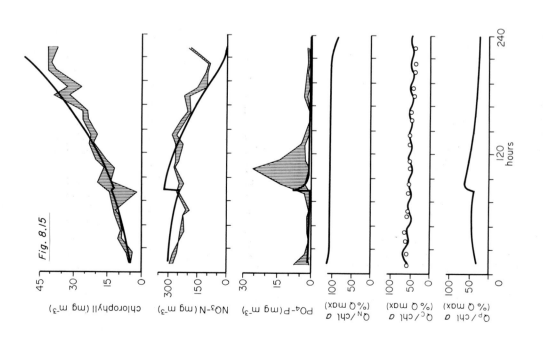

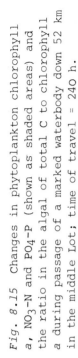

Fig. 8.15 Changes in phytoplankton chlorophyll a, NO₃-N and PO₄-P (shown as shaded areas) and the ratio in the algal of total C to chlorophyll a during passage of a marked waterbody down 52 km in the middle Lot; time of travel = 240 h.

The curves corresponding to the numerical simulation of changes in biomass, nitrate and phosphate and of contents of algal N (Q_N), C (Q_C) and P (Q_P) for the initial conditions were:

$Q_C : Q_N : Q_P = 60 : 8 : 1$ mg per mg chl a; $R/A_{opt} = 0.1$

Fig. 8.16 Results of simulations showing the importance of respiration of change in biomass:

R/A_{opt} = 0.05   (—·—·—)

0.1   (————)

0.2   (— — —)

Fig. 8.17 Results of simulations showing the influence of the initial chemical composition of the phytoplankton:

$Q_C : Q_N : Q_P$ = 30 : 8 : 4   (————)

60 : 8 : 1   (— — —)

60 : 2.5 : 0.4   (—·—·—)

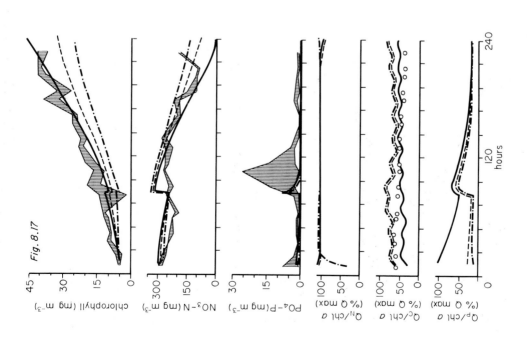

Fig. 8.17

therefore, less available for the algae. When the concentration of available dissolved phosphorus falls below a threshold value, with up-take no longer compensating for cellular reserves, algal multiplication continues at the expense of internal reserves. The exhaustion of these last leads progressively to cessation of growth and, with losses through mortality becoming preponderant, to a reduction in the popu-lation. This continues until a new source of nutrient elements - from urban effluents for example - permits cellular stocks to be recons-tituted.

8.6 FAUNA

The Lot contains various species of fish:

Carnivores: *Salmo trutta fario, Esox lucius, Micropterus salmoides, Lucioperca lucioperca, Perca fluviatilis, Lepomis gibbosus, Anguilla anguilla, Cottus gobio.*

Omnivores: *Ameiurus nebulosus, Alburnus alburnus, Leuciscus cephalus, Rutilus rutilus, Scardinius erythrophthalmus, Abramis brama, Blicca björkna.*

Bottom-living: *Gobio gobio, Leuciscus leuciscus, Barbus barbus, Phoxinus phoxinus, Cobitis barbatula, Chondrostoma nasus.*

Burrowing: *Cyprinus carpio, Tinca tinca, Chondrostoma toxostoma.*

Migratory: *Mugil* sp., *Petromyzon marinus, Lampetra planeri, Alosa alosa.*

The construction of large dams on the Haut-Lot and the Truyère, as well as the weirs in the middle and lower basin, have transformed the distribution of fish in the Lot (Tourenq & Dauba, 1978). Upstream the drop in water temperature has led to a rapid diminution of numerous white fish, with *Salmo trutta fario* replacing *Leuciscus leuciscus* as the dominant. Several species have been introduced into the reservoirs of the Haut-Lot and Truyère: *Cyprinus carpio, Tinca tinca, Rutilus rutilus, Esox lucius, Lucioperca lucioperca.* The weirs further down-stream have stopped the passage of migrators, in particular *Salmo salar* which, although abundant until the middle of the 18th century, has disappeared today.

In spite of these important changes, however, the fish population of the Lot is still abundant and the benthos, especially of chironomids and oligochaetes, relatively rich.

8.7 HYDRAULIC MANAGEMENT OF THE LOT

The studies made on the Lot have served to define the objectives for improving water quality, taking into account the diverse interests of tourism, agriculture, industry and the valley (Décamps, 1978). The demands on the water are varied and often antagonistic. It is essential to make sure that water resources are adequate in quantity and quality to cover present and future needs.

Nowadays a lack of water usually appears in July and August as a result of the increase in use, notably for irrigation (Table 8.6) and the management of reservoirs for hydroelectric purposes. These inadequate flows lead to slow passage of the water and, as a consequence, changes in ecological conditions. The residence time of the water increases, temperatures rise, phytoplankton increases and domestic and industrial pollution are concentrated. At the same time, hydroelectric requirements disturb the regularity of flows.

Table 8.6 Requirements for irrigation water in the valley of the Lot. Calculation based on estimate of 1500 m^3 ha^{-1}. (Anonymous, 1979)

	irrigated surface (ha)	water used (m^3)	maximum rate of pumping (m^3 s^{-1})
1979	11100	17×10^6	5.75
projected increase for following 10 years	8000	12×10^6	4.20
predicted total requirement	19100	29×10^6	9.95

The 1975 investigations revealed the changes likely to influence river communities: (i) floating wastes, (ii) inadequate flows, (iii) pulses in flow, (iv) pollution by domestic and industrial wastes, (v) turbidity due to suspended materials, (vi) cleanliness and accessibility of the banks. In fact, the most serious nuisance is the presence of excessive heavy metal loads below the entry of the Riou Mort.

These data have led to various objectives for water for drinking, industry and irrigation, compensation for water removed, fish life, environmental needs and tourism (Anonymous, 1976). A set of actions have been proposed: (i) combat industrial and domestic pollution, (ii) maintain flows by the creation of a reservoir ensuring that discharge does not drop below 12 m^3 s^{-1} at the Haut-Lot/Truyère confluence, (iii) regulate flows by a more rational management of hydroelectric uses, (iv) control the exploitation of gravel, (v) eliminate floating wastes, (vi) strengthen the banks.

This series of complementary actions was drawn up to reconcile the diverse uses of a river long domesticated by man.   This programme for the hydraulic management of the Lot is now being realized.

8.8   CONCLUSION

The hydroelectric management of the Lot has led to reduced discharges in a river already transformed by the needs of navigation.   These conditions have favoured the development of phytoplankton: algae produced in spring in the reservoirs of the Haut-Lot and the Truyère are continuously inoculated into the middle and lower sections of the Lot.   These algae proliferate when conditions of light and temperature are favourable.   Above all, low flows, which lead to slow passage of the water, favour the increase of algal populations.   These also assist the sedimentation of suspended materials.   In certain stretches a thermal stratification can even appear, with a reduction in dissolved oxygen at depth.

For practical purposes it is apparent that the improvement of flows is essential to improve the quality of the water.   The example of the Lot shows in effect that the mode of functioning of river and lake ecosystems can be very similar.   In the middle and lower sections of the Lot, more or less pronounced periods of stagnation are frequent in July, August and September;   they are accompanied by a simultaneous increase of detrital and phytoplankton organic matter.   The principal problem posed is thus to determine the minimum discharge that can maintain biological equilibrium in the river.   On what criterion should this be made?   Developments occurring in some stretches of the Lot during summer provide one approach: the minimum discharge permissible should be that which prevents the appearance of a stratification in temperature and oxygen in the stretches most threatened by eutrophication.   It still remains to determine precisely the geomorphological conditions which produce this phenomenon.

Acknowledgements

We thank A. Dauta for calculations made using the numerical simulation model.   The figures were produced by H. Casanova and A. Païs.

REFERENCES

Angelier E., Bordes J.M., Lucchetta J.C. & Rochard M. (1978)   Analyse statistique des paramètres physico-chimiques de la rivière Lot. *Annls Limnol.* 14, 39-57
Angelier E. & Décamps H. (1978)   Le fonctionnement des ecosystèmes dans les eaux aménagées. *Acad. Agricult. fr.* 741-48

Anonymous (1973) *Livre Blanc de la Vallée du Lot*, 74 pp. Association
    pour l'Aménagement de la Vallée du Lot.

Anonymous (1976) *Propositions pour un Schéma d'Aménagement Hydraulique
    de la Vallée du Lot*, 123 pp.   SODETEG.

Anonymous (1979) *Propositions d'Actions en Vue de l'intégration de la
    Vallée du Lot au Plan Décennal de Développement du Grand Sud-Quest*,
    31 pp. Mission Interdépartementale et Association pour l'Aménagement
    de la Vallée du Lot.

Bordes J.M., Lucchetta J.C. & Rochard M. (1973) Etude d'un écosysteme
    d'eau courante : le Lot. Thèse 3e cycle, Toulouse, 152 pp.

Capblancq J. & Dauta A. (1978) Phytoplancton et production primaire
    de la rivière Lot. *Annls Limnol.* 14, 85-112.

Capblancq J., Dauta A., Caussade B. & Décamps H. (1982) Variations
    journalières de la production du phytoplancton en rivière :
    modélisation d'un bief du Lot. *Annls Limnol.* 18, 101-32.

Capblancq J. & Décamps H. (1978) Dynamics of the phytoplankton in the
    river Lot. *Verh. int. Verein. theor. angew. Limnol.* 20, 1479-84

Capblancq J. & Tourenq J.N. (1978) Hydrochimie de la rivière Lot.
    Annls Limnol. 14, 25-37.

Casanova-Batut Th. (1977) Etude préliminaire en vue de la modélisation
    de l'écosysteme Lot. Thèse 3e cycle, Toulouse, 109 pp.

Dauta A. (1975a) Etude du phytoplancton du Lot. Thèse 3e cycle, Tou-
    louse, 107 pp.

Dauta A. (1975b) Etude du phytoplancton du Lot. *Annls Limnol.* 11,
    219-38.

Dauta A. (1982a) Conditions de développement du phytoplancton. Etude
    comparative du comportement de huit espèces en culture. I. Deter-
    mination des parametres de Croissance en fonction de la lumière
    et de la température. *Annls Limnol.* 18, 217-62.

Dauta A. (1982b) Conditions de développement du phytoplancton. Etude
    comparative du comportement de huit espèces en cultures. II. Rôle
    des nutriments: assimilation et stockage intracellulaire. *Annls
    Limnol.* 18, 263-92.

Dauta A., Brunel L. & Guerri M.M. (1982) Détermination expérimentale
    des parametres liés à l'assimilation de l'azote et du phosphore par
    Scenedesmus crassus. *Annls Limnol.* 18, 33-40.

Dauta A., Décamps H., Denat D., Galharague J., Giot D., Massio J.C. &
    Trolet J.L. (1975) Etude de la turbidité et des matières en
    suspension dans les eaux du Lot et de leurs relations
    avec les écoulements. 3. Evolution des matières en suspension et
    de la turbidité dans une masse d'eau marquée. Rapport B.R.G.M., 75
    S.G.N. 388 M.P.Y., 31 pp., 31 plates.

Décamps H. (1978) Qualité des eaux et développement de la vallée du
    Lot. *Annls Limnol.* 14, 163-79.

Décamps H. & Casanova-Batut Th. (1978) Les matières en suspension et
    la turbidité de l'eau dans la rivière Lot. *Annls Limnol.* 14, 59-84.

Décamps H., Capblancq J., Casanova H. & Tourenq J.N. (1979) Hydro-
    biology of some regulated rivers in the Southwest of France. In:
    Ward J.V. & Stanford J.A. (Eds) *The Ecology of Regulated Streams*,
    pp. 273-88. Plenum Press, N.Y.

Décamps H., Capblancq J., Casanova H., Dauta A., Laville H. & Tourenq J.N. (1981) Ecologie des rivières et developpement: l'expérience d'aménagement de la Vallée du Lot. In: Lefeuvre J.C., Long G. & Ricou G. (Eds) *Ecologie et Développement*, pp. 219-34.

Décamps H., Massio J.C. & Darcos J.C. (1976) Variations des teneurs en matières minérales et organiques transportées dans une rivière canalisée, le Lot. *Annls Limnol.* 12, 215-37.

Delclaux F. (1980) Production primaire en milieu thermiquement stratifié: modélisation et application au Lot. Thèse Docteur-Ingénieur, Toulouse, 177 pp.

Delclaux L. & Guerri M.M. (1980) Biocinetiques de la production primaire, étude expérimentale des fonctions de base de la production algale. Thèse 3e cycle, Toulouse

Denat D. (1977) Etude de la production primaire dans le réservoir de Cajarc (Lot). Thèse 3e cycle, Toulouse, 94 pp.

Droop M.R. (1973) Some thoughts on nutrient limitation on algae. *J. Phycol.* 9, 274-82.

Gagneur J. (1976) Etude des diptères du Lot et étude de la retenue de Cajarc. Thèse 3e cycle, Toulouse, 189 pp.

Galharague J. (1974) Exploitation des sables et graviers de la vallée du Lot. Rapport B.R.G.M., 74 S.G.N. 239 M.P.Y., 30 pp.

Galharague J. & Giot D. (1974) Etude des matières en suspension dans les eaux du Lot et de leurs relations avec les écoulements. 1e partie: résultats des études qualitatives. Rapport B.R.G.M., 74 S.G.N. 392 M.P.Y. 33 pp., 2 annexes, 8 plates.

Galharague J. & Giot D. (1975a) Etude des matières en suspension dans les eaux du Lot et de leurs relations avec les écoulements. 2e partie: résultats des études quantitatives. Rapport B.R.G.M., 75 S.G.N. 154 M.P.Y. 44 pp., 6 annexes, 21 plates.

Galharague J. & Trolet J.L. (1975b) Etude de la turbidité et des matières en suspension dans les eaux du Lot et de leurs relations avec les écoulements. 1. Suivi de la turbidité dans les eaux du Lot et liaison avec le débit. Rapport B.R.G.M., 75 S.G.N. 386 M.P.Y. 17 pp., 33 plates.

Galharague J., Giot D. & Trolet J.L. (1975c) Etude de la turbidité et des matières en suspension dans les eaux du Lot et de leurs relations avec les écoulements. 2. Evolution des profils en long de Saint-Parthem à Castelfranc à différents débits régularisés. Rapport B.R.G.M., 75 S.G.N. 387 M.P.Y. 12 pp., 16 plates.

Galharague J., Giot D. & Trolet J.L. (1975d) Etude de la turbidité et des matières en suspension dans les eaux du Lot et de leurs relations avec les écoulements. 4. Etude des éléments en traces dans les vases pendant l'été 1975. Rapport B.R.G.M., 75 S.G.N. 389 M.P.Y. 6 pp., 7 plates.

Jørgensen S.E. (1976) An eutrophication model for a lake. *Ecological Modelling* 4, 253-78.

Labat R., Roqueplo C., Ricard J.M., Lim P. & Burgat M. (1977) Actions écotoxicologiques de certains métaux (Cu, Zn, Pb, Cd) chez les poissons dulcaquicoles de la rivière Lot. *Annls Limnol.* 13, 191-207.

Laur C. (1975) Etude faunistique du Lot et étude du sédiment de la retenue de Cambeyrac (Truyère). Thèse 3e cycle, Toulouse, 122 pp.

Lehman J.T., Botkin D.B. & Likens G.E. (1975) The assumptions and
    rationales of a computer model of phytoplankton population dynamics.
    Limnol. Oceanogr. 20, 343-64
Massio J.C. (1976) Facteurs d'évolution des matières en suspension
    minérales et organiques dans les eaux du Lot. Thèse 3e cycle,
    Toulouse, 146 pp.
Say P.J. (1978) Le Riou-Mort, affluent du Lot pollué par métaux lourds.
    I. Etude préliminaire de la chimie et des algues benthiques. *Annls
    Limnol.* 14, 113-31
Strathmann R.N. (1967) Estimating the organic carbon content of phyto-
    plankton from cell volume or plasma volume. *Limnol. Oceanogr.* 12,
    411-18
Talling J.F. (1957) The phytoplankton population as a compound photo-
    synthetic system. *New Phytol.* 56, 133-49
Tourenq J.N., Capblancq J. & Casanova H. (1978) Bassin versant et
    hydrologie de la rivière Lot. *Annls Limnol.* 14, 9-23
Tourenq J.N. & Dauba F. (1978) Transformation de la faune des poissons
    dans la rivière Lot. *Annls Limnol.* 14, 133-38
Vollenweider R.A. (1965) Calculation models of photosynthesis depth
    curves and some implications regarding day rates estimates in
    primary production measurement. *Memorie Ist. ital. Idrobiol.*,
    Suppl. 18, 427-57

9 · Estaragne

P. LAVANDIER & H. DÉCAMPS

9.1   INTRODUCTION

The Estaragne shows the rigorous conditions of a torrential high mountain stream: a long period of snow cover, low temperatures, rapid
changes in flow. As it is situated for the large part above the forest
line, it receives light energy directly when not covered by snow. It
shelters a rich and diverse community of benthic invertebrates specialized for montane conditions. The Estaragne drainage basin is characterized also by having only a very slight human influence, a feature of
particular interest at a time when numerous mountain streams have
succumbed to pollution resulting from the development of tourism.

A range of studies have been devoted to the ecology of the Estaragne
and other high mountain streams in the Pyrenees (Angelier, 1980;
Berthélemy, 1966; Dawson, 1973; Décamps, 1967a,b, 1968; Décamps &
Elliott, 1972; Décamps & Lafont, 1974; Décamps & Lavandier, 1972;
Décamps & Pujol, 1977; Dumas, 1976; Elliott, 1973; Giani & Lavandier,
1977; Lavandier 1974, 1975, 1976, 1979, 1981, 1982a,b; Lavandier & Mur,
1974; Lavandier & Pujol, 1975; Laville & Lavandier, 1977; Thomas, 1970).
This chapter is based on the results of these studies.

Fig. 9.1 Upstream Estaragne.

9.2   GENERAL DATA

DIMENSIONS   LOCATION

  Area of catchment   6.5 km^2
  Length of stream   2.8 km   42° 48'N   0° 9'E
  Highest point in catchment 3006 m
  Drop in height   600 m

METEOROLOGY

  Snow covers the upper part of the stream for 6 to 8 months and the
  lower part for 4 to 5 months

LAND USE

  Population: none. Sheep grazing in summer. Tourist visits. Basin
  is included in the Parc National des Pyrénées.

PHYSICAL FEATURES OF STREAM

Discharge at selected points in summer 1972 (in $1 \ s^{-1}$):

	minimum	maximum
km 0.05	50	350
km 1.1	80	1050
km 2.8	90	1350

Seasonal changes in flow - for 1969-1971 mean daily flow at km 1.1:

  in winter, under the snow $20 \ 1 \ s^{-1}$
  at snow-melt $1200\text{-}1400 \ 1 \ s^{-1}$

Substratum - ranges from sheets or rocks to pebbles, with local pockets
of sand and under low-flow conditions.

Temperature - maximum between 1971 and 1973:

km 0.05	1.1	2.8
4.5 $^{\circ}$C	8 $^{\circ}$C	13 $^{\circ}$C

Fig. 9.2
Downstream Estaragne
at Lac Oredon

9.3   GEOGRAPHICAL, PHYSICAL AND CHEMICAL FEATURES

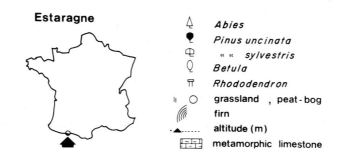

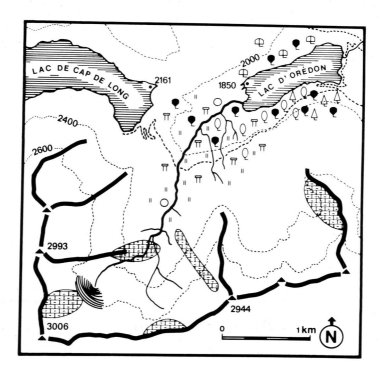

Fig. 9.3 The Estaragne and its catchment.

The Estaragne is part of the Neste d'Aure system and thus ultimately of the Garonne. It lies in the high mountains of the central Pyrenees. It rises at 2450 m and ends at 1850 m at Lac Oredon.   It is only 2.8 km long, with an average slope of 20%. The width of the stream is mostly between 1.5 and 5 m. Of the 6.5 km^2 catchment area, 75% lies above the 2200 contour (Fig. 9.3).

The basin consists largely of schists and granites, but a calcareous strip forms a break on the slope shortly after the source. The vegetation cover which is reduced by the source, increases progressively to form a turf of *Festuca eskia*, with *Rhododendron ferrugineum* and *Abies pectinata* in the lower part. The basin lies within the Pyrenees National Park and human influence is confined to the passage of walking tourists and to the pasturing of a few sheep.

The high altitude and orientation towards the north lead to a rigorous climate. Thus a permanent bank of consolidated snow (névé) feeds the source and snow covers the upper part of the stream for 6 - 8 months of the year and the lower part for 4 - 5 months. This snow can reach 5 m in depth towards 2200 m.

Between 1969 and 1971, discharge during winter was as low as 20 l s^{-1} under the snow and frequently exceeded 600 l s^{-1} during snow-melt (Fig. 9.4). In the absence of snow cover, rain has an immediate, often violent, effect on discharge; its influence on stream water level continues for several days.

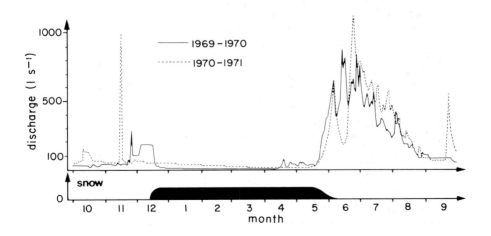

Fig. 9.4 Mean daily discharge of the Estaragne at 2100 m altitude over two hydrological years.

The chemistry of the Estaragne was studied regularly from 1971 to 1973. The mean values discussed here are taken from the paper by Lavandier & Mur (1974). Temperature (Fig. 9.5) is close to 0 °C throughout winter. During summer, it remains close to 5 °C at the

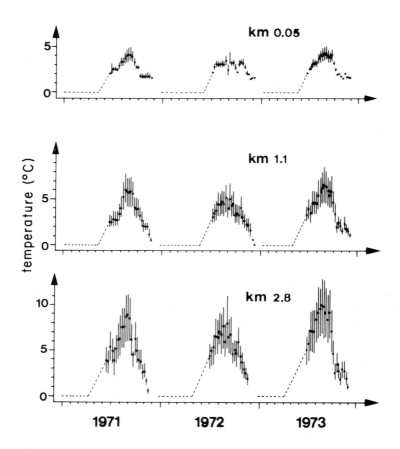

Fig. 9.5 Mean temperature and weekly range at three stations on the Estaragne.

source and rarely exceeds 12 °C downstream although daily variations can be more than 7 °C. Between 1971 and 1973 the number of degree-days above 0 °C ranged from 450 to 540 near the source and from 940 to 1160 at the mouth.

The concentration of dissolved oxygen is always near saturation. The ionic composition of the water is dominated by Ca^{2+} among the cations and by HCO_3^- and SO_4^{2-} among the anions. Conductivity drops steadily on passing from upstream to downstream (means from 114 µS cm^{-1} to 80 µS cm^{-1} in 1972). The total content of dissolved salts drops because of the input of less mineralized water as soon as the basin becomes granitic (means from 78.7 mg l^{-1} to 53.7 mg l^{-1} in 1972).

The content of most of the elements studied thus drops on passing down-stream (Table 9.1).

Table 9.1 Mean concentrations of water chemistry variables (mg l^{-1}) in the Estaragne during summer 1972.

distance from source (km)	Na	K	Mg	Ca	Fe total
0.05	0.48	0.34	0.37	22	0.140
2.8	0.49	0.30	0.31	15.5	0.015

distance from source (km)	HCO_3	NO_3-N	PO_4-P	SiO_2-Si	SO_4-S
0.05	57.5	0.188	0.004	1.20	2.2
2.8	42.9	0.171	0.0022	1.20	1.3

The concentrations of individual ions - except K^+ - are significantly correlated with flow. Concentrations of HCO_3^-, SiO_2, Ca^{2+}, Na^+, Mg^{2+} at all stations, and in addition SO_4^{2-} and NO_3-N near the source, vary inversely with flow. In contrast, PO_4-P at all stations, as well as SO_4^{2-}, NO_3-N and Fe well away from the source, vary directly with flow. Concentrations of Cl^- (of the order of 0.5 to 0.75 mg l^{-1}) appear to vary inversely with flow.

These correlations have made it possible to estimate the quantity of dissolved substances transported by the stream. For the hydrological year 1970-71, the volume of water which flowed was 7.9 x 10^6 m^3 (7.2 x 10^6 m^3 in 1969-70) and the total content of dissolved substances 497 tonnes (451 tonnes in 1969-70). About 65% of this load was trans-ported during snow-melt, 30% during the period without a snow cover and 5% in winter. When calculated per unit area, these values correspond to 765 kg ha^{-1} yr^{-1} in 1970-71 and 694 kg ha^{-1} yr^{-1} in 1969-70. Estimates for individual components for the hydrological year 1970-71 (kg ha^{-1} yr^{-1}) are:

	mean	confidence limits
HCO_3^-	487	(338 - 769)
Ca^{2+}	177	(123 - 246)
SO_4-S	17	(5 - 38)
SiO_2-Si	14.5	(10.7 - 21.5)
Na^+	5.4	(3.8 - 7.7)
Mg^{2+}	3.8	(2.3 - 5.4)
K^+	4.6	(1.5 - 7.7)
NO_3-N	2.3	(1.2 - 4.6)
Fe total	0.2	(0.15 - 0.4)
PO_4-P	0.03	(0.01 - 0.06)

9.4 VEGETATION

Right up to the melting snow the bed of the Estaragne is covered with
Hydrurus foetidus and diatoms, among which the dominants are *Diatoma
hiemale, Ceratoneis arcus, Meridion circulare* and *Cymbella ventricosa*.
Mosses, such as *Brachythecium rivulare, Cratoneuron commutatum* and
Hygrohypnum molle, are locally very abundant, especially in the tributaries. This aquatic vegetation plays a fundamental role, since it is
an endogenous source of food for herbivores and detritivores.

Hydrurus foetidus, the most abundant alga proliferates in the main
stream. During its growth, fragments of varying length are stripped by
the current, especially in August when the algal population is most
developed and in October when it completes its vegetative cycle. Part
of the aquatic vegetation of the Estaragne and its tributaries is thus
regularly moved by the current. However, after fragmentation, an
important part accumulates on stones or is deposited in pools. The
fine vegetable debris thus mixes with the mineral particles and is
progressively incorporated into the substratum.

In total, the plant biomass transported by the current in 1972 was
estimated to be:

> 80 kg fresh weight, 50 m from the source (2370 m)
> 210 kg fresh weight, 1.1 km from the source (2150 m)
> 270 kg fresh weight, 2.8 km from the source (1850 m)

At the two higher stations, in the subalpine meadow, the majority of
this biomass is endogenous in origin; at the third station, in the
Rhododendron - pine zone, it also includes exogenous leaves and woody
debris.

9.5 BENTHIC INVERTEBRATES

9.51 COMPOSITION AND DISTRIBUTION

The benthic community of the Estaragne includes nearly 200 species of
invertebrates (Lavandier, 1979). Ephemeroptera (5 species) and chironomids (60 species) form more than 80% of the total number of larvae
(Table 9.2). *Baetis alpinus* is by far the most abundant species (Table
9.2)*. Among the more than 600000 individuals collected to provide
samples representative of the various conditions in the stream, the two
dominant species make up over 50% of the population and the nine most
abundant species over 70%.

The temperature regime, length of snow cover and current speed all

---

*Nomenclature used is that of Illies (1978)

Table 9.2 Relative abundance of the various groups and main species in the benthic community of the Estaragne.

	no. of species	percentage
Groups:		
Ephemeroptera	5	47
Diptera Chironomidae	60	35.2
Plecoptera	28	6.7
Oligochaeta	17	4.7
Trichoptera	27	3.1
Planaria	2	1.3
Diptera Empididae	7	0.8
Diptera Simuliidae	8	0.6
other Diptera	17	0.4
other groups	24	0.2
total	195	100.0

Main species:		
Baetis alpinus	Ephemeroptera	39.6
Diamesa hamaticornis	Chironomidae	10
Rhithrogena loyolaea	Ephemeroptera	7
Eukiefferiella calvescens	Chironomidae	3
Micropsectra bidentata	Chironomidae	3
Trichodrilus macroporophorus	Oligochaeta	2.2
Orthocladius rivicola	Chironomidae	2.9
Corynoneura lobata	Chironomidae	1.6
Isoperla viridinervis	Plecoptera	1.5

determine the distribution of invertebrates in the Estaragne. The coldest upstream stations harbour a stenothermal high montane fauna, while downstream the fauna is more eurythermal. Species typical of middle or low montane regions appear only in the downstream Estaragne or in some tributaries where thermal and flow conditions are less severe. Thus, three tendencies can be recognized when interpreting species distribution.

(i) Certain species decrease in number progressively on passing downstream and disappear altogether at the lowest stations e.g. the stone-fly larva *Arcynopteryx compacta*, the caddis-fly larvae *Drusus rectus* and some chironomids mainly in the subfamily Diamesinae.

(ii) Certain species colonize the whole stream. These include the may-fly larvae *Baetis alpinus* and *Rhithrogena loyolaea*, the oligochaete *Trichodrilus macroporophorus*, the stone-fly larva *Isoperla viridinervis*, all species which are among the most abundant of the Estaragne benthic community (Table 9.2).   Others are the planarian *Crenobia alpina*, the caddis-fly *Drusus discolor* and chironomids of the subfamily Orthocladiinae.

(iii)   Species of the mid-montane zone increase progressively downstream: *Polycelis felina, Leuctra lamellosa, Allogamus auricollis, Nais alpina, Protonemura vandeli.*

Overall the fauna increases in diversity on passing downstream (Fig. 9.6).   Its density, everywhere around or above 10000 individuals m^{-2}, is greatest in the subalpine zone (about 2000 m).

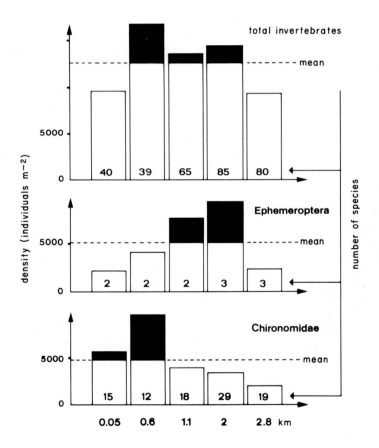

Fig. 9.6 Mean population density and number of species at five stations on Estaragne, showing total fauna, Ephemeroptera and Chironomidae. Parts of populations which exceed the mean for all stations are in black.

The adults play an essential role in the distribution of the benthic insects. Reproductive females provide a real mixing of the community within the modest dimensions of this hydrographical system. This distribution of egg-laying tends each year to enlarge the territory of the species: the young larvae invade the whole of the network, but then disappear at stations less suited for their development.

Sandy substrata, which form where the current is moderate during periods of low flow, but disturbed frequently when the flow increases, shelter only forms capable of burying themselves in the interstices without harm: small larvae, oligochaetes, chironomids, empidids. The mobility of the substratum prevents the development of algae and food is restricted to fine exogenous vegetable debris. At downstream sites some detritivores of larger size appear in pools where sand and vegetable debris accumulate. This is the situation with the (Plecoptera) Nemouridae.

The underside of pebbles and rocks subject to rapid currents form more populated biotopes and shelter the principal rheophilous invertebrates of the Estaragne. The stability of the substratum permits the growth of *Hydrurus foetidus* on various materials. Both intra- and inter-specific competition is severe, especially among carnivores. Thus territorial needs appear to limit the number of large individuals of *Perla grandis* to only one or two in the small cascades of one of the Estaragne tributaries. Similarly, certain sets of species, which have close ecological niches, tend to exclude each other during the course of their development e.g. the caddis larvae *Rhyacophila intermedia* and *R. evoluta* or the stone-fly larvae *Isoperla viridinervis* and *Arcynopteryx compacta*.

Altogether the population is diverse, and enriched in moderate to rapid currents. Of course the distribution - and consequently the interrelations between individuals - changes constantly in space and time as a function of the biology of the species and environmental conditions. Thus the majority of species become more rheophilous during growth. The fauna is however concentrated in the axis of the flowing water during the low winter flows under the snow, but when the cover breaks up, the animals shelter under rocks, in crevices or in the deeper layers of the stony bed. In each of these situations new relationships are established, with the formation of more or less ephemeral micro-communities.

9.52 AUTECOLOGY OF THE MOST ABUNDANT SPECIES

9.521 *Baetis alpinus*

This species colonizes the whole hydrographical network. The females lay their eggs indiscriminantly in the Estaragne and its tributaries, including zones unfavourable for the development of the species. After hatching the young larvae move in large numbers into the deeper layers of the stony bed. They then colonize various biotopes and tend to move towards faster currents during their growth; they are thus frequently transported downstream. At any given station the biomass lost in this way can exceed one-third of the annual production. These losses are compensated for by a high reproductive ability (700 - 2300 eggs per female) and by migration of reproductive females upstream.

The length of the life of a larva depends on local conditions of temperature and snow cover. At the lowest stations, the life cycle takes one year; the full-grown larvae are small (5 mm mean length) and carry a maximum of 1800 eggs. At the highest stations development takes two years; the larvae reach a mean length of 7 mm at emergence and can lay more than 2500 eggs.

In fact, over the whole of the Estaragne the number of larvae diminishes strongly during the course of the summer. Downstream, where the females are small, of 1000 eggs hatched in autumn, 100 larvae survive to the end of winter and about ten get as far as the instar preceeding emergence after one year of larval existence. Upstream, where females are larger, of 1000 eggs laid by one female in autumn, 30 larvae survive the end of the first winter and two to the instar preceding emergence after two years of larval life. Thus upstream population are maintained only thanks to the eggs deposited by females flying from further downstream. In the neighbourhood of the source, the population of *Baetis alpinus* can even depend exclusively on this input since the larvae appear unable to reach maturity here. '

9.522 *Rhithrogena loyolaea*

This, the other may-fly abundant in the Estaragne, has a longer period of larval development, namely two years in downstream and three years in upstream stretches. The mean length of mature larvae is 10 mm and there is no significant difference in size between upstream and down-stream larvae. Virtually all growth takes place during the period without snow cover. The population of *R. loyolaea* thus appears stable from one year to the next. Upstream, from a mean clutch of 4000 eggs laid, about one-third hatch (Humpesch & Elliott, 1980). 400 larvae survive after one year, 50 after two years and less than 10 reach the end of the last instar after three years. Drift losses are much less than with *Baetis alpinus* and represent only 15% of the annual larva production. This last is greatest in the middle region of the stream, at about 2100 m, where it ranges from 0.9 to 1.8 g m^{-2} yr^{-1}.

9.523 *Isoperla viridinervis*

This stone-fly develops over three years upstream and two years down-stream. Unlike the preceding species, it is essentially carnivorous, capturing for 9/10 of its life the larvae of chironomids and *Baetis*

alpinus. Its behaviour varies markedly during the course of the growth
of the animal. Young larvae move into the sediment, regularly ingesting
mineral particles and capturing diverse first instar larvae and small
oligochaetes. At an older stage the larvae of *I. viridinervis* become
very rheophilous and change their food. They capture larvae of
chironomids and *Baetis alpinus* on the *Hydrurus foetidus* or, temporarily
during the snow melt, larvae of *Rhithrogena loyolaea* which have a similar habitat.

Drift, predominantly nocturnal, affects all larval stages of *Isoperla
viridinervis*. The resulting loss of biomass can exceed one-fifth of
the annual production. In compensation the mature larvae tend to move
actively against the current, but it is mainly because of fertile
females flying upstream that the larval density is maintained at the
highest altitudes.

Each female lays about 600 eggs between June and September, depending
on altitude. Maximum numbers of larvae occur in September-October,
corresponding to when the young larvae climb to the bed of the flowing
water. The number decreases strongly during the period without snow,
especially in upstream populations.

The production of larvae is greatest (0.84 g m^{-2} yr^{-1}) at 2190 m
where the population is most abundant. Downstream populations are less
productive ($0.09 - 0.16$ g m^{-2} yr^{-1}), but, in spite of their lower
density, can produce as much or slightly more than those living near
the source (0.11 g m^{-2} yr^{-1}). The mature larvae, which essentially
determine production measurements, are in effect relatively more frequent downstream because of a less rapid loss of individuals (lower
mortality, shorter life cycle).

9.524 *Chironomids*

Several chironomids are among the most abundant species (Table 9.2).
They are much affected by the current. During the whole period lacking
a snow cover one can observe a larval drift, this being greatest in the
middle of the day. The extent of this phenomenon is augmented by the
constant stripping of *Hydrurus* filaments to which are attached numerous
Diamesinae larvae. For these last, the annual biomass lost in this way
can exceed one-fifth of the annual larval production.

The dynamics of the Diamesinae populations reflect those of the
dominant species, *Diamesa hamaticornis*, whose eggs, laid in autum mainly
hatch at the end of winter. The Orthocladiinae and the Tanytarsini, in
contrast, hatch in abundance from autumn onwards, the numbers of larvae
decreasing much more among the former than the latter during the early
instars. In fact some Diamesinae larvae and all the Orthocladiinae
larvae hatched in autumn disappear into basal water during winter;
they reappear in large numbers after the snow melt in the *Hydrurus*
tufts. It is probable that these larvae move into the interstices of

the substratum. The larvae of Tanytarsini, which are more rheophilous,
stay in the axis of the stream in winter where they remain vulnerable
to carnivores.

After the flood due to snow melt, the density of chironomid larvae
diminishes in an irregular fashion. This is accelerated, particularly
for the Diamesinae and the Orthocladiinae, by the drift of old *Hydrurus*
tufts at the end of summer. At the same time the larvae increase in
weight very rapidly, especially *Diamesa hamaticornis*. Altogether the
larval production of Diamesinae is 1.3 g m^{-2} yr^{-1} near the source,
1.5 g m^{-2} yr^{-1} at 2150 m and 0.5 g m^{-2} yr^{-1} downstream. This population constitutes the majority of chironomid production in the
Estaragne.

9.525 *Taxa other than insects*

Two groups of worm reach a noteworthy abundance, the oligochaetes and
the planarians (Table 9.2). The oligochaete populations appear quite
stable. The most abundant species, *Trichodrilus macroporophorus* is
almost alone in colonizing the most rapid and coldest waters. Downstream the higher temperatures and accumulation of vegetable debris
favour a variety of oligochaetes. As for the planarians, which are
more numerous towards the end of summer, they have frequently been
discovered to feed on adult females of *Baetis alpinus* after oviposition
and also on caddis larvae. For both the oligochaetes and the planarians,
production is higher downstream, where it reaches about 0.2 g m^{-2} yr^{-1}.

Two groups of benthic invertebrates of small size have not been
countered. *Hydracarina* occurs particularly in mosses in these high
altitude waters. The copepod *Bryocamptus zschokkei* is very abundant
in the under-flow and is certainly important as food for young insect
larvae.

9.53   POPULATION DYNAMICS

9.531 *Life cycles*

The majority of Estaragne species develop in one year. This, for
example, is the case with Chironomidae, Simuliidae and Leuctridae. It
is also the case with species which come to maturity only in lower parts
of the stream e.g. the caddis *Allogamus auricollis*. With other species
development takes one year downstream and two years upstream. One can
distinguish, for example in *Baetis alpinus*, two types of population
as a function of altitude; these can co-exist at the same station with
relative proportions differing from year to year as a result of climatic
conditions.

Various Estaragne species are characterized by a life cycle which is
always longer than one year. Some, as the caddis *Drusus rectus* and *D.*

discolor, have a two-year cycle over the whole stream system; others develop over two or three years, depending on the altitude. This is the situation with the four largest species: *Isoperla viridinervis, Rhyacophila intermedia, R. evoluta* and *Rhithrogena loyolaea*. The caddis *Micrasema morosum*, which lives in mosses of the cold tributaries, has the same type of life cycle.

Three other carnivorous stone-fly larvae always develop over three years. These are *Perla grandis* restricted to the warmest tributary, *Arcynopteryx compacta* living in the highest parts of the system and *Chloroperla breviata* present along the whole stream. Finally the detritrivore stone-fly larva, *Pachyleuctra benlocchi*, has a four-year cycle in the Estaragne, one of the longest known for an aquatic insect.

It should be pointed out that oligochaetes, planarians and aquatic beetles certainly survive after their reproductive period for several years.

9.532 *Growth and period of development*

Growth, which is negligible or nil in winter under the snow, takes place essentially during the period without snow cover. It is then approximately exponential. It is slow for many species as the snow melts, then accelerates with the summer rise in temperature. The growth rate of an individual is related principally to the mean temperature of its surroundings and the lower temperatures near the source are most often responsible for the lower growth rate here. The length of biological cycle is thus also tied to length of snow cover and to temperature.

9.533 *Loss of individuals*

During the period without snow cover, there are successive population explosions due to the rapid biological cycle and high reproductive ability. The greatest changes in numbers observed between two successive cohorts are in the populations with an annual cycle, like the chironomids (5-fold increase) and those resulting from females coming from downstream, like *Baetis alpinus* in the neighbourhood of the source (13-fold increase). For species with a longer life cycle, a more or less identical number of individuals reach the end of each cycle. These are the species which constitute the stable and permanent inhabitants.

Over the year, the reduction in numbers which is relatively slow in winter, reaches its maximum in the period without snow cover when downstream drift is important. It is difficult to assess the mortality of young larvae some of whom shelter in the interstices of the substratum after hatching. Further, the rate of hatching varies according to the station as a function of mean temperature (Elliott & Humpesch, 1980). As a consequence estimates made of the numbers of eggs carried by the female are 5 to 10 times higher than the numbers of young larvae estimated by direct sampling from the stream.

After the first instar is accomplished, quite a large proportion of the fauna always seems to come to the end of its growth. The lowest mortality rates are found among the caddis with an annual cycle, the highest among the may-fly larvae, but these last benefit from the high reproductive abilities of the species. For those species, the duration of whose life cycle is variable, the elongation of their development time reduces by a factor of 3 to 8 the proportion of individuals capable of completing their cycle.

9.534 *Movement of the benthos*

The benthos of the Estaragne gains periodically new habitats as a result of flow conditions under flood or snow melt. The majority of the larvae particularly mature specimens, maintain themselves between rocks, pebbles or under banks. Some move into the top ten centimetres of the stream bed. These are mainly larvae of the early instars, such as *Rhithrogena loyolaea* (30%), Orthocladiinae (20%), *Isoperla viridinervis*, *Baetis alpinus*, Diamesinae and Tanytarsinae. These species together with *Bryocamptus zschokkei*, are abundant in the under-flow.

During the year larval density in the underflow is greatest during the snow-melt; it decreases at the end of the flood and increases again towards winter when new generations hatch. The interstices of the substratum thus constitute a refuge from which the stream can be colonized.

Downstream larval drift is particularly pronounced in a mountain torrent such as the Estaragne. The proportion of the biomass moved being greater than that in calmer waters. Thus, at 2150 m, the biomass transported in one year (2 kg dry weight = approximately 7.5×10^6 individuals) is equivalent to the whole production from an area of 300 m^2; at this altitude this is 100 m length of stream. According to species, the surface area needed for the production of the biomass transported ranges from 30 to over 1000 m^2 (Lavandier, 1979):

	annual drift (g)	surface needed to produce equivalent biomass (m^2)
Orthocladiinae	6	30
Drusus rectus	80	130
Rhithrogena loyolaea	173	200
Isoperla viridinervis	240	320
Diamesinae	410	380
Drusus discolor	300	390
Baetis alpinus	470	500
Rhyacophila intermedia	150	1150

The last species is particularly vulnerable to drift at the end of the larval stage.

Overall the number of individuals in the drift reflects the composition of the benthic community; chironomids and may-fly larvae form more than 85% of the drift fauna, stone-fly and caddis larvae less than 3%. The likelihood of entering the drift varies according to group: Diamesinae and caddis are always relatively more frequent in the drift than in the benthos; Orthocladiinae, may-fly and stone-fly larvae are the converse, except for *Chloroperla breviata* which, at the end of its life cycle, drifts abundantly.

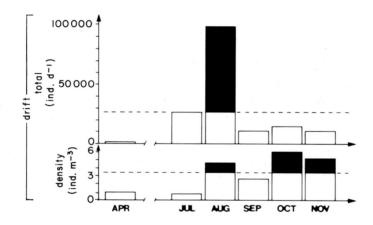

Fig. 9.7 Seasonal changes in total drift and drift density at 2150 m. Parts which exceed the mean for all the months shown are in black.

During the course of the year, drift of the most species is greatest at the end of the snow-melt flood, at the time of recolonization of the stream beds (Fig. 9.7). Another maximum can occur in autumn when the new eggs hatch and when *Hydrurus* breaks loose. During floods, especially in spring, all benthic species can be moved. At lower water, transport by the current is related more to the behaviour of the larvae, notably their reaction to changes in light flux; it is during this period that the density of drift individuals in the water is greatest (6 individuals m^{-3}).

The upstream movement of the larvae concerns only mature individuals and is important for only one species, *Isoperla viridinervis*. At the time of its maximum activity, the numbers moving upstream reach one-tenth of the number and one-quarter of the biomass of those in the drift. In contrast, the flight of reproductive females upstream plays an important role in compensating drift losses for species, particularly

Baetis alpinus and *Rhithrogena loyolaea*.

9.6 FUNCTIONING OF THE ECOSYSTEM

9.61 INVERTEBRATE FEEDING

The majority of Estaragne invertebrates feed on vegetation more or less altered, often associated with mineral particles. At 2000 m and above, feeding is based on *Hydrurus* and diatoms. Detritus of endogenous or exogenous origin augments these algae at lower stations and when the stream is covered by snow.

The principal primary feeders in the Estaragne are chironomid and ephemeropter larvae. Some species which are primarily herbivores like *Allogamus auricollis* and *Drusus* spp. can also capture invertebrates, whilst others, primarily carnivores like *Arcynopteryx compacta*, can feed on algae and detritus to an extent which is quantitatively impor- tant. Even the most markedly carnivorous species in the stream, *Isoperla viridinervis* and *Rhyacophila* spp., feed partly on *Hydrurus* at certain stages of their cycle.

Chironomid and ephemeropter larvae form the base for the nutrition of carnivores in the Estaragne, in varying proportions according to season. The following percentages of prey consumed at 2150 m give an idea of the differences observed between periods without and with snow cover.

prey	summer	winter
Chironomidae	70%	13%
Baetis alpinus	15	68
Rhithrogena loyolaea	10	17
others	5	2

9.62 INVERTEBRATE PRODUCTION

The mean annual production of benthic invertebrates was studied from 1971 to 1973 (Lavandier, 1979, see Table 9.3). It ranged from 4.9 g m^{-2} yr^{-1} at the source to 7.6 g m^{-2} yr^{-1} at the lowest station. This pro- duction depends upstream on species whose cycle last two or more years; below 2000 m more than 80% of production depends on species which are univoltine. This difference is not reflected in the values for the ratio P/$\bar{\text{B}}$, which always remains close to 4 (Table 9.3). The two stations located on tributaries (slightly above their confluence with the Estaragne) present a contrast with one another. The first, at 2150 m, has a low production; while being covered by snow too long for species living at lower altitudes, it reaches summer temperatures too high for high mountain species. The second tributary at 1920 m, is more productive than the lowest station on the main stream: conditions

ameliorated by its position make it a refuge for numerous middle montane
species.

Table 9.3 Mean biomass, production and the ratio $P/\bar{B}$ for all benthic invertebrates at five stations on the Estaragne and two on tributaries for the 1971-1973 study period. P, production (g yr^{-1});
B, mean biomass (g).

Estaragne

altitude (m)	2370	2190	2150	1920	1850
distance from source (km)	0.05	0.6	1.1	2	2.8
snow cover (month)	6-8	5-8	5-7	4-6	4-5
mean width (m)	1.5	2.5	4	4	5
max. temperature (^{o}C)	4.5^{o}	7°	8.5^{o}	11°	13°
mean biomass (g m^{-2})	1.18	1.67	1.63	1.55	1.92
production (g m^{-2} yr^{-1})	4.90	6.70	6.55	6.50	7.60
$P/\bar{B}$	4.2	4.0	4.0	4.2	4.0

Tributaries

altitude (m)	2150	1920
snow cover (month)	5-6	4-6
mean width (m)	0.7	0.9
max. temperature (^{o}C)	15°	17°
mean biomass (g m^{-2})	0.65	1.90
production (g m^{-2} yr^{-1})	3.30	8.10
$P/\bar{B}$	5.1	4.3

Benthic invertebrate production in the Estaragne over 1971-73 was
estimated to be 65.2 kg dry weight ha^{-1} yr^{-1} or 35 Kcal m^{-2} yr^{-1}. 36%
of this was due to the may-fly larvae *Baetis alpinus* and *Rhithrogena
loyolaea*, 31% to the caddis *Allogamus auricollis, Drusus* spp. and
Rhyacophila spp. and 16% to Diptera.   Production by carnivores was
22% of the total, a high percentage, which we believe reflects the
nature of energy transfer in the Estaragne community.

9.63   ENERGY FLOW

The flow of energy at three stations on the Estaragne (Fig. 9.8)
reflects the differences in the environment and benthic communities on
passing from upstream to downstream.

(i) The endogenous production of *Hydrurus* and diatoms, important

(A)

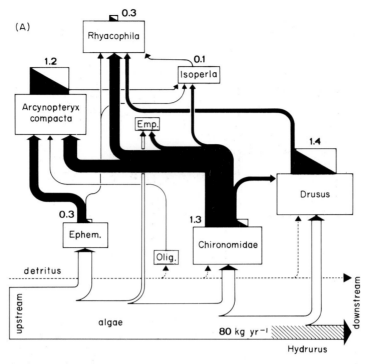

(B)

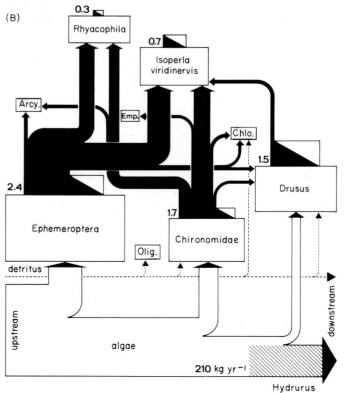

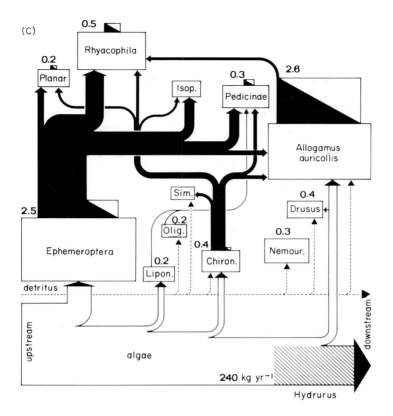

(C)

Fig. 9.8A Energy flux 0.05 km from the source. Taxa mentioned are those whose mean biomass is equal or more than 0.02 g m^{-2}; areas are in each case proportional to the mean biomass. Production is shown for values above 0.1 g m^{-2} yr^{-1}; flux width is proportional to the amount of production.

Rhyacophila	*Rh. contracta + Rh. evoluta*
Isoperla	*I. viridinervis*
Emp.	Empididae
Ephem.	*Baetis alpinus + Rhithrogena loyolaea*
Olig.	Oligochaeta
Drusus	*D. rectus + D. discolor*

Fig. 9.8B Energy flux km 1.1 from the source.

Arcy.	*Arcynopteryx compacta*
Chlo.	*Chloroperla breviata*

Fig. 9.8C Energy flux km 2.8 from the source.

Sim.	Simuliidae

throughout the course, is supplemented downstream by exogenous material.

(ii) The composition of the community changes at each trophic level. Among herbivores chironomids replace may-fly larvae, among omnivores *Allogamus auricollis* replaces *Drusus* spp.; among carnivores, *Arcynopteryx compacta*, which is dominant near the source, is gradually replaced by *Isoperla viridinervis*, which in turn diminishes, while *Rhyacophila* spp., planarians and Pedicinae become important.

(iii) At the same time as the number of species increases, the trophic web increases in complexity and diversification of the detritivores into micro- and macro-forms, leads to more niches being occupied, greater mean biomass and greater production (Table 9.4).

Table 9.4 Features of the benthic invertebrate community at three Estaragne stations for the 1971-1973 study period.

distance source (km)	0.05	1.1	2.8
number of species	40	65	80
diversity	2.96	2.78	2.85
evenness	0.80	0.67	0.65
mean biomass (g m^{-2})	1.18	1.63	1.92
production (g m^{-2} yr^{-1})	4.90	6.55	7.60
P/$\bar{B}$	4.2	4.0	4.0
mean carnivore biomass as fraction of mean total biomass	0.4	0.2	0.2
carnivore production as fraction of total production	0.4	0.2	0.2
proportion of species whose cycle is longer than one year	0.23	0.11	0.10

In the course of three kilometres benthic colonization of the Estaragne changes from one dominated by its proximity to consolidated snow to one typical of a mountain stream. However herbivores remain predominant among the primary producers and detritivores do not attain the production characteristic of streams fed by lakes or ones lower down among forest. Overall the Estaragne has the autotrophic nature of a mountain stream coming directly from consolidated snow (névé) situated in grassland above the tree zone.

9.7   GROWTH OF TROUT IN THE TORRENT OF ESTIBÈRE

As trout are practically absent in the Estaragne, their ecology has been studied in a neighbouring stream, the Estibère (Elliott, 1973;

Dumas, 1976). This stream is rather similar to the Estaragne in length
(3.5 km), altitude (2850 - 2450 m), mean slope (17%) and duration of
snow cover (7 months). It differs especially in its passage through a
series of lakes and by its S-E orientation (Décamps, 1967b). The
chemistry of the Estibère also differs from that of the Estaragne, the
former having a more granitic catchment. Values of 0.304 to 0.45 meq
l^{-1} has been recorded for the ionic content of the Estibère, as compared with 1.140 to 0.537 meq l^{-1} on passing down the Estaragne. Alkalinity (HCO_3^- approx. 20 mg l^{-1}) and Ca (approx. 6 mg l^{-1}) are both
lower than in the Estaragne (Table 9.1).

The trout population of the Estibère has been studied at 2040 m in a
zone where the slope is reduced considerably (around 2%). This population includes indigenous *Salmo trutta* and *S. gairdneri* introduced
between 1958 and 1960 in large numbers.

The *S. gairdneri* population of the Estibère is characterized by a
good density, but a biomass and productivity mediocre in comparison
with mean values obtained for *S. trutta* at less elevated altitudes
(Table 9.5). The structure of the Estibère population, with 41 and 60%
of alevins in two consecutive years, makes an estimate of recruitment
important. This last is essentially due to a high reproductive potential (3.2 eggs m^{-2}) and the absence of more mature competitors in zones
where the alevins live. In effect, after reproduction, the alevins
remain in shallow zones with a brisk current; they do not move into
deeper zones with a slow current until the beginning of their second
year. These deeper zones contain 96% of trout aged one year or more
and which are for the most part sedentary. It is in these zones in
contact with strongly undermined banks that the trout shelter during
winter.

Table 9.5 Comparison of features of *Salmo gairdneri* in the Estibère
(data of Dumas, 1976) with means for populations of *S. trutta* in
France.

features of the populations	unit	*Salmo gairdneri* in Estibère	*Salmo trutta* in France
biomass	kg ha^{-1}	50	123
density	(animals ha^{-1})	4300	3200
annual survival		31%	26%
length at 3 years	cm	16.5	23.5
net annual production (trout >18 cm)	kg ha^{-1} yr^{-1}	21	46.5

At the end of the first year, the annual rate of survival (31%) remains high as a consequence of the absence of disease, pollution and predation. However growth remains feeble, and in spite of very rapid growth in summer (1 cm per month), the length of 3-year old trout is only 16.5 cm and the net annual production is 65 kg ha^{-1} yr^{-1}. These low values are due to the rigorous climatic conditions of the Estibère: the water temperature is around 0.1 °C for seven months and growth is limited to summer months (Fig. 9.9).

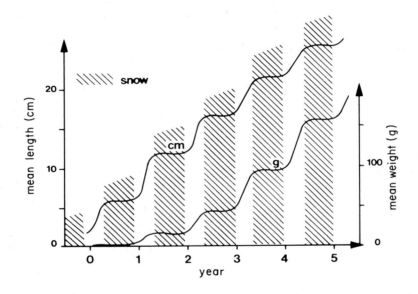

Fig. 9.9 Growth (length, weight) of *Salmo gairdneri* in the Estibère. The stream is free of snow-cover from June to October; reproduction of the trout takes place in July.

Studies have been made on the feeding by trout in the Estibère in relation to the abundance of invertebrate drift (Elliott, 1973). For three different mean temperatures (4.7 °C, 7.3 °C, 10.8 °C), the invertebrate populations in the drift and stomach contents of trout were analyzed over a 24-hour period. These showed that *S. gairdneri* and *S. trutta* were feeding principally on drift invertebrates. The two species showed a similar behaviour. At all three temperatures the principal feeding period took place in the first hours of the night, that is, when numerous benthic invertebrates were in the drift. However at 10.8 °C there was also a period of feeding during the day, at times when terrestrial invertebrates and emergent aquatic insects reach an important proportion of the drift.

In view of the abundance of invertebrates in the drift, it is probably not the availability of food but the temperature of the water which plays a major limiting role. In effect, the amount of feeding by the trout during one day was related to the mean temperature. This supplement is due to a larger stomach capacity of the trout, but especially to a more rapid gastric evacuation at a higher temperature. Using the exponential model for the rate of gastric evacuation as a function of temperature which has been established for *S. trutta* (Elliott, 1972), one can estimate the time after which no more than 15 - 10% of the stomach is filled; it is at this stage that *S. trutta* feeds once more. Applied to the observations for Estibère, these estimates show that trout with a full stomach can eat only once a day at mean temperature of 4.7 °C and 7.3 °C, but twice a day at a mean temperature of 10.8 °C. Consequently, as can be seen in Table 9.6, trout in the Estibère obtain sufficient energy for growth only when temperature conditions permit a sufficiently rapid gastric evacuation that they can feed twice a day.

Table 9.6 Energy budget for trout at three mean temperatures in Estibère, with energy value of food consumed (Q_c), energy of egestion and excretion ($Q_w = Q_c/5$), energy of resting metabolism (Q_r), additional energy for swimming activity (Q_a), and energy for growth (Q_g). (After Elliott, 1973)

mean temperature of water	gastric evacuation (h) [*]	energy budget	comments
4.7 °C	21 - 27	$Q_c - Q_w = Q_r$	one meal/day provides energy for Q_r only
7.3 °C	16 - 21	$Q_c - Q_w = Q_r + Q_a$	one meal/day provides energy for Q_r but non max Q_a
10.8 °C	11 - 13	$Q_c - Q_w = Q_r + Q_a + Q_g$	two meals/day provide energy for Q_r, Q_a and Q_g

[*]time at which 15% and 10% of the food is left in the stomach and therefore at which a new meal is possible: see text.

9.8 CONCLUSIONS

Situated in the high mountains, the Estaragne (and also the neighbouring Estibère) is cold and covered by snow for eight months of the year. Flows are extremely variable and the bed of the stream regularly disturbed by floods. The absence of trees along the banks reduces considerably input from exogenous materials.

Under such difficult conditions one might expect to find only a
sparse benthos, limited to a few species. This is far from the situation: the benthic invertebrate community of the Estaragne survives not
only by its diversity and complexity, but also by its permanence in
time.

How does this come to be? Two responses can be distinguished.

(i) Certain species show an annual life cycle; they come in successive waves and exploit to the maximum the available resources. Their
growth is rapid, their abundance can vary, but recruitment assures the
maintenance of populations in the more extreme conditions.

(ii) Other species show a life cycle extended to two or more years
when conditions become more rigorous. Several generations of the same
species can thus coexist, this permits a greater occupation of diverse
habitats and acts as a protection against environmental fluctuations.
Thus a sudden flood will cause especial harm only to the most active
age-class. Besides, the species are capable of entering into a condition of reduced metabolism at various instars of their larval development, these are adjusted permanently to the conditions in their environment and constitute the stable part of the population. The coexistence
of these two strategies confers a dynamic equilibrium in the benthic
community of the Estaragne.

Another important factor for survival in the Estaragne is the
spatial heterogeneity of the hydrographic system. Some tributaries
with temperature and flow conditions which are more clement serve as
refuges for the more fragile species. The lower part of the torrent
has the more diverse fauna. Adults of many species extend their habitat range by flying upstream to deposit their eggs. This results in a
mixing which ensures the unity of the community and the possible
occupation of all aquatic sites within the catchment.

The arrival of winter determines the rhythm of life in this community.
Each year, following the snow melt flood, the benthos once again under-
goes its development. Herbivores then occupy a predominant place because
of the abundant algal cover on the bed and the rarity of detritus of
exogenous origin.

In fact the Estaragne constitutes an ecosystem maintained in a
juvenile state by its ambient conditions. It is a fragile ecosystem,
with species often at the survival limit, as shown by the example of
trout in the neighbouring Estibère.

The high mountain torrents thus correspond with ecosystems whose
equilibrium is precarious. Protection turns out to be a delicate problem since it depends on the whole hydrographical system; any action at
one point in the system could have repercussions elsewhere in it.

Thus, because of the upstream flight of fertile females, destruction of downstream habitats could affect the colonization of stations close to the source.

Acknowledgements

We thank J.M. Elliott for helpful suggestions. The figures were drawn by H. Casanova.

REFERENCES

Angelier M.L. (1980) Recherches sur l'écologie des hydracariens dans les eaux courantes, 96 pp. Thèse 3e cycle, Toulouse.

Berthelemy C. (1966) Recherches écologiques et biogéographiques sur les plécoptères et coléoptères d'eau courante (*Hydraena* et Elminthidae) des Pyrénées. *Annls Limnol.* 2, 227-458.

Besch W.K., Backhaus D., Capblancq J. & Lavandier P. (1972) Données écologiques sur les algues benthiques de haute montagne dans les Pyrénées. I. Diatomées. *Annls Limnol.* 8, 103-18.

Cuinat R. (1971) Principaux caractères démographiques observés sur cinquante rivières à truites francaises. Influence de la pente et du calcium. *Annls Hydrobiol.* 2, 187-207.

Dawson F.H. (1973) Notes on the production of stream bryophytes in the high Pyrenees (France). *Annls Limnol.* 9, 231-40.

Décamps H. (1967a) Introduction à l'étude écologique des Trichoptères des Pyrénées. *Annls Limnol.* 3, 101-76.

Décamps H. (1967b) Ecologie des Trichoptères de la vallée d'Aure (Hautes-Pyrénées). *Annls Limnol.* 3, 399-577.

Décamps H. (1968) Vicariances écologiques chez les Trichoptères des Pyrénées. *Annls Limnol.* 4, 1-50.

Décamps H. & Elliott J.M. (1972) Influence de la mesure chimique du débit sur les invertébrés d'un ruisseau de montagne. *Annls Limnol.* 8, 217-22.

Décamps H. & Lafont M. (1974) Cycles vitaux et production des *Micrasema* Pyrénéennes dans les mousses d'eau courante (Trichoptera, Brachycentridae). *Annls Limnol.* 10, 1-32.

Décamps H. & Lavandier P. (1972) La faune du torrent d'Estaragne. In: (Eds) Kajak Z. & Hillbricht-Ilkowska A. *Productivity Problems of Freshwaters, Proceedings of the IBP-UNESCO Symposium on Productivity Problems of Freshwaters*, 918 pp.

Décamps H. & Pujol J.Y. (1977) Influences humaines sur le benthos d'un ruisseau de montagne dans les Pyrénées. *Bull. Ecol.* 8, 349-58.

Dumas J. (1976) Dynamique et sédentarité d'une population naturalisée de truite arc-en-ciel (*Salmo gairdneri* Richardson) dans un ruisseau de montagne, l'Estibère (Hautes-Pyrénées). *Annls Hydrobiol.* 7, 115-39.

Elliott J.M. (1972) Rates of gastric evacuation in brown trout, *Salmo trutta* L. *Freshwat. Biol.* 2, 1-18.

Elliott J.M. (1973) The food of brown and rainbow trout (*Salmo trutta* and *S. gairdneri*) in relation to the abundance of drifting invertebrates in a mountain stream. *Oecologia* 12, 329-47.

Elliott J.M. & Humpesch U.H. (1980) Eggs of Ephemeroptera. *Rep. Freshwat. Biol. Ass.* 48, 41-52.

Giani N. & Lavandier P. (1977) Les Oligochètes du torrent d'Estaragne (Pyrénées-Centrales). *Bull. Soc. Hist. Nat. Toulouse* 113, 234-43.

Humpesch U.H. & Elliott J.M. (1980) Effect of temperature on the hatching time of three *Rhithrogena* spp. (Ephemeroptera) from Austrian streams and an English stream and river. *J. Anim. Ecol.* 49, 643-61.

Illies J. (Ed.) (1978) *Limnofauna Europaea*, 2nd Edn. xviii + 532 pp. Gustav Fischer Verlag, Stuttgart.

Lavandier P. (1974) Ecologie d'un torrent pyrénéen de haute montagne. I. Caractéristiques physiques. *Annls Limnol.* 10, 173-210.

Lavandier P. (1975) Cycle biologique et production de *Capnioneura brachyptera* D. (Plécoptères) dans un ruisseau d'altitude des Pyrénées Centrales. *Annls Limnol.* 11, 145-56.

Lavandier P. (1976) Cycle vital de *Pachyleuctra benlocchi* (Navas) dans un ruisseau d'altitude des Pyrénées Centrales (Plecoptera). *Annls Limnol.* 12, 1-5.

Lavandier P. (1979) Ecologie d'un torrent pyrénéen de haute montagne : l'Estaragne, 532 pp. Thèse Doctorat d'Etat, Toulouse.

Lavandier P. (1981) Cycle biologique, croissance et production de *Rhithrogena loyolaea* Navas (Ephemeroptera) dens un torrent pyrénéen de haute montagne. *Annls Limnol.* 17, 163-79.

Lavandier P. (1982a) Evidence of upstream migration by female adults of *Baetis alpinus* Pict. (Ephemeroptera) at high altitude in the Pyrenees. *Annls Limnol.* 18, 55-59.

Lavandier P. (1982b) Developpement larvaire - Régime alimentaire, production d'*Isoperla viridinervis* Piclet (Plecoptera, Perlodidae) dans un torrent froid de haute montagne. *Annls Limnol.* 18, 301-18.

Lavandier P. & Mur C. (1974) Ecologie d'un torrent pyrénéen de haute montagne. II. Caractéristiques chimiques. *Annls Limnol.* 10, 275-309.

Lavandier P. & Pujol J.Y. (1975) Cycle biologique de *Drusus rectus* (Trichoptera) dans les Pyrénées Centrales : Influence de la température et de l'enneigement. *Annls Limnol.* 11, 255-62.

Laville H. & Lavandier P. (1977) Les Chironomides (Diptera) d'un torrent pyrénéen de haute montagne : l'Estaragne. *Annls Limnol.* 13, 57-81.

Thomas A.G.B. (1970) Taxonomie et répartition des Ephémèroptères et de quelques Diptères aquatiques (Tipulidae et Psychodidae) des Pyrénées, 105 pp. Thèse 3e cycle, Toulouse.

10 · Rhine

G. FRIEDRICH & D. MÜLLER

10.1   INTRODUCTION

The Rhine (Rhein) is a river of the superlative; there is surely no other stream about which so much has been spoken, sung and written. The recognition of its significance goes back to ancient Graeco-Roman times. As early as this, the Rhine was a navigation way of great importance, with an extensive chain of Roman and Teutonic settlements. Some of these have remained important cities to this day (Basel, Strasbourg, Koblenz, Köln). The Rhine is the only river which connects the Alps with the North Sea.

Thanks to its beautiful scenery and vineyard slopes, it is above all the region of the Middle Rhine whose praises poets have so often sung and that is now, as before, a favourite recreation and holiday region. During recent decades the Rhine basin has become one of the most important industrial agglomerations in Central Europe, which has led to the term "Rhine rail". With this recent development the river has acquired the role of a great supplier, providing drinking water for about 20 million people, as well as industrial and cooling water for Switzerland, France, Germany and the Netherlands. Rhine water contributes considerably to the sprinkling irrigation of special crops, of particular importance to the Netherlands. The Rhine is the world's most densely used inland navigation way with the busiest shipping traffic. Notwithstanding the extraordinarily high water quality requirements, the Rhine has to take up wastes to a much greater extent than most rivers. It is due to this immense burden that the river is also called "Europe's cloaca".

Due to its dominating role among the rivers of Central Europe, the Rhine was the object of intensive investigations as far back as the turn of the century, at the time of the first researches in hydrobiology and limnology. The works of Lauterborn (1905, 1916, 1917, 1918) are of particular interest, representing landmarks in the history of limnology. The river has since been investigated over and over again. The main emphasis has always been on determining and documenting the degree and load of pollution. A comprehensive limnological study on the Rhine, such as written recently on the Danube (Liepolt, 1967), is still awaited. As more recent and extensive descriptions on the Rhine the following may be mentioned: "Umweltprobleme des Rheins" (Environmental Problems of the Rhine : Bundesminister des Innern, 1976); this is a study of the Council of Experts. This work, prepared on the initiative of the Federal Minister of the Interior, lays the main emphasis on problems resulting from disturbances in the ecological equilibrium and the pollution by waste waters. The hydrology of the Rhine is presented comprehensively and discussed in "Das Rheinstromgebiet - Hydrologische Monographie" published by the Internationale Kommission für die Hydrologie des Rheins (1978). The regions of the Rhine between Basel and the German-Dutch frontier which have preserved their natural landscale to some degree, and as such are worth particular protection, have been listed by Solmsdorf et al. (1975); this work includes many maps.

Owing to the manifold and intensive utilization of the Rhine there
is a great number of national and international organizations created
specially for its protection and improvement. Here are some which are
especially important.

IKSR - Internationale Kommission zum Schutze des Rheins
       gegen Verunreinigungen (Member States: France, Germany,
       Luxembourg, The Netherlands, Switzerland).
IAWR - Internationale Arbeitsgemeinschaft der Wasserwerke im
       Rheineinzugsgebiet (International Working Group of the
       Waterworks in the Rhine Basin) with the participating
       working groups:

       AWBR - Arbeitsgemeinschaft der Wasserwerke Bodensee –
              Rhein (Working Group of Waterworks, Bodensee-Rhine)
       ARW - Deutsche Arbeitsgemeinschaft der Rheinwasserwerke
              (German Working Group of Rhine Waterworks)
       RIWA - Niederlandische Rijnkommissie Waterleidingbedrijven
              (Dutch Rhine Commission of Waterworks)

83 waterworks from five countries belong the IAWR. They supply 20
million people with drinking water and a considerable part of the
industry with service water.

Regular publications on the results of Rhine water analyses are
issued both by IAWR and IKSR. The German Working Group of Rhine Waterworks (ARW) presents additionally regular information on the most
essential requirements from the standpoint of water supply enterprises.
In addition, there have been special international agreements concluded
in order to protect the Rhine against the introduction of dangerous and
harmful substances (Rhine-Chemistry Agreement of the Neighbouring
States in 1976) and to mitigate the chloride load (Chloride Agreement
between France, Switzerland, Germany and the Netherlands in 1972).
The guidelines worked out by the European Community on the protection of
surface waters against the release of dangerous and toxic substances to
the Rhine may also contribute to the water quality improvements in the
river. This Guide includes a "black list" forbidding the release of
particularly dangerous matter to the rivers and lakes, as well as a
"grey list" of substances which ought to be avoided.

In the following presentation it is possible to deal with only a few
important aspects of the ecology of the river. The account is
restricted to the region between the Bodensee in the south and the
approach area of the Rhine-Delta in the north; that is essentially the
stretch flowing through the territory of the Federal Republic of
Germany. The Rhine downstream of the Dutch border is discussed in
Chapter 17.

10.2   PHYSIOGRAPHY AND HYDROLOGY

With an overall length of 1320 km, the river extends over 700 km from
south to north (Fig. 10.1). The source is in the Alps with its two
headstreams "Vorder Rhein" and "Hinter Rhein". The Vorder Rhein rises
from Lake Toma (2344 m); the Hinter Rhein has its source in the
"Paradies Glacier" (2216 m). The two headstreams which join at an
elevation of about 600 m, have already flowed 70 km (Vorder Rhein) and
60 km (Hinter Rhein), respectively. The river then goes north, under
the name "Alpine Rhine", towards the Bodensee. Before entering the
Bodensee, the Rhine slopes only slightly, which would result in considerable sedimentation if the river bed had not been narrowed by dams.
Sedimentation in fact takes place in the lake, with the mouth of the
river "growing" into the lake, so that Bregenz bay will get cut off
more and more from the rest of the Bodensee.

The origin of the whole Rhine basin dates back to the glacial epoch.
Accordingly, the Rhine in its present-day form can be regarded as
geologically young. The Bodensee, with its 542 km^2 surface area and
276 m depth, represents a huge balancing reservoir counteracting discharge fluctuations. In addition, the lake has a significant influence
on water quality. From the Bodensee the Rhine flows westwards and
breaks through the Jura Mountains, where it bears the name "Hochrhein".
The swift and turbulent flow in this stretch, which includes the Rhine
Falls near Schaffhausen with a drop of 24 m, indicate that erosion is
still dominant here. The Swiss river Aare flows into the "Hochrhein".
The Aare drains an even larger area than the Rhine at this point. At
Basel there is a northward change in direction; here the river enters
the Upper Rhine plain with a length of 300 km. The entry of the Neckar
and the Main subjects the Rhine to a heavy pollution load. The Neckar,
Main and the Upper Rhine have all been made navigable by weirs and
transformed into a chain of river impoundments (10.3, 11.33). In the
region of the Neckar and Main there are important settlement and
industrial agglomerations. The Nahe joins the northern end of the
Upper Rhine. This is the start of the Middle Rhine (see frontispiece),
where the river flows through the Rhenish Slate Mountains both westwards
and eastwards. At the narrowest place, with its famous Lorelei Cliff,
the whole valley is no more than 200 m wide. At the city of Koblenz,
two tributaries, the Lahn and Mosel, discharge into the Rhine. The
Mosel, with a basin of 28000 km^2, represents the largest tributary of
the Rhine. It is an important international waterway, transformed into
a chain of river impoundments. After passing the "Mittelgebirge", the
Rhine enters the lowlands at Bonn. The wide meanders of the Lower
Rhine have formed a "garland" of old branches here. The main tributaries of the Lower Rhine are the Wupper, Ruhr and Lippe. In addition,
the Rhine meets the Emscher, a river heavily loaded by wastes from the
industrial "Ruhr District".

The Rhine then flows with wide curves in a northerly direction to
the German-Dutch frontier. Immediately after leaving the frontier it
branches into three rivers, the Waal (Fig. 10.3), Neder-Rijn and

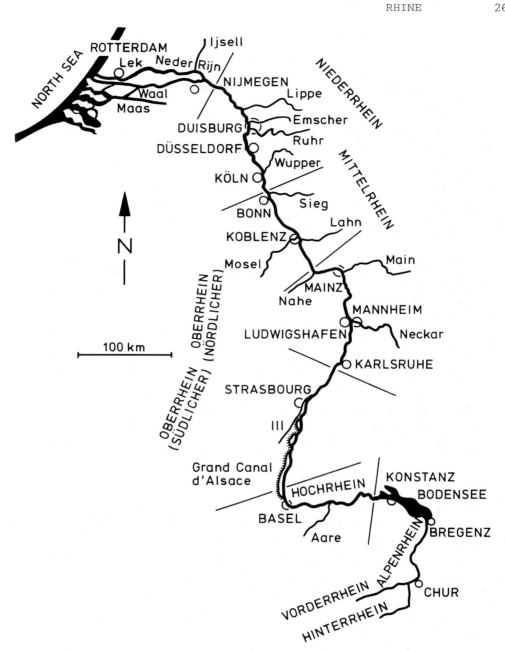

Fig. 10.1 Map of Rhine (Rhein), showing the Bodensee, the main towns and sections of the River.

Table 10.1 General data on Rhine catchment and river.

DIMENSIONS
    Area of catchment   185300 km^2
    Source of river   2344 m altitude
    Length of river (i) total (ii) in chapter   (i) 1320 km
                                                 (ii) 850 km to Dutch border

CLIMATE
Precipitation
    Alpine and Prealpine Region: summer wet, up to 3000 mm, mostly 1200 - 1800 mm,
        dry in winter. Above 1800 m more than 50% of precipitation as snow.
    Mountains and Upper Rhine valley: mountains wet in summer, 800 - 1200 mm.
    Upper Rhine valley, 400 - 800 mm.
    Lower Rhine typical maritime climate, 600 - 800 mm.

Temperature
    Upper Rhine valley is warm in summer and holds the German record for a maximum
        temperature (39.2 °C).

GEOLOGY
    The recent Rhine is young. Until the Oligocene the "Urrhein" flowed in the
older tectonic valley, the Aare belonged to the Rhone-system and the "Alpenrhein"
went to the Danube-system. The recent system of Rhine and tributaries formed during
the Pleistocene glaciation.

POPULATION
    Population and industry are concentrated along the "Rheinschiene" from Basel to
Rotterdam, with many large cities: Basel, Strasbourg, Karlsruhe, Ludwigshafen/
Mannheim, (Frankfurt am Main), Mainz, Koblenz, Bonn/Köln, Düsseldorf, Krefeld,
Duisburg, (Ruhrgebiet), Wesel.

LAND USE
    Above Bodensee: agriculture and forestry are dominant.
    Upper Rhine valley: intensive agriculture, wine, fruits, agriculture, forestry.
    Middle Rhine: agriculture, wine, forestry in the mountains.
    Lower Rhine: intensive agriculture.
    There is intensive industrial activity from Basel to the Lower Rhine and also
        intensive gravel mining in the flood beds of the Upper and Lower Rhine regions.

PHYSICAL FEATURES OF RIVER
    On-stream lakes and weirs
    Bodensee and three lakes in Switzerland (Aare-catchment). Several weirs
in the Upper Rhine: Grand Canal d'Alsace, Dam Gambsheim and Iffezheim.

    Discharge at selected points (based on data for 1931-1960)

                                        Kaub   Rees

    lowest ever recorded   482   590
    mean annual minimum   700   990
    mean   1560   2210
    mean annual maximum   4040   6660
    highest ever recorded   6150   9500

    Influence of regulation
    Marked in connection with landscape use, especially in Upper Rhine valley
and Lower Rhine, mainly to protect land against high floods.

    Substratum
    High Rhine   :   rocks, limestone
    Upper Rhine   :   gravel
    Middle Rhine :   rocks, gravel
    Lower Rhine   :   gravel, mud in breakwater fields

    Temperature (1971-1980):   selected points - from Deutsches
    Gewässerkundliches Jahrbuch (1983)

                    km   minimum   mean   maximum

    Walshut   102.8   1.2   10.6   23.2 °C
    Worms   443.4   2.7   13.7   26.2
    Köln   688.0   0.0   12.5   25.2
    Rees   837.4   1.8   12.8   26.2

Fig. 10.2 Rhine at Mannheim/Ludwigshafen, looking north. On left, Ludwigshafen with industrial plants; on right, Mannheim with port and mouth of Neckar. (Kommission für die Hydrologie des Rheingebietes; photo Landesbildstelle Rheinland-Pfalz)

Fig. 10.3   Bifurcation of the Rhine at Millingen and Pannderden. Left branch = Waal; right branch = Lek, Neder-Rijn (Ernst, 1972).

IJssel, which flow mostly in a western direction and form the so-called "Rhine Delta". The Rhine here comes into contact with the Maas river several times.

The hydrological divisions of the Rhine are shown in Table 10.2 and a detailed hydrological description can be found in a monograph of the Rhine Basin (Internationale Kommission für die Hydrologie des Rheingebietes, 1978). The hydrological regime of the Rhine is characterized by a well-balanced stream-flow. The ratio between lowest low water and highest high water amounts to 1 : 16 on the Upper Rhine (Karlsruhe) and 1 : 20 on the Lower Rhine (Emmerich) (Hess, 1959). This may be attributed to the superimposition of various effects. In summer there is much snow water from the Alpine region as well as runoff from precipitation. Runoff in the "Mittelgebirge" (uplands) is influenced more intensively by precipitation than in the Alpine region. A further balancing effect on the stream-flow regime is that due to the lakes in the Alpine foothills, primarily the Bodensee, with its large volume and annual water level fluctuations of about 2 m. Inflows from the uplands are particularly high in winter and spring. This leads to a more uniform discharge, especially in the Middle Rhine. On the upper Lower Rhine there are further considerable inflows from the uplands which increase the discharge maximum in spring.

Fig. 10.4 shows the changes in the flow regime of the Rhine in its course from Basel to Köln. It is transformed in this long course from a river with an Alpine stream-flow regime, i.e. high discharge in summer, into a river that is influenced considerably by runoff from the uplands. Fig. 10.5 shows the characteristic discharges of a long-term annual series.

Current velocities are less than 1 m s^{-1} on the entire "Hochrhein" and range approximately from 1 to 1.5 m s^{-1} on the free-flow reaches of the Upper Rhine and on the Lower Rhine. These values are not only higher with increasing discharge, but also fluctuate strongly from place to place. At sections with river impoundments speeds may drop down to 30 cm s^{-1} or less. Water movements caused by shipping exert an additional effect on the banks.

As a result of the well-balanced stream-flow regime, water depths remain relatively constant. The maximum depths on the Upper Rhine range at mean water level between 3 and 6 m, on the Middle Rhine between 4 and 7 m, and on the Lower Rhine between 5 and 10 m (Internationale Kommission für Hydrologie des Rheins, 1978). The mean depths over the whole river width are equal to about two-thirds of these river depths.

Table 10.2 Hydrological divisions of the Rhine (modified from Solmsdorf *et al.*, 1975)

main section	section	zone	boundary points	km	altitude (m)	slope in %
	Alpine Rhine		sources of Vorder Rhein and Hinter Rhein			
1 Alpine Rhine			entry to Bodensee		395	0.0
	Lake Rhine		Konstanz	0	395	0.04
			Stein	25	394	
	Hochrhein		Basel	170	244	1.0
	Upper Rhine	furcation zone	Karlsruhe	360	100	0.7
2 Upper Rhine		meander zone	Mainz	500	80	0.1
	Rhein		Bingen	530	77	0.1
		transverse valley	Koblenz	590	59	0.2
3 Middle Rhine	Middle Rhine	Neuwied basin	Andernach	615	52	0.3
		broadening valley	Bonn	655	45	0.1
4 Lower Rhine	Lower Rhine		Pannerden	870	10	0.1
	estuary, inclusive of branches in the Netherlands		North Sea	1000		0.1

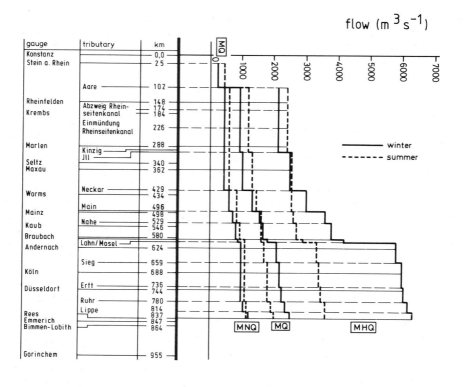

Fig. 10.4 Winter and summer discharge in Rhine for period 1936–1965 (Bundesminister des Innern, 1976). MHQ = mean of annual maxima; MQ = mean flow; MNQ = mean of annual minima.

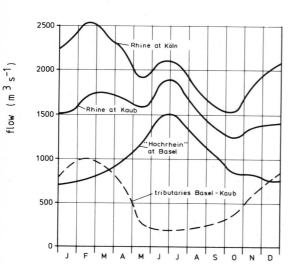

Fig. 10.5 Monthly discharge in Rhine and tributaries, based on long-term observations on the Basel–Kaub stretch (Deutsches Gewässerkundliches Jahrbuch, Rheingebiet, 1974).

10.3 WATER ENGINEERING MEASURES

Until the early years of the 19th century the Rhine was relatively free
from human interference. Where the stream was not forced by mountains
into a narrow channel, alluvial plans were formed, with a width of
several kilometres. These valley lowlands were either divided into
several river branches (furcations) or the river flowed through them in
large meanders. Extensive alluvial meadows existed mainly on the Alpine
Rhine, the Upper Rhine plain and the Lower Rhine. In these regions the
Rhine often changed its bed, devastating agriculture, forestry and
settlements near the stream. This could even happen on some partial
reaches where the "thalweg" (the line following the lowest part of the
riverbed, which is used as the downstream shipping route) was the
national frontier. Such changes in the channel dislocated the
nationality of important riparian places and handicapped shipping
traffic. Mosquito plagues occurred in the Upper Rhine plain, where the
lowland forests were flooded for long periods in summer. The relative
importance of towns and settlements was also influenced considerably in
the Lower Rhine region by natural displacements of the river course
(Hoppe, 1970).

Since the mid-19th century the Rhine and its tributaries have been
subjected to changes by engineering measures, which have partly led to
serious consequences in the ecological system. The most significant of
these interventions was the so-called Tulla Rhine rectification (1817-
1876), whose main objective was to extend the agricultural use of the
alluvial valley plain. The first stages in this project were to
concentrate the stream flow in a confined channel, cutting straight the
meander loops (Bensing, 1966). In the Upper Rhine plain these rectification measures caused an increase in current velocity which led to
intense bottom erosion, amounting at Basel to about 7 m and on the
Basel-Mannheim stretch to about 1.5 m (Schäfer, 1973). This has
resulted, especially in the southern part of the Upper Rhine plain, in
a considerable lowering of the ground-water table and in the withering
of some riparian forests. Thus, the original natural unity of the Rhine
river and the Rhine valley has become destroyed. Several old branches
and backwaters have ceased to exist. Suggestions for restoring the
landscape have come from Schäfer (1973), among others, and work is now
in progress.

Subsequent river training operations have dealt with flood control,
generation of hydroelectric power and shipping traffic. In the Swiss
river basin the retention capacity has been increased by constructing
reservoirs for hydro-power utilization and by diverting the Aare river
and one of its tributaries through three lakes in the Alpine foothills.
The Aare and the Hochrhein in the stretch between Schaffhausen and
Basel have been developed for hydro-power utilization and, as a result,
considerable changes in water depth and flow velocity have taken place.

In the stretch from Rheinfelden to Basel the Rhine has been developed
for shipping traffic using up-to-date ship units. In the section up to

Fig. 10.6 Upper Rhine at Philippburg and Speyer (in background).
The character of landscape is determined by the old Rhine branch of
the former meandering system with its alluvial forest and by the
regulated river.

Fig. 10.7 Backwater of the Gamsheim river dam downstream of Strasbourg. Structures from left to right: fixed transverse dike, power
stations, locks and the Ill River diverted into backwater of dam
(Internationale Kommission für die Hydrologie des Rheingebietes (1978).

Breisach there is in French territory a lateral canal with navigation
locks and hydro-power plants. In the old Rhine channel, the "Rest
Rhine", the minimum discharge amounts to not more than 15 to 30 m^3 s^{-1}.
In the section up to Strasbourg the shipping route and the main stream
flow run alternately in the old Rhine channel and in short lateral
canals equipped with hydro-power plants and navigation locks. Here, too,
there are extremely low flows in the sections of the "Rest Rhine". The
final stretch up to Rastatt has also been subjected to river impoundments. In order to maintain the ground-water level, to preserve the
remaining areas of natural lowland forests and to retain floodwater,
large-scale engineering operations have been carried out. Currently, a
project to establish large flood retention spaces is envisaged. Fig.
10.8 indicates the alterations in the Rhine channel over the
"Taubergiessen" area since 1827. This last region of well-preserved
lowland forest is to be placed under nature protection.

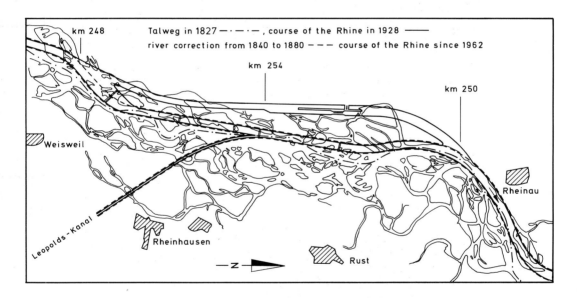

Fig. 10.8 Changes in the channel of the Rhine since 1827 in the
Taubergiessen region (near Freiburg).

In order to prevent scour in the river bed downstream of the last
dam, erosion activities are counterbalanced by addition of gravel from
riverside gravel pits ('bed load supply'). As compared with the
construction of further impounding weirs, this method has proved to be
particularly economical. Further, ecological effects resulting from
this technique are less than those entailed by dam construction.

On the Middle Rhine and the German Lower Rhine, the channel has been deepened in order to facilitate navigation. The resulting ecological effects are relatively insignificant compared with the engineering measures taken further upstream; they have led, nevertheless, to a considerable impoverishment of the fields and meadows. The Dutch section of the Lower Rhine and some of its river mouth branches have suffered in particular. The intensive gravel dredging operations in the Upper Rhine valley and on the Lower Rhine can be regarded as drastic interventions in the relations of water and environment. An area of 61 km^2, that is about 2.7% of the alluvial plain of the German Rhine, had been "used up" by excavations by 1973 (Solmsdorf *et al.*, 1975).

As a consequence of river impoundments on the southern reach of the Upper Rhine, flood waves proceed more rapidly nowadays. This means an augmented flood danger downstream of the development section (Karlsruhe, Mannheim, Worms). To reduce the flood danger again, new hydraulic engineering measures are needed. The aim is to make the probable maximum 200-year flood pass without damage, as was the case before intensive river development. This objective is to be achieved by deliberately inundating parts of the alluvial plains which had earlier served as flood retention areas. The total area to be subjected to artificial flooding extends over 75 m^2.

The floods to be attenuated by retention measures usually occur during the winter semi-annual periods. The flood-retaining areas (polders) will be inundated for several days in every single flooding operation. This is expected to favour the alluvial forest vegetation by deposition of sediment as well as a rise of the ground-water level beneficial to the formation of moist biotopes. Such control measures, however, are not expected to cause essential changes in ecological conditions in the river itself.

10.4 ECOLOGICAL ZONATION

10.41 INTRODUCTION

In conformity with its length and the multiform character of the land-scale through which it flows, the Rhine has a well-marked subdivision into biological zones (Lauterborn, 1916-1918; Knöpp, 1957; Schwoerbel, 1959). Whereas Lauterborn could report on a relatively natural biological zonation, Knöpp had to conclude that all sensitive species had gone from the Upper Rhine (in the region of Basel) down to the Lower Rhine (German-Dutch frontier). Only some few forms which are definitely tolerant to pollution have survived; a "forced" biological zonation has come into being, governed and determined mostly by the point-source effluents and the subsequent zones of self-purification.

10.42 PLANTS

10.421 Benthic algae

As with most rivers, there is no comprehensive account of the benthic algae. Besides the usual taxonomic difficulties, problems arise due to the use of rivers for navigation and due to the changing water level; in particular it is difficult to use artificial substrates for periphyton studies. The macroscopic species of the Hochrhein and Upper Rhine are known from the studies of Lauterborn (1910), Jaag (1938) and Krause (1976). Krause investigated the waters of the Upper Rhine valley with respect to the impoverishment of the landscape by river regulation and eutrophication. He concluded that originally there was probably a considerable diversity of aquatic environments and their algal communities. He recognised (p. 227) five types of community, based partly on modern observations made by him on the relatively few undisturbed waters.

(a) *Rivularia haematites* in waters of 20-30 cm depth and fast currents, but which generally lack water in winter. The water-courses are exposed to sunlight.

(b) *Hildenbrandia rivularis, Lithoderma fontanum, Cladophora glomerata* in moderately fast streaming water of 50 - 100 cm depth. The water-courses are shaded and rarely fall dry.

(c) *Batrachospermum ectocarpum, Draparnaldia glomerata, Bangia atropurpurea* in very fast currents under narrow bridges.

(d) *Chara hispida, C. aspera, C. contraria* in ponds coming from ground-water and with high light penetration, but nearly anaerobic conditions on the bottom.

(e) *Tolypella intricata* in pools drying out periodically.

Waters of types a - c have more or less disappeared due to anthropogenic causes. Nevertheless the rare brown alga *Pleurocladia lacustris* could be found by intensive collecting. The species grows in the so-called "Taubergiessen", in the last almost undisturbed rivulets of the Upper Rhine valley.

Only a few typical examples of the Characeae-ponds remain and the *Tolypella*-vegetation is much impoverished. In dammed waters nowadays, *Cladophora crispata, Hydrodictyon reticulatum* and *Enteromorpha* are visible signs of eutrophication. In the river bed itself *Hildenbrandia* and *Lithoderma* were conspicuous till 1961. The flat incrusting layers of these algae have been overgrown, especially by the diatom *Melosira varians*. *Thorea ramosissima*, in earlier times already a rare species, is known from recent times only from a restricted area.

According to Lange-Bertalot (in press), *Ulothrix zonata* is one of the dominant benthic algae of the Upper and Middle Rhine. *Bangia atropurpurea* is rare, but Chantransia- forms are frequent down to as far as the Lower Rhine. *Schizomeris leibleinii* occurs in the splash zone from the middle to the upper Lower Rhine (Lange-Bertalot, in press) and also in the lower part of the Lower Rhine (G.F., unpublished). After construction of some sewage plants, the species disappeared, though it is uncertain whether or not this was due to the improved water quality. Lange-Bertalot & Lorbach (1979) investigated diatom communities from the mouth of the Aare to Bonn. Benthic diatoms often formed massive growths, mainly widespread species of *Navicula, Gomphonema, Cymbella* etc. The use of glass slides exposed in the Lower Rhine has shown that the incrusting algae *Protoderma* sp. and *Sphaerobotrys fluviatilis* together with *Stigeoclonium* sp., are abundant; *Chamaesiphon incrustans* and the iron bacterium *Siderocapsa treubii* were frequent as well (Schiller, pers. comm.).

10.422 *Macrophytes*

The macrophytic vegetation of the Rhine cannot be studied properly without taking the whole alluvial valley into consideration. It was predominantly in the old branches of the Upper and Lower Rhine plain and in their surrounding bottom-land regions that large populations of helophytes and hydrophytes developed, greatly dependent upon the river. As a result of the extensive river regulation, many species have been destroyed and others are threatened by silting. Solmsdorf, Lohmeyer and Mrass (1975) investigated the whole of the Rhine between the Bodensee and the German-Dutch frontier and divided the present-day alluvial valley, which is characterized by embankments, into three main zones:

> river bed with the islands
> flood zone (dike foreshore) with intact (i.e. periodically
>   flooded) old branches
> flood-free zone with old branches tending to silt up.

As mentioned in 10.2, floods occur on the Hochrhein in June/July and November and on the Lower Rhine mainly in February and somewhat less intensively in July. In the flood zone the high water stages may exceed the mean water stages by 7 metres or more. It was in this zone that the large riparian forests once grew.

A vegetation close to the natural state, and still dependent on the Rhine today, can be found only on the so-called "Restrhein" stretch of the Upper Rhine, on some islands in the Rheingau and on the Middle Rhine, as well as in a few places scattered along the whole river. Thanks to the recognition of nature reserves, a number of old river branches with natural vegetation are still preserved, though even these are endangered in many places, as on the Lower Rhine, by sludge accumulation or by the hazards of being trampled or bitten by cattle.

The alluvial meadow is of decisive importance as living space for the animal world as well. The old Rhine branches which remain are therefore indispensable biotopes in every respect. Two such regions are the Taubergiessen in the Upper Rhine plain and the Xanthen old river branch on the Lower Rhine. Besides these wetlands of international importance on the Rhine are the regions around Bodensee Untersee, the Upper Rhine area between Lörrach and Kehl, the Rhine meadows at Bingen-Erbach and the lower Lower Rhine region from Bislich to the frontier (Solmsdorf *et al.*, 1975).

Hochrhein and Upper Rhine

According to the early studies of the Upper Rhine by Lauterborn (1910, 1917), the "flowing" Rhine was almost free from phanerogams. Only aquatic bryophytes, predominantly *Fissidens grandifrons* in the Hochrhein and *Fontinalis antipyretica* in the Upper Rhine, were present in the clear and cool water, forming luxuriant growths. It was not before the Mainz region that *Potamogeton* (*P. pectinatus* and *P. perfoliatus*) also appeared. Koch (1926) and Kummer (1934) referred to the occurrence of *Ranunculus fluitans*. Since 1970 this weed has become a great problem, particularly upstream and downstream of the Schaffhausen Falls and led to a total overgrowth in some reaches of the Rhine. As a result of excessive weed growth the river bottom once covered by gravel has silted up very rapidly. To make it worse, the abundance of aquatic weeds has caused water level rises on the Upper Rhine of up to 30 cm. In one quite short stretch of the Swiss Rhine, 1000 tonnes of *R. fluitans* was removed in 1974 with a submersible mowing unit within 4 weeks, in order to prevent operation troubles at a down-stream water power plant (Knecht, 1982). According to Thomas (1975), this mass growth is due to aqueous phosphate exceeding a threhold of about 0.07 mg l^{-1} PO_4-P.

Of particular scenic attraction are the abandoned river courses in the Upper Rhine region with their diverse vegetation. Lauterborn had already described them enthusiastically in his reports as sites of botanical treasures. In spite of considerable losses, due to river training and deep erosion as a result of desiccation, and notwith-standing the filling up of old beds, there are still large areas with remarkable aquatic and alluvial vegetation (Philippi, 1973, 1977, 1978a, 1978b). Since the beginning of the century species which have been lost include a large number of *Potamogeton* and Characeae species e.g. *Tolypellopsis stelligera*. Other species possess an extremely limited ecological amplitude and may, therefore, easily disappear as a consequence of even relatively slight disturbances e.g. *Naias marina, N. minor, Trapa natans, Nymphoides peltata* and *Salvinia natans*. These still exist today, but usually in very restricted areas.

As stated by Philippi (1978, p.132), in just the Upper Rhine region of Baden fifty species of aquatic plants may now be considered as extinct or threatened. He recognised the following categories:

extinct or disappeared
> e.g. *Marsilea quadrifolia, Oenanthe fluviatilis, Stratiotes aloides*

acutely endangered
> e.g. *Salvinia natans, Potamogeton coloratus, P. gramineus, Utricularia bremii*

endangered
> e.g. *Potamogeton densus, P. acutifolius, P. angustifolius, Hydrocharis morsus-ranae, Sparganium minimum, Nymphaea alba, Trapa natans* (acutely endangered?), *Hottonia palustris.*

In contrast to the losses resulting from pollution, development of the Rhine and construction of gravel pits, the following aquatics have increased:
> *Potamogeton pectinatus, P. crispus, P. perfoliatus, P. lucens* (?), *Elodea nuttallii, Lemna gibba, Ceratophyllum demersum, Myriophyllum spicatum, Utricularia neglecta* (?). (*E. nuttallii* is frequent in Upper Rhine and Nordrhein-Westfalia, but *E. ernstiae* grows in restricted places where water does not freeze: G. Philippi, pers. comm.)

Distributional data have also been obtained for *Phragmites*, whose retrogression is due primarily to water engineering measures. Under particularly favourable conditions, man-made waters such as gravel pits may become replacement biotopes for plants in danger, as shown by Krause (1975) for some *Chara* species, different Vaucheriaceae and *Myricaria germanica*. Favourable conditions could be created in many places by using suitable materials to cover the bottom and edges of such ponds.

Middle Rhine

As a result of its confined river bed, the Middle Rhine has apparently always been poor in aquatic vegetation; Lauterborn's description (1918) includes only some species of aquatic bryophyte. The literature on the river flora also includes records for *Butomus umbellatus, Potamogeton compressus, P. densus, P. pectinatus, Ranunculus fluitans* and *Zannichellia palustris*. *Butomus* and *Potamogeton pectinatus* have been found in recent times (A. Krause, pers. comm.). If the construction of the river bank were to leave a few gaps above the line of the mean water level of the river, a rich perennial vegetation would be likely to develop, even under present-day conditions; this might contribute, even if only moderately, to the ecological regeneration of the river basin (Krause, 1981).

Lower Rhine

In the region of the Lower Rhine, the river bends which dry periodically, the oxbows and backwaters, are all habitats with a rich flora and vegetation. In addition, dry areas between the bottom of the Rhine-valley and the lower terrasse of the slopes are well known habitats for so-called "Stromtalpflanzen", thermophilic species which extend north only in the Rhine valley e.g. *Koeleria pyramidata, Poa bulbosa, Cynodon* dactylon, *Thalictrum minus*, *Silene conica*, *Holosteum umbellatum*, *Isatis tinctoria*, *Diplotaxis tenuifolia*, *Brassica nigra*, *Ononis spinosa*, *Euphorbia segueriana*, *Eryngium campestre* and *Peucedanum chabraei*. *Cucubalus baccifer*, a liane relative abundant in the south, can also be estimated as a "Stromtalpflanze" in the Lower Rhine region as well (Burckhardt & Burgsdorf, 1964, 1965; Knörzer, 1960, 1963).

The forest stands rich in elm trees, which were once a significant part of the vegetation, have been almost totally destroyed; the rare persistant elms are now endangered by disease. Endangered as well are the thermophilic grasslands, because of the growing intensity of agriculture and camping activities.

Backwaters, oxbows and "Kolke" of the Lower Rhine, especially in the lowest part near Wesel, contain some well developed stands of reeds and other macrophytic vegetation rich in very rare and endangered species. *Nymphoides peltata* is a typical example which lives only directly by the river. These waters beside the Rhine are endangered especially by cattle and fertilisers and by the smaller water bodies being filled in deliberately. The waters may also be filled by intensive sludge growth, an effect of eutrophication (Burckhardt & Burgsdorf, 1964, 1965; Hild, 1960, 1968; Hild & Rehnelt, 1965a, 1965b; Solmsdorf *et al.*, 1975).

Of special interest are the plains of sand and gravel between middle and low water, which regularly fall dry in summer. This is a zone where a scattered vegetation of the Polygono-Chenopodietum-Association grows, rich in nitrophilous and thermophilic species, the most abundant of which belong to *Polygonum*, *Chenopodium* and *Bidens*. This association is also found in other big rivers of central Europe, which have a changing water level, like the Elbe and Weser (Lohmeyer, 1970a, 1970b).

In harbours and their vicinity there can occur a lot of adventive plant species imported with shipping and goods from overseas e.g. *Solanum nititibacatum*, *Abutilon theophrasti* and *Ricinus communis*. Tomatoes (*Lycopersicon esculentum*) and thorn-apple (*Datura stramonium*) are regularly to be found (Stieglitz, 1980). On sludge, which dries in summer but remains wet for the rest of the year, stands of a Nanocyperion-vegetation grow more or less regularly, especially *Limosella aquatica* and *Potentilla supina*. Where the sludge is very fine and remains wet all the time, the vegetation closest to the water develops a green cover of the macrophytic alga *Botrydium granulatum*.

10.423 Phytoplankton

In a river like the Rhine, flowing through several lakes and impoundment zones, a correspondingly great variety in plankton species is to be expected. Yet, relatively few studies have been made of the plankton of the river (Lauterborn, 1910; Kolkwitz, 1912; Seeler, 1938; Czernin-Chudenitz, 1958).

In any case, the question often discussed in earlier times, whether
the Rhine plankton represents a real "potamoplankton" or not, can be
seen today to be merely a dispute about words. Eutrophication of the
river has led to so great a phytoplankton growth that even over the
relatively short stretch from Bonn to Emmerich a considerable increase
in chlorophyll can be measured, which is not likely to be due exclusively to the plankton-rich affluents. Table 10.3 gives information
on maximum phytoplankton densities since 1912. The phytoplankton plays
a very important role in the ecosystem of the Rhine, as already pointed
out impressively by Knöpp (1957). On the other hand, zooplankton has
never occurred in the stream in significant amounts. Both the earlier
and the more recent studies indicate a zooplankton density, which for
Crustacea, is far below one individual per litre (Seeler, 1938; Seibt,
1980) and for Rotatoria less than 500 individuals per litre. The
following description is therefore restricted to phytoplankton.

Table 10.3 Maximum standing crop of phytoplankton in Rhine.

section	date	individuals $(\times\ 10^3\ 1^{-1})$	month	reference
Hochrhein	1912	22–23	August	Kolkwitz 1912
	1955	1800	March	Czernin-Chudenitz 1958
Upper	1912	44	August	Kolkwitz 1912
Rhine	1936	500	September	Seeler 1936
	1955	1800	March	Czernin-Chudenitz 1958
	1974	11000–15000	April	Backhaus & Kemball 1978
Lower	1936	4000	September	Seeler 1936
Rhine	1955	2000	March	Czernin-Chudenitz 1958
	1970/74	14000	August	Heuss 1975
	1979	59000	August	G. F. (unpublished)
	1980	46000	May	" "
	1981	56000	April	" "
	1982	63200	May	" "

Hochrhein and Upper Rhine

Until the late 1930s the phytoplankton growth and occurrence in the
Hochrhein and Upper Rhine had been governed by the plankton of the
great lakes through which the Rhine and Aare rivers flow. This
plankton consisted almost exclusively of diatoms, chrysophytes and
cryptomonads. Abundant species were: *Cyclotella bodanica, C. melosi-
roides, C. socialis, C. comta,* cryptomonads, *Ceratium hirundinella,
Asterionella formosa, Fragilaria crotonensis, Diatoma elongatum,
Synedra acus* including var. *angustissima, S. delicatissima, Dinobryon
sertularia* and *Sphaerocystis schroeteri; Oscillatoria rubescens,
Tabellaria fenestrata* and *Melosira islandica* var. *helvetica* had come
into the Rhine from the Zürichsee through the R. Aare (Lauterborn,
1917; Backhaus & Kemball, 1978).

The first investigations at the beginning of the 20th century
(Marsson, 1907-1911; Kolkwitz, 1912) showed very low cell numbers with
less than 100 cells ml^{-1}. Seeler (1936) had already found cell
numbers up to 1000 ml^{-1} in the Upper Rhine and distinct indications of
pollution in form of *Sphaerotilus* flocs drifting with the current. At
that time, however, such pollution loads had been local phenomena,
counterbalanced again by the self-purification of the stream. By 1941
Fragilaria crotonensis and *Asterionella* were largely displaced in the
Hochrhein by *Tabellaria fenestrata*. In addition, there was an increase
in the following diatoms: *Diatoma vulgare, Nitzschia acicularis,
Synedra ulna* var. *oxyrhynchus, Melosira varians, M. islandica* ssp.
helvetica and *Fragilaria crotonensis*. These results were confirmed by
Czernin-Chudenitz (1958).

The Upper Rhine had earlier shown a plankton greatly similar to that
of the Hochrhein both in composition and density. It was only in the
old branches that plankton could develop to high densities. That is
why these waters were of vital importance to the fishery. At km 173
(approximately Franco-Swiss border) Czernin-Chudenitz (1958) found
preponderantly *Melosira varians, Nitzschia acicularis, Cyclotella* spp.,
Chroomonas sp., *Oscillatoria rubescens, Synedra ulna* and *Tabellaria
fenestrata*.

In the Upper Rhine from about Germersheim (km 384) there is already
a population of *Stephanodiscus hantzschii,* which takes over the role of
an "indicator form for the Rhine plankton" here (Backhaus & Kemball,
1978). Its occurrence is associated with the saline waste discharges
from the Alsace. Large populations of chlorococcalean green algae and
µ-algae were also found. *Rhodomonas minuta* var. *nanoplanctica,
Melosira islandica* ssp. *helvetica, Gloeotilopsis planctonica, Gonio-
chloris pulchra* and *Nitzschia fruticosa* all have originated from the
lakes through which the river had passed. *Nitzschia acicularis,
Asterionella formosa* and *Fragilaria crotonensis* are also characteristic
species in the Upper Rhine today.

Middle Rhine

Lauterborn (1918) did not find obvious changes in the plankton of
the Middle Rhine as compared with the Upper Rhine. Seeler (1936),
however, could show that the planktonic biocoenoses were changed in
this section under the influence of the waste load. Czernin-Chudenitz
(1958) also found that centric diatoms were dominant, but the overall
species composition had become impoverished as a result of heavy
pollution. The entrance of the Mosel leads to new planktonic elements
in the Rhine. Different bodies of water flow side-by-side with a
different content. Such stripe formation is characteristic of the
Rhine and makes detailed studies difficult.

Lower Rhine

At the beginning of the century, "The plankton of the Lower Rhine shows far into the Delta region a great similarity in its essential features to the plankton of the Middle Rhine ..." (Lauterborn, 1918). The main constituents of the plankton were diatoms such as *Fragilaria crotonensis*, *Asterionella formosa*, *Synedra delicatissima*, *Tabellaria fenestrata* var. *asterionelloides*; somewhat less abundant were *Synedra actinastroides*, *Melosira*, *Stephanodiscus hantzschii*, *Cyclotella* as well as, sporadically, *Coscinodiscus lacustris*, *Attheya zachariasi* and *Rhizosolenia longiseta*. Among the Cyanophyceae the occurrence of *Oscillatoria rubescens* is not very rare ...". The list continues with a long series of Chrysophyceae and some Chlorophyceae. Cell densities were very low at that time. During the last decade extensive studies on the phytoplankton of the Lower Rhine have been conducted until 1974 by Heuss (1975) and subsequently by G. Friedrich. The species composition of phytoplankton has not changed markedly during these years, but populations are much higher with maxima well over 4×10^7 cells l^{-1} in 1979-1982. The plankton of the Lower Rhine is dominated by centric diatoms e.g. *Stephanodiscus hantzschii*, *S. tenuis*, *S. subsalsus*, *S. astraea* inc. var. *minutula*, *S. dubius*, *Melosira granulata* var. *angustissima*, *M. islandica* and *Cyclotella meneghiniana*. The first two are overwhelmingly predominant. Even *Asterionella formosa* may attain densities of 1.5×10^6 cells l^{-1}. The seasonal changes during 1981 of the most important groups are shown in Fig. 10.9. Filamentous Cyanophyceae, and in particular *Oscillatoria rubescens,* which had often been mentioned in the earlier literature, have disappeared. This is another clear indication of intensive changes in the "Rhine water" biotope. These changes are without doubt due to the influence of the saline waste load and to eutrophication. The considerable increase in plankton density has also been favoured by decreases in concentrations of toxic substances and to the heavier preloading of plankton because of impoundments in the Upper Rhine and from the major tributaries, the Neckar, Main (Lange-Bertalot, 1974) and Mosel.

Although there is a considerable net increase in plankton density between Bonn and the German-Dutch border, there is, nevertheless, a significant depression in the Greater Duisburg area, which may be due to toxic waste. This interpretation has been corroborated by measurements on oxygen production potential and other biochemical parameters. As soon as this toxic component is eliminated through further developments in waste treatment, a still more intensive plankton growth may be expected in the lower Lower Rhine (Friedrich & Viehweg, in press).

Seasonal periodicity

Investigations on benthic algae in different rivers have shown that the seasonal periodicity of algal coenoses decreases with increasing waste load; this occurs apparently because species with distinct periodicity are affected more intensively than those without it (Backhaus, 1968; Friedrich, 1973). In the case of the Rhine the widely different origins of the water are an additional factor to be considered. In the far past it can be assumed that the seasonal periodicity of the plankton was governed and affected by that prevailing in the Alpine and

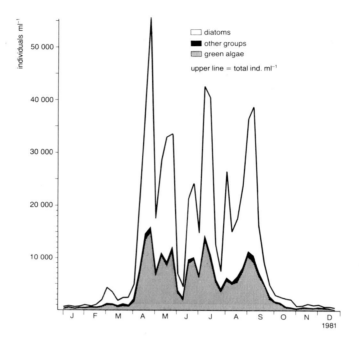

Fig. 10.9 Changes in phytoplankton density and relative proportions
of main taxa at Bimmen during 1982, based on weekly sampling. Relative proportions of groups are indicated by areas between lines
(i.e. not total height in case of diatoms).

and Alpine foothill lakes. Backhaus & Kemball (1978) measured a winter
minimum in March in the Upper Rhine, followed by rapid growth in April
leading to an annual maximum. Subsequently the crop fell slowly,
decreasing irregularly until August. The Lower Rhine shows a somewhat
different picture. There are generally two maxima here (Heuss, 1975;
G. F. unpublished; Fig. 10.10) a spring maximum from March to May and
a summer maximum from June/July to October. It is due mainly to the
intensive rainfall in sub-catchments, and the subsequent increased
discharges in the Lower Rhine that the plankton fluctuation is affected.
Centric diatoms are permanently predominant, Chlorophyceae may almost
reach the cell density of diatoms, though a high proportion of these
may be μ-algae.

10.43   ANIMALS

Headwaters and the Alpine Rhine extending to the Bodensee

As is characteristic of high-mountain streams, both the headwaters and
the upper section of the Alpine Rhine are typical salmonid waters with

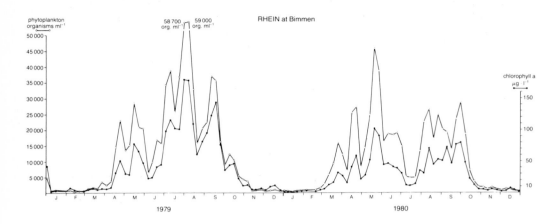

Fig. 10.10 Changes in phytoplankton density and chlorophyll *a* concentration at Bimmen in 1979-1980, based on weekly sampling. (Compare with observations on individual groups in 1981 shown in Fig. 10.9)

rheophile and rheotolerant species. In the Swiss Vorarlberg valley, the Rhine forms a lowland stream with heavy bed load transport, which has been still further increased by canalization. This has resulted inevitably in an impairing effect on the environment for benthic organisms. According to Bloesch (1977), because of the gravelly-sandy bottom, this Rhine section is considerably poorer in species than the R. Aare which has a similar water quality but a more silty substratum. The Aare has more insect larvae and molluscs.

The dominant fish in this section are trout and grayling. The abandoned river channels in the alluvial area and the small watercourses flowing towards the Rhine ("Giessen") are biotopes of their own.

Hochrhein

After flowing through the Bodensse, the Rhine reaches a section (Hochrhein) whose aquatic bryophytes and calciphilic algae provide a favourable environment for a diverse fauna (Caspers, 1980a) consisting predominantly of rheophile insect larvae. Caspers found 158 macro-zoobenthic species in 1978/79 (Table 10.4). Owing to the great progress in taxonomy it is difficult to compare faunistic lists with ones made at the beginning of the century, but the totals in 1908, 1910 and 1916 for Ephemeroptera, Plecoptera and Trichoptera were 19, 13 and 54, respectively, indicating obvious losses in all three orders. Examples of individual species which have been lost are the Trichoptera *Chimarra marginata, Glossosoma boltoni* and *Stactobiella risi* and the Ephemeroptera *Oligoneuriella rhenana,Rhithrogena* spp., *Ecdyonurus* spp.

and *Heptagenia* spp. There has been an increase in species characteristic of eutrophic still-water sections, such as the Ephemeroptera *Cleon dipterum* and *Caenis luctuosa*, the Trichoptera *Limnephilus lunatus* and a large number of chironomid species. Lauterborn (1916-1918) discovered only three species of Gastropoda (*Lymnaea peregra, Ancylus fluviatilis, Bithynia tentaculata*) in 1916. In 1978 and 1979 Caspers found 18 such species, whose appearance he attributed to the transformation of the Hochrein into a chain of impoundments, as well as to a growing eutrophication of the Bodensee.

Table 10.4 Composition of benthic fauna in Hochrhein near Bad Säckingen in 1978/1979 (Caspers, 1980a).

	number of species		number of species
Porifera	2	Ephemeroptera	7
Hydrozoa	1	Plecoptera	0
Tricladida	4	Odonata	1
Gastropoda	10	Heteroptera	2
Lamellibranchiata	8	Coleoptera	4
Oligochaeta	5	Trichoptera	21
Hirudinea	3	Diptera	85
Isopoda	2	(Chironomidae only	73)
Amphipoda	2	Bryozoa	1

Upper Rhine

The increased uniformity of the river bed and the increasing waste load have led to a considerable change in the fauna of the Upper Rhine (Kinzelbach, 1978). The impoverishment characterized by total disappearance of pollution-sensitive species reached its peak during the period 1970/1971. Insofar as permitted by the impact of waste load, epilithic species such as *Ancylus fluviatilis* and *Glossiphonia complanata* usually inhabiting the Hochrhein and the Middle Rhine had spread over the whole stretch. In recent years (since 1975) more species have been observed again. In addition to the reappearance of indigenous species, however, an increasing "alienation process" caused by the penetration of species new to the Rhine, has become apparent. Of the 32 new species found by Kinzelbach (1978), some come from other rivers in the region, but some from other continents.

Middle Rhine

The Rhine breaks through the Rhenish Slate Mountains here and flows in a narrow valley with steep banks and cragged rocks. The dominant

epilithic species are typical mountain-species forms, because of the
rise in water temperatures during the summer period. Barbel and nase
are the prevalent first species. The effects of effluents have led to
a considerable decline in fauna species diversity here as well
(Schwoerbel, 1959). Quite recently, however, the conditions for
colonization have become somewhat better again, as with the Upper
Rhine. These improvements are due primarily to the reduction in waste
load and the resulting higher oxygen concentrations in the Rhine
(Conrath *et al.*, 1977). An impressive example spotlighting the biological situation in the Rhine is given by the phenomenon of the
appearance of abundant caddis flies (*Hydropsyche contubernalis*) in the
Koblenz region during 1975-1977. Increased oxygen concentration, the
existence of flat-water zones, favourable substratum and the inflow of
Mosel water rich in phytoplankton led to the widespread propagation of
this insect. Disturbed trophic conditions (absence of consumers and
nutritional rivals for the caddis flies) probably played an equally
great role in this process (Tittizer, 1977). The general regeneration
in the ecological state of the Rhine has led to an extension of the
distribution range for this species (Caspers, 1980b). As a result of
growing diversity in trophic relations, however, a decrease in the mass
occurrence of this species has taken place. Further information on
the distribution of chironomids in the Rhine is given by Wilson &
Wilson (in press), who analyzed the composition of exuviae collected
from sites down the whole river.

Soppe (1983) has conducted several studies with the help of a
diving bell on the benthos in the zone of the Lorelei Cliff where the
Rhine has only a width of 130 m but its deepest spot with a depth of
23 m. Of the two types of substrata distinguished, solid boulder
gravel was almost free from faunal colonization, whereas the native
bedrock served as a biotope mainly for *Hydropsyche* larvae and *Ancylus
fluviatilis*. In current-free places, however, the boulder gravel
substratum had about 30 species of macroinvertebrate. Similar data
have been obtained near Bonn, i.e. the transition from Middle Rhine to
the Lower Rhine (Caspers, 1980b). The disappearance of pollution-
sensitive species, caused by the increasing pollution load and bank
development, has become obvious here as well. The river regulation
measures have led to the disappearance of macrophytes as microbiotopes
e.g. for the dragon-flies *Calopteryx splendens* and *C. virgo*. Bless
(1981) reports on the reappearance of the lamellibranchs *Unio pictorum*
and *Anodonta cygnea* in the lower section of the Middle Rhine and in
the upper section of the Lower Rhine. Both have now spread again far
into the Lower Rhine (G.F., unpublished).

Lower Rhine

The Lower Rhine is characterized by an extreme uniformity in
environmental conditions for the fauna. The whole of the bank is
protected, mainly with basalt stones. The "Greater Köln area" adds a
heavy point-source pollution load to the upper section of the Lower
Rhine. Modern studies are not yet complete enough to make a detailed

comparison with those of Lauterborn. However, it is known that the Plecoptera and Ephemeroptera, as well as about half of the Mollusca species, had disappeared between the beginning of the century and 1955 (Knöpp, 1955). On the 200 km stretch between Köln and Emmerich, there were vast depopulated zones until the mid-1970s in which no macroinvertebrate community could live (Herbst *et al.*, 1969, 1971; Miegel, 1963; Schiller, 1974). Only in the abandoned river branches through which the Rhine was flowing at times of flood could some species survive.

With the decrease in pollution and the improvement in oxygen supply a gradual recolonization has taken place (Landesamt für Wasser und Abfall Nordhein - Westfalen, 1982). The present species include for the most part pollution-insensitive forms with a wide ecological valency, forming a remnant biocoenosis from the inhabitants of the hard substratum. The most abundant macroinvertebrates in the Lower Rhine today are (W. Schiller, pers. comm.): *Ancylus fluviatilis, Asellus aquaticus, Bithynia tentaculata, Dreissena polymorpha, Dugesia lugubris, D. tigrina, Glossiphonia complanata, Fredericella sultana, Erpobdella octoculata, Hydropsyche* sp., *Plumatella* sp., *Radix peregra, Rheotanytarsus* sp., *Sphaerium corneum, Spongilla lacustris*. Less frequent are: *Acroloxus lacustris, Dendrocoelum lacteum, Gammarus pulex, Physa fontinalis, P. acuta, Helobdella stagnalis*.

Bed load transport also plays a role in the spatial and temporal distribution of benthic macroinvertebrates. Petran & Kothé (1978) have shown that the majority of organisms of the bottom fauna (Porifera, Platyhelminthes, Gastropoda, Bivalvia, Hirundinea, Crustacea, Isopoda, Insecta, Bryozoa) is rather sensitive to bed load transport. In contrast to these, mud sabellariid worms and chironomid larvae, which are able to dig into the river bottom, are found even in stream sections with intense bed load transport. A biological inventory executed by Tittizer (1980) on part of the Lower Rhine with the help of a diving bell, indicates that near the left bank, where redeposition of the river bottom does not occur, the greatest richness in species and density of individuals can be found. Towards the middle of the river, where frequent or permanent bed load movements may be expected, there is a rapid decrease in both species and individuals. The right bank zone is colonized by macrofauna coenoses poor in species and individuals (Fig. 10.11); bed load transport takes place here only at higher stream-flow.

10.5 FISH FAUNA AND FISHING

The fish fauna of a river like the Rhine cannot be understood without knowing the history of fishing in the region. The fish population has always indicated the water quality better than anything else and fishing had been the basis of existence for whole towns and villages on the Rhine throughout the centuries. Most of the information about the fish fauna has been gathered by professional fishermen and sport anglers. Changes in the fish fauna show seven phases of the history of the Rhine (chiefly according to Trahms, 1954).

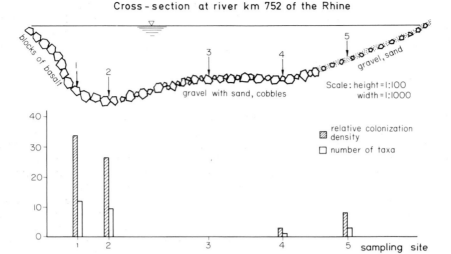

Fig. 10.11 Cross-section of Rhine km 752. 'Relative colonization density' gives a very approximate estimate of density of individuals (T. Tittizer, pers. comm.).

Phase 1 - before 1765

   The Rhine was clean and flowed freely in its natural channel, with a main stream and a large number of lateral and old branches. Extensive flood plains provided important spawning grounds. Mature salmon migrated upstream in hundreds of thousands each year and found spawning grounds in the tributaries, which were accessible without any obstruction. Despite intensive fishing, the fish population and production remained constantly high, some place names still remind us of that time and it is reported in old documents that the servants were complaining about having to eat salmon too often.

Phase 2 - after 1765

   As a result of river engineering measures taken to make effective agricultural use of the alluvial meadow and in the interest of navigation, the tributaries of the river were cut off and through the construction of groynes the stream bed became confined. Several tributaries became inaccessible for migratory fish. As a result of dike construction important spawning and resting places got lost. First of all, spawning grounds for the salmon became restricted. In their place, due to the newly produced groyne fields, some new spawning grounds were available for other fish. Because of the intensive deep erosion, these biotopes mostly lost in the meantime

their significance for the aquatic fauna. The introduction of steam-
ship navigation caused heavy damage to the fish fauna by disturbing the
fish and especially by destroying the fry. The propellers of ships
caused a considerable proportion of the young fish living in the flat
groyne fields to be thrown on land. In addition to these changes in
the living space the first serious waste loads attached to the "founder
years" contributed to an impairment of conditions.

Phase 3 - from 1915

The growing urbanization and the rapidly developing large-scale
industry at the close of the 19th century had already given rise to
pollution which led to a sudden decrease in the fish production about
1915. Besides the decrease in the salmon stock, there was also a
reduction in the "mayfish" (*Alosa alosa*) population, whose migration
upstream had been until then a special occasion for public festivals.
In 1924, after a 12-year period, the rents for the lease of fishing
sections on the Lower Rhine had to be fixed anew and reduced to a half-
or even to one-third, because of the lower fish yields. The last
sturgeon (*Acipenser sturio*) was caught in the Rhine in 1931. By then
the "mayfish" had also disappeared and in the most polluted part of the
Lower Rhine the remaining salmon became inedible because of their strong
phenolic taste. The eel, however, began to grow in importance, becoming
the "bread and butter fish" of the fishermen.

Phase 4 - from 1933

Industrial production increased by leaps and bounds but without
adequate waste treatment, leading to great and rapid damage to the fish
fauna. There was a short recovery after the end of the war, due to the
postwar stagnation in production and shipping.

Phase 5 - from about 1948

As the towns and industries destroyed during the war were reconstruc-
ted, pollution increased rapidly. Long stretches began to show zones of
biological devastation. The fish became inedible because of bad taste.
Fish mortality and large oil carpets on the surface of the water
occurred so often that long sections of the Lower Rhine were practically
free of fish.

Phase 6 - from about 1954

In this phase intensive efforts were started in order to improve
water quality on the Rhine; a characteristic measure was the rule
obliging ship-owners to discharge their waste oil only at special
collecting places. Construction of clarification plants was intensified.
Nevertheless, the waste load of the river increased to such an extent
that it practically paralysed the Rhine fishery as a branch of trade.
Only in the section above Karlsruhe could professional fishermen still
exist. The most serious fish mortality ever, extending from the Main

to the Netherlands in the summer of 1969 (Ant, 1969), falls also in this phase of the history of the Rhine.

Phase 7 - since about 1975

The first signs of a recovery in the fish stock in the Rhine have been apparent since 1975. At least sport fishing has improved considerably. Single catches have been reported of some sensitive species that were thought to have been extinct in the Rhine (Bless, 1981; Böving, 1981).

Originally, there were about 40 species of fish and Cyclostomata living in the Rhine. Out of these 8 species have disappeared (state in 1978): *Acipenser sturio, Salmo salar, S. trutta, Coregonus oxyrhynchus, Alburnus bipunctatus, Leuciscus souffia agassizi, Leucaspius delineatus, Alosa alosa*. *Petromyzon marinus* occurs only as a few specimens in the Netherlands Rhine section. (According to Bless, 1981, some salmon have sporadically been caught again in the Rhine.) In contrast to this loss, there has been an introduction of foreign species. This occurred first with the North American species, *Oncorhynchus tshawytscha, Salmo gairdneri, Ictalurus nebulosus, Micropterus salmoides* and *Lepomis* sp. More recently, the "grass carp" (*Ctenopharyngodon idella*) from Eastern Asia has appeared as well. Similar changes in the fauna have been observed in other large river basins; according to Lelek (1978) there are now twelve naturalized foreign species in the Danube, seven in the Elbe and eight in the Main. In a 100-km long section on the northern Upper Rhine and southern Middle Rhine Lelek found that of the 39 species originally present, not more than 19 could be detected in 1976. Still water regions were of great importance as residence areas for the remaining fish. The European roach (*Rutilus rutilus*) was the most frequent fish and is now also the dominant on the Lower Rhine with some 40 to 50% of individuals (Böving, 1981).

10.6   CHEMISTRY AND WASTE LOAD

Knowledge of the chemistry of the Rhine is especially important for an understanding of its ecology, because of the presence of toxic substances from industrial and household sources and also from numerous non-point sources originating from streets or agriculture.

Inorganic salts

The water of the Rhine is moderately hard, neutral to slightly alkaline and with a high buffering capacity. As a result the relatively high pollution load of inorganic acids has much less effect than it would do in a poorly buffered water. Table 10.5 shows the changes in concentrations of various elements in 1982 on passing down the river.

Table 10.5 pH range and mean concentrations (mg l^{-1}) of various elements in the Rhine during 1982. (After DKRR, 1983)

	Hochrhein Oehningen km 22.9	Upper Karlsruhe km 362.3	Middle Koblenz km 590.3	Lower Rhine Kleve-Bimmen km 865
pH	7.9 - 8.8	7.7 - 8.5	7.3 - 7.7	7.1 - 7.6
Na	5	63	61	68
K	2	5	5	6
Mg	8	8	10	12
Ca	41	48	64	66
Cl	7	100	99	136
SO$_4$-S	9	10	17	22

Since the Rhine is a very important source of freshwater for the Netherlands, its loading with salts discharged from potash works, mining drainage and the chemical industry has considerable significance. The first important increase in the chloride load is due to the waste discharge of the Alsatian potash industry, whereas increase on the Lower Rhine is a result of mining drainage operations and effluents from the chemical industry. A reduction in the chloride loading is economically feasible only in the case of potash industry; an international agreement was made to deal with this problem (see 10.1).

Oxygen content and organic load

Because of its key role in the river ecosystem, the oxygen content of the Rhine water has been a focus of public interest. After the second world war the concentration decreased until 1969 due to the high loads of poorly or untreated sewage effluent. Since that time there has been a tremendous programme of building or improving sewage works and the oxygen content has increased again (Fig. 10.12). In 1982, for instance, the lowest concentration reported (by Deutsche Kommission zur Reinhaltung des Rhein, 1983) was 4.9 mg l^{-1} O$_2$. There has been a comparable decrease in the concentration of slightly degradable organic compounds, measured as BOD (Fig. 10.13). The improvement in water quality during 1970 to 1980 (Fig. 10.14) cannot, however, be attributed entirely to developments in waste disposal. It is due also to the downward economic trend in industrial production.

The waste load of the Rhine should be reduced in the coming years as a result of further developments at waste treatment plants and by internal preventive measures in industry. The aim is to arrive at a stage where biochemical oxygen demand and oxygen concentration meet the required quality standards (quality class II, BOD <5 mg l^{-1}, oxygen > 6 mg l^{-1}). Problems associated with the stretch below the Neckar (still evident in 1980) have recently been solved; particular problems that still exist are the stretch below the Main and the whole of the lower Rhine.

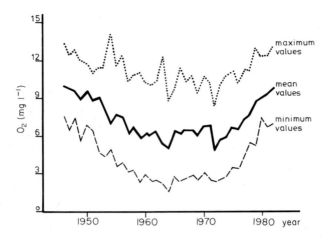

Fig. 10.12 Dissolved oxygen in the Lower Rhine at the German–Dutch border. (Modified from RIWA, 1983)

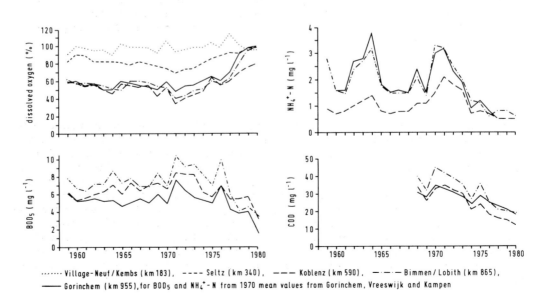

Fig. 10.13 Changes in water quality at five sites on Rhine during 1959–1980.

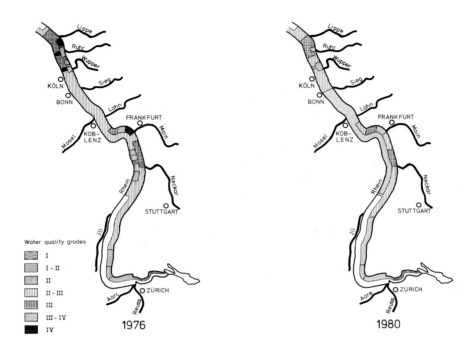

Fig. 10.14 Water quality maps of the Rhine for 1976 and 1980 (LAWA, 1976, 1980). Grades (in terms of pollution): I, very low; I-II, low; II, moderate; II-III, critical; III, strong; III-IV, very strong; IV, overloaded.

Nutrients

An indication of the changes in phosphate which have taken place may be gained by considering the Bodensee. Up to about 1950 the concentrations of phosphate were near the detection limit of the method used for routine monitoring, but since that time there has been a very rapid increase. This is due to agricultural, industrial and domestic uses of phosphate. At the end of winter stratification in 1970 phosphate-P reached 40 µg l^{-1} (Lehn, 1972: see also Fig. 10.15). Increases in phosphate are evident all down the Rhine, as well as increases in combined nitrogen (nitrate and ammonium) (IAWR, 1981; RIWA, 1983). Correlated changes shown by plants are the great increases in *Ranunculus fluitans* in the Hochrhein (10.422) and phytoplankton from the Upper to the Lower Rhine (10.423).

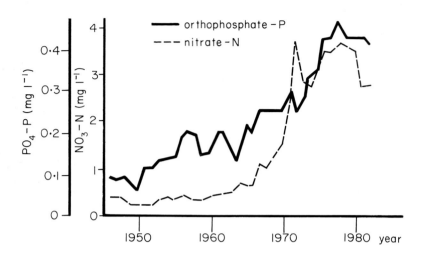

Fig. 10.15 Orthophosphate-P and nitrate-N concentrations in the Lower Rhine at the German-Dutch border. (Modified by RIWA, 1983)

Heavy metals

The natural content of heavy metals in the Rhine is low, but because of industrial and other sources concentrations in the water and especially the sediments can reach potentially harmful levels. Because of the multiform interactions between liquid and solid phases (water - seston - sediment) and because of problems at all steps in sampling and analysis, data on heavy metals need to be interpreted with particular care. Dissolved heavy metals play a less important role than those associated with suspended particles and sediments. Heavy metals in sediments are of great economic importance because it is usually impossible to dispose of dredged materials on to agricultural land. Reduction in heavy metal pollution depends largely on industry, since waste and water treatment do not give fully satisfactory results. An extensive body of data and interpretation can be found in the works of Förstner & Müller (1974), Müller (1979) and Malle & Müller (1982). In general many of the heavy metals showed peak concentrations and loads at the end of the 1960s and beginning of the 1970s. Since that time the loads for most elements have remained little changed, but those for chromium and cadmium have shown a marked decrease. Table 10.6 gives some indication for 1982 of the increases taking place on passing down the river.

Table 10.6 Heavy metals in the Rhine during 1982: mean concentrations (μg l^{-1}) and mean load (g s^{-1}). (After IKSR, 1983)

	Hochrhein Oehningen km 22.9		Upper Karlsruhe km 362.3		Middle Koblenz km 590.3		Lower Rhine Kleve-Bimmen km 865	
	conc.	load	conc.	load	conc.	load	conc.	load
Cr	<2.0	0.6	<2.0	2.0	5.3	11.9	7.9	22.3
Ni	5.3	2.0	9.6	14.1	<5.0	8.1	9.2	25.4
Cu	2.8	1.2	2.9	4.4	7.4	16.5	25.7	69.0
Zn	<30.0	6.3	<30.0	23.8	45.0	99.1	108.8	291.5
Cd	<0.3	0.1	<0.3	0.2	<0.3	0.3	0.6	1.6
Hg	<0.05	0.02	<0.05	0.06	0.08	0.2	<0.2	0.59
Pb	<5.0	1.0	<5.0	4.0	<5.0	8.8	11.6	32.4

Micropollutants

As the input of easily degradable substances and obvious sources of coarse pollution have been much decreased, increasing attention has been paid to substances which are poorly degradable, some of which are subject to bioaccumulation and some of which are known to be toxic, carcinogenic or mutagenic. Because of their long-term effects, organo-halogens are of special interest. A tremendous range of these substances which enter the river either as industrial wastes or following their use of pesticides etc. For instance, during assessment of Rhine water quality in the Nordrhein-Westfalia region, there are 65 compounds which are detected regularly. These include:

> chlorohydrocarbons of low molecular weight, e.g. chloroform;
> chlorohydrocarbons of high molecular weight, e.g. dichlorbenzol;
> polycyclic, aromatic hydrocarbons, e.g. 3,4 benzpyren;
> chlorophenols;
> biocides, e.g. hexachlorcyclohexane.

Because of this complexity, only some compounds can be assessed quantitatively, so 'total organic chlorine' is used as an additional summarizing parameter. As with many other pollutants, its concentration increases in passing downstream (Table 10.7).

The concentrations of these substances are in general - looking at the Rhine as a whole - low and below toxic thresholds. Further, there is some indication of a decrease. The mean concentrations in 1982 were lower than in previous years, though peak loadings were still similar (KIWA, 1983). Nevertheless there is a considerable risk that substances of biological activity are overlooked (Sontheimer, 1983). The use of trout as test fish can still demonstrate chronic toxic effects and the AMES-Test, mutagenic effects (KIWA, 1983).

Table 10.7 Total organic chlorine (μg l^{-1} Cl) in the Rhine during 1975 (IAWR, 1978).

	km	1975		1976		1977		1978	
		$\bar{x}$	max	$\bar{x}$	max	$\bar{x}$	max	$\bar{x}$	max
Basel	160	50	130	40	100	50	140	20	60
Karlsruhe	360	40	120	30	80	45	150	50	140
Köln	680	100	340	40	100	60	170	100	270
Wittlaer	750	100	290	80	270	120	330	100	240

Waste heat

Since the end of the sixties the waste heat load has also become an urgent problem. The situation at present is as follows (M. Wunderlich, pers. comm.). Due to the construction of more and higher-capacity power plants, the limits of a straightforward once-through system of cooling have been reached. The concentrated discharge of large heat loads can lead to restriction on other uses of the water and affects the ecological state of the surface waters.

The loading of the German Rhine section with waste heat discharges is now regulated with the aid of recommendations by the Working Community of the German Federal States (LAWA, 1977). The maximum temperature rise permitted generally in rivers is equal to $\Delta T = 5$ K; the maximum water temperature should not exceed 28 $^{\circ}$C in summer-warm waters and 25 $^{\circ}$C in summer-cold waters. These temperatures represent calculated values obtained with the assumption of a complete mixing of cooling water and river water. In the Rhine, however, a stronger margin is set for the temperature rise ($\Delta T = 3$ K). This is justified by the assumption that, owing to the large width of the river, complete mixing is not possible over long stretches.

The five states bordering the Rhine decided in 1972 to take measures to prevent the aggravation of the thermal load in the river. A resolution was agreed that only power plants with cooling towers or similar cooling systems should be permitted. To the best of present knowledge, it is possible to avoid adverse consequences for the ecology and water resources management of the Rhine with the existing temperature limit values and by adequate associated measures (artificial aeration of cooling water).

In assessing the effects of the waste load on the ecology of the Rhine an important aspect is the insignificant transverse mixing. Waste water discharges and tributaries can be regarded in some sections of the Rhine as completely mixed only after at least 30 to 100 km. It is on the mountainous stretch between Koblenz and Bingen that faster mixing may occur. There is a similar situation on the impounded parts of the Hochrhein and the Upper Rhine, where most of the mixing takes place in the turbines of the power stations.

10.7 SELF-PURIFICATION AND MICROBIAL ACTIVITY

For a stream so heavily loaded with wastes as the Rhine, self-purification is of decisive ecological and economic importance. Considering
self-purification in the light of sewage engineering, its economic
value can be quantified. During the period of heavy pollution a waste
load of 50 x 10^6 inhabitants (P) and population equivalents (PE) was
discharged into the Rhine before it reached the Dutch frontier. At
Bimmen the BOD$_5$ load corresponded to 35 x 10^6 P + PE; this means that
15 x 10^6 P + PE could be eliminated by self-purification. Taking a
yearly average expenditure of 40 DM per P + PE on preventive measures,
this corresponds to a saving of about 600 million DM per year
(Bundesminister des Innern, 1976). Without the influence of toxic
substances the effects of self-purification would have been about 30%
higher (Knöpp, 1970).

Summation parameters like TOC (total organic carbon), COD (chemical
oxygen demand) and BOD (biochemical oxygen demand) are particularly
suited to characterize the self-purification process. Sampling for the
longitudinal profiles of the Rhine between Karlsruhe and Emmerich
shown in Fig. 10.16 was carried out by helicopter for the Upper and
Middle Rhine on 22 October 1979 and for the Lower Rhine on 25 October
1979; this assured a delay of only a few hours before analysis. The
natural discharge of the Rhine is usually quite low in the second half
of October, which makes the effects of waste effluents particularly
serious. Nevertheless, discharge was relatively high during the
October 1979 study, amounting to about 1250 m^3 s^{-1} at Kaub (between
Mainz and Koblenz). This corresponds to a value between MLQ (mean low
discharge) and MQ (mean discharge). The pollutant concentrations
depending on the waste discharge were about one third lower than those
to be expected at MLQ, because of the relatively high degree of
dilution. However it is the fundamental changes along the course of
the river that are emphasized here, rather than the magnitude of
individual parameters.

The longitudinal profiles of the various waste parameters in the
Rhine are not at all similar (Fig. 10.16). The degradable organic
matter content, characterized by C-BOD$_5$ (determined by applying a
nitrification inhibitor) is relatively low along the whole stretch.
The maxima on the Lower Rhine are about twice as high as the minima on
the Middle and the Upper Rhine. The concentrations of degradable
organic substances in the Main, Wupper and Emscher are considerably
higher than in the Rhine; those in the Mosel are less than in the Rhine.
In contrast, the effect of effluents from large-scale industry is much
higher judged by COD (oxidation with bichromate). The Neckar, Mosel
and Ruhr show similar concentrations of COD to those in the Rhine.
On the whole, there is an increase in COD on passing down the river.

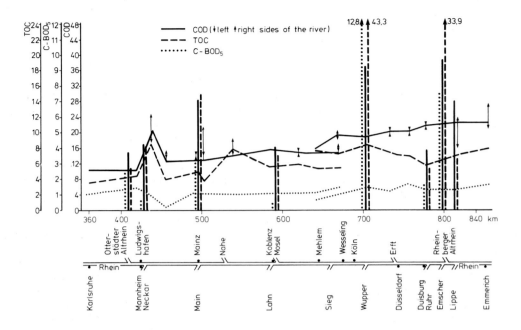

Fig. 10.16 Changes in COD (chemical oxygen demand), TOC (total organic carbon) and C-BOD$_5$ (5-day BOD in presence of allyl thiourea) along the Rhine. Measurements made on 22 and 25 October 1979; the break in the lines from km 640 - 670 is due to observations at the lower sites being made on the second day; all concentrations in mg l^{-1}.

In general, differences similar to those observed with COD are found for TOC in the Upper and Middle Rhine. It is remarkable, however, that in the Lower Rhine there is no similar increase in TOC corresponding to that which occurs with COD. Correlations between these three parameters permit conclusions to be made concerning the "biochemical degree of purification" (C-BOD$_5$/COD) and the "specific oxygen demand of organic substances " (COD/TOC). The biochemical degree of purification, which is characterized in the Rhine section investigated mostly by values between 0.05 and 0.1, indicates that self-purification is well-advanced. For fully degradable substances the quotient of the biochemical degree of purification is usually equal to 1; fresh domestic sewage and wastes subjected to biological treatment are generally characterized by quotients equal to 0.7 and 0.25, respectively. Only the Wupper and Emscher rivers show an essentially lower degree of purification than the Rhine.

The quotient for the specific oxygen demand of organic substances in the Upper and Middle Rhine also indicates no peculiarities. It does

show, however, a considerable increase in the Lower Rhine. From this
it can be inferred that, although the substances present in the Lower
Rhine are largely oxidized in a biochemical respect, this is not so
from a chemical point of view.

In addition to the oxidation of organic substances, nitrification
takes an essential part in the oxygen consumption in the rivers. The
ammonium content comes from waste waters or gets released from protein
compounds during their degradation. Whereas C-BOD_5 increases relatively
little on passing down the river, the oxygen consumption by nitrification
increases along the whole stretch. Upstream of Karlsruhe oxygen
consumption by nitrification amounts to not more than 20% of the
total BOD, whereas on the Lower Rhine it is 60 to 80%. If the yearly
average is considered, however, the proportion of nitrification in the
BOD_5 may not be quite so high.

The increase in oxygen consumption by nitrification with the
extension of the river reach is influenced by the ammonium concentration
in the water of both ammonium and nitrifying bacteria. At the
beginning of the test section the former is mostly 0.3 - 0.5 mg l^{-1} and
on the Lower Rhine mostly 0.5 - 1.0 mg l^{-1}. However, the increase in
nitrifying bacteria is likely to be of great influence, this increase
being characterized by "the maximum oxygen consumption due to nitrification under standard laboratory conditions"(Müller & Kirchesch,
1981a). The biomass of nitrifying organisms increases more than ten-
fold by the time the river reaches the German-Dutch border. Tributaries with a high degree of self-purification, like the Neckar, Mosel
and Ruhr, also contribute to the biomass of nitrifying bacteria. In
the Main and Wupper, which are loaded with fresh waste waters, the
concentration of nitrifying bacteria is very low. The R. Emscher
differs from all the other tributaries because it combines a high
concentration of nitrifying organisms with a low degree of purification.
This is due to the fact that the Emscher is the effluent of a large-
scale treatment plant (river sewage works), where the sludge is
regulated so as to facilitate the development of nitrifying bacteria.

The diel fluctuations in the oxygen content determined by the primary
production range up to 3.0 mg l^{-1}, depending on the stretch, season and
radiation intensity. Oxygen contents below 2 mg l^{-1} at Mainz and below
4 mg l^{-1} on the Lower Rhine were not unusual during the period of the
maximum waste load. Conditions are far more favourable today and they
need to be improved only on the Lower Rhine.

A river like the Rhine, receiving large amounts of waste waters from
the chemical industry, may always be suspected of containing toxic
substances which have inhibitory effects on the biota. It is usually
very difficult to produce causal-analytical evidence for such interference, but this becomes easier with individual point sources of
pollution. Knöpp (1970) was the first to deal with this problem. His
results for September 1966 showed that in the section of greatest
importance for water management, there was a marked discrepancy between

the load profile computed from emission data and the longitudinal profile of BOD determined from water samples. This was probably due to the toxic effects of effluents. This interpretation has been con-firmed by data on algal photosynthetic activity (see below). Knöpp (1968) found that among the samples of water tested, six were highly toxic for microbial degradation and thirteen reduced algal photosynthesis.

In investigations on the effects of rise in water temperature in October 1972 and 1979, Müller (1975) and Wunderlich (1978) applied a quite different method to determine the inhibition ratios of self-purification on the Lower Rhine. In order to study the effects of calefaction, Rhine water was incubated both at river temperature and 10 °C higher for seven hours. For the Upper and the Middle Rhine there was an increase in oxygen consumption at the higher temperature. In the Lower Rhine, where Knöpp's investigations had already revealed an inhibitory influence, the rate of increase in oxygen consumption was either very low or even negative. This suggests that the effects of poisons increases when the water is warmed. As the waste load of the Rhine has strongly decreased during recent years, a reduction in toxic wastes may also be assumed. This has been corroborated by studies on the rate of increase in oxygen consumption when water temperature rises due to heat discharge (Wunderlich, 1980). Nevertheless there is still a measurable inhibitory influence in parts of the Lower Rhine (Friedrich & Viehweg, in press).

Thanks to the strict control regulations, the calefaction of Rhine water by cooling water discharges has not led to serious ecological effects. Nevertheless, the results of some studies may be of interest. As a result of the rise in temperature, the rate of metabolic processes will be increased. The degree of increase plays an important part in determining the substances contained in the river water. Wunderlich (1980) concluded for the Rhine that the increased rates of oxygen consumption due to biochemical degradation and nitrification are subject to both marked spatial and temporal fluctuation. Low rates of increase often occur in the zones loaded with waste discharge from fresh communal sewage. Particularly high rates of increase are typical of zones where self-purification is well advanced.

In the degradation of organic substances, psychrophilic and meso-philic bacteria i.e. those active even at low (6 °C) or high tempera-tures (35 °C), respectively, play an essential part. Plate techniques have shown that psychrophilic bacteria are dominant during winter. A short period of warming, as often occurs when water passes once through power plants with continuous cooling, exerts no detrimental effect on psychrophilic bacteria.

Some of the hydro-electric power plants on the Rhine are operated with cooling towers having an outlet cooling system. Power plants of this type may lead to a considerable rise in oxygen consumption, even if the outflow and inflow temperatures are equal (e.g. oxygen consump-tion doubles within 24 hours during incubation at 20 °C after a short

period at 40 °C (Wunderlich, 1977; Friedrich, 1979). This presumably
results from short-term activation of bacteria in quiescent stages
(hypnospores, hunger forms). As only a relatively slight part of the
discharge is used for cooling purposes and water is subjected to
aeration while flowing through cooling towers and outlet structures,
this increase in oxygen consumption has not led to special problems.

Outlet cooling systems and closed-circuit cooling systems show
positive features from the viewpoint of water resources management,
because they act as fixed bed reactors in the nitrification process.
In a cooling system with outlet the ammonium contained in the river
water will be oxidised up to about 40% and in a closed-circuit cooling
system this oxidation is almost complete. Less important but still
quantifiable is the decrease in organic carbons after the water has
passed through a cooling system (Wunderlich, 1980).

Although heterotrophic organisms play the major role in a river
with heavy waste load, autotrophic organisms must also be considered.
The remainder of this chapter deals with phytoplankton metabolism,
based on investigations of the "oxygen production under laboratory
conditions" referred to as "OPL" (DIN 38412, Part 14).

Fig. 10.17 shows longitudinal profiles of OPL and chlorophyll *a* in
late October 1979. On the Upper Rhine at Karlsruhe there is almost
no active phytoplankton at this season, although in the anabranches
(e.g. the Otterstadt old Rhine branch), OPL is still high. Oxygen
production increases more or less constantly along this stretch of
river, with the impounded tributaries making an important contribution.
From the longitudinal profiles of OPL and chlorophyll, it is possible
to estimate to what extent primary production or phytoplankton biomass
are involved in C-BOD_5 or COD. In the Upper Rhine the contribution of
algae to the BOD is more or less insignificant; it is about 20% in the
Middle Rhine in autumn, eventually rising to about 30% in the Lower
Rhine near the German-Dutch border. The corresponding algal contribution to the COD is less. Phytoplankton is more important in summer.

Before dealing with annual changes in phytoplankton activity, two
terms need to be explained. "Direct algal BOD" denotes the biological
oxygen demand needed for the degradation of a momentarily present
biomass of algae. "BOD of algal origin" denotes all products resulting
from algal primary production (breakdown stages, extracellular products).
The latter is higher than the former and the relative difference between
these two quantities increases further downstream.

The annual course of algal activity may be characterized by the OPL,
measurements of which are made for the most important sections of the
Rhine. Of the two years shown in Fig. 10.18, 1979 was relatively "dry",
whereas 1980 was "wet". The dynamics of algal activity on the Hochrein
at Reckingen still correspond well with those of the Bodensee. A
characteristic feature is the spring maximum, which then breaks down,
presumably due to nutrient deficiency and grazing.

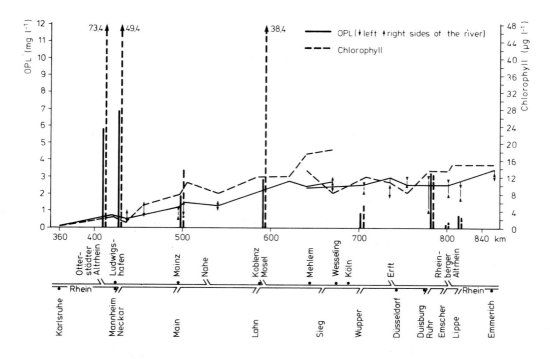

Fig. 10.17 Changes in algal activity measured as OPL (oxygen production in laboratory) and chlorophyll *a* along Rhine. Measurements
made on 22 and 25 October 1979 ; the break in the lines from km 640 -
670 is due to observations at the lower sites being made on the
second day. Assay made in laboratory using light (200 µmol photon
m^{-2} s^{-1})-dark bottle technique at 20 °C.

Algal activity shows a more or less constant increase, any drops
being correlated mostly with hydrological events. For example, flow at
Koblenz amounted to about 2000 m^3 s^{-1} in May 1980, a period with high
OPL values in the Middle and Lower Rhine. It then rose to 4000 m^3 s^{-1}
in August 1980 and this resulted in a reduction in the OPL to 25% of its
initial value.

Studies on actinomycetes, which are of interest as producers of
odorous substances, have suggested (Müller, 1980) that, at least in the
Rhine, these belong to the "sediment bacteria". Owing to their slow
growth and adherence of their filaments to particles, they are hardly
capable of colonizing the flowing water body actively. Consequently,
their concentration in waters appears to depend to a large extent on
the content of resuspended sediment. Bacteria responsible for cresol

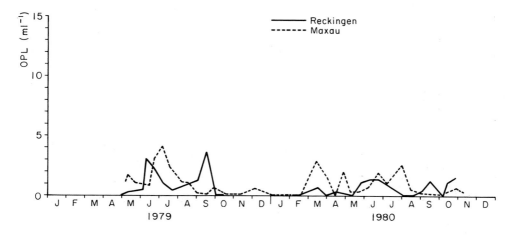

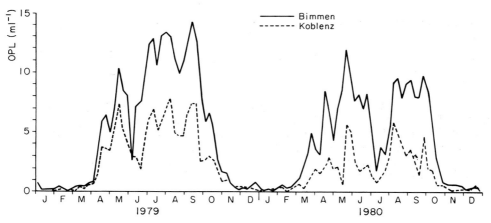

Fig. 10.18 Changes in algal activity measured as OPL (oxygen production in laboratory) in 1979-1980 based on weekly measurements at four sites. (Details given in legend to Fig. 10.17)

degradation appear to have a similar behaviour. During the course of the year, the concentration of such sediment bacteria is correlated negatively with that of the algae. It is clear however that a great deal more research needs to be carried out on the sediment bacteria. Knowledge about microbial activities and self-purification in the flowing water of the Rhine may now be considered quite good. Knowledge about the contribution of the sediment bacteria and sessile organisms in general to the metabolism of the Rhine should be expanded.

10.8 CONCLUDING REMARKS

Although the Rhine and the land along its banks is heavily used for
settlements, industry and traffic, the river has conserved or regained
some of its original biological characteristics and has not lost its
attraction for tourists from all over the world. Its ecosystem is on
the way of recovery from very severe damage, but the possibility of
future danger remains because of accidents connected with a river used
so intensively.

There are some special measures which would help ensure better
conditions in and around the river. In the Upper Rhine plain the
ecology of the alluvial valley could be reactivated by the construction
of connections between old river branches, gravel pits and tributaries
with high water quality. These measures would augment the ground water,
support the typical biotope of the alluvial forest, improve the quality
of the side waters and make some improvement to water quality of the
main river. Similar measures could be taken along the Lower Rhine, but
the situation there is more difficult due to the deep erosion of the
river and the dense settlements. The gravel workings in the alluvial
plain of the Upper Rhine and of the Lower Rhine are especially
important. Future operations should include much better reclamation
measures to improve the ecological situation in the river and the
alluvial plain.

When the river has reached the quality objectives (German water
quality class II) which are based on biological analyses of the benthos,
water quality management has then to evaluate and describe the waste
load of the river by methods capable of differentiating various components. Such methods should permit the concentrations of substances
hazardous for drinking and other water uses to be assessed. They
should also monitor components which influence adversely the ecology
of the river and its self-purification capacity. Water quality
management has to handle micropollutants with more care. Better
measures to avoid pollution are needed; this means that industry has to
develop waste water-free techniques and to take care that the consumers
of their products do not spread so many harmful substances into the
environment and the waters.

With the progress of mechanical-biological waste water treatment,
the input of nutrients for algae is getting more important, as these
nutrients cause eutrophication problems in the slow-flowing stretches
of the Rhine in the Netherlands. In a river, which has been so much
improved, it will be worthwhile to evaluate its suitability for bathing
from the hygienic point of view.

Looking back in history, one can see a correlation between good
times for man and bad times - pollution - for the Rhine. Only when
the ecological crisis of the Rhine from 1950-1970 caused severe problems
in the use of the water by man, did sanitation programmes start to
uncouple this correlation. Let us hope that the national and inter-

national efforts for improving the condition of the Rhine will continue till the end of the century, so that the condition of the river will match its importance as the artery of central Europe.

Acknowledgements

We have to thank our many co-workers who helped us to compile this paper and in particular, Frau Marlies Viehweg and Frau Ingrid Busch. Thanks also to Dr D. Backhaus (Landesanstalt für Umweltschutz, Baden-Württemberg) for unpublished data.

REFERENCES

Anhut U. (1977) Untersuchungen zur Molluskenfauna des rechten unteren Niederrheins im Raum zwischen Wesel, Rees und Emmerich. *Gewässer und Abwässer* 62/63, 17-62

Ant H. (1969) Biologische Probleme der Verschmutzung und akuten Vergiftung von Fliessgewässern, unter besonderer Berücksichtigung der Rheinvergiftung im Sommer 1969. *Schriftenreihe für Landschaftspflege und Naturschutz* 4, 97-126

ARGE Rhein (Arbeitsgemeinschaft Rhein) (1968) Die Verunreinigung des Rheins und seiner wichtigsten Nebenflüsse in der Bundesrepublik Deutschland - Arbeitsgemeinschaft der Länder zur Reinhaltung des Rheins. *Denkschrift, Stand* 1965. Ministerium fur Ernährung, Landwirtschaft und Forsten, Düsseldorf

Backhaus D. (1968) Ökologische Untersuchungen an den Aufwuchsalgen der obersten Donau und ihrer Quellflüsse, II. Die räumliche und zeitliche Verbreitung der Algen. *Arch. Hydrobiol., Suppl.* 34 1/2, 24-73

Backhaus D. & Kemball A. (1978) Gewässergüteverhältnisse und Phytoplanktonentwicklung im Hochrhein, Oberrhein und Neckar. *Arch. Hydrobiol.* 82, 166-206

Bensing W. (1966) Gewässerkundliche Probleme beim Ausbau des Oberrheins. *Gewässerkundliche Mitteilungen* 10, 85-102

Bless R. (1981) Beobachtungen zur Muschelfauna des Rheines zwischen Köln und Koblenz. *Decheniana (Bonn)* 134, 234-43

Bloesch J. (1977) Bodenfaunistische Untersuchungen im Aare und Rhein. *Schweiz. Z. Hydrol.* 39, 46-68

Böving M.-P. (1981) Die Fischfauna des Rheinstromes und seiner direkt angrenzenden Altwässer im Niederrheingebiet. *Decheniana (Bonn)* 134, 360-73

Bundesminister des Innern (1976) Umweltprobleme des Rheins. *Sondergutachten 1976 des Rates von Sachverständigen für Umweltfragen, herausgegeben vom Bundesminister des Innern (BMI).* Verlag Heger, Bonn-Bad Godesberg

Burckhardt H. & Burgsdorf H.L. (1964, 1965) Flora des Altrheins bei Xanten und seiner Umgebung. I Floristische Untersuchung. II. Pflanzengesellschaften des Xantener Altrheins. *Gewässer und Abwässer* 37/38, 7-45 ; 43, 7-47

Caspers N. (1980a)   Die Makrozoobenthos-Gesellschaften des Hochrheins
    bei Bad Säckingen.   *Beiträge Naturk. Forsch. Sudw. Dtl.* 39, 115-42.
Caspers N. (1980b)   Die Makrozoobenthos-Gesellschaften des Rheins bei
    Bonn. *Decheniana (Bonn)* 133, 93-106.
Conrath W., Flakenhage B. & Kinzelbach R. (1977)   Übersicht über das
    Makrozoobenthos des Rheins im Jahre 1976. *Gewässer und Abwässer*
    62/62, 63-84.
Czernin-Chudenitz C.W. (1958)   Limnologische Untersuchungen des
    Rheinstromes.   III Quantitative Phytoplanktonuntersuchungen. *Forsch-
    ungsberichte des Wirtschafts und Verkehrsministeriums Nordrhein-
    Westfalen* 536, 422 pp.   Köln & Opladen.
Deutsches Gewasserkundliches Jahrbuch, Rheingebiet (1974)   Landesamt
    für Gewässerkund.   Mainz.
Deutsches Institut fur Normung (1982)   Bestimmung der Sauerstoffproduk-
    tion mit der Hell-Dunkelflaschen-Methode unter Laborbedingungen, SPL.
    DIN 38 412 Teil 14.   Deutsches Institut fur Normung e. V., Berlin.
Ernst E. (1972)   *Der Rhein - eine europäische Stromlandschaft im Luft-
    bild* 146 pp.   Konkordia-Verlag, Bühl.
Förstner U. & Müller G. (1974)   *Schwermetalle in Flussen und Seen als
    Ausdruck der Umweltverschmutzung.*   Springer Verlag, Berlin.
Friedrich G. (1973)   Okologische Untersuchungen an einem thermisch
    anomalen Fliessgewässer (Erft/Niederrhein). *Schriftenreihe der
    Landesanstalt fur Gewässerkunde und Gewässerschutz der Landes
    Nordrhein-Westfalen* 33, 1-125 + 16 Tables.
Friedrich G. (1979)   Biologische Untersuchungen am Rhein in Nordrhein-
    Westfalen.   Vortrag auf der Jahrestagung der deutschsprachigen
    Limnologen 1979 in Schlitz (Hessen).   (Unpublished: data available
    from author.)
Friedrich G. & Viehweg M. (in press)   Recent developments of the phyto-
    plankton and its activity in the lower Rhine. *Verh. int. theor.
    angew. Limnol.* 22.
Herbst H.V., Friedrich G. & Heuss K. (1969)   *Biologische und chemische
    Untersuchungen zur Kennzeichnung der Beschaffenheit des Rheines am
    Ende einer extremen Niedrigwasserperiode,* 90 pp.   Landesanstalt für
    Gewässerkunde Nordrhein-Westfalen, Krefeld.
Herbst H.V., Friedrich G. & Heuss K. (1973)   Biologische und chemische
    Untersuchungen des Rheines bein Niedrigwasser im Oktober 1971.
    *Schriftenreihe der Landesanstalt fur Gewasserkunde und Gewasserschutz
    Nordrhein-Westfalen* 35, 1-44.
Hess F. (1959)   Probleme des Hochwasserschutzes am Niederrhein.
    Gewässer und Abwässer 23, 1-88.
Heuss K. (1975)   *Das Phytoplankton der Ströme Rhein und Donau - ein
    Vergleich. 18. Jahrestagung der Internationalen Arbeitsgemeinschaft
    Donauforschung der Societas Internationalis Limnologiae. Kurzrefe-
    rate* 1, pp. 217-229.   Bundesanstalt für Gewässerkunde, Koblenz.
Hild H.J. (1960)   Hydrobiologische Untersuchungen in einigen abwasser-
    belasteten niederrheinischen Gewässern. *Hydrobiologia* 16, 203-14.

Hild H.J. & Rehnelt K. (1965a) Öko-soziologische Untersuchungen an einigen niederrheinischen Kolken. *Ber. deutsch. bot. Ges.* 78, 289-304.

Hild H.J. & Rehnelt K. (1965b) Hydrobiologische Untersuchungen an niederrheinischen Gewässern. *Hydrobiologia* 25, 422-65.

Hild (H.) J. & Rehnelt K. (1967) Oko-soziologische Untersuchungen am Boetzelaerer Meer (Niederrhein). *Ber. deutsch. bot. Ges.* 79, 647-68.

Hoppe (1970) Die grossen Flussverlagerungen des Niederrheins in den letzteon 2000 Jahren und ihre Auswirkung auf Lage und Entwicklung der Siedlungen. *Forschungen zur deutschen Landeskunde* 189.

IAWR (Internationale Arbeitsgemeinschaft der Wasserwerke im Rhein-einzugsgebiet) (annual) *1 Arbeitstagungen. 2 Jahresberichte.* Sekretariat der IAWR, Postfach 8169, NL-1005 AD, Amsterdam, The Netherlands.

IKSR (Internationale Kommission zum Schutze des Rheins gegen Verun-reinigungen) (1977) *Bericht über die physikalisch-chemische Untersuchung des Rheinwassers* VII, 1972-1975. Koblenz. Also annual reports: *1 Zahlentafeln. 2 Tätigkeitsberichte.*

Internationale Kommission für die Hydrologie des Rheingebietes (1978) *Das Rheingebiet Hydrologische Monographie*, 279 pp. Staatsuitgeverij, Christoffel Plantÿnstraat 1, The Hague, Netherlands.

Jaag O. (1938) Die Kryptogamen-Vegetation des Rheinfalls und des Hochrheins von Stein bis Eglisau. *Mitt. Naturf. Ges. Schaffhausen* 14, 1-158.

Kinzelbach R. (1978) Veränderungen der Fauna des Oberrheins. Beitr. Veröff. Naturschutz. *Landschaftspflege Baden-Würtemberg* 11, 291-301. (Karlsruhe).

KIWA (1983) Toxikologische Untersuchungen des Rheinwassers. The Netherlands Waterworks Testing and Research Institute KIWA Ltd., Rijkswijk, Netherlands.

Knecht A. (1982) Das Wachstum von Wasserpflanzen im Hochrhein. *Deutsche gewässerkundliche Mitteilungen* 26(4), 109.

Knöpp H. (1957) Die heutige biologische Gliederung des Rheinstromes. *Deutsche Gewässerkundl. Mitt.* 1(3), 56-63. Koblenz.

Knöpp H. (1960) Untersuchungen über das Sauerstoff-Produktions-Potential von Flussplankton. *Schweiz. Z. Hydrol.* 22, 152-166.

Knöpp H. (1968) Stoffwechseldynamische Untersuchungsverfahren für die biologische Wasseranalyse. *Int. Rev. ges. Hydrobiol.* 53, 409-41.

Knöpp H. (1970) Neuere Studien zur Biologie und Biochemie des Rheins. *Gewässerschutz Wasser - Abwasser* 3, 67-82. Institut für Siedlungs-wasserwirtschaft der Rhein, Westf. Techn. Hochschule, Aachen.

Knörzer H.K. (1960) Die Salbei-Wiesen am Niederrhein. *Mitt. flor. soz.* N.F. 8, 209-21.

Knörzer K.H. (1963) Pflanzenwanderungen am Niederrhein. *Der Niederrhein* 30, 134-37. (Krefeld).

Koch W. (1926) Die Vegetationseinheiten der Linthebene. *Jb. St. Gallener Naturf. Ges.* 61, 1-144.

Kolkwitz R. (1912) Quantitative Studien über das Plankton des Rhein-
    stroms von seinen Quellen bis zur Mündung. Mitteilungen aus der
    Kgl. *Prüfungsanstalt für Wasserversorgung und Abwasserbeseitigung*
    16, 167-209. (Berlin). See also *Ber. deutsch. bot. Ges.* 30, 205-26.
Krause A. (1981) Über die Flora und Vegetation der dem Uferschutz
    dienenden Bruchsteinmauern - Pflastern und - Schüttungen am
    nördlichen Mittelrhein. *Natur und Landschaft* 56(7/8), 253-60
    (Stuttgart).
Krause W. (1975) Siedlungen gefährdeter Pflanzen in Baggerseen der
    Oberrheinebene. *Beitrage zur naturkundlichen Forschung Südwest-
    deutschlands* 34, 187-199. Karlsruhe.
Krause W. (1976) Veränderungen im Artenbestand makroskopischer
    Süsswasseralgen in Abhängigkeit vom Ausbau des Oberrheins.
    Schriftenreihe f. Vegetationskunde 10, 227-37. (Bonn-Bad Godesberg).
Krebs F. (1978) *Vergiftung der Selbstreinigung durch toxische Abwässer*
    182 pp. Bundesanstalt fur Gewässerkunde, Koblenz.
Kummer G. (1934) Die Flora des Rheinfallgebietes. *Mitt. Naturf. Ges.
    Schaffhausen* 11. (see pp. 68-73, Die Wasserflora des Rheines und
    seiner Ufer).
Landesamt für Wasser und Abfall Nordrhein-Westfalen (1982) *Gewässer-
    güteberich* 1981. (1983) Gewässergütebericht 1982, Düsseldorf.
Lange-Bertalot H. (1974) Das Phytoplankton im unteren Main unter dem
    Einfluss starker Abwasserbelastung. *Cour. Forsch. Inst. Senckenberg*
    12, 1-88.
Lange-Bertalot H. (in press) The algal communities of the Rhine. In:
    Kinzelbach R. *The Rhine.* Junk, The Hague.
Lange-Bertalot H. & Lorbach K-D. (1979) Die Diatomeenbesiedlung des
    Rheins in Abhängigkeit von der Abwasserbelastung. *Arch. Hydrobiol.*
    87, 347-63.
Lauterborn R. (1905) Die Ergebnisse einer biologischen Probenunter-
    suchung des Rheins. *Arbeiten a.d. kaiserlichen Gesundheitsamt* 22,
    630-52. (Berlin).
Lauterborn R. (1910) Die Vegetation des Oberrheins. *Verh. naturhisto-
    risch-medizinischen Vereins Heidelberg N.F.* 10, 450-502.
Lauterborn R. (1916-1918) Die geographische und biologische Gliederung
    des Rheinstromes. *Sitzungsbericht d. Heidelberger Akad. Wissensch.
    Math.-Nat.* Abt. 1: 6, 1-61; 2: 5, 1-70; 3: 7, 1-87.
LAWA (Länderarbeitsgemeinschaft Wasser) (1976) *Die Gewässergütekarte
    der Bundesrepublik Deutschland.* 16 pp. + map. Published by LAWA
    under chairmanship of: Minister für Landwirtschaft, Weinbau und
    Umweltschutz des Landes Rheinland-Pfalz.
LAWA (1977) *Grundlage für die Beurteilung der Wärmebelastungen von
    Gewässern.* Verlag Krach, Mainz.
LAWA (1980) *Die Gewässergütekarte der Bundesrepublic Deutschland.*
    16 pp. + map. Published by LAWA under chairmanship of: Minister
    fur Ernährung, Landwirtschaft, Umwelt und Forsten des Landes Baden-
    Württemberg.
Lehn H. (1972) Zur Trophie des Bodensee. *Verh. int. Verein. theor.
    angew. Limnol.* 18, 467-74.
Lelek A. (1976) Veränderungen der Fischfauna in einigen Flüssen
    Zentraleuropas (Donau, Elbe und Rhein). *Schriftenreihe für
    Vegetationskunde* 10, 295-308.

Lelek A. (1978) Die Fischbesiedlung des nördlichen Oberrheins und des südlichen Mittelrheins. *Natur und Museum* 108, 1, 1-9.

Liepolt R. (1967) *Limnologie der Donau. Eine monographische Darstellung.* Schweizerbart, Stuttgart.

Lohmeyer W. (1970a) Zur Kenntnis einiger nitro- und thermophiler Unkrautgesellschaften im Gebiet des Mittel- und Niederrheins. *Schriftenreihe fur Vegetationskunde* 5, 29-62. (Bonn-Bad Godesberg.)

Lohmeyer W. (1970b) Über das Polygono-Chenopodietum in Westdeutschland unter besonderer Berücksichtigung seiner Vorkommen am Rhein und im Mündungsgebiet der Ahr. *Schriftenreihe für Vegetationskunde* 5, 7-28. (Bonn.)

Malle H.G. & Müller G. (1982) Metallgehalt und Schwebstoffgehalt im Rhein. *Wasser Abwasser* 15, 11-15.

Marsson M. (1907-1911) Bericht über die Ergebnisse der vom 14. bis zum 21.10.1905 ausgeführten biologischen Untersuchung des Rheins auf der Strecke Mainz bis Koblenz. *Arbeiten aus dem kaiserlichen Gesundheitsamt (Berlin)* 25, 140-63 (1907); 28, 29-61 (1908); 28, 92-124 (1908); 28, 549-71 (1908); 30, 543-74 (1909); 32, 59-88 (1909); 33, 473-99 (1910); 36, 260-90 (1911).

Miegel H. (1961) Die Molluskenfauna des Xantener Altrheins. *Gewässer und Abwässer* 29, 7-12.

Miegel H. (1963) Untersuchungen zur Molluskenfauna linksrheinischer Gewässer im Niederrheinischen Tiefland und des Rheingebietes. *Gewässer und Abwässer* 33, 1-75.

Müller D. (1975) Zum Einfluss der Aufwärmung auf die mikrobiologen Stoffumsetzungen in Gewässer. *Deutsche Gewässerkundliche Mitteilungen* 19, 76-82.

Müller D. (1980) Die Bedeutung der Actinomyceten für die Geruchstoffbelastung des Rheins und des Rheinuferfiltrats, 122 pp. Dissertation, Universität Stuttgart.

Müller D. & Kirchesch V. (1981a) Bestimmung des "maximalen Nitrifikations-Sauerstoffverbrauchs unter standardisierten Laborbedingungen, MNSL" als Analogwert für die Biomasse der Nitrifikanten. *Vom Wasser* 56, 145-51.

Müller D. & Kirchesch V. (1981b) Nitrifikation in Fliessgewässern (Bedeutung - Messung - Berechnung). *Vom Wasser* 57, 199-214.

Müller G. (1979) Schwermetalle in den Sedimenten des Rheins. Veränderungen seit 1971. *Umschau in Wissenschaft und Technik* 79, 778-83.

Petran M. & Kothé P. (1978) Influence of bedload transport on the macrobenthos of running water. *Verh. int. Verein. theor. angew. Limnol.* 20, 1867-72.

Philippi G. (1973) Zur Kenntnis einiger Röhrichtgesellschaften des Oberrheingebietes. *Beitr. naturk. Forsch. Südw-Deutschland* 32, 53-95. (Karlsruhe).

Philippi G. (1977) Vegetationskundliche Luftbildinterpretation als Mittel zur Erfassung von Trophiestufen im Gewässerbereich am mittlere Oberrhein im Vergleich zur mittleren Saar. In: *Gewässerüberwachung durch Fernerkundung. Der mittlere Oberrhein im Vergleich zur mittleren Saar,* pp. 33-48. Bundesanstalt für Landeskunde und Raumordnung, Bonn-Bad Godesberg.

Philippi G. (1978a) Die Vegetation des Altrheingebietes bei Russheim. In: *Der Russheimer Altrhein, eine nordbadische Auenlandschaft. Natur- u. Landschaftsschutzgebiete Bad.-Württ.* 10, 103-267. (Karlsruhe).

Philippi G. (1978b) Veränderungen der Wasser- und Uferflora im badischen Oberrheingebiet. *Beih. Veröff. Naturschutz u. Landschaftspflege Bad.-Württ.* 11, 99-134. (Karlsruhe)

RIWA (Rijncommissie Waterleidingbedrijven) (1983) Jahresbericht 1982. Teil A. Der Rhein. Den Haag, Netherlands.

Schäfer W. (1973) Altrheinverbund am nördlichen Oberrhein. *Cour. Forsch.-Inst. Senckenberg* 7, 63 pp. (Frankfurt/Main)

Schiller W. (1974) *Der Gütezustand des Rheines in Nordrhein-Westfalen 1972-1974.* Internal publication of Landesanstalt für Wasser und Abwasser Nordrhein-Westfalen, Krefeld.

Schmitt A. (in press) Das Makrozoobenthos des Rheinkanals. Verh. d. 11. *Jahreshauptversammlung deutschen Gesellschaft Ökologie,* März 1981.

Schwoerbel J. (1959) Die biologische Gliederung des Rheinstromes. *Das Gas - und Wasserfach* 100(44), 1-6.

Seeler T. (1938) Uber eine quantitative Untersuchung des Planktons der deutschen Ströme unter besonderer Berücksichtigung der Einwirkung von Abwässern und der Vorgänge der biologischen Selbstreinigung. II Der Rhein. *Arch. Hydrobiol.* 30,

Seibt D. (1980) *Erfassung der organischen Drift im Rhein - ein Beitrag zur faunistischen Bestandsaufnahme am Niederrhein.* Diplomarbeit, Universität Köln.

Solmsdorf H., Lohmeyer W. & Mrass W. (1975) Ermittlung und Untersuchung des schutzwürdigen und naturnahen Bereiche entlang des Rheins (Schutzwürdigen Bereiche im Rheintal). *Schriftenreihe für Landschaftspflege und Naturschutz* 11, 186 pp. (Bonn-Bad Godesberg).

Sontheimer H. (1983) *Der Rhein im Jahre 1982.* Vortrag auf der 9 Arbeitstagung der IAWR in Köln, April 1983.

Soppe E. (1983) Verteilung des Makrozoobenthos im Querprofil des Rheins bei der Loreley. Verh. d. 11. *Jahreshauptversammlung der deutschen Gesellschaft für Ökologie,* März 1981.

Stieglitz W. (1980) Bemerkungen zur Adventivflora des Neusser Hafens. *Niederrhein, Jahrb.* 14, 121-28.

Thomas E.A. (1975) Kampf dem zunehmenden Wasserpflanzen bewuchs in unseren Gewässern. *Wasser. u. Engergiewirtschaft* 1/2, 1-8.

Tittizer T. (1977) *Untersuchung über die Verbreitung und die Ursachen des Larvenvorkommens der Köcherfliege* Hydropsyche contubernalis *McLachlan im Rhein bei Koblenz,* 4 pp. Ergebnisbericht Bundesanstalt für Gewässerkunde Koblenz.

Tittizer T. (1980) Die Biologie als Hilfsmittel bei der Erfassung des Geschiebetransportes in Gewässern. *BfG-Informationen* 2, Koblenz.

Trahms O.K. (1954) Der Rhein: Abwasserkanal oder Fischgewässer. *Mitt. Rhein. Ver. Denkmalpflege und Heimatschutz* 3/4, 1-8.

Wilson R.S. & Wilson S.E. (in press) A reconnaisance of the River Rhine using Chironomidae pupal exuviae (Diptera, Insecta). *Proc. 8th International Symposium on Chironomidae,* Jacksonville, Florida.

Wunderlich M. (1977)   Bewirken die herkömmlichen Kühlverfahren von
    Kraftwerken eine Aktivierung bakterieller Ruhestadien?   In:
    (Ed. I. Daubner) *II. Int. Hydrobiol. Symp. Smolenice 1975,* pp. 287-300
    Verlag slowak. Akad. Wissenschaften, Bratislava.
Wunderlich M. (1978)   Introduction of waste heat and effects on
    microorganisms.   *Verh. int. Verein. theor. angew. Limnol.* 20,
    1855-60.
Wunderlich M. (1980)   *Kühlwasserbelastung der Gewässer - Forschungs-
    bericht UFO-PLAN-Nr. 10204033,* 94 pp.   Bundesanstalt fur Gewässer-
    kunde, Koblenz.

11 · Neckar

I. PINTÉR & D.BACKHAUS

11.1 INTRODUCTION

The Neckar and the Main are the largest tributaries on the right bank
of the Rhine. The position of the Neckar in central western Europe
makes it a river typical of those with a moderately atlantic climate -
that is, its flow regime is not subject to extreme climatic factors.
The Neckar is also a typical middle mountain river; it lacks both high
mountains and a long lowland region. The downstream stretch is
however canalized, a feature shared with two neighbouring rivers, the
Main and the Mosel. The Neckar has a relatively large catchment area
in relation to river length, so that a large part of south-west Germany
belongs to it.

The lower valley of the Neckar has been populated for a very long
time. In the neighbourhood of Heidelberg there are remains of Stone
Age settlements and in the upper stretches there are numerous places
whose origin starts with Roman times. The beauty of the Neckar valley
is told in many stories and songs. It is, above all, the city of
Heidelberg, which has spread the fame of the river throughout Europe.

11.2 GENERAL DATA

DIMENSIONS

Area of catchment	14000 km^2
including: Enz	2228 km^2
Kocher and Jagst	3840 km^2
Length	370 km
Highest point in catchment (Hohloh-Berg)	980 m
Source	706 m
Mouth	86.5 m (i.e. drop = 619.5 m)
Slope	1.67‰

CLIMATE

Precipitation
The catchment mostly lies in the rain shadow of the Schwarzwald (Black
Forest), with annual totals in the lower regions in the range 600-800 mm
and reaching 900-1200 mm locally (Schwarzwaldrand, Frankish Schild).
The climate is more or less of Atlantic type, with SW winds predominating.

GEOLOGY
Predominantly in the "Südwestdeutsches Schichtstufenland".
(See Fig. 11 and text)

LAND USE
About three-quarters of the catchment is used for agriculture, and
the remainder (excluding conurbations) for forestry or similar activities.

POPULATION

Whole catchment	4.6 x 10^6 people
including central Neckar region	2.36 x 10^6 people

The main conurbations of Stuttgart, Heilbronn, Pforzheim, Heidelberg
and Mannheim have a density of 1000 people km^{-2}.

PHYSICAL FEATURES OF THE RIVER

Discharge ($m^3 s^{-1}$) at selected points

km from mouth	place	catchment (km^2)	absolute minimum	mean lowest	mean	absolute maximum
318	Oberndorf	694.7	0.76	1.98	7.9	288
202.6	Plochingen	4002	3.70	9.7	44.1	1150
125.1	Lauffen	7910	14.1	23.4	84.7	1270
61.4	Rockenau	12710	20.0	35.4	126.0	2150

Mean monthly discharge ($m^3 s^{-1}$) at Rockenau gauging station for 1951-1979

		influence
Nov	71	autumn rain
Dec	141	
Jan	163	
Feb	430	snow-melt
Mar	340	
Apr	220	spring
May	149	
Jun	91	
Jul	64	rain — poor summer and early autumn
Aug	58	
Sep	53	
Oct	54	

Regulation

  27 weirs on canalized stretch between mouth and river - km 200
  (upstream from mouth)

Substratum

  In uncanalized stretch: blocks, boulder, cobbles, pebbles

  In canalized stretch: silt and gravel

Temperature

  In uncanalized stretch: annual range from 0 to 22 ^{O}C

  In canalized stretch downstream of influence of cooling waters:
     winter temperatures never below 4-5 ^{O}C
     summer temperatures reaching 28-30 ^{O}C or 26-27 ^{O}C after complete
     mixing

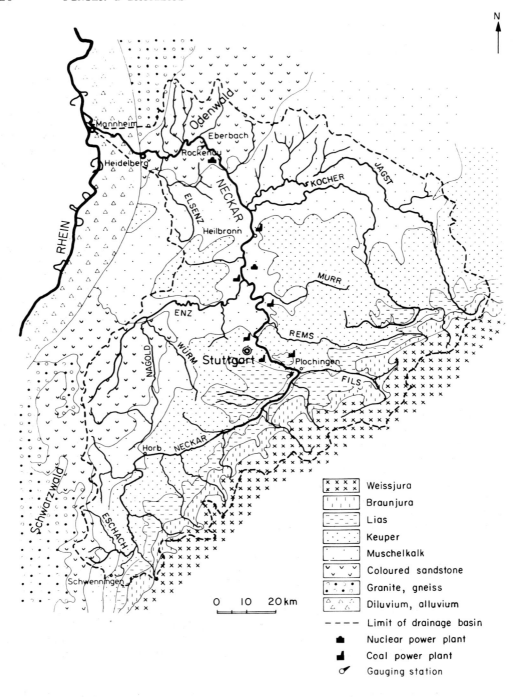

Fig. 11.1 Neckar, showing various geological formations, gauging stations and power stations.

11.3 PHYSICAL AND GEOGRAPHICAL BACKGROUND

11.31 PHYSIOGRAPHY AND GEOLOGY

The Neckar arises in a small area of woodland not far from the town of Schwenningen. Geographically its origin belongs with the eastern edge of the Black Forest, itself in the southern part of the 'Schwäbisch - Fränkisches Stufenland'. It is also a region well known geologically due to the Triassic and Jurassic rocks whose various layers come to the surface within a narrow region (Fig. 11.1). From its source the river first flows north-east with a mean slope of $0.28^{\circ}/oo$, but after 170 km it bends sharply to the north-west, eventually reaching the Rhine at Mannheim after 370 km, with a mean slope of $0.55^{\circ}/oo$. The Enz, Kocher and Jagst all have important effects on flow.

During its course the Neckar successively flows through, among other rock types, limestone, the Liassic covered Keuper rocks, Odenwald red sandstone, limestone again and, finally, the alluvial deposits of the Rhine valley. The river valley changes according to type of rock from V-shaped to broad U-shaped or sometimes with cliffs on one bank. Soils and land use also depend on the geology and these in turn affect the river. The predominantly calcareous materials of the catchment have an influence on river chemistry.

11.32 HYDROLOGY

Of the 14000 km^2 catchment, about 4000 km^2 belong to the uncanalized stretch down to Plochingen. The entry of the Enz doubles the catchment area, while the catchments of the Kocher and Jagst together total about the same as the Enz. The remainder of the catchment consists of small tributaries, mostly from the Odenwald. The main collecting ground of the Neckar lies therefore partly in the north-east part of the Black Forest and adjacent 'Schichtstufenland' as far as the Schwäbische Alb (upper Neckar, Enz) and partly in mountain and hill country of the so-called 'Fränkisches Schild' (Internationale Kommission für die Hydrologie des Rheingebietes, 1978).

The mean discharge per unit area as a result of precipitation within the catchment down as far as the gauging station at Plochingen)km 202.6 from mouth, catchment area = 4002 km^2) is only about 0.011 m^3 s^{-1} km^{-2}. The contributions per unit area from the Enz, Kocher and Jagst are similar. This may be compared with values for the western part of the Black Forest, which are in the 0.02 - 0.04 m^3 s^{-1} km^{-2}. The mean discharge of the Neckar rises from 44 m^3 s^{-1} at Plochingen to 126 m^3 s^{-1} at Rockenau and remains similar as far as the mouth (Fig. 11.2).

The Neckar shows very variable flows. At Rockenau gauging station discharge ranges from 20 to 2150 m^3 s^{-1}, with a mean of 126 m^{-3} s^{-1} (Landesanstalt für Unweltschutz, 1980: see Fig. 11.3). Low waters

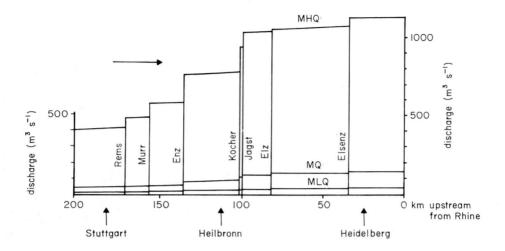

Fig. 11.2 Discharge in the canalized part of the Neckar for 1966–
1975. (Distance measured from entry in to Rhine). MHQ = mean of
annual maxima; MQ = mean; MLQ = mean of annual minima)

are the most important ecologically. These occur regularly in summer
and autumn. During this period the ratio between Neckar water and
water from outside sources (effluents, cooling water) is in the region
of 1 : 1 or even as low as 0.8 : 1 (Schaal, 1981).

11.33 CANALIZATION

In 1920 the first steps were taken towards the development of a
shipping route, which was finally completed in 1968. This affects
200 km of river from Plochingen (where the R. Fils enters) to the
mouth (Fig. 11.4). This stretch of the river forms a class IV waterway
("Waterway of international importance"). The 160.7 m drop in height
requires 27 locks, each of which has a double sluice and a weir;
almost all also have a small hydro-electric plant. The shipping canal
is at least 2.5 m deep and mostly 3 m.

    The length of time the water takes to pass down the canalized
stretch depends markedly on flow conditions. At the mean annual flow
minimum this is 800 hours, but only 95 hours at the mean annual flow
(Schmitz, 1971). This would result theoretically in a change of
current speed from 7 to 50 cm s^{-1}, if local variations could be neglected.

    Construction along the lower Neckar was carried out without fore-

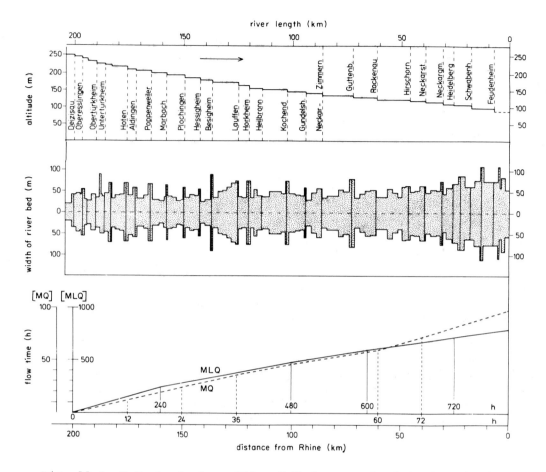

Fig. 11.3 Hydrological profile of Neckar.   A) altitude; B) width
of river bed, based on measurements; C) mean time for passage of
water under different flow conditions in 1966-1975. MQ = mean flow;
MLQ = mean of annual minima. (C, after Schmitz (1971))

thought for the likely effects on the river ecosystem. Such effects
are likely to be particularly pronounced, in view of the high input of
waste and warm water effluents. Thus river management nowadays has to
pay a lot of money to remove or ameliorate the problems caused (see
11.4).

11.34 LAND USE AND POPULATION

As a rough approximation almost one-quarter of the Neckar catchment is
covered by forest. The largest continuous areas are in the Black

Fig. 11.4 Aerial view of Neckar at km 47 (near Hirschhorn). Note canalized banks of river, weir with sluice for ships and turbulence created by ships. (Photo by permission of Regierung von Oberbayern Nr. G 7/88 048 Hersteller "Photogrammetrie GmbH")

Forest and Odenwald - that is, in the periphery of the region. Further, the Keuper heights are forested over a large area. In contrast, the central Lias and limestone are covered with good, heavy soils. These are the favoured area for settlements. Except for a few stretches with steep rocky slopes, the river flows here through a landscape which is used intensively.

The use of the land depends partly on soil quality and climate, partly on the type and size of farm. In the Neckar region there are various agricultural zones. The upper region as far as the vicinity of Stuttgart and also the higher limestone slopes along the Jagst are characterized by cereals. In the district north-east of Stuttgart there is a zone of root crops, which reaches as far as Odenwald. There are large areas of intensive farming (vegetables, fruit, berries, vine-yards) around Stuttgart and Heilbronn which contribute an additional nutrient input into the river (Fig. 11.4). In the downstream area (below Heidelberg) there are more root-crops, especially sugar-beet.

The structure of the settlements also differs in the various types of landscape. Large villages predominate in the regions which are favoured for farming, with up to about 2000 inhabitants. Such villages are often very old. The Black Forest and Keuper area were colonized later and the villages are mostly smaller and often little more than a cluster of farmhouses. The great part of the Neckar region is however transformed by industrialization.

There is an industrialized strip already in the upper Neckar, while other industrial regions follow the valleys of the Fils, Rems, Murr, Kocher and Enz. Some of them come together in the broader conurbation of Stuttgart with its industrial concentration and dense population. In hardly any other part of Germany are towns packed so densely (Fig. 11.5). Other industrial zones are around Heilbronn and Pforzheim, as well as in the Rhine - Neckar triangle at Heidelberg and Mannheim. 4.6 million people live in the Neckar catchment, of whom about half live in the middle Neckar region.

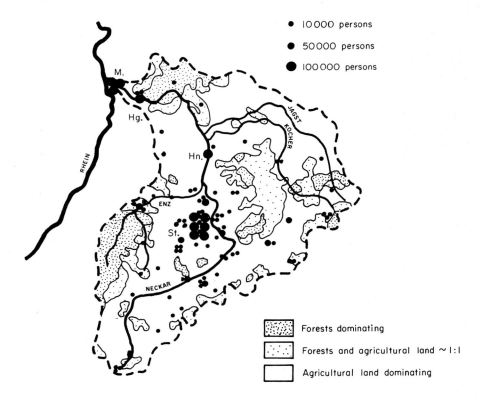

Fig. 11.5 Forests and agricultural land in the Neckar region. Hg, Heidelberg; Hn, Heilbronn; M, Mannheim; St, Stuttgart. (Modified from Internationale Kommission für die Hydrologie des Rheingebietes, 1978)

The characteristics of the economic structure of the Neckar region is the broadly scattered mix of agriculture and industry and the fact that the latter is specialized - heavy industries like coal and steel are lacking. Key industries are machine tools, car manufacture, electrical components, clocks, jewellery and textiles. Many of these industries have a high requirement for electrical power, which is supplied from the middle region of the Neckar itself. The need for cooling water leads to many problems in water management. Other critical problems resulting from the density of population and industry are the need for drinking and industrial water and waste disposal. The need for water is too great for the Neckar catchment alone, so drinking water is brought from the Bodensee and other districts at the rate of 8 - 10 m^3 s^{-1} at peak demand. Most of this extra water eventually reaches the Neckar as waste water.

11.4   WATER QUALITY PROBLEMS

Water quality in the Neckar is dominated by the need to use it for removal of wastes, its role as a coolant for power plants and the control of flow for energy supply and shipping. At times of low flow, sewage effluent contributes 30% of total flow below Stuttgart. What is more, periods of low flow coincide with warm weather. Heating is increased in the canalized stretch both because of damming and the requirement for cooling water. It is not unusual for temperatures to reach 25 $^\circ$C here. The higher temperatures favour microbial, oxidative breakdown of organic materials, thus reducing the oxygen content of the water. The natural oxygen supply from the air is reduced because of lower current speeds and the water depth of 3 m. Depletion of oxygen is particularly marked in autumn when temperatures are still high and algal photosynthesis is less effective.

A further effect of damming is the marked deposition of sediments which takes place in the middle stretch of the Neckar. The often relatively low efficiency of the sewage treatment plants leads to a relatively high organic content in these sediments (something up to 20%) and this results in an additional demand. Every year 1.8 x 10^6 m^3 of sediments are deposited in the canalized stretch in addition to the sediments which are already there and are estimated to be about 2 x 10^6 m^3. The extra sediments have to be removed by dredging, especially in the middle part, and their quality leads to further problems of how to get rid of the material which has been dredged. In addition to a high organic content, the sediments are enriched by heavy metals from the metal industry, cadmium being a particular problem. In recent times the manganese content has also increased at some sites, perhaps due to efforts to increase the oxygen content of the water. An increased lead content may also result due to its co-precipitation with MnII (G. Müller in symposium discussion: see Institut für Sedimentforschung der Universität Heidelberg, 1981).

The use of sediments for agriculture is impossible in view of its potentially toxic composition. It has been suggested that they should be deposited and the metals recycled. A favourite suggestion is to incorporate the dredged material with the materials normally used to make bricks, but up to now research has not shown how this can be done satisfactorily.

In order to deal with the problems of water quality, a cleaning-up programme was decreed by the Landesregierung of Baden-Württemberg in 1976. Priorities are put forward to make optimal use of financial resources and improve the situation in the water. Among possible approaches to improvement are:

    (i)   reduction in pollution;
   (ii)   addition of oxygen;
 (iii)   addition of extra water to the river.

The first was chosen for the initial goal up to 1980. The second approach has been used in some cases, with the oxygen enrichment of water used by hydro-electric plants; this approach is however adopted only at times of low flow, because of its high energy requirement.

The daily load of organic effluents added to the river is shown in Table 11.1. Most of the waste is directed through sewage treatment plants, there being only forty-two firms with their own treatment plant and separate effluent. These consist of ten metal works, three aluminium surface treatment works, seven chemical works, four cement and gravel works, eleven paper manufacturers, two glue factories and five concerned with food manufacture. The development of efficient sewage treatment (judged by BOD_5) in the effluent demands a high annual capital investment for building sewage treatment plants (Table 11.2). 1765×10^6 deutschmarks have been spent to get a 40% improvement in capacity, which has resulted in a 56% improvement in effluent quality.

Table 11.1 Daily addition of BOD_5 to various sections of the Neckar (Ministerium für Ernährung, Landwirtschaft Umwelt Baden-Württemberg, 1976).

section	BOD_5 (tonnes d^{-1})
above Fils	96.1
Fils to mouth of Enz	181.5
Enz	62.9
below Enz to mouth of Kocher	29.8
Kocher + Jagst	55.8
below Jagst to mouth	45.7

Table 11.2 Annual expenditure in the Neckar region on treatment works over the period 1975-1980.   (P, person; PE, person equiv.)

year	capacity $(P + PE) \times 10^6$	expenditure $(DM \times 10^6)$	mean quality of waste $(mg\ l^{-1}\ BOD_5)$
1975	7.3	185	54
1976	8.1	290	46
1977	8.7	250	36
1978	9.3	340	33
1979	9.4	360	25
1980	10.2	340	24

The potential energy in the dammed region is used to generate electricity. There are 24 hydro-electric plants, each with a fall in river height of 3 - 8 m. The flow requirements of individual plants range from 46 to 100 $m^3\ s^{-1}$. The annual output of all plants is around 55 MW (Neckar AG und Wasser-und Schiffahrsdirektion Stuttgart, 1971). Besides this direct requirement for water energy from the river, cooling water is needed for eight oil, coal or nuclear power plants with a maximum output of 3500 MW. These plants are confined to a stretch of only 122 km. The two nuclear plants lie at km 129 (Neckarwestheim: 858 MW) and km 77 (Obrigheim: 345 MW). If all the heat had to be removed by water alone, it would require 130 $m^3\ s^{-1}$ at Rockenau gauging station. Such a flow is not reached during 270 days a year, so quite apart from the ecological impact of heating, it is essential to use cooling towers.

Some of the power plants have already got cooling towers, which allow up to 70% of the additional heat energy to be lost to the atmosphere when meteorological conditions are favourable. The nuclear power plant at Neckarwestheim, which commenced in 1976, is fitted with cooling towers which use water in three different ways to maximize the efficiency of heat loss. At present the older power plants have been ordered to improve their cooling systems. This is very urgent if the present plans are to be fulfilled by the end of the 1980's, because by then the maximum output of all plants will reach 6000 MW. The maximum temperature permitted for warm water effluents is 30 - 35 $^{\circ}C$ (depending upon origin), the maximum rise in river temperature is 5 K and the maximum temperature in the river itself is 28 $^{\circ}C$. State government regulations control the rise in temperature permitted by each of the eight plants. A complication of these plants is that the extent of pollution is determined by changes in oxygen (see below) as well as temperature.

For the cooling requirements of the Neckarwestheim power plant (40 $m^3\ s^{-1}$), there are both temperature and oxygen regulations. Sufficient oxygen must be added by the power plant to compensate for the decrease that would occur due to increased microbial activity as a result of the temperature rise. When the oxygen concentration in the Neckar is very low, then an exception to the rules is permitted. The whole river at Neckarwestheim is allowed to rise by an extra 2 K, but the oxygen must be increased by the extra amount corresponding to this temperature rise. This regulation was drawn up as a result of ecological considerations.

Energy planning on the Neckar has involved voluminous studies and calculations (Ministerium für Ernährung, Landwirtschaft and Umwelt, Baden-Württemberg, 1973). These have resulted in decisions that power plants built in the future must have circulating cooling water and that the existing power plants must reduce their heat waste at critical times. Because cooling towers cause a considerable loss of water by evaporation, energy-producing firms in Baden-Württemberg are obliged to compensate for such loss. This will probably require further reservoirs to be built.

11.5 PHYSICO-CHEMICAL FEATURES

The annual amplitude in water temperature in the Neckar is marked; it follows the natural yearly changes in radiant energy and air temperature. For the non-canalized Neckar a mean annual temperature around 13 $^{\circ}$C (at Plochingen) is characteristic. This rises to about 16 $^{\circ}$C near the entrance to the Rhine. If it is considered that flow increases four-fold along this stretch and the flows from tributaries tend to reduce the temperature of the main river, then it will become clear that the additional heat load due to warm-water effluents is considerable. There are no effluents in the last 77 km of the river and therefore this part can be regarded as a cooling stretch. Another indication of the effects of the heat inputs is that the winter temperature seldom drops below 5 $^{\circ}$C in the canalized stretch. In summer temperatures reach 24 - 25 $^{\circ}$C here, or exceptionally even 27 $^{\circ}$C, as during the extremely hot summer of 1976.

The ground level of the main mineral components (Ca^{2+}, Mg^{2+}, Na^+, Cl^-, SO_4^{2-}, HCO_3^-) is high as a result of the geology of the catchment. This results in the high conductivity value of 750 μS cm^{-1} where the Neckar enters the canalized stretch (Fig. 11.6). This is reduced slightly by the rivers from the Odenwald which have a low mineral content, but their flows are too little to have much effect. More important for the chemistry of the river are the effluents from Stuttgart sewage treatment plant (km 174) and predominantly those from the salt industry at Heilbronn (km 111). These latter lead to marked increases in Cl^-, Na^+ and Ca^{2+}, with resulting mean conductivities of at least 1250 μS cm^{-1}. Most of these contents are conservative, with little loss on passing downstream due to chemical or biological activities.

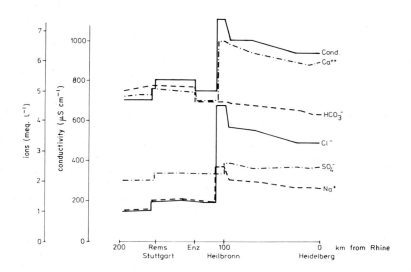

Fig. 11.6 Changes in conductivity and concentration of various ions on passing down the canalized Neckar in 1978-1979.

The changes in pH taking place annually in the canalized stretch are strongly influenced by primary production, with a range of 7.0 - 8.6. The mean value along this stretch varies only little (7.6 - 7.8). The diverse sewage effluents have little effect on the strong buffering capacity of Neckar water.

The Neckar is strongly influenced by the addition of organic wastes. Because of the unfavourable hydrological conditions in the canalized stretch (little turbulence, great mean depth, strong sedimentation), this leads to a marked reduction in oxygen content, as discussed in more detail below. The load of persistent organic compounds is however low in comparison with the Rhine and lower Main, because chemical industry is sparse in the Neckar region. In contrast, the concentrations of most heavy metals are - especially in the sediments - higher than in the Rhine and Main, as a result of the metal treatment industries.

The Neckar is in practice free of effluents for only a few kilometres. It is polluted as soon as it enters the town of Villingen-Schwenningen where it receives effluent from the town sewage treatment plant (115000 persons + person-equivalents). In the 170 km of the upper stretch, and indeed until the beginning of the canalized stretch, the tributaries, Eschach, Eyach and Ammer contribute a considerable organic load in addition to effluents reaching the Neckar directly from the towns of Rottweil, Horb and especially Tübingen (165000 persons + person equivalents). The situation was particularly critical until sewage treatment plants were built at Villingen-Schwenningen (in 1979) and Tübingen (in 1978). BOD_5 values of 15 mg l^{-1} were no exception when river flows were low; dissolved organic carbon was around 8 mg l^{-1}. However self-purification is generally adequate because of the good oxygen supply, small river depth and strong slope. The improved effluents resulting from the new treatment plants have halved the organic load in the river. A further drop in BOD_5 (to values less than 6 mg l^{-1}) is predicted when tertiary treatment is added to the Villingen-Schwenningen plant. In the canalized stretch the critical points are in the region of Stuttgart and Heilbronn, but the entry of the Fils, Rems and Murr are also quantitatively important. The sewage treatment plant of Stuttgart was for long particularly inadequate for the 1.25 million persons + person equivalents and the 3 - 4 m^3 s^{-1} effluents which were always a very heavy load for the Neckar until sewage treatment was extended in 1981.

In the lower stretch of the Neckar from the entry of the Kocher to the mouth the density of population is less. Both because of this and because the necessary improvements to sewage treatment plants have mostly been made already, the additional load added to the river is lower. Heidelberg is still the critical part on this stretch. The results for dissolved organic carbon in 1973-1976 and 1979 (Fig. 11.7) show this clearly, though there was a considerable decrease in 1979 due to improvements made in the intervening years. This change was due mainly to the reduction in biologically degradable substances.

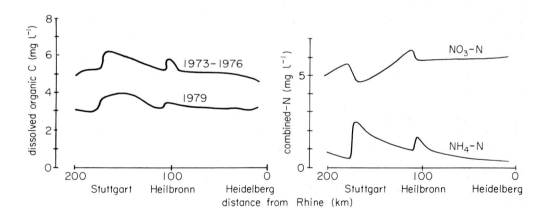

Fig. 11.7 Comparison for dissolved organic carbon (DOC) between the mean for 1973-1976 and that for 1979. (Data based on water taken continuously, frozen and analyzed once every two weeks)

Fig. 11.8 Changes in NO_3-N and NH_4-N on passing down the canalized Neckar in 1979. (Data based on water taken continuously, frozen and analyzed once every two weeks)

Ammonium reaches a mean concentration of 25 mg l^{-1} N in some effluents and there are marked rises in the river below Stuttgart and Heilbronn (means of 2.5 and 2 mg l^{-1} NH_4-N, respectively). There is intensive nitrification (Fig. 11.8) below these effluents and denitrification probably also plays an important role below Stuttgart. The NH_4-N load in the river shows characteristic annual changes, being lower in summer and higher in winter, because nitrification is less at lower temperatures.

The concentration of inorganic phosphate is strongly influenced by anthropogenic factors. The mean average concentrations in the canalized stretch lie between 1 and 1.5 mg l^{-1} PO_4-P. Phosphate in the river shows a quasi-conservative character. Although it is removed by organisms, especially plants, the amount removed is not enough to have a marked effect on aqueous concentrations. Light is the main factor limiting plant growth in the canalized stretch of the Neckar, not nutrients.

11.6 BIOLOGICAL COMPOSITION

11.61 BIOCOENOTIC ZONATION

Because of its geographical position in the south-west German middle mountain region, the slope and temperature regime of the Neckar mean that it belongs naturally to the group of summer-warm rivers, with a mountain stream character in the upper and middle stretches. On passing downstream about half the river would fall into the trout and grayling zones. In the upper and middle stretches, the natural biocoenotic zonation is disturbed by human and industrial effluents and, although the condition of the river has been improved in recent years, it is probably still the sewage load which is most responsible for the absence of benthic species that would otherwise be expected. In particular, many species of stenoecious Plecoptera, Ephemeroptera and Coleoptera have either disappeared completely or are present in very low populations. Fish populations were disturbed for a long time, but are now beginning to build up again.

The physical features of the present-day canalized stretch (Fig. 11.9) indicate that it would belong formally to the barbel zone, but in practice this species is not present. This stretch of the river has changed even more than those upstream, with its lentic conditions, raised temperatures and high sediment load all being important biotic factors.

11.62 OPEN WATER

As would be expected in a dammed river, there is a marked development of phytoplankton. Interpretation of the physical and chemical data

Fig. 11.9 Typical canalized stretch (km 137, near Besigheim).

for this part of the river suggest that nutrients should never be a limiting factor for algal growth (Pintér & Bauer, 1980). Algal populations, especially those dominated by diatoms, are often sufficiently dense to colour the water, but surface blooms do not form. Populations are apparently limited by the poor light penetration. Secchi depths are mostly less than 50 cm and the compensation point is around 2 m on bright summer days, less under other conditions. The possibility, that the concentration of toxic materials in effluents is sometimes sufficiently high to influence phytoplankton growth, requires research.

There is usually a distinct development of planktonic algae in March, with populations reaching around 1000 individuals ml^{-1} (Fig. 11.10). By April this is about one order of magnitude larger and in summer densities are still higher, with extreme records of about 60000 individuals ml^{-1} (Backhaus & Kemball, 1978).

Besides the high nutrient concentrations, other factors which affect phytoplankton growth are input of light energy, water temperature and flow. Phytoplankton density shows a positive correlation with the first two factors and a negative one with the last (Backhaus & Kemball, 1978). Increases in flow above 50 m^3 s^{-1} (at Plochingen) bring about a marked reduction in population density.

It is possible to distinguish three different phytoplankton regions

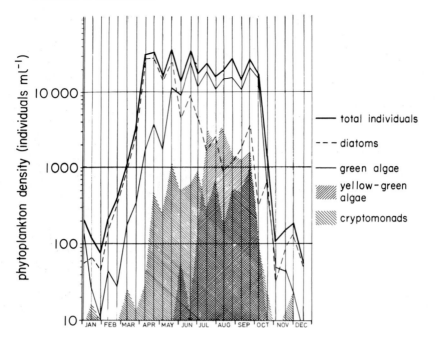

Fig. 11.10 Temporal changes through 1974 in phytoplankton density in Neckar at km 94.2 (near Gundelsheim). (Data based on samples taken at 2-weekly intervals)

on passing down the canalized stretch. Above Stuttgart there is a relatively high biomass with 2 - 2.3 mg fresh weight l^{-1}. From Stuttgart to Heilbronn this drops to about 1 mg l^{-1}, but below Heilbronn it increases again, with values locally as high as 3.5 mg l^{-1}. The decrease between Stuttgart and Heilbronn is probably due to the heavy sewage effluent loading. Chloride and conductivity are two obvious variables which increase below Heilbronn, but it is an open question whether these are in any way responsible for the increase in algal biomass.

Some 300 algal taxa have been recorded for the Neckar (Backhaus, 1980), but only a few make up the bulk of the biomass. Among these are *Stephanodiscus hantzschii*, *S. subsalsus*, *Scenedesmus* spp. and various other Chlorococcales. *Stephanodiscus* forms almost 90% of the biomass in spring, but by July green algae form more than 80% of the biomass (based on fresh weight) (Fig. 11.11). Coenobial forms typically grow as single cells, a feature well known in algal culture studies.

Among the grazers, rotifers and ciliates hold first place. Only during periods of summer low flow, when the residence time of the water behind the dams is long, do large populations of cladocera and

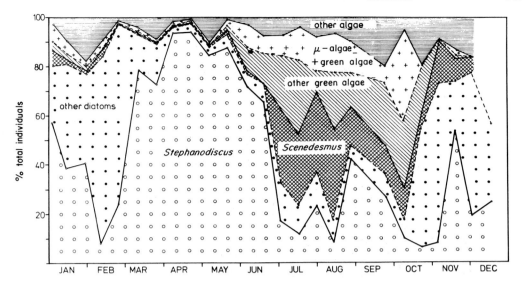

Fig. 11.11 Relative proportions of various algal groups in Neckar phytophyton at km 94.2 (near Gundelsheim). (Data based on samples taken at 2-weekly intervals; compare with Fig. 11.10)

copepods build up. Relatively little is known about the species composition of the zooplankton, apart from a detailed study of the rotifers (Rapp, 1973). Over 20 species of rotifer have been recorded, but only *Rotaria rotaria, R. neptunia, Brachyonus calyciflorus, Chlanis dilata, Asplanchna brightwelli* and some species of *Synchaeta, Polyarthra* and *Filina* build up high populations. Although there have been no detailed studies, it seems probable that grazing plays a relatively minor role in reducing phytoplankton populations.

As mentioned above, fish populations have started to increase since 1978 when improvements in water quality were made. During the period of worst pollution, 1960-1970, fish-kills were a regular happening. The critical oxygen conditions and toxic materials permitted only 'white fish', a species-rich group predominantly of cyprinids. Today the fish populations of the upper Neckar approach those of forty years ago, before the period of economic expansion. So now at least in the upper stretch there are again carp, tench, eel, pike, zander, as well as trout and nase (Schoch, 1981). Most of them are stocked periodically by angling associations.

11.63 RIVER BED

Few studies have been made of either the macro- or micro-vegetation of the river bed, so it is impossible to give a general picture. There is sometimes a strip of *Potamogeton pectinatus* in the canalized stretch (Hoffman, 1980), mostly between 9.5 and 2 m wide. Stones in the wave zone of the river bank are often covered with *Cladophora glomerata* and

Vaucheria sp. or a thick layer of filamentous blue-green algae. The bulk of this last is made of species of *Phormidium*, such as *P. autumnale*, *P. favosum* and *P. tenue*.

The periphyton consists largely of diatoms, with about a hundred species recorded. Of these, the following become abundant at times: *Achnanthes lanceolata*, *Cocconeis placentula*, *Diatoma vulgare*, *Gomphonema angustatum*, *Navicula avenacea*, *N. mutica*, *Nitzschia capitellata*, *N. palea*, *N. paleacea*, *Surirella ovata*, *Synedra ulna*. A re-invader is the red alga *Thorea ramosissima*; although known from old records this had disappeared until recently (D.B., unpublished).

The macrobenthos of the upper Neckar shows no obvious differences from those of other large river systems in the region, such as the headwaters of the Danube. The species composition of the littoral region of the middle and lower stretches is that typical for lowland rivers. The communities differ relatively little whether they are subject to strong or weak influences of sewage. *Asellus aquaticus*, various bryozoa, sponges, Hirudinea, Gastropoda and Turbellaria are abundant everywhere. Communities can be differentiated only by differences among the less common species. Heterogeneity arises because of differences in substratum. If the larval forms of nematodes, which are recognizable only with difficulty are excluded, then the numbers of species range in various situations from 12 to 24. Especially typical of the shipping stretch of the Neckar are *Asellus aquaticus*, masses of *Spongilla lacustris*, *Ephydatia fluviatilis*, *Plumatella repens*, *P. fungosa*, *Fredericella sultana*, *Bithynia tentaculata*, *Acroloxus lacustris*, *Planaria torva*, *Dendrocoeleum lacteum* and - recently - *Dugesia tigrina*.

The benthic organisms of the sediments in the middle of the river are known less well. There is no stretch of water which is totally still and some of the above species also occur here. The most obviously different community is that of the soft sediments just above the dams and in harbours. Such sediments, with their high organic content, are dominated by members of the *Chironomus thummi* group. This is especially the case where the influence of sewage effluents is somewhat stronger, such as below Stuttgart and Heilbronn. Tubificids dominate where the organic component of the sediment is less, while *Lumbriculus variegatus* can form large masses in intermediate conditions. In contrast to the other two groups, *L. variegatus* is influenced particularly by the chemical condition of the sediment, while mineral composition, size and structure of particles are less important. The mussel *Dreissena polymorpha* has recently invaded the river, following the improvements in water quality. Insect larvae of the genera *Hydropsyche* (Trichoptera) and *Ischnura* (Odonata) appear in noticeable numbers only in lateral channels with faster flows.

11.7 BIOLOGICAL PROCESSES AND THE OXYGEN CONTENT
OF THE RIVER

11.71 INTRODUCTION

Canalization, heavy loading with sewage effluents and the requirements
for cooling water all have a marked influence on biological processes
in the river. Optimal conditions for biological activity occur when
three factors operate together:

    (i)   long residence time
   (ii)   high content of organic and inorganic nutrients
  (iii)   high water temperature.

The oxygen supply is much influenced by processes of O_2 consumption,
which is sometimes hardly compensated by physical and biological O_2
input.

11.72 PHYSICAL INPUT OF OXYGEN

In the canalized stretch the passage of oxygen from air to water is
slow, especially under low flow conditions when turbulence is also low.
Hoffmann's (1980) calculations showed an oxygen input of about 0.03 mg
l^{-1} h^{-1}. Accordingly the velocity constant for natural reaeration
(K_2) is also very small at times of low flow, with values between 0.15
and 0.25 d^{-1} (Pintér, 1980). When flows are higher, reaeration is
better because of the increased turbulence and because more water
passes over the weirs. This last takes place in flow reaches from 45
to 100 m^3 s^{-1}, depending on the size of the turbines. The strong
influence of turbulence on the Neckar has been shown by correlations
between oxygen load on the one hand and flow and water temperature on
the other (Pintér, 1980). Multiple correlations show high values for
this calculation ($r^2 \geqslant 0.76$).

11.73 BIOLOGICAL INPUT OF OXYGEN

Among processes adding oxygen to the water, phytoplankton phytosynthesis
is predominant in summer and oxygen production depends on algal biomass
and factors influencing its photosynthesis.

The relationship between the amount of alga per unit water column
determined by the standard laboratory photosynthesis test (Ph_{OPL}), the
OPL test: see Fig. 11.12, photosynthetic active radiation (PAR) and
the rate of photosynthetic input of oxygen to the Neckar (Ph_{Neckar})
are given by the empirical equation (Hoffmann, 1980):

$$Ph_{Neckar} = F \cdot Ph_{OPL} \cdot a \cdot PAR^b$$
where F is a calibration factor

        a, b are calculated constants.

(Hoffmann's symbols have been changed for use with English text). For
fine days in summer, Ph_{Neckar} was calculated to be 1.75 mg O_2 l^{-1} h^{-1}
in the upper 50 cm; this value dropped to 0.125 mg l^{-1} h^{-1} at a depth
of 1 m. On dull days the oxygen input drops rapidly to zero (Hoffmann,
1980).

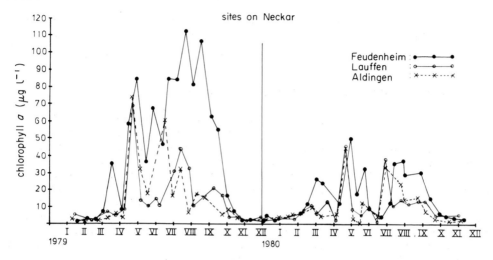

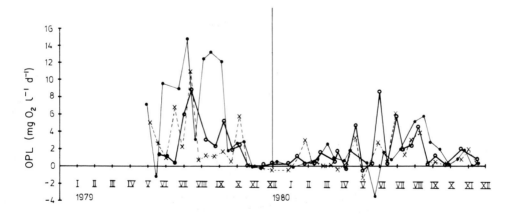

Fig. 11.12 Changes in chlorophyll *a* and OPL (oxygen production in
laboratory) through 1979-1980. (Data based on samples taken at 2-
weekly intervals; oxygen production in laboratory measured at 20 °C,
using light and dark bottles)

Measurement of the net oxygen production-potential i.e. photosynthesis under standard laboratory conditions (Ph_{OPL}) showed a peak of

15 mg O_2 l^{-1} d^{-1} during a two-year study (Fig. 11.12). With very dense algal populations, this value can reach 18 mg l^{-1} d^{-1}. There are marked differences not only between months but also between years. In the cold and wet summer of 1980, the peak value algal biomass and Ph_{OPL} reached half that of the previous warm, dry year.

In comparison with phytoplankton the contribution of macrophytes to the the overall oxygen load of the river can be neglected: Hoffmann (1980) calculated a mean value of 0.1 - 0.2 mg O_2 l^{-1} d^{-1}.

11.74 MICROBIAL OXIDATION OF ORGANIC COMPOUNDS

The high concentrations of organic substances in the Neckar result from the degradation both of the biomass which is produced autochthonously and also of allochthonous materials such as waste effluents. These substances are in general broken down according to a first order reaction (Pintér, 1980). A large part of the easily oxidizable organic material is broken down in sewage treatment plants before it reaches the river. The mean value for BOD_5 in the river in 1979/80 was about 8 mg l^{-1} over much of its length (Fig. 11.13). Peak values were much higher, especially in 1980. These can result from storm run-off disturbance of the bottom sediment or the failure of a sewage treatment plant. The results are quite similar to those found for the Main (Tobias, 1976).

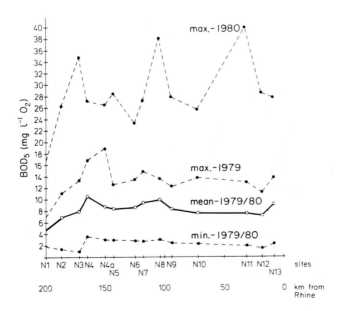

Fig. 11.13 Changes in BOD_5 on passing down the canalized Neckar in 1979-1980. (Data on samples taken at 2-weekly intervals). Maxima for 1979 and 1980 shown separately to indicate of high flow in 1980

Large areas of the river bed, especially upstream of lock-gates, are covered with fine sediment. A heavy deposition of sediments is continually taking place. As well as old sediments, which can be re-suspended by the movements of ships, sediments recently brought into the river become deposited as the river passes downstream. Microbial oxidation in these sediments adds to the overall oxygen-deficit in the water column. The intensity of oxygen consumption by these sediments has been found to be correlated positively with the oxygen concentration in the free water (Hoffmann, 1980). Oxygen consumption was found experimentally to range between 1 and 9 g m^{-2} d^{-1}, depending on the oxygen concentration of the water. This leads to decreases in oxygen concentration in the free water of between 0.35 and 3.0 g m^{-3} d^{-1}. Because of this relationship between oxygen consumption and oxygen in the environment, the efficiency of artificial aeration is reduced considerably.

11.75 AMMONIUM OXIDATION AND NITRATE REDUCTION

As can be seen from Fig. 11.8, the ammonium content of the river water is especially high between Stuttgart and Heilbronn. This is influenced to a marked extent by the rate at which bacterial nitrification takes place. This process proceeds according to the equations:

$$NH_4 + 1\tfrac{1}{2}\,O_2 \xrightarrow{\;Nitrosomonas\;} NO_2^- + H_2O + H_2 + 76\ kcal.$$

$$NO_2^- + \tfrac{1}{2}\,O_2 \xrightarrow{\;Nitrobacter\;} NO_3^- + 24\ kcal.$$

Noticeable nitrification takes place only when water temperature exceeds 15 °C and the rate increases markedly with a further rise in temperature (Q_{10} = 2-3). In the middle Neckar, where the natural physical supply of O_2 is particularly inadequate, ammonium oxidation stresses the O_2 budget in the warm months, especially when flows are low. It is difficult to quantify the relative demands for O_2 of the oxidation of organic materials and of the oxidation of ammonium, but the extremely high ammonium concentration in some stretches suggests the latter process is sometimes even more important than the former.

Reductive reactions can also occur, whereby nitrate is reduced to ammonium or N_2 (denitrification):

$$NO_3 \longrightarrow NO_2 \longrightarrow NO \underset{\longrightarrow NH_2OH \longrightarrow NH_3}{\overset{\longrightarrow N_2O \longrightarrow N_2}{\Big\langle}}$$

The Neckar appears to resemble other rivers with a high suspended matter and detritus content in that denitrification occur while oxygen is still present in the water. This takes place on oxygen-free microzones around particles of sediment. Denitrification probably plays an important role below Stuttgart, where there is a marked drop

in the nitrate concentration of the water (Fig. 11.8); partially
anaerobic conditions exist in the lowermost part of the water column.

11.76 PLANT AND ANIMAL RESPIRATION

Because of the high phytoplankton density in the Neckar, account
must be taken of its potential for high respiratory activity. Negative
balances are most likely to occur at times (as in autumn) when high
temperature is combined with low light intensity. On the other hand
the respiratory requirement for oxygen of the littoral and benthic fauna
is so slight that it can be ignored when drawing up an oxygen budget.

11.77 OXYGEN BALANCE IN THE RIVER

The balance of O_2-producing and O_2-requiring processes differs along
the length of the Neckar. The balance shifts towards the latter,
particularly under the influence of effluents in the greater Stuttgart
region (km 150 - 180), where summer oxygen-deficits are frequent and
extend well into the water column. The occurrence of low flows during
the warmer months is particularly likely to lead to deficits (Fig.
11.14). The high rate of oxygen production in summer during periods of
high light intensity hinders O_2 concentrations from dropping to zero,
but there is an obvious alteration in concentrations. Completely
anaerobic conditions used to occur regularly in autumn; photosynthetic
activity is low at this time of year, yet water temperatures are still
high enough for effective biodegradation of dead plankton. Such
conditions seldom occur nowadays because of the general improvements
in effluent quality.

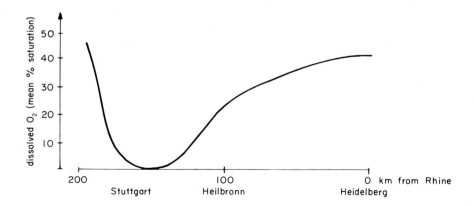

Fig. 11.14 Changes in dissolved oxygen on passing down the canalized
Neckar in 1972-1976.

There is a notable improvement in oxygen conditions if flows at Plochingen gauging station exceed about 50 $m^3 s^{-1}$. Besides natural aeration along this stretch and the higher flow/effluent ratio, aeration during passage over weirs is also important. Most power stations on the middle Neckar have been designed to work for a flow of 45 - 50 $m^3 s^{-1}$. Any surplus is led downstream by a weir overfall, where considerable re-oxygenation can occur (Pintér, 1979). As a result, at times of high flow in winter oxygen concentrations approach saturation.

11.8 CONCLUDING COMMENTS

Even among European rivers, the influence of man on the Neckar has been especially pronounced. The long canalized stretch has a profound effect on the ecology of the river, with its locks, shipping channel, power stations and associated thermal enrichment, and municipal and industrial effluents. When development was taking place along the middle and lower Neckar, the effects of effluents and calefaction were either unknown or underrated. The many problems have led to steps to eliminate, or at least reduce, harmful features along the canalized stretch, although the question of economic needs versus the ecological health of the river must always be kept in the foreground. In order to maintain conditions which are at least reasonable it is important that the ratio between volumes of freshwater and of the various effluents does not drop below about 1 : 1.

Among the most important problems is the extremely high rate at which sediment is deposited in the canalized part of the river (180000 $m^3 yr^{-1}$). This not only reduces the cross-sectional area of the river, but also leads to greatly increased microbial activity, which is itself enhanced by the warm water, and hence is an important factor leading to reduced oxygen concentrations. It is particularly difficult to deal with these sediments because of their relatively high content of heavy metals, making it unsuitable for use on agricultural land. Up to now no satisfactory solution has been found for this problem.

The drop in oxygen concentrations during the warmer months is another serious problem. This results from the influence of microbial activity on organic materials and ammonia. Although there has been a considerable improvement in recent years in the quality of organic materials entering from effluents, levels still remain high. Ammonia is also important because of its potential toxicity to fish, especially if pH values are high. Further measures are needed to deal with this problem.

Other aspects may become increasingly important in the future. A reduced sediment load would lead to increased light penetration, which, combined with high concentrations of plant nutrients, would in turn favour very high phytoplankton crops. The first signs are already at hand. A stage may even be reached when the main biological problem is not a saprobic one, but one resulting from eutrophication.

Acknowledgements

We thank Dr H. Buck (Stuttgart) for help with the account of zoobenthos.

REFERENCES

Backhaus D. (1980) Phytoplanktonuntersuchungen im Freiwasser des Neckars. In: Studien zum Gewässerschutz 4, 312 pp. *Landesanstalt für Umweltschutz Baden-Württemberg - Institut für Wasser- und Abfallwirtschaft*, pp. 170-229, Karlsruhe.

Backhaus D. & Kemball A. (1978) Gewässergüteverhältnisse und Phytoplanktonentwicklung im Hochrhein, Oberrhein und Neckar. *Arch. Hydrobiol.* 82, 166-206.

Gradmann R. (1931) Süddeutschland. 2. Die einzelnen Landschaften, 553 pp. Verlag J. Engelhorns Nachf., Stuttgart.

Hoffmann M. (1980) Biologischer Sauerstoffumsatz im Neckar/Photosynthese, tierische Atmung, mikrobielle Sauerstoffzehrung im Sediment. In: Studien zum Gewässerschutz 3, 276 pp. *Landesanstalt für Umweltschutz Baden-Württemberg - Institut für Wasser- und Abfallwirtschaft*, pp. 2-195, Karlsruhe.

Internationale Kommission für die Hydrologie des Rheingebietes (1978) Das Rheingebiet. - Hydrologische Monographie, Kartenteil. Bonn, Paris (no fixed office)

Landesanstalt für Umweltschutz Baden-Württemberg (1980) Reglement für den Kühlbetrieb des KKW Neckarwestheim, 42 pp. *Landesanstalt für Umweltschutz Baden-Württemberg*, Karlsruhe.

Landesanstalt für Umweltschutz Baden-Württemberg - Institut für Wasser- und Abfallwirtschaft (1980) Deutsches gewässerkundliches Jahrbuch, Sonderheft Land Baden-Württemberg, Donau- und Neckargebiet Abflussjahr 1979, 215 pp. *Landesanstalt für Umweltschutz Baden-Württemberg - Institut für Wasser- und Abfallwirtschaft*, Karlsruhe.

Ministerium für Ernährung, Landwirtschaft und Umwelt Baden-Württemberg (1973) Wärmelastplan Neckar. Wasserwirtschaftsverwaltung Heft 3, 18 pp. *Ministerium für Ernährung, Landwirtschaft und Umwelt Baden-Württemberg*, Stuttgart.

Ministerium für Ernährung, Landwirtschaft und Umwelt Baden-Württemberg (1976) Sanierungsprogramm Neckar, Plochingen bis Mannheim. Wasserwirtschaftsverwaltung Heft 5, 45 pp. + 13 maps. *Ministerium für Ernährung, Landwirtschaft und Umwelt Baden-Württemberg*, Stuttgart.

Neckar AG und Wasser- und Schiffahrtsdirektion Stuttgart (1971) Der Neckarausbau zwischen Mannheim und Plochingen - Fünfzig Jahre Neckar-Aktiengesellschaft, 39 pp. *Neckar AG und Wasser- und Schiffahrtsdirektion Stuttgart*, Stuttgart.

Pintér I. (1979) Naturmessungen zum Sauerstoffeintrag durch verschiedene Belüftungsverfahren. In: Natur- und Modellmessungen zum künstlichen Sauerstoffeintrag in Flüsse, Symposium Darmstadt 1979. Schriftenreihe des Deutschen Verbandes für Wasserwirtschaft und Kulturbau (DVWK), Heft 49, pp. 119-25.

Pintér I. (1980) Mikrobielle Sauerstoffzehrung und physikalischer Sauerstoffeintrag im Freiwasser des Neckars. In: *Landesanstalt für Umweltschutz Baden-Württemberg - Institut für Wasser- und Abfallwirtschaft*, Karlsruhe. Studien zum Gewässerschutz Nr. 3, 276 pp., pp. 196-254.

Pintér I. & Bauer L. (1980) Gütezustand des Freiwassers des Neckars.
    In: Studien zum Gewässerschutz Nr. 4, 312 pp., pp. 73-169.
    *Landesanstalt fur Umweltschutz Baden-Württemberg - Institut für
    Wasser- und Abfallwirtschaft,* Karlsruhe.
Rapp R. (1973) Ökologische Untersuchungen über Rotatorien in
    Stauhaltungen des Neckars von Cannstadt bis Marbach zur Beurteilung
    der Gewässerbelastung. Zulassungsarbeit fur das Lehramt, päda-
    gogische Hochschule Esslingen, 53 pp. (Report to Pedagogische
    Hochschule Esslingen)
Schaal H. (1981) Der Neckar als Lebensader einer Industrieregion -
    Wasserwirtschaft im Kraftfeld vielseitiger Interessen. In:
    Referate zum 1. Neckar-Umwelt-Symposium, 16/17 Februar 1981 in
    Heidelberg, 135 pp., pp. 1-13. Institut für Sedimentforschung der
    Universität, Heidelberg.
Schmitz W. (1971) Die Ausbreitung von Spaltprodukten bei Reaktorun-
    fällen in Fliessgewässern. In: Radioaktive Stoffe und Trinkwasser-
    versorgung bei nuklearen Katastrophen. Bericht der Arbeitsgruppe
    "Trinkwasser-Kontamination", Anhang Nr. 9, 1-15. Bundesministerium
    des Innern, Bonn.
Schoch E. (1981) Fischbestand und Wasserqualität des Neckars. In:
    Referate zum 1. Neckar-Umwelt-Symposium, 16/17 Februar 1981 in
    Heidelberg, 135 pp., pp. 75-78. Institut für Sedimentforschung der
    Universität, Heidelberg.
Tobias W. (1976) Kriterien für die ökologische Beurteilung des
    unteren Mains. II. Untersuchungen über den organischen Stoff-
    haushalt. Courier Forschungsinstitut Senckenberg 18, 137 pp.

Useful sources on information not quoted in text

Dornier System (Ed.) (1972-1977) Prognostisches Modell Neckar,
    Berichte 1-25. Postfach 1360, D-7990 Friedrichshafen. Dornier
    System GmbH.
Institut für Sedimentforschung der Universität Heidelberg (1981)
    Referate zum 1. Neckar-Umwelt-Symposium 16./17. Februar 1981 in
    Heidelberg, 135 pp. Heidelberg.
Landesanstalt für Umweltschutz Baden-Württemberg (Ed.) Studien zum
    Gewässerschutz. Heft 1: Kontrollverfahren und Massnahmen zum
    Schutze der Gewässer und des Wassers gegen Vereinigung im Rahmen
    des Umweltschutzes, 53 pp. (1971). Heft 3: Untersuchungen über den
    Sauerstoffhaushalt des Neckars, 276 pp. (1980). Heft 4: Unter-
    suchungen über den Gewässergütezustand des Neckars, 312 pp. (1980.
    Heft 5: Die Verfahren der biologischen Beurteilung des Gütezustandes
    der Fliessgewässer (systematisch-kritische Übersicht, 125 pp. (1980).
    Griesbachstr. D-7500 Karlsruhe.
Landesanstalt für Umweltschutz Baden-Württemberg (1979) Umweltqualitäts-
    bericht Baden-Württemberg 1979, 217 pp. Griesbachstr. D-7500
    Karlsruhe.

12 · Shannon

K. HORKAN

12.1   INTRODUCTION

The Shannon is the longest river in the British Isles, with a main
channel length (including the estuary) of some 359 km. It is a mature
river (Nevill, 1972).   It flows not much above base level on a gentle
gradient and cuts laterally instead of down, producing broad meanders
and frequent lake-like expanses. The source is a spring-fed pool known
as the "Shannon Pot" which is located only 45 km from Donegal Bay at an
elevation of 104 m on hills which rise to 600 m (Fig. 12.1).   However,
because of high ground between there and the coast, the river is forced
to flow south, widening into many lake-like expansions as it meanders
slowly through the Central Plain. The first of these is Lough Allen
(3500 ha).   From the outflow of this lake, the main channel drops only
18 m over a river distance of 230 km, passing through further large
lakes, the most important of which are Lough Ree (10457 ha) and Lough
Derg (11635 ha).   Downstream of the latter the Shannon falls 30 m in
only 32 km to reach tidal waters at Limerick. The river then continues
its course through 99 km of a broadening submerged valley to reach the
Atlantic Ocean at Loop Head.

   The catchment, including that of the estuary, drains some 14100 km^2
or approximately one-sixth of the area of the Republic of Ireland
(McCumiskey, 1977). The more important tributaries are: the Boyle
River which flows through Lough Key (897 ha), a scenic amenity area;
the River Suck; the River Brosna which drains the larger Midland lakes;
the Little Brosna River and the River Inny. The population of the
whole catchment is about 500000.

12.2   GEOLOGY

The Shannon rises on hills which consist of rocks of the Yoredale and
Pendleside Series (Upper Avonian Shales and Sandstones; Fig. 12.1).
The river flows off these hills into Lough Allen which is a glacial
lake, continuing from there through a narrow band of Upper Carboniferous Limestone. Further downstream the river briefly crosses Avonian
Shales and Sandstones followed by Carboniferous Limestone, then an area
of Lower Avonian Shales with Carboniferous Slate and Calciferous Sandstone. The main influence is, however, the Limestone of the Central
Plain, which dominates the catchment downstream as far as the hills on
the southern end of Lough Derg. Here the rock formations are mainly
Silurian with some Old Red Sandstone hills surrounded by a narrow band
of Lower Avonian Shales. West of Limerick the submerged valley that is
now the estuary is floored by an expanse of Limestone. On the western
side of the Fergus Estuary the channel crosses Coal Measure uplands
which are in turn replaced by Millstone Grit and Flagstones, particularly on the north bank (Co. Clare). On the south bank (Co. Kerry)
there are some Upper Avonian Shales and Sandstones.

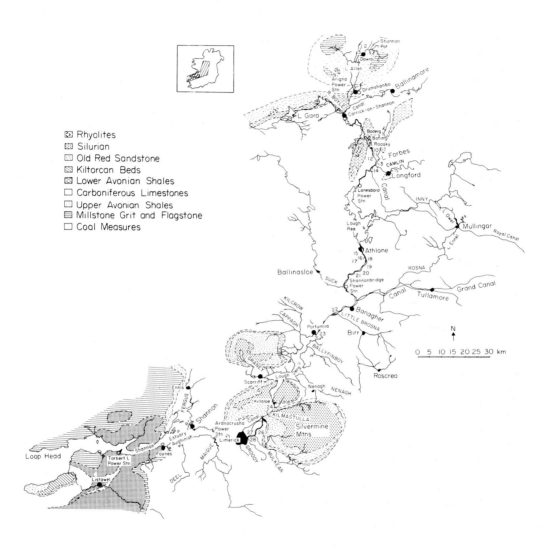

Fig. 12.1 Shannon Catchment, showing main tributaries, general geology and larger towns. Numbers refer to sampling stations in Table 12.3.

Most of the important tributaries arise on either Upper or Lower Carboniferous Limestone, apart from the Boyle River which originates on hills formed of Lower Avonian Shales and Old Red Sandstones. The Shannon Lakes, apart from Lough Allen, whose origins are glacial, originated as great solution expanses of the main channel (Praeger, 1974).

12.3 SOILS AND LAND USE

A variety of soils are found in the catchment. Because some are poorly drained and because the river is broad and slow-flowing through a frequently waterlogged plain of low relief, flooding is common, particularly in winter.

Upstream of Lough Allen the mountainy and hilly areas are dominated by peaty gleys with their associated climatic peats. Further downstream these change to drumlin gleys with some associated peaty gleys, interdrumlin peat and acid brown earths. Around Rooskey and again south of Athlone there is an area of basin peat with organo-mineral soils. Lough Ree and the northern end of Lough Derg are mostly surrounded by grey brown podzolics with associated gleys. Between Lough Ree and Athlone there are also some grey brown podzolics but here they are associated with basin peat and regosols. West of Lough Derg there are drumlin soils, while around Killaloe there are acid brown earths and associated brown podzolics. Further downstream towards Limerick these soils are replaced by gleys and associated peaty gleys. West of Limerick there are mudflats, especially on the Co. Clare (north) side and these are principally regosols with some gleys. Small areas of grey-brown podzolics occur on the north bank while shallow brown earths and grey brown podzolics are found on the south side. West of Foynes on the south shore and Ennis on the north side the principal soils are mainly gleys and associated acid brown earths and peats.

Before farming developed, Ireland was covered in oak forests. These were often felled to make room for grasslands, the growing of which is particularly suited to the mild Irish climate. The forests may also have declined due to climatic changes with increased rainfall which resulted in peat formation. In the low-lying Midland region adjoining the Shannon large areas of peat bogs are found which are suited only to sheep grazing and peat production. Cattle rearing and mixed farming are most common on the richer pastures of the catchment.

12.4 CLIMATE

Because of the influence of the warm Atlantic currents and mild sea breezes the Irish climate lacks extremes. The mean annual temperature range is only about 10°C and varies from a mean in January of 4.4 °C in

the Midlands (5.5 $^{\circ}$C along the west coast) to a mean in July of 15 $^{\circ}$C.
Frost is relatively rare along the west coast but is more common in the
Midlands. The Shannon and its lakes do not freeze, although ice may
form in sheltered bays of the larger water bodies; this seldom lasts
for more than a few days.

The prevailing winds from the Atlantic bring in rainfall which is
greatest along the western seaboard and declines eastwards. On the
west side of the Shannon annual precipitation may be as high as 1750 -
2250 mm in hilly areas, whereas on the eastern side it averages 750 -
1000 mm, with a mean annual 30 year average over the Shannon catchment
of 1024 mm (An Foras Forbartha, 1977). At Killaloe where the mean
annual rainfall over a 15 year period is 1029 mm, the average run-off
(inflow) was estimated at 539 mm and total losses amounted to 490 mm.
Losses due to evapotranspiration and deep seepage constituted 37.6% of
the mean annual rainfall for this period (McCumiskey, 1977).

12.5 TOWNS AND INDUSTRY

Due to the low-lying boggy nature of the surrounding land and the tendency towards winter flooding few towns are found along the main river
channel. The more important ones are: Carrick-on-Shannon (population
2208), an important tourist centre containing the largest inland marina
in Ireland; Athlone (9778) which has manufacturing and electronics
industries as well as boat-building; Limerick (60665) which is a shipping port, as well as having numerous industries including manufacturing
and electronics; Shannon New Town (7000) which has a wide range of
industries, which are mostly light manufacturing and electronic plants
geared for export through the adjoining International Airport.

The Shannon Estuary is rapidly becoming an important industrial
development area. The small port of Foynes is an oil rig supply base
as well as the export point for barytes. Nearby, at Aughinish Island,
a £500 million aluminium production plant has recently been completed.
Electricity is generated at Tarbert (oil-fired) and a new coal-fired
power station is being constructed at Money Point (12.62). Future
developments planned for the estuary include oil refineries and their
associated industries.

Anthracite is mined at Arigna and supplies the energy source for a
local power station. Milled peat for fuel and horticulture is also
important in the catchment. Silver, lead and zinc-bearing ores are
extracted from the Silvermine Hills. The fishing industry in the area
is based mainly on commercial salmon and eels (the latter at Killaloe
and Athlone), with some shell-fishing in the estuary; a herring fishery
is active off Loop Head (see 12.93).

12.6   UTILIZATION OF THE RIVER

12.61   NAVIGATION

The first work on the development of the Shannon as a navigation channel began at Killaloe in 1755. By 1787 the river had been made navigable as far upstream as Carrick-on-Shannon (Carty, 1977). The more difficult Limerick to Killaloe stretch was not completed until 1814. The latter, known as the Errina-Plessey Canal, by-passed different stretches of the river and contained a total of 11 locks. This canal was abandoned in 1930 with the completion of the hydro-electric works (see below) containing a double lock with a lift of 30.5 m.

The Lough Allen Canal at Drumshanbo came into operation in 1814 but was little used and was closed in 1936. Navigation was linked to the Erne when the Ballinamore and Ballyconnell Canal was opened in 1860. However, by 1880 this canal was derelict. Now enthusiasts are trying to have this link with the Erne system re-opened. The Shannon was connected to Dublin by 1806 with the completion of the Grand Canal. A rival waterway, the Royal Canal, was opened in 1809. Steam power came to the river in 1826 and regular cargo-carrying barges plied between Dublin and Limerick. Periodically improvements were made to the navigation system, these including the construction of a number of weirs and deepening of the locks to accommodate a minimum draft of 2.0 m. However, the coming of the railways caused the demise of the water borne cargo trade. Barges which were moving 100000 tons of cargo annually by the turn of the century found their trade declining.

The first signs of a revival in interest in the Shannon as a waterway occurred in 1954, when two passenger launches were purchased by C.I.E. (National Transport Authority) for day trips from Athlone. The first hire cruiser for family holidays became available in 1960. Now there are some 560 boats for hire and at least a similar number of private craft on the water. At present over 7600 passages are made by boat through Athlone Lock during a summer season and this number is rising steadily (see 12.63).

12.62   POWER PRODUCTION

The Shannon became important for power production during the 1920s when the Dail (Irish Government) passed an act in 1925 enabling the first development of the Shannon hydro-electric works at Ardnacrusha (310 MW) (O'Leary, 1977). This involved raising Lough Derg by about 60 cm (2 feet), which concentrated the entire fall of the river over a length of some 80 km at one point. The average annual flow of the Shannon past Killaloe is 190 m^3 s^{-1}, of which 180 m^3 s^{-1} is available for power generation. Since that time several other power stations have been built in the catchment including Arigna (15 MW, coal fired from local supplies), Lanesboro (60 MW, peat fired from local supplies), Shannonbridge (80 MW, peat fired from local supplies), Tarbert (620 MW,

oil-fired) and Money Point (900 MW, coal fired, under construction).
The capacity of some of the smaller plants is being expanded at
present.

12.63 AMENITIES

Due to the fact that much of the Shannon is navigable, the river and
its lakes are becoming very popular with tourists, particularly in
terms of pleasure cruising holidays. The navigation channels are well
marked and boating facilities, fuel, water and other provisions are
available in towns along the river and in small villages adjacent to
sheltered bays on the lakes. By European standards the river is relatively uncrowded and unpolluted. If the Shannon had the same density
of boats as the Norfolk Broads there would be 200000 boats (Ransom,
1971), compared to the 1100 or so craft found today. The monetary
return from this type of tourism along the Shannon is valued at a
minimum of Irf10 million (data of Irish Tourist Board).

Angling is also very important on the Shannon, as mentioned in 12.5.
Excellent salmonid (*Salmo salar* and *S. trutta*) and coarse fishing are
available with special attention paid to large pike, bream, rudd, rudd/
bream hybrids, perch, tench and eels. The thermal effluent canal at
Lanesboro Power Station is a renowned haunt for specialist coarse
fishermen, as it supports large tench and bream.

Other aquatic based pastimes are enjoyed on the river including
canoeing, water-skiing and duck shooting. There are safe beaches on
some of the lakes, picnic areas, a Forest Park on Lough Key, historic
sites and some holiday village developments are also found.

12.7 WATER RESOURCES AND FUTURE MANAGEMENT

The water resources of the Shannon, including tributaries, constitute
about 21% of the total available in the Republic of Ireland. Of the
15 year mean annual rainfall of 1029 mm at Killaloe, the average run-
off is 539 mm and total losses due to evapotranspiration and deep
seepage accounts for the remaining 490 mm or 47.6% of the total
(McCumiskey, 1977).

Values for low flow at various points along the Shannon are given in
Table 12.2, together with the 99.5 percentile values at Limerick. The
annual average flow at Killaloe is included for comparative purposes.
On the tributaries there are wide variations in the ratio of low flows
to average flows ranging from 1 to 10%, the higher the ratio the larger
the tributary. Values for these low flows range from less than 100000
l km^{-2} d^{-1} to as low as 3000 l km^{-2} d^{-1}. The significance of the magnitude of such low flows is that they limit the extent to which the water
can be used for receiving effluents before quality problems develop.

Table 12.1 Distribution of water resources in the Republic of Ireland (from McCumiskey, 1977). Population estimates are based on the 1971 Census. The population of the country was 3.3 million in 1979 compared to 2.9 million in 1971.

region	population ($\times 10^3$)	average run-off ($m^3\ s^{-1}$)	low flow run-off ($m^3\ s^{-1}$)	average run-off per capita (l person^{-1} d^{-1})	low-flow run-off per capita (l person^{-1} d^{-1})
1 Eastern	1160	114.0	5.8	8592	432
2 South-Eastern	413	198.8	15.2	41600	3180
3 Southern	430	235.5	11.5	47320	2320
*4 Shannon, excluding tributaries joining estuary	259	178.5	10.0	59550	3340
5 Mid-Western	253	124.8	5.0	42600	1696
6 Western	183	186.2	4.7	87920	2218
7 North-Western	221	231.7	6.2	90560	2410

*
The entire catchment, including tributaries joining the estuary, has a population of 470000.

Many public water supplies in the Shannon catchment use the main channel, the tributaries and the lakes for their abstractions. This implies that future demands for surface water abstraction in the catchment requires careful control of discharges into the river system in order to maintain satisfactory water quality conditions. To date the various developments in the freshwater reaches of the Shannon system have not resulted in widespread pollution. This position reflects the low degree of industrialization and small population size rather than good management practices. However, some localised problem areas occur on a few of the tributaries and a number of lakes (12.82). Recent trends show that industrial growth, intensification of agriculture and the expansion of towns in the Shannon catchment are now in progress.

As various water uses are inter-related they cannot be properly managed in the catchment as a whole without adopting a comprehensive approach. This should take into account all possible uses, with special emphasis on abstraction, discharge and amenity. In achieving such objectives the attainment and maintenance of sufficiently good

Table 12.2 Minimum flows over the 44-year period, 1932-1975, at various sites in the main channel of the Shannon and the larger tributaries (after McCumiskey, 1977).

location	catchment (km^2)	minimum flow (m^3 s^{-1})
Shannon		
outlet of Lough Allen	425	1.7
Carrick-on-Shannon	1255	2.3
Rooskey	1981	3.0
Athlone	4455	8.0
Limerick	11250	12.4 (99.5 percentile)
		26.4 (95 percentile)
Tributaries		
Graney - Scariff	296	0.23
Killimor - Moate	194	0.039
Suck - Ballinasloe	1373	1.37
Mountain - Strokestown	197	0.156
Boyle - Boyle	500	0.395
- Ballaghaderreen	207	0.05 (95 percentile)
Camlin - Longford	207	0.09
Inny - Ballymahon	1046	0.24
Brosna - Tullamore	106	0.1
- Clara	324	0.85
- Ferbane	1186	2.0
Little Brosna - Birr	479	0.84
- Roscrea	26	0.026
Nenagh - Nenagh	142	0.22

water quality to safeguard public health is a priority. To achieve these standards requires the collection of good basic data, which must take into account present and future water uses, waste water discharges, and the hydrological, physical, chemical and biological characteristics of the catchment.

The standards laid down in the E.E.C. Directive (Council of the European Communities, 1978) which covers salmonid waters is perhaps the most appropriate for the Shannon and its main tributaries. If such standards were adopted and complied with they would ensure adequate protection for other uses also.

12.8   WATER QUALITY

12.81   THE RIVER

Much of the main channel of the Shannon is inaccessible by road. As a
result the river has not been monitored as regularly as some of the
other large Irish rivers. However, a number of surveys have been
carried out and the main findings are discussed here.

In 1971 the chemistry of the river upstream of Dowra in the Upper
Shannon to Thomond Bridge (Limerick) in the Lower Shannon was investigated at 22 stations. The results showed that the greater portion of
the length surveyed was in a 'satisfactory' condition (Flanagan & Toner,
1972), but the number of stations sampled was relatively few and the
possibility of the survey missing localised polluted stretches was
realised. Hence, a more intensive survey of the main channel, the
lakes and the main tributaries was carried out in the summer of 1972.
37 stations on the Shannon and 13 locations on the tributaries were
examined (Table 12.3). These results confirmed that over most of its
length the Shannon was indeed in a 'satisfactory' condition but there
were some stretches where slight impairments were noted. Orthophosphate levels were relatively high downstream of Carrick-on-Shannon,
Athlone and below the polluted Camlin tributary confluence. Ammonia
levels were slightly elevated at the former location and also below
Rooskey. These results were attributed to organic waste discharges
from these towns and also to small discharges, floating solids and
debris from cruising boats, the problem being most notable near
marines and quays. The localised enrichment observed below the towns
mentioned above was mild in nature and the water quality in the river
was generally 'fair' to 'good'. Filamentous algal growths including
Cladophora were common in the enriched stretches, but macroinvertebrate
diversity was reduced only slightly.

The Lower Shannon was found to have good water quality conditions
as far as Limerick City, in the vicinity of which occasional elevated
BOD values were recorded. No evidence was found of any contamination
by waste discharges along this lower river reach (Flanagan, 1974).
One tributary in this stretch, the Kilmastulla River, has been deterio-
rating in quality in recent years as a result of toxic discharges from
mining activities in the nearby Silvermine Mountains. Heavy metal
levels have caused denudation of the fauna in this tributary. In
October 1980, an overspill from the tailings pond at the mines resulted
in a toxic waste load reaching the Shannon. This was a 'one off'
event, but points to the need for vigilance at all times in areas
where potentially serious problems exist. The mine has now closed.

One problem associated with some sections of the main river channel
is that of peat-silt deposits, which build up particularly below
Lanesboro, Shannonbridge and at points in the stretch from Banagher to
Portumna (Toner *et al*., 1973). This results in the virtual absence of
macroinvertebrates in these areas due to the fine deposits of peat-

Table 12.3 Water chemistry of the Shannon (main channel), together with biological quality rating. u/s = upstream; d/s = downstream; tr = trace. Quality ratings: Q5, good; Q4, fair.

sta-tion	location	distance from st. 1 (km)	dissolved oxygen (% saturation)	BOD (mg l⁻¹ O₂)	NH₃-N (mg l⁻¹)	NO₃-N (mg l⁻¹)	reactive phosphate-P (mg l⁻¹)	hardness (mg l⁻¹ CaCO₃)	quality rating (Q)
	Upper Shannon, Dowra Area								
1	Metal Br. N. of Corratober	0	98 - 119	0.8 - 1.2	0.01 - 0.03			55 - 75	5
2	Bridge in Dowra	6.0							5
3	1 km d/s Dowra	7.0	98 - 122	0.7 - 1.4	0.01 - 0.02			30 - 50	5
	Carrick-on-Shannon Area								
4	Hartley Bridge	36.0	98 - 118	0.8 - 1.8	0.01 - 0.03			45 - 65	5
5	2 km u/s Carrick-on-Shannon	38.0				0.12	0.015		5
6	Carrick-on-Shannon Bridge	39.8	100 - 120	0.8 - 1.7	0.03 - 0.04	0.12 - 0.19	0.025 - 0.035	45 - 100	5
7	0.5 km d/s Carrick-on-Shannon	40.5	95 - 122		0.02 - 0.04				4
8	u/s Jamestown Bridge	49.5	100 - 118	0.9 - 1.4	0.01 - 0.03			75 - 105	5
	Middle Shannon, Rooskey Area								
9	u/s Rooskey Bridge	67.0	100 - 120	1.0 - 2.0	0.02 - 0.05			90 - 140	5
10	0.2 km d/s Rooskey Bridge	67.2							5
11	u/s Lock and d/s Canal	67.6	94 - 116		0.025	0.11	0.055		5
12	u/s Lough Forbes	72.0							5
13	u/s Camlin River, N. Branch	76.0	94 - 112	1.1 - 1.7	0.02 - 0.06	0.10 - 0.12	0.025 - 0.065		5
14	u/s Camlin River, S. Branch	77.0	94 - 116			0.19 - 0.33			4
	Athlone Area								
15	d/s Athlone Weir	124.0	90 - 104	1.4 - 5.5	0.02 - 0.30	0.22 - 0.34	0.015 - 0.125	90 - 145	4
16	d/s disused canal	125.0	100 - 112		tr - 0.08	0.17	0.010	165 - 195	5
17	u/s Wrens Island	127.0	100 - 112		0.07		0.035		5
18	u/s Long Island	129.5	104 - 107			0.44			5
19	d/s Long Island	132.0	104 - 108		tr	0.42	0.005		5
20	Inchalee Island	134.0	102 - 104		tr	0.20 - 0.41	0.030		5
21	u/s Clonmacnoise	137.5	80 - 89	1.1 - 1.6	0.02 - 0.06	0.22 - 2.50	0.015 - 0.050		5
	Lower Shannon, Lough Derg Area								
22	Meelick	174.0	78 - 118	0.5 - 3.5	tr - 0.15	0.25 - 2.50	0.003 - 0.060	160 - 205	-
23	Portumna Bridge	187.0	76 - 114	0.8 - 4.4	<0.01 - 0.19	0.28 - 1.90	tr - 0.065		-
24	Killaloe Bridge	222.0	86 - 110	0.5 - 3.9	tr - 0.20	0.28 - 1.90	tr - 0.033		-
25	O'Briens Bridge	229.5	88 - 117	1.0 - 3.5	0.01 - 0.06	0.36 - 1.20	0.010 - 0.100		-
26	Castleconnell	234.5	90 - 125	1.0 - 3.4	0.01 - 0.07	0.35 - 1.20	0.010 - 0.140		-
27	2.5 km WNW Lisnagry Station	238.5	96 - 100	1.0 - 2.7	0.01 - 0.07	0.40 - 1.35	0.010 - 0.110		-
28	2.0 km W Mount Shannon	239.0	100 - 120	1.0 - 2.8	0.01 - 0.11	0.45 - 1.40	0.010 - 0.140		-
29	Athlunkard Bridge	244.5	92 - 114	1.0 - 4.1	0.01 - 0.08	0.43 - 1.30	0.010 - 0.310		-
30	Thomond Bridge	248.0	84 - 109	1.0 ->7.0	0.02 - 0.37	0.31 - 1.20	0.010 - 0.150		-

silt covering the substratum, although chemically the water quality is
not altered. One tributary, the deep slow-flowing River Suck, also
receives peat-silt deposits, but its water quality is 'good'.

12.82 THE LAKES

The Shannon Lakes also have some influence on the main channel. The
uppermost lake (Lough Allen) has soft water with significant coloration
(Table 12.4). From the upstream to the downstream lakes there is a
gradual increase in mineral content and pH because of the change from
harder rocks to more soluble limestone. An exception to this is Lough
Key which is more similar to the two lower lakes, Ree and Derg, because
the Boyle River flows through a limestone area (Fig. 12.1). Lough
Allen is chemically the poorest lake, the other lakes having a relatively high conductivity and hardness. Most of them are quite shallow
apart from Ree and Derg, where troughs of 30 m or more are found
(Charlesworth, 1963) and tend mostly to be low-lying and exposed,
apart from hilly areas adjoining Loughs Key and Derg.

   In terms of trophic status the lakes range from oligotrophic (Lough
Allen) to eutrophic (Lough Derg). Lough Derg is of particular interest
as data were collected by Southern & Gardiner (1926, 1932, 1938) and
Southern (1935) from November 1921 to March 1923. These workers were
based in a freshwater laboratory near Portumna and this limnological
station was the first of its kind to be built in the British Isles.
Unfortunately, the station was closed down because of a lack of funds.
Examination of the plankton data of Southern and Gardiner show that
some changes have taken place, with increased numbers of some clado-
ceran and rotifer species which prefer slightly enriched conditions,
as well as an increase in the blue-green algal populations. There
has been some reduction in water clarity. These changes are slight
rather than excessive and as yet no serious decline in water quality
has occurred.

12.9 BIOTA

12.91 PLANTS

According to Praeger (1974) the fact that almost all of the larger
Irish rivers have no marked watersheds and have at least part of their
courses in the Limestone Plain has made it possible for an easy migration of species from one river system to the next resulting in a
comparative uniformity of flora and fauna. Yet a few species are
restricted to only one or two river basins. Examples of localised
species occurring in the Shannon system include *Inula salicina* which
is found only on the shores of Lough Derg, *Teucrium scordium* found
along the shores of Loughs Ree and Derg (Webb, 1963) and *Sium latifo-
lium* found along the river bank of the Shannon and also in the Erne

Fig. 12.2 Lough Key, showing Castle Island and reed-covered shore-line in sheltered bays.

Fig. 12.3 The Shannon upstream of Lough Rae, showing typical low-lying banks and scattered beds of emergent plants in shallow water. (Photos: courtesy Irish Tourist Board)

Table 12.4 Mainstream lakes on the Shannon: selected morphological, physical, chemical and biological characteristics, showing ranges recorded during various surveys made during the period 1973-1982. tr = trace.

lake	altitude m	size ha	conductivity μS cm^{-1}	colour hazen units	hardness mg l^{-1} CaCO$_3$	pH	ortho-phosphate mg m^{-3} P	oxidised N mg m^{-3} N	ammonia mg m^{-3} N	chlorophyll a mg m^{-3}	Secchi disc transparency m	trophic status
Allen	50.0	3500	90 – 100	50	30 – 50	7.1 – 7.4	tr – 40	100 – 190	tr – 40	1.4 – 4.2	1.8 – 2.5	oligotrophic
Bofin	42.0	408	190 – 315	40 – 60	80 – 140	7.8 – 8.1	10 – 40	140 – 590	10 – 30	1.9 – 5.5	1.2 – 2.3	mesotrophic
Boderg	42.0	430	155 – 310	50 – 60	70 – 145	7.7 – 8.0	10 – 50	140 – 650	10 – 20	1.7 – 7.5	1.0 – 2.0	mesotrophic
Key	44.5	897	310 – 360	50 – 100	150 – 200	8.0 – 8.3	5 – 20	80 – 880	10 – 50	<1.9 – 9.6	1.5 – 4.5	naturally eutrophic
Ree	38.0	10457	300 – 395	30 – 60	165 – 195	7.8 – 8.4	10 – 140	90 – 900	10 – 170	0.9 – 25.0	1.0 – 5.5	slightly enriched
Derg	33.0	11635	410 – 510	20 – 70	205 – 210	7.8 – 8.6	tr – 85	100 – 2000	tr – 70	2.2 – 29.0*	0.9 – 4.0	moderately enriched

*A special survey on Lough Derg in August 1982 reported concentrations within the 30 – 50 mg m^{-3} range.

system; *Chara tomentosa* which is unknown in Great Britain, occurs in
the Shannon basin, particularly in Lough Ree and in the Midland lakes
which drain into the Shannon. Praeger (1974) suggested that the distribution of these rare species may signify that they are late arrivals
rather than glacial relicts and little dispersal has yet occurred.
The red alga *Bangia atropurpurea* was recorded in Lough Derg near
Portumna only recently by Scannell (1972). It is known to have occurred
previously in the two canals which drain into the Shannon and from
whence it probably spread to the lake. Several other aquatic plants
have spread from the Royal and Grand Canals to the Shannon basin
including *Sagittaria* spp., *Glyceria aquatica, Equisetum variegatum*
and several charophytes.

12.92   MACROINVERTEBRATES

Information on the macroinvertebrate fauna in the catchment is sparse
and is based mostly on limited sampling carried out in connection with
water quality assessments. The Upper Shannon above Lough Allen, which
varies from torrential reaches to fast-flowing riffle areas, has a
diverse fauna typical of a clean upland stream. Baetidae tend to dominate, with Gammaridae and Ecdyonuridae numerous. Other taxa well
represented include *Isoperla* spp., *Leuctra* spp., *Astacus* sp., *Elmis*
spp., *Potamopyrgus*, Ceratopogonidae, *Hydropsyche* spp., and *Simulium*
spp. (An Foras Forbartha, unpublished data).

Below Lough Allen movement of water in the main channel of the river
becomes sluggish for most of the remainder of its course to the sea.
The fauna changes to suit these conditions and is dominated in most
cases by Lamellibranchs, Oligochaeta, Chironomidae and Gastropods
(Toner *et al.*, 1973). Other groups commonly encountered include
Asellus spp., Polycentropidae and other Trichoptera, Ephemeroptera
(excluding Ecdyonurids) and Hirudinea. Less frequent are Plecoptera,
Simuliidae, other 'Diptera', Ecdyonuridae and sponges. Some groups
are well represented but confined to particular stretches; these include
Gammarus spp. which occur mainly in the upper part of the river and
Mysis spp. mainly below Athlone. Detailed species lists can be found
in the account of Toner *et al.* (1973).

In general, the faunal composition reflects the deep slow-flowing
nature of the river, with burrowers and crawling types best suited to
the silted conditions. Differences between stations can be attributed
to the nature of the substratum. Even where slight organic enrichment
has been noted (e.g. below Athlone), both sensitive and tolerant groups
are abundant. Larger populations of Lamellibranchs and Gastropods
featuring such genera as *Sphaerium, Anodonta, Bithynia, Potamopyrgus*
and *Theodoxus* are abundant at stations downstream of lakes. Where the
substratum is covered in peat and silt deposits few fauna are encountered, although no adverse chemical effects have been observed at
such locations (see 12.8).

12.93 FISH

According to Went (1940) the Shannon, prior to the initiation of the hydro-electric works, was an outstanding salmon fishery with a world-wide reputation. Following the construction of the dam near Ardna-crusha conditions were altered dramatically with the result that the salmon fishery went into serious decline. Net catches up to 188000 kg were recorded before the river was harnessed for electricity supply (O'Dowd, 1977). Following completion of the works returns declined to only 19000 kg (Table 12.5). Fish passes were installed to improve the situation and catches increased again to over 100000 kg. Commercial eel fishing also declined on the river for the same reason and is recovering again. Reduced catches of salmon in recent years are attri-buted to the effects of Ulcerative Dermal Necrosis (U.D.N.) disease in the late 1960s and drift netting for salmon at sea in the 1970s.

Table 12.5 Salmon and eel net landings from the Shannon. The value for 1927 is typical of annual returns before the construction of Ardnacrusha Power Station. In 1939 catches were still badly affected by the hydro-electric works and fish passes were construc-ted to improve the situation. Catches then returned to 'normal'. (Data from O'Dowd (1977) and Annual Sea and Inland Fisheries Reports, 1963-1980)

year	salmon landings (kg)	eel landings (kg)
1927	188244	65382
1939	19051	–
1963	162822	33536
1964	157468	40365
1965	156282	61723
1966	110058	54184
1967	141771	87171
1968	127177	51823
1969	173043	36034
1970	169676	35054
1971	214889	31131
1972	132840	8038
1973	199581	40232
1974	162778	39200
1975	190193	34996
1976	118638	60000
1977	96948	38100
1978	76064	24000
1979	48752	42411
1980	43332	33459

Some excellent brown trout fishing is found on the lakes with the best catches usually in May and June when *Ephemera danica* is emerging in large numbers. The main river channel and the larger lakes provide good coarse fishing especially for pike, bream, tench, perch, rudd and rudd/bream hybrids. Whitefish are not well represented in Irish waters but one exception is the pollan (*Coregonus pollan elegans*) which is found in the Shannon lakes. This fish is of commercial importance in Lough Neagh but is not found in sufficiently large quantities to be a viable commercial proposition on the Shannon. The smelt (*Osmerus eperlanus*) has been recorded from the Shannon lakes by Kennedy (1948) and is also mentioned by Went (1948) but is usually not seen by anglers.

REFERENCES

An Foras Forbartha (1971) *Preliminary Report on Surface Water Resources in the Shannon Water Resources Region.* An Foras Forbartha, Water Resources Division.

Carty J.J. (1977) The management of the River Shannon in the 1980s. *J. Inst. public Admin. Ireland* 25(2) Special issue, 220-30.

Charlesworth J.K. (1963) The bathymetry and origin of the larger lakes of Ireland. *Proc. R. Ir. Acad.* 63 B(3), 61-71.

Clabby K.J. (1981) *The National Survey of Irish Rivers. A Review of Biological Monitoring, 1971-1979*, 308 pp. An Foras Forbartha, Water Resources Division, WR/R12.

Council of the European Communities (1978) *Council Directive of 18 July 1978 on the Quality of Fresh Waters needing Protection or Improvement in Order to Support Fish Life.* (78/659/E.E.C.). Official Journal of the European Communities. No. L222/7.

Flanagan P.J. (1974) *The National Survey of Irish Rivers. A Second Report on Water Quality*, 98 pp. An Foras Forbartha, Water Resources Division.

Flanagan P.J. & Toner P.F. (1972) *The National Survey of Irish Rivers. A Report on Water Quality*, 213 pp. An Foras Forbartha, Water Resources Division.

Kennedy M. (1948) Smelt in the Shannon. *Ir. Nat. J.* IX, 151-52.

Lennox L.J. & Toner P.F. (1980) *The National Survey of Irish Rivers. A Third Report on Water Quality*, 104 pp. An Foras Forbartha, Water Resources Division.

McCumiskey W. (1977) The management of the River Shannon in the 1980s. *J. Inst. public Admin. Ireland* 25(2) Special issue, 192-210.

Nevill W.E. (1972) *Geology and Ireland*, 263 pp. Allen Figgis & Co. Ltd., Dublin.

O'Dowd P. (1977) The management of the River Shannon in the 1980s. *J. Inst. public Admin. Ireland* 25(2) Special issue, 179-87.

O'Leary D.P. (1977) The management of the River Shannon in the 1980s. *J. Inst. public Admin. Ireland* 25(2) Special issue, 160-80.

Praeger R.L. (1974) *The Botanist in Ireland*, 587 pp. EP Publishing Ltd.

Ramson P.J.G. (1971) *Holiday Cruising in Ireland. A Guide to Irish Inland Waterways.* David and Charles, Newton Abbot, England.

Scannell M.J.P. (1972) "Algal paper" of *Oedogonium* sp., its occurrence in the Burren, Co. Clare. *Ir. Nat. J.* 17(5), 147-52.

Sea and Inland Fisheries Reports (1963-1980). Stationery Office, Dublin.

Southern R. (1935) The food and growth of brown trout and the River Shannon. *Proc. R. Ir. Acad.* XXII, 87-151.

Southern R. & Gardiner A.C. (1926) Reports from the Limnological Laboratory. I. The seasonal distribution of the Crustacea of the plankton in Lough Derg and the River Shannon. *Fisheries, Ireland, Sci. Invest.* 1.

Southern R. & Gardiner A.C. (1932) Reports from the Limnological Laboratory. II. The diurnal migrations of the Crustacea of the plankton in Lough Derg. *Proc. R. Ir. Acad.* 40 B, 121-59.

Southern R. & Gardiner A.C. (1938) Reports from the Limnological Laboratory. IV. The phytoplankton of the River Shannon and Lough Derg. *Proc. R. Ir. Acad.* 45 B, 89-129.

Toner P.F., Clabby, K.J. & Bowman J.J. (1973) *Report on the Results of a Biological and Chemical Survey of the Main River Channel and Lakes of the Shannon System, between Battle Bridge and Killaloe undertaken in July 1972*, 89 pp. An Foras Forbartha, Water Resources Division.

Webb D.A. (1963) *An Irish Flora*, 261 pp. Dundalgan Press Ltd., Dundalk.

Went A.E.J. (1940) Salmon of the River Shannon. *J. Dept Agriculture, Dublin* 37, No. 2.

Went A.E.J. (1948) Irish freshwater fish. Some notes on their distribution. *The Salmon and Trout* 118, 9 pp.

13 · Caragh

H. HEUFF & K. HORKAN

13.1 INTRODUCTION

The Caragh, which drains part of south-west Ireland, is an excellent
example of a river system which is relatively unpolluted from headwaters
to estuary. This is unusual in Western Europe and, for precisely this
reason, the river was chosen for detailed biological investigation. The
survey, on which most of this chapter is based, took place between October 1974 and February 1977 and was financed by a grant from the National
Board for Science and Technology to the Botany and Zoology Departments
of University College, Dublin. The chapter describes the general ecology of this non-calcareous, fast-flowing, clean water system.

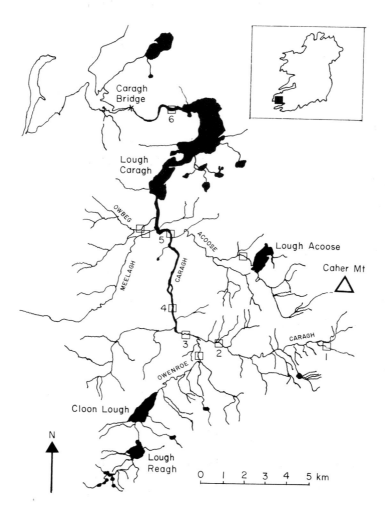

Fig. 13.1 Caragh River catchment.

13.2 GENERAL DATA

DIMENSIONS

Area of catchment   160 km^2
Highest point in catchment   975 m
Length of stretch described (stations 1-6)   22.4 km
Drop in height down this stretch   104 m (= 4.6 m km^{-1})

CLIMATE (Rohan, 1975 and Irish Meteorological service, pers. comm.)

Precipitation

Data obtained at Valencia Observatory, the nearest major one to
the catchment and Glenvickee inside the catchment.

Mean annual values:

Valencia   (1931-1960)   1398 mm
Glenvickee   (1941-1960)   2364 mm

Temperature

Mean daily value at Valencia (1931-1960)

February   6.8 $^{\circ}$C
August   15.4 $^{\circ}$C

Mean daily sunshine at Valencia (1931-1960)   3.72 h

Relative humidity

mean at Valencia   82%

GEOLOGY

Old Red Sandstone (Walsh, 1968)

LAND USE

Rough grazing (Agricultural statistics, 1975)

POPULATION (Census of populations 1971, 1979)

1144 people = 7.15 people km^{-2}

PHYSICAL FEATURES OF RIVER

Discharge at Caragh Bridge (Data from Electricity Supply Board charts
for 1975 and 1976)

mean daily value 7.29 m^3 s^{-1} = 45.6 s^{-1} km^{-2}
range   0.24-55.1 m^3 s^{-1}

River regime: highest flows generally in October-February
              lowest flows   April-September
              flash floods   throughout the year

Fig. 13.2 A headstream on slopes of Caher mountain at about 350 m. This rocky and treeless landscape is typical of the upper catchment.

Fig. 13.3 Stunted trees cover the sides of streams in a few areas of the upper catchment. The steep-side valley shown lies upstream of the more typical landscape in Fig. 13.2.

13.3   TOPOGRAPHY

The Caragh River system occupies glacially deepened valleys on the northern side of the mountains which form the backbone of the Iveragh peninsula.   The valley floors are relatively broad and flat and are flanked by steep slopes rising 200 to 700 m above the floor of the valley.   The source of the Caragh is on the southern slopes of the Macgillycuddy Reeks at an altitude of approximately 400 m, where several rivulets join to form a relatively permanent stream.   This, in a distance of just over one kilometre, then descends the rocky and peaty slopes to the valley floor at approximately 100 m.   From here it first flows westward for 9 km through the Bridia valley and at station 3 (Fig. 13.1) is 7.3 m wide.   The current speed in this stretch is moderate to fast and the substratum consists mostly of small stones and gravel.   On leaving the Bridia valley it turns northwards, and is characterised by steep riffles over bedrock and deep sandy pools (width at station 4 is 12.5 m).   Just before entering Lough Caragh the river becomes deep and slow-flowing with a peaty-muddy bottom.   Below Lough Caragh the river flows westward to enter the sea (width at station 6 is 47 m).   The substratum of this stretch consists of large boulders, coarse pebbles and some gravel.

It should be noted that the Caragh River study was mainly concerned with the river on the valley floor.

13.4   CLIMATE AND HYDROLOGY

South-west Ireland has a pronounced oceanic climate.   High humidity, cool summers and mild winters are typical.   The distribution of precipitation is fairly even over the year, but in general winters are somewhat wetter than summers.   Dry spells (15 or more days without rain) are extremely rare.   At Valencia Observatory, 30 km WSW of the catchment, evaporation (Irish Meteorological Service, monthly weather report, Part I) and potential evapotranspiration (Connaughton, 1967: computed using Penman formula) exceed rainfall for only a short period during each year, usually in June and July, while actual evapotranspiration is always smaller (Irish Meteorological Service, pers. comm.); this applied even during the relatively dry years (1975, 1976) when the Caragh investigation took place.   Strong winds from the south-west are commonplace.   Snowfall is infrequent and short-lived.   In this type of climate primary production can take place the whole year around. The median dates of the beginning and end of the grass growing season lie, respectively, between the first and fifteenth of February and the fifteenth of December and the first of January (Atlas of Ireland, 1979).

The Caragh catchment covers 160 km^2.   The Caragh river reaches the sea as a 109th order stream sensu Shreve (1966), using the Irish Ordnance Survey map, scale 1 : 126720.   The total length of the main channel is 27.6 km.   Since 1945 the Electricity Supply Board (E.S.B.) has maintained a water-level gauging station at Caragh Bridge (National

Fig. 13.4 Rocky cascades in the Caragh just above station 2 at about 60 m altitude.

Fig. 13.5 Stretch just below boulders in Fig. 13.4, with deep pool in middle ground and pebbly riffle in foreground. All three types of environment can be found for most of the length of the river.

Grid Reference V712 926), downstream from Lough Caragh. However most
of the data sheets have not been analysed and hence the river regime
cannot be described. Fig. 13.6 shows daily discharge at Caragh Bridge
(data sheets of E.S.B., analysed by An Foras Forbartha) and the daily
mean rainfall for four weather stations in the catchment (data of Irish
Meteorological Service) for a 6-month period.

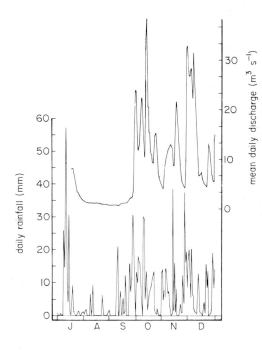

Fig. 13.6 Daily rain-
fall in the Caragh
catchment and mean daily
discharge at Caragh
Bridge from July to
December 1976.

From this figure it is clear that river discharge shows a rapid
response to rainfall in the catchment. This is due to the low ground
water storage capacity of the area, the steep slopes (mean slope 170 m
km^{-1}) and the generally impermeable nature of the soils. Flash floods
occur at frequent intervals throughout the year. For example, the
greatest rise in water level measured during the study period was 1.80 m
in 22 h on the 17/18 October 1974, corresponding to an increase in dis-
charge from 3.8 to 70 m^3 s^{-1} (23.8 l s^{-1} km^{-2} to 438 l s^{-1} km^{-2}). The
mean rainfall on the 17 October for four rainfall stations in the catch-
ment was 37.8 mm. This is 38% of the total rainfall for October of
that year. The mean total rainfall for the previous 10 days was 20.1 mm
and was spread more or less evenly.

13.5 GEOLOGY, SOILS AND VEGETATION

The rock type in the catchment is Old Red Sandstone with conglomerates
of the Quartz-Jasper type. Three types of Old Red Sandstone have been

recognised in the area (Walsh, 1968):

(a) Green Sandstone, with a high content of chloride

(b) Purple Sandstone, containing iron

(c) Grey Sandstone, containing relatively little chloride
     or iron.

The area was covered during the last Ice Age by a radiating ice cap
which had its centre to the south (Lewis, 1977). Evidence of this
glaciation can be clearly seen from the ice-scouring of the hills and
the presence of cirques and moraines. The glacial material which has
been deposited in the valleys of the catchment is derived from the Old
Red Sandstone and is, therefore, of similar chemical composition.
Although the Old Red Sandstone meets Lower Carboniferous Limestone at
the northern shore of Lough Caragh this has no effect on the ecology of
the catchment since the limestone is deeply covered by till of Old Red
Sandstone origin and nowhere does the limestone reach the surface
(Warren, pers. comm.).

The soils in the catchment are poor. They consist of peaty podzols,
lithosols and climatic peat (Soil Map of Ireland, 1974). These soils
are strongly leached and an iron pan is often present. In the case of
the peaty podzols and lithosols the parent material is Old Red Sandstone, which although a sedimentary rock, is highly resistant to
weathering. It consists mostly of silica and is, therefore, acidic and
a poor supplier of nutrients. The glacial till derived from this rock
is low in calcium; only $0.07 - 0.24\%$ $CaCO_3$ was found in the C-horizon
(Warren, pers. comm., Schotter method). A high percentage of unde-
composed organic material is present in the soils, which are waterlogged for a considerable part of the year. The land use potential of
these soils is low. The vegetation in the area is typical of western
lowland and mountain blanket bog, flush, wet heaths and unmanured
permanent lowland wet meadow: i.e. Sphagnetalia magellanici,
Scheuchzerietea, Parvocaricetea, *Molinia* and *Erica tetralix* heaths and
Molinietalia coeruleae. There is also a small amount of woodland
present, mostly conifer plantations (Vegetation Map, Atlas of Ireland,
1979).

13.6 LAND USE AND POPULATION

The population density in the area is low. The majority of the people
at work are engaged in farmwork (Census of populations, 1971, 1979
and Agricultural Statistics, 1975). Settlements are dispersed and
consist of small farms with small irregular fields (pasture, hay and
crops) and largely unenclosed peat and upland areas. The main arable
crops are potatoes and oats. Fertiliser input seems to be low,
although detailed figures are not available (agricultural adviser,
pers. comm.).

Table 13.1 Comparison for (a) land use, and (b) livestock density
of Caragh catchment with that of the whole of Ireland (based on
Agricultural Statistics, 1975).

(a) land use	% of total land area	
	Caragh	Ireland
rough grazing	59.1	14.9
pasture	14.4	45.9
meadow	3.7	15.5
ploughed area	0.6	6.7
woods, plantation	4.3	4.3
other land	17.9	12.7
(b) livestock	density (animals per ha)	
	Caragh	Ireland
cattle	0.26	1.03
sheep	0.81	0.53
poultry	0.18	1.37
horses	0.0054	0.013
donkeys	0.0061	0.0054
pigs	0.0013	0.11

Table 13.1 gives details of the agricultural land use and demonstrates the importance of grazing and sheep in the economy of the area.
There are no towns or villages in the catchment. 52.8% of permanent
housing units have no piped water supply of any kind (inside or outside, piped or mains), 69% have no toilet (flush, chemical or dry)
(Census of Populations, 1971). From the above it is clear that
nutrient input from human activities in the catchment must be on a
very small scale.

In Ireland land units used in the results from the more recent agricultural statistics do not follow catchment boundaries. The above
figures have been calculated from the 1975 agricultural census for an
area approximately twice the size of the catchment, but this larger
area can be considered to have a similar composition to that of the
catchment itself. However, if a precise nutrient budget for the river
were to be calculated it would be of considerable use if statistical
information were made available in land area units that follow natural
boundaries, as townlands often do in Ireland.

13.7   WATER CHEMISTRY

13.71   INTRODUCTION

The chemistry of the water was investigated at ten sampling stations in the catchment at approximately monthly intervals from December 1974 to February 1977, with six stations along the main channel and four on the tributaries (Norton et al., 1978).   The water in the Caragh river is low in electrolytes, extremely soft, slightly acidic, low in nitrate and moderately poor in phosphate. The major ions present are chloride and sodium. The water is clear and low in suspended solids, even during floods. Table 13.2 summarizes both data on water chemistry and also rainwater and air from Valencia Observatory for 1975 and 1976 (Monthly Weather Report, Part I). An unexplained imbalance exists between the sum of cations and the sum of anions (0.51 and 0.81 meq l^{-1} respectively) in the data collected on Caragh water chemistry. Care has been taken not to draw any conclusions from the data that require absolute figures, but it was felt that comparisons could still be made if it is assumed that systematic errors are involved.

13.72   ORIGIN OF DISSOLVED SALTS

In the Caragh catchment, the resistant rock, peaty soils, low human population and marginal agriculture are all poor sources of mineral ions. Comparison of data for rain and river water (Table 13.2), especially those for chloride, sodium and potassium, indicates that rain probably plays an important role as a supplier of electrolytes into the system. The high proportion of chloride and sodium compared to other ions indicates the marine origin of the former. In addition to the rainfall a further supply of marine derived ions could be carried in by the air on the prevailing south-westerly winds, as was shown for Connemara (Groenendael et al., 1979).

13.73   SEASONAL PATTERNS

Seasonal patterns of chemistry are influenced strongly by the fact that the river is subject to high flow conditions frequently and at irregular intervals throughout the year. Mean values of conductivity, total hardness, calcium hardness, magnesium and calcium are significantly and inversely correlated with discharge at Caragh Bridge. This indicates that there is a constant input (Golterman, 1975) of magnesium and calcium, probably originating mainly from rock and till, diluted at different times to varying degrees by the rain. No significant correlation was found between discharge and chloride, which is to be expected as most of the chloride is of marine origin and supplied by the rain.

Table 13.2 Chemistry of Caragh River Water (Norton *et al.*, 1978), precipitation and air at Valencia Observatory (Irish Meteorological Service, Weather Report, Part I). (Data on Si from An Foras Forbartha; data on SO_4-S, Cu, Zn, Cd and Pb from Dowling *et al.*, 1981)

parameters measured monthly	units	Caragh Dec. 1974–Feb. 1977		rainfall 1975–1976		air 1975–1976	
		$\bar{x}$	range	$\bar{x}$	range	$\bar{x}$	range
conductivity	$\mu S\ cm^{-1}$	70	47 – 123	70.4	27 – 208		
pH			5.8 – 7.6		4.6 – 6.8		
alkalinity	$meq\ l^{-1}$ HCO_3	0.25	0.05 – 0.90	0.14	0.01 – 1.04		
calcium hardness	$mg\ l^{-1}$ $CaCO_3$	8.2	2 – 22				
total hardness	$mg\ l^{-1}$ $CaCO_3$	16.8	4 – 38				
Cl	$mg\ l^{-1}$	15.2	6.0 – 28.0	13.0	3.1 – 51.3	15.8	3.7 – 29.4
Na	"	6.1	0.8 – 11.3	8.5	2.3 – 31.9	7.4	2.3 – 31.2
Ca	=	2.15	0.5 – 6.0	0.6	0.30 – 1.30	1.5	0.4 – 6.2
Mg	=	1.37	0.15 – 5.25	0.9	0.10 – 3.60		
K	=	0.5	trace – 3.5	0.5	0.14 – 1.30	1.11	0.43 – 4.4
PO_4-P	=	0.011	0.001 – 0.045	0.036	0.01 – 0.80		
NO_3-N	=	0.050	0.011 – 0.200	0.14	0.01 – 0.68	1.83	0.70 – 5.4
NH_4-N		0.03	trace – 0.10	1.34	0.61 – 2.8	7.1	4.8 – 9.6

parameters measured occasionally in Caragh River

	SiO_2-Si $mg\ l^{-1}$	SO_4-S $mg\ l^{-1}$	Cu $\mu g\ l^{-1}$	Zn $\mu g\ l^{-1}$	Cd $\mu g\ l^{-1}$	Pb $\mu g\ l^{-1}$
$\bar{x}$	0.3	6	6	20.3	0.5	3
range	trace – 1.0	5 – 7				

13.74 DIFFERENCES WITHIN THE CATCHMENT

In general in the Caragh, contrary to the more usual situation in which
the concentration of dissolved salts increases downstream, pH, alkalinity, total hardness, calcium hardness, magnesium and calcium actually
decrease between stations 2 and 6. Against this general trend and in
agreement with the usual situation there is a significant increase for
calcium and calcium hardness in the Bridia Valley (between stations 1
and 2). The decrease in concentration of these ions in the main channel
below station 2 is due to dilution by the two major tributaries, the
Owenroe and Acoose rivers, whose catchments contain a larger proportion
of upland peatland than the Bridia Valley. Both the Owenroe and Acoose
flow through lakes, where, physico-chemical and biological processes
may perhaps favour a further decrease. Likewise, the decrease in concentration which occurs between stations 5 and 6 is probably due to
Lough Caragh.

    The pattern for chloride is different: a general increase is seen
with proximity to the sea and exposure to westerly winds. The Owbeg
River is significantly richer in chloride than all other sampling
sites except station 4. This subcatchment is the nearest to the sea,
separated from it only by a relatively low saddle (365 m). The
Meelagh River is significantly higher in chloride than the Owenroe and
Acoose and significantly lower than the Owbeg. The Owenroe and Acoose
rivers, both relatively low in chloride, are furthest away from the
sea and the Owenroe catchment is also sheltered by a ridge of relatively high hills (600 m).

13.8 BIOTA

13.81 ALGAE

The Caragh is rich in species of algae, mainly blue-greens, greens and
diatoms: approximately 185 taxa have been identified. Table 13.3
lists dominant and abundant taxa as well as some rarer species. In
general the diatoms are quantitatively the most important group, the
major genera present being *Achnanthes, Eunotia* and *Gomphonema*. Blue-
green algae are common, including several heterocystous forms (and
thus presumed nitrogen-fixers). Most algal species can be found
throughout the year although their relative abundance varies; for
example, the filamentous green are most conspicuous in summer and
autumn. The standing crop of algae varies considerably between
different sites, being greater in shallow areas with fast currents
and a fairly stable substratum than in pool areas with lower current
speeds. A crude measure of algal biomass was obtained by scrubbing
a 1 m^2 area and measuring the volume of wet, preserved algae sedimented in a measuring cylinder. Values ranged at station 1 from 815
ml m^{-2} in April to 10 ml m^{-2} in November. Fig. 13.4 shows mean values
of biomass from 5 stations (1, 2, 3, 4, 6), all riffle areas, and
discharge (E.S.B. data, analysed by Heuff) from December 1974 to
November 1975. Peak biomass was found during low flows, but does not
seem to be related solely to discharge, as low values of biomass

Table 13.3 Some algal taxa in the Caragh River.

abundant or common	other
blue-greens	
Chaemaesiphon confervicola	Aphanocapsa spp.
C. incrustans	Aphanothece spp.
C. minutus	Calothrix braunii
C. polonicus	C. parietina
Dichothrix baueriana	Chroococcus sp.
D. orsiniana	Coelosphaerium sp.
Hydrococcus cesatii	Desmosiphon maculans
H. rivularis	Gloeocapsa sanguinea
Lyngbya aestuarii	Homoeothrix sp.
L. epiphytica	Lyngbya spp.
L. kuetzingii	Merismopedia sp.
Nostoc spp.	Oscillatoria spp.
Phormidium retzii	Phormidium spp.
Stigonema mamillosum	Plectonema sp.
Tolypothrix penicillata	Schizothrix sp.
reds	
Lemanea fluviatilis	Batrachospermum atrum
Rhodochorton violaceum	B. moniliforme
diatoms	
Achnanthes spp.	Achnanthes flexella
A. minutissima var. cryptocephala	A. minutissima
A. saxonica	Amphora ovalis
A. exilis	Anomoeoneis exilis
A. affinis	Cocconeis placentula
Ceratoneis arcus	Cyclotella spp.
Cymbella spp.	Diploneis ovalis
Diatoma spp.	Epithemia zebra var. saxonica
Denticula sp.	Melosira sp.
Didymosphenia geminata	Meridion circulare
Eunotia spp.	M. circulare var. constrictum
Fragilaria vaucheriae	Navicula gregaria
Frustulia rhomboides	Nitzschia spp.
F. rhomboides var. saxonica	N. palea
f. undulata	Pinnularia spp.
Gomphonema acuminatum	P. borealis
G. acuminatum var. coronata	Rhopalodia gibberula
G. constrictum	Stauroneis phoenicenteron
G. olivaceoides	S. anceps
G. parvulum	Stenopterobia intermedia
Tabellaria flocculosa	Surirella spp.
T. fenestrata	
Navicula spp.	

cont'd.

Table 13.3 continued ..

abundant or common	other

greens

Bulbochaete spp.	*Aphanochaete repens*
Coleochaete scutata	*Botryococcus braunii*
C. orbicularis	*Chaetosphaeridium globosum*
Gongrosira fluminensis	Desmidiales (12 genera)
Microspora amoena	*Dictyosphaerium pulchellum*
Mougeotia spp.	*Draparnaldia* sp.
Oedogonium spp.	*Hormidium* cf. *rivulare*
Rhodoplax schinzii	*Microspora tumidula*
Spirogyra spp.	*M. wittrockii*
Ulothrix zonata	*Schizochlamys gelatinosa*
Zygnema sp. (18 - 26 µm)	

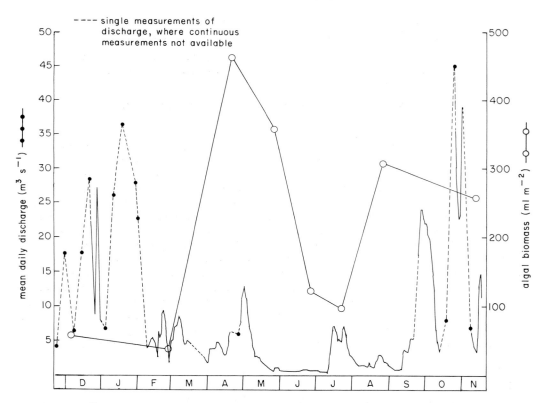

Fig. 13.7 Mean daily discharge at Caragh Bridge and mean algal biomass for 5 sampling stations from December 1974 to November 1975.

occurred in June and July, yet relatively high levels of biomass still occurred in November, after extensive flooding had occurred. It is possible that extremely low flows and high temperatures during the exceptionally dry summer of 1975 may have limited growth after the spring peak. That high biomass still occurred in November must mean that the species present are well adapted to high flow conditions. In the upper reaches of the river (stations 1 and 2) peak biomass was due mainly to diatoms. In the middle reaches it was due mainly to, in spring, *Ulothix zonata* (station 3) and *Microspora amoena* (station 4) and in August to co-dominance of diatoms and green filamentous forms (mainly *Oedogonium*). Below Lough Caragh (station 6) the spring maximum was due to a mixture of diatoms, *Lemanea fluviatilis* and *Rhodochorton violaceum*, while the August maximum was mainly due to *Zygnema* sp. Several species in the Caragh form conspicuous macroscopic growths on stones and bedrock, others are common epiphytes. The macroscopic growths can be divided into five groups according to their growth form:

(a) Mixtures of encrusting species, closely and firmly attached to the rock.

(b) Sheets of gliding filaments, without any obvious means of attachment (e.g. *Phormidium*).

(c) Thick felt-like masses, attached to rocks and stones (e.g. *Didymosphenia geminata, Gomphonema* spp.).

(d) Long trailing filaments, attached by the base to rocks, stones and plants (e.g. green filamentous forms and *Lemanea fluviatilis*).

(e) Short isolated tufts (e.g. *Tolypothrix penicillata, Dichothrix* spp.).

Encrusting species cover rock, stones and gravel and are well adapted to fast currents, some even surviving unstable substrata. For example, mixtures of the following encrusting species are commonly found in the Caragh: *Gongrosira fluminensis, Coleochaete* spp., *Stigeoclonium* bases and others, creeping Chaetophorales sheets and the lichen *Verrucaria* cf. *rivularis*. *Rhodoplax schinzii*, a green alga containing a brilliant scarlet pigment, is well represented on water-falls. Unlike the Upper Rhine (Jaag, 1932) and Tweed (Holmes & Whitton, 1975) the crusts are not conspicuous enough to make the rock crimson and the alga not obvious macroscopically. This species has not been found before in Ireland, but because it is so inconspicuous may well have been overlooked previously. Another encrusting species is *Desmosiphon maculans*, a genus new to Ireland and Britain. It forms small, hard, dark green spots on stones in fast-flowing water and as it was found only near a bridge it may prefer shady conditions.

Phormidium retzii forms strong, dark green sheets on stones and plants. It can be co-dominant with the diatoms in early spring (station 4) and is a striking feature in the river.

Didymosphenia geminata forms thick felt-like masses on rocks in riffles. The rock is covered with a layer of *Didymosphenia* attached by their stalks, while these individuals have other *Didymosphenia*'s growing epiphytically on them. In this way a dense structure is built up which also supports a community of other, smaller epiphytes, e.g. species of *Gomphonema*, *Achnanthes* and *Eunotia* spp.; most of these are also attached by stalks. These felt-like masses usually contain a population of invertebrates, mainly chironomid larvae. A considerable amount of fine sand is caught within the structure. It seems possible that these diatoms may obtain silica directly from the sand, rather than from the water (see Patrick, 1977), which is relatively poor in silica (mean 0.3 mg l^{-1} Si, range trace - 1.0 mg l^{-1} Si).

Long trailing filaments become obvious in the river during periods of low flow. These consist of green algae, e.g. *Spirogyra* spp., *Oedogonium* spp., *Mougeotia* spp., *Zygnema* sp. (18 - 26 μm). However, these algae are widespread and moderately abundant in the river the whole year around, especially in areas with lower current speeds and in pools. Few of these taxa can be identified to species level since no fruiting material has been found. It is estimated that the 10 genera involved probably represent 24 species, using criteria like width of filament, chloroplast characteristics etc.

An interesting zonation is present on riffles with a lot of bedrock and large boulders (e.g. station 5). *Stigonema mamillosum* covers rocks just above the average water level. In its dry condition it has the appearance of short, coarse, dark grey hair. Below this zone, usually just submerged, colonies of *Nostoc* a few millimetres in diameter cover the rock. Below the *Nostoc* zone *Lemanea fluviatilis* grows abundantly on the river bed, at least in spring and autumn. Only a few differences were found between stations based on presence or absence of species. For example, *Batrachospermum* spp. although not abundant anywhere, were absent from the shallower stations (1, 2) and from station 5, which is a cascade. *Microspora amoena*, present at most stations, was absent from station 1 and only occasionally present at station 6. Some taxa were only found downstream of Lough Caragh, e.g. *Oedogonium undulatum*, *Chaetosphaeridium globosum*, *Coelosphaerium*, while *Stigeoclonium* sp. was more common here than elsewhere. *Nitella opaca* was found only at station 3, growing in deep water on a sandy bottom.

The composition of the algal flora is, as might be expected, indicative of the oligotrophic nature of the Caragh system. Species present which are characteristic of slightly acidic water with low salt concentrations (Patrick, 1977) are *Eunotia* spp., *Frustulia rhomboides*, *Pinnularia* spp. and *Tabellaria flocculosa*. Some are calciphobe and can tolerate high levels of iron, e.g. *Eunotia* spp., *Pinnularia* spp., *Frustulia*, *Gomphonema acuminatum*, *Stauroneis phoenicenteron* (Patrick, 1977). Twelve genera of desmids have been recorded in the river. These, while never abundant, are widespread. Many live loosely attached amongst other algae or are washed out from surrounding bogs. The presence of *Ulothrix zonata*, *Microspora amoena*

and *M. tumidula* is thought to indicate oxygen rich water (Printz, 1964).
Members of the genus *Zygnema*, with a cell width less than 30 μm are
restricted to streams with water poor in electrolytes, at least in
Scandinavia (Israelson, 1949). In the Caragh a *Zygnema* sp. (18 - 26 μm)
is common throughout the river. It is interesting to note that *Cladophora*, which is common elsewhere in Ireland, is absent from this river,
as the freshwater members of this genus are generally considered to be
characteristic of eutrophic conditions (Van den Hoek, 1963).

    In recent times few algal studies have been carried out in oligo-
trophic calcium-poor rivers. In Ireland the only case reported is a
headwater stream of the Inver system at Recess (National Grid Reference
L867 459) (McDonnell & Fahy, 1978). When sampled in July 1976, this
stream was dominated by *Batrachospermum* spp. with the filamentous green
algae *Mougeotia*, *Spirogyra* and *Zygnema* next in abundance. Israelson
(1949) classifies Scandinavian rivers into two major types: "*Zygnema*
type" and "*Vaucheria* type", corresponding respectively to oligotrophic
and eutrophic waters. The algal vegetation in the Caragh is comparable
to oligotrophic streams in Scandinavia ("*Zygnema* type"). Several of
the differential species for streams poor in minerals mentioned by
Israelson are also present in the Caragh: *Mougeotia*, *Spirogyra*,
Zygnema, *Bulbochaete*, *Stigonema mamillosum*, *Batrachospermum moniliforme*,
Schizochlamys gelatinosa. Israelson distinguishes two sub-types within
the *Zygnema* type: brown water streams with much *Batrachospermum* and
clear water streams with many encrusting forms. The Caragh would fit
in with the latter sub-type while the Inver stream would presumably fit
the first sub-type. The Caragh also shows similarities to the Kaltis-
jokk, an oligotrophic stream in Sweden (Klasvik, 1974) in which
Eunotia spp. are mentioned as abundant. Klasvik found that *Achnanthes
minutissima* var. *cryptocephala* is much more abundant than *A. minutissima*,
as is also the case in the Caragh. Round (1959) noted that the diatom
flora of the upper sediment layers of acid lakes in Ireland is similar
to the flora of oligotrophic Scandinavian lakes. As many of the
species mentioned by Round have also been recorded from the Caragh,
this is further evidence that the biology of the Caragh is probably
similar to Scandinavian oligotrophic waters. Other streams comparable
to the Caragh are described by Douglas (1958) and Whitford &
Schumacher (1963).

13.82 MACROPHYTES

The overall cover of macrophytes in the river is extremely low and
they are entirely absent from many fast-flowing stretches. 51 species
of vascular plant and 15 species of bryophyte were recorded. The
vascular plants can be divided into three groups based on the habitat
they occupy:

    (a) Gravel beds which are exposed during periods of low flow;

    (b) Shallow sandy areas near the river margin;

(c) Deeper water with moderate flow.

On the gravel banks *Oenanthe crocata*, *Apium inundatum*, *Juncus bulbosus*
f. *fluitans* and *Callitriche* spp. are common and can form a dense cover.
As these plants trap sand and silt, banks may build up until swept
away by flash floods. Shallow sandy areas near the shore, especially
in the lower stretches of the river, are sparsely vegetated by plants
such as *Eriocaulon aquaticum*, *Lobelia dortmanna*, *Littorella uniflora*
and *Elatine hexandra*. In deeper water in the upper reaches of the
river *Scirpus fluitans* occurs, while in the middle and lower reaches
Juncus bulbosus f. *fluitans* and *Myriophyllum alterniflorum* are common,
accompanied very occasionally by a batrachian *Ranunculus* species.
Marsh and floating leaf zones are absent except in a few slow-flowing
stretches between station 4 and Lough Caragh, where occasional plants
of *Nuphar lutea* occur and a narrow fringe of *Phragmites australis*,
Equisetum fluviatile and *Carex rostrata* is developed.

Bryophytes are not an important component of the aquatic vegetation,
except in areas of bedrock, where the major species are *Fontinalis
antipyretica*, *Fissidens* sp., *Grimmia alpicola* var. *rivularis* and
Cinclidotus fontinaloides. Luxurious growths of bryophytes occur on
big boulders in the river which project above the average water-level.

The species composition of the macrophytes is similar to that given
by Haslam (1978) as typical of mountain and moorland streams in S.W.
England and Scotland. Of the trophic level indicators cited by Haslam
only those species characteristic of dystrophic and oligotrophic
conditions occur in the Caragh.

13.83 MACROINVERTEBRATES

Macroinvertebrates from riffle areas (stations 1, 2, 3, 6) were sampled
quantitatively on four occasions (December 1974, February, August,
December 1975). Numerous qualitative samples were taken throughout
the survey. Over 200 taxa were identified including 160 species
(Dowling *et al.*, 1981). The more important observations are outlined
here.

Sponges were confined to station 6 below Caragh Lake. Low numbers
of triclads (*Phagocata vitta*) occurred at stations 1 - 3 and 6. Five
oligochaete families were identified, the most important species being
Lumbriculus variegatus and *Stylodrilus heringianus*, which were widespread. Hirudinea were found only in low numbers. *Gammarus duebeni*
occurred throughout the system, being most common below the lake.
Asellus aquaticus occurred only in silted pools near stations 4 and 5.
There were eight species of Plecoptera; of these, *Leuctra inermis*
was most plentiful at station 1, being replaced by *Chloroperla
torrentium* and *Perla bipunctata* at station 2. *Isoperla grammatica*
and *Leuctra hippopus* were common downstream of Caragh Lake. Fourteen
species of Ephemeroptera were identified and of these *Baetis rhodani*

was the most important with largest numbers occurring at station 1. Other important mayflies included *Rhithrogena semicolorata* (upper), *Siphlonurus lacustris* (peaty pools), *Ecdyonurus torrentis* (upper), *E. venosus* (all stations), *Caenis moesta* and *C. rivulorum* (below Caragh Lake). Only one odonate species was found (*Coenagrion puella*, station 5). Hemiptera consisted of *Aphelocheirus montandoni* (station 6) and small numbers of other species including *Gerris najas*.

Trichoptera are well represented, especially below the lake. The combined observations of O'Connor (1975) and Dowling *et al.* (1978) give a total of 43 species. The latter survey reported that the following were most important in the headwaters: *Agapetus fuscipes*, *Diplectrona felix*, *Philopotamus montanus*, *Tinodes machlachlani*, *Wormaldia* sp., *Hydropsyche siltalai*. The last was the most common net-spinning form in the upper stretch. Trichopteran numbers were highest below the lake where *Neureclipsis bimaculata*, *Hydroptila* spp., *H. pellucida*, *Cheumatopsyche lepida*, *Ithytrichia lamellaris*, *Psychomyia pusilla*, *Polycentropus flavomaculatus*, *P. kingi* and *Chimarrha marginata* were all well represented.

Of the six species of Coleoptera recorded *Esolus parallelipipedus* was dominant, while *Limnius volckmari*, *Oulimnius tuberculatus* and *Elimis aenea* were less important. The Chironomidae were investigated thoroughly and yielded 48 species. Two genera were numerically dominant, *Cricotopus* and *Eukiefferiella*, both being well represented at all stations. Simuliidae were relatively important in the middle reaches. Of the other Diptera only *Dicranota* and *Ephydra* occurred regularly with highest numbers of both at station 2. Hydracarina were encountered everywhere but did not represent a high proportion of the biomass. Of the six species recorded, one, *Pergamasus brevicornis*, is new to the British Isles and is not listed by Gledhill & Viets (1976). Mollusca were poorly represented presumably because the low ionic content of this river system. Of seven genera recorded only *Pisidium/Sphaerium* spp. and *Ancylus fluviatilis* were ever found in numbers; they were most common below the lake.

It is clear that the Caragh River has a wide diversity of species at all stations, including a good representation of the most sensitive forms. Such results are indicative of a clean unpolluted river along the entire stretch examined and would be represented as Q5 (i.e. good quality) by the Irish classification method (Flanagan & Toner, 1972). The specific differences found at station 6 are caused by a lacustrine influence carrying organic matter composed of detritus, phyto- and zooplankton from Caragh Lake and do not show any signs of a reduction in water quality. Other differences between stations in the system are attributed to flash flooding, which may occur in any season, resulting in movement of the substratum and a paucity of suitable food. Evidence in support of this is the fact that at station 1 the fauna is poorer than elsewhere being located just downstream of a cascade. The further downstream the organisms are located in this system the wider the river and the more capable it is of dealing with sudden floods without denuding the fauna.

13.84 FISH

The Caragh is renowned as a salmonid river, with a plentiful supply of both salmon and sea-trout. Brown trout are common also but they are usually small except for a few larger specimens in the lake (E. Evans, pers. comm.). However, commercial fishing, including some illegal netting may be actually reducing stocks migrating upstream to spawn. Apparently rod and line catches had been in decline in recent years until an improvement occurred in 1980. The latter is mainly attributed to a successful run of salmon during a very wet winter.

13.9 CONCLUSIONS

The only old biological records available for the river system concern Lough Caragh. Plankton was collected and described from the lake in 1906 (West & West), and a list of some macrophytes was given by Praeger (1934). Recent surveys of plankton and vascular plants in Lough Caragh (Flanagan & Toner, 1975, and H. Heuff, unpublished data) have found the lake to be practically unchanged. In general it appears that the Caragh system is not only relatively unpolluted from headwaters to the estuary, but also has apparently remained hardly changed for three-quarters of a century. Because of this special efforts should be made to preserve the present situation and where possible future develop-ments of intensive farming, industry or excessive afforestation should be avoided.

REFERENCES

Agricultural Statistics (1975) Central Statistics Office, Dublin, Ireland.
Atlas of Ireland (1979) Royal Irish Academy, I.S.B.N. 0 901714 13 5
Census of Population (1971, 1979) Central Statistics Office, Dublin, Ireland.
Connaughton M.J. (1967) Global Solar Radiation, Potential Evapo-transpiration and Potential Water Deficit in Ireland, Irish Metero-logical Service, Technical Note No. 31.
Douglas B. (1958) The ecology of the attached diatoms and other algae in a small stony stream. *J. Ecol.* 46, 295-322.
Dowling C., O'Connor M., O'Grady M.F. & Clynes E. (1981) A base-line survey of the Caragh, an unpolluted river in South-West Ireland: topography and water chemistry. *J. Life Sci. R. Dubl. Soc.* 2(2), 137-45.
Dowling C., O'Connor J.P. & O'Grady M.F. (1981) A base-line survey of the Caragh, an unpolluted river in South-West Ireland: observations on macroinvertebrates. *J. Life Sci. R. Dubl. Soc.* 2(2), 147-59.
Dowling C., O'Connor J.P. & O'Grady M.F. (1978) The Caragh River Survey 1974-1977, Final Report, section 3: Studies on the macro-invertebrates of the River Caragh. National Science Council of Ireland.

Flanagan P.J. & Toner P.F. (1975) A preliminary survey of Irish lakes, An Foras Forbartha, Water Resources Division.

Gledhill T. & Vietz V.O. (1976) A synonymic and bibliographic check-list of the freshwater mites (Hydrachnellae and Limnohalacaridae, Acari) recorded from Great Britain and Ireland. *Sci. Publs Freshwater Biol. Ass. U.K.* 1.

Golterman H.L. (1975) Chemistry. In: Whitton B.A. (Ed.) *River Ecology* pp. 39-80. Blackwell, Oxford.

Groenendael J.M. van, Hochstenbach S.M.H., Mansfeld M.J.M. van & Roozen A.J.M. (1979) The influence of the sea and of parent material on wetlands and blanket bog in West Connemara, Ireland. Thesis, Catholic University Nijmegen, Netherlands.

Haslam S.M. (1978) *River Plants*. Cambridge U. P., Cambridge.

Hoek C. van den (1963) *Revision of the European species of* Cladophora. E.J. Brill, Leiden, Netherlands.

Holmes N.T.H. & Whitton B.A. (1975) Notes on some macroscopic algae new or seldom recorded for Britain. *Vasculum* 60, 47-55.

Israelson G. (1949) On some attached Zygnemales and their significance in classifying streams. *Bot. Notiser* 4, 313-58.

Jaag O. (1938) Die Kryptogamenflora des Rheinfalls und des Hochrheins von Stein bis Eglisau. *Mitt. Naturgesellschaft Schaffhausen* 14, 1-158.

Klasvik B. (1974) *Computerised Analysis of Stream Algae*. Vaxtekol. Stud. 5, Uppsala.

Lewis C.A. (1977) *South and South West Ireland, Guidebook for Excursion Al5*. International Union for Quaternary Research. I.S.B.N. 0 902246 83 6.

McDonnell J.K. & Fahy E. (1978) An annotated list of the algae comprising a lotic community in Western Ireland. *Ir. Nat. J.* 19(6), 184-7.

Monthly Weather Reports (1975, 1976) Part I, general weather report. Irish Meterological Service, Dublin.

Norton M.A., Dowling C., O'Grady M.J. & Clynes E. (1978) The Caragh River Survey, 1974-1977, Final Report, Section 2: Physico-chemical Investigation Appendix 1, physico-chemical results. National Science Council of Ireland.

O'Connor P.J. (1975) Freshwater studies part II: an investigation of the trichopterous fauna and its distribution in various habitats in Ireland. Ph.D. Thesis, University College, Dublin.

Patrick R. (1977) Ecology of freshwater diatoms and diatom communities. In: Werner D. (Ed.) *The Biology of Diatoms*. Blackwell, Oxford.

Praeger R.L. (1934) *The Botanist in Ireland*. Hodges Figgis, Dublin

Printz H. (1964) *Die Chaetophoralen der Binnengewasser, Hydrobiologia* 24, 1-376.

Rohan P.K. (1975) *The Climate of Ireland*. Stationery Office, Dublin.

Round F.E. (1959) A comparative survey of the epipelic diatom flora of some Irish loughs. *Proc. R. Ir. Acad.* 60 (B,5), 193-215.

Shreve R.L. (1966) Statistical law of stream numbers. *J. Geol.* 74, 17-37.

Ordnance Survey of Ireland (1974) Soil Map of Ireland.

Walsh P.T. (1968) The Old Red Sandstone West of Killarney, Co. Kerry, Ireland. *Proc. R. Ir. Acad.* 66 (B,2), 9-26.

West W. & West G.S. (1906) A comparative study of the plankton of some Irish lakes. *Trans. R. Ir. Acad.* <u>23</u> (B,2), 77-116.

Whitford L.A. & Schumacher G.J. (1963) Communities of algae in North Carolina streams and their seasonal relations. *Hydrobiologia* <u>22</u>, 133-96.

Whittow J.B. (1974) *Geology and Scenery in Ireland.* Penguin Books.

14 · Suir

K. HORKAN

14.1   INTRODUCTION

The River Suir has been chosen as the third representative of Irish
rivers, as it provides a useful contrast with the River Caragh
(Chapter 13). Although it rises in the mountains and the quality of
the water in its upper stretches is excellent, the downstream part is
much more nutrient-rich, with local problems due to massive growths of
macrophytes. The chapter attempts to give an overall picture of
current knowledge of the river.

The ecology of the Suir and its various tributaries is much influenced by the geology of the catchment. Like other rivers in Ireland
its ancestral course was formed about 60 million years ago (Nevill,
1972) during the emergence of Europe and Asia from the Chalk Sea.
Rainfall resulted in erosion of the new land surfaces, washing silt
down the slope seawards, thus creating channels within which the first
river valleys developed. Many rivers in Ireland, including the Suir,
originally flowed from north to south, following the gradient of the
chalk surface, but as erosion progressed deeper into the valley floor,
the waters reached harder layers such as shales. In the case of the
Suir, this resulted in the course of the river altering to an east-
west direction wearing through synclines before turning south again
near the sea (Fig. 14.1).

The Suir rises on Borrisnoe Mountain, just north of the well-known
landmark known as the Devil's Bit, in an area of Silurian deposits.
The river falls rapidly through a narrow band of Old Red Sandstone
and from there onto Lower Carboniferous Limestone, which is the main
geological feature of the catchment. The gradient is quite steep at
first, falling 320 m in the first 9 km, but then becomes more gradual,
descending only 60 m over the next 58 km to Clonmel. East of Clonmel
the river skirts along the Comeragh Mountains, where the area adjacent
to the south bank consists mainly of Avonian Shales. Near the estuary
the rocks are a mixture of Intrusive and Volcanic silica-poor rocks.

Some of the tributaries such as the Drish and Rossestown Rivers
originate in Upper Carboniferous limestone descending from there through
Lower Carboniferous Limestone. Two other important tributaries, the
Clodiagh and Multeen Rivers, flow over Silurian rocks through narrow
bands of Old Red Sandstone and Lower Avonian Shales and Sandstones
before continuing onto the Lower Carboniferous Limestone.

Near the headwaters there are acid brown earths with peaty podzols
below these. Much of the Suir catchment consists of flat to undulating
lowlands which are mostly grey-brown podzolics and associated gleys.

The moderate, moist climate, lacking extremes, renders the country
suitable for farming. Apart from the hilly headwaters, the Suir catchment includes very rich grasslands with beef cattle and dairy farming
and also some cereals.

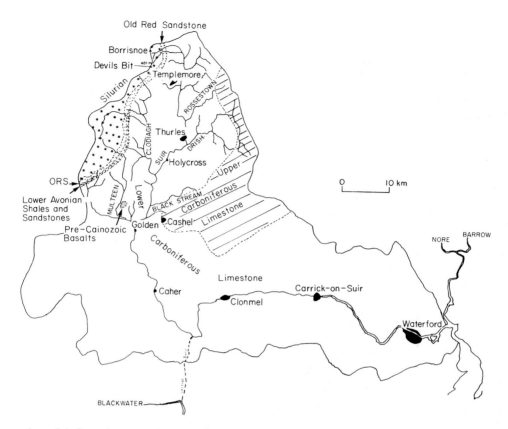

Fig. 14.1 River Suir catchment, showing upper tributaries and general geology. The Initial Suir is also shown flowing southwards before it eventually changed to its present course (see text).

Because of the rich farming hinterland in the Suir Valley there are several important provincial towns. Agriculture related industries are to be found in these towns and include sugar refining, food processing, meat and protein production, creameries, ice cream manufacture and brewing. The historic city of Waterford lies on the estuary below the confluence of the three "sister" rivers, the Barrow, Nore and Suir. Industries to be found here include glassware, chemicals and shipping.

Many parts of the Suir Valley are popular with motoring tourists, but the main attraction is salmonid angling. Commercial salmon fishing occurs along the lower reaches of the River Suir, where best catch returns are from drift nets. However, in recent years overfishing by the latter is regarded as being a cause of decline in catches of salmon and sea-trout by anglers in the freshwater reaches. In some stretches of the Suir, usually below towns, angling is poor because of

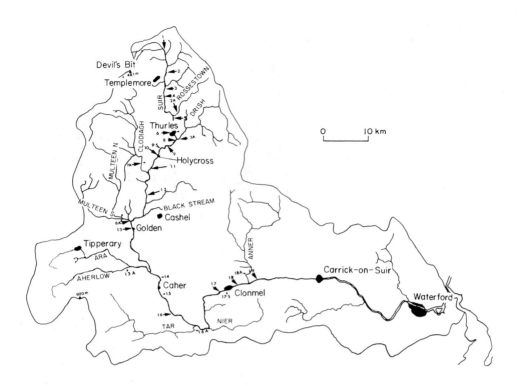

Fig. 14.2 River Suir catchment, showing main towns, tributaries and the more important sampling stations.

the effects of domestic and industrial discharges (mostly organic, biodegradable) on water quality. The value of salmonid catches in the River Suir during 1978 was estimated to be about IR £0.4 million (An Foras Forbartha, 1980). Other amenities and recreational activities are best developed in the lower reaches of the main channel. These include boating, sailing, water-skiing and other aquatic pastimes and also river walks and picnicing.

Fig. 14.3 (opposite) R. Suir sampling station upstream of Holycross Bridge. This is a recovery zone dominated by *Ranunculus* spp. and *Potamogeton pectinatus*. The emergent is *Oenanthe* sp. (Photo courtesy of Bord Failte-Irish Tourist Board)

14.2   GENERAL DATA

DIMENSIONS
    Area of catchment   3610 km^2
    Length of main channel   184 km
    Altitude at source   449 m, on Borrisnoe Mountain

CLIMATE

Precipitation
    mostly 800-1000 mm, reaching 1800 mm in upper catchment (An Foras Forbartha,
    1980). Snow sometimes falls in the mountains, but never persists for long
    enough to permit a concentrated period of snow-melt in spring.

Temperature
    The mean values for Ireland (Praeger, 1974) provide a reasonable indi-
    cation for the Suir catchment:
                January 4.4 $^\circ$C;   July 15.5 $^\circ$C

POPULATION
    Templemore   2556   Clonmel   12418
    Thurles   7439   Carrick-on-Suir   5511
    Tipperary   4929   Waterford   32617
    Cashel   2588

PHYSICAL FEATURES OF RIVER

Hydrometric data
Selected stations on Suir and some of its larger tributaries. Data include
long-term mean flows, highest recorded mean monthly flows per unit of time and
per unit of catchment area, lowest recorded mean monthly flows per unit of time
and per unit of area. (Data from An Foras Forbartha, unpublished.)

river		station	long-term mean flow ($m^3\ s^{-1}$)	highest recorded mean monthly flow		lowest recorded mean monthly flow	
				($m^3\ s^{-1}$)	($m^3\ s^{-1}$ 100 km^2)	($m^3\ s^{-1}$)	($m^3\ s^{-1}$ 100 km^{-2})
Suir	Thurles	6	4.26	7.75 (Jan)	3.28	0.10 (Aug)	0.65
Suir	Breakstown Lodge	9.5	9.97	19.87 (Jan)	3.88	0.46 (Oct)	0.73
Suir	New Br.	14	20.32	38.56 (Dec)	3.44	1.55 (Aug)	0.62
Suir	Caher Park Br.	15	29.41	55.32 (Jan)	3.47	4.19 (Oct)	0.78
Suir	Clonmel	17.5	54.20	92.05 (Jan)	4.24	8.83 (Aug)	1.34
Clodiagh	Rathkennan Br.	7A	5.26	10.75 (Jan)	4.37	0.24 (Aug)	0.67
Multeen (S)*	Aughnagross	8A(S)	2.17	3.76 (Dec)	4.33	0.21 (Aug)	1.13
Multeen (N)*	Ballinaclough	8A(N)	1.55	3.33 (Jan)	4.44	0.26 (Aug)	0.74
Aherlow	Killardry	13A	7.98	14.12 (Jan)	5.17	1.25 (Oct)	1.59
Tar R.	Tar Br.	16A	5.92	10.83 (Jan)	4.75	1.43 (Oct)	1.36
Anner	Anner	18A	6.18	11.57 (Jan)	2.74	0.91 (Aug)	0.58

S South Branch   N North Branch

14.3 PHYSICO-CHEMICAL FEATURES

14.31 HYDROLOGY AND WATER RESOURCES

Maximum flows have almost always been recorded in January and minimum flows in late summer or early autumn (see 14.2). Hydrological studies of the time of occurrence of maximum and minimum flows in the U.K. have shown that they become later towards the south and east. Beaumont (1975), for instance, quotes examples for the north-west of Scotland and East Anglia, where the mean maximum flows occur in December and March, respectively. The Suir, its location in south-east Ireland, falls in between. The detailed behaviour of flows within the catchment is influenced by the geology, storage capacity of the groundwater and the proportion of precipitation reaching the rivers.

Table 14.1 summarizes two important sets of statistics important for planning. The 95 percentile flow is perhaps the most useful guide to decision makers of the potential of receiving water to assimilate waste discharges from urban and industrial zones. For the assessment of the water resources of a catchment when planning abstractions, it is also necessary to know the dry weather flow. Detailed investigations are now under way to improve the basic hydrological information in the catchment with the construction of weirs, monitoring of groundwater and more accurate gauging of the system in order to plan for careful industrial and urban expansion in the future.

Table 14.1 Catchment area, estimated 95 percentile flow, and estimated dry weather flow at selected stations on the Suir and some larger tributaries.

river	location	stn. no.	catch- ment km^2	95 per- centile flow $m^{-3} s^{-1}$	dry weather flow $m^3 s^{-1}$	$1 s^{-1} km^{-2}$
Suir	Thurles Br.	6	236	0.20	0.09	0.40
Suir	Beakstown Lodge	9.5	512	1.10	0.30	0.60
Suir	New Br.	14	1120	2.73	1.10	1.00
Suir	Caher Park	15	1602	6.10	3.20	2.00
Clodiagh	Rathkennan	7A	246	0.45	0.17	0.70
Multeen (S)*	Aughnagross	8A(S)	87	0.30	0.10	1.15
Multeen (N)*	Ballinaclogh	8A(N)	75	0.30	0.10	1.33
Aherlow	Killardry	13A	273	1.68	0.90	3.30
Tar	Tar Br.	16A	228	1.68	1.00	4.39

*S = South branch;   N = North branch

14.32 WATER ABSTRACTION

The location of all surface abstractions in the catchment are shown in
Fig. 14.5. Most of the large abstractions are from surface waters,
mainly from rivers and streams, although a few impoundments are also
involved. Smaller abstractions tend to be provided from wells and boreholes. In general, most of the abstractions in the Suir catchment occur
in areas remote from point waste discharges. However, a few industrial
abstractions, notably those for the Sugar Co. below Thurles, protein and
meat plants below Caher and a chemical factory just upstream of Carrick-
on-Suir could be abstracting water of doubtful to poor quality at all
times. Hence, regular monitoring particularly at these points is
essential to avoid providing sub-standard water for such industrial
concerns. It would appear that in planning for the future it will be
necessary to adopt stringent water quality standards along the main
channel in order to maintain or improve present conditions to ensure
suitable water to meet future industrial need and also to allow upstream
migration of salmonids. Such standards would be ensured by adopting
the Council of the European Communities (C.E.C.) Directive for salmonid
waters (1978) and would require a high standard of sewage treatment by
all towns presently discharging into the upper portion of the main
channel and on the tributaries of the catchment.

14.33 WATER QUALITY

The Suir experiences a wide range of water qualities over the length
examined (Table 14.2). In addition there is often a considerable range
in concentrations of different chemical constituents, which is at least
in part due to differences between summer and winter flows. The effects
upon water quality of domestic or industrial wastes show up immediately
downstream of their discharge. However, nutrients reaching the river
from non-point sources such as runoff from fertilized agricultural land
can also have a considerable affect on water quality.

Water quality is excellent in the upper reaches because there is
little industry, towns or extensive farming. Initial impairment
commences below Templemore, where organic effluent is responsible for
increased BOD and also, at times, low dissolved oxygen concentrations.
Further downstream at stations 4 and 5 (Fig. 14.2), the river improves
in quality; the large weed growths sometimes lead to oxygen super-
saturation in summer. The river then receives both industrial and
domestic organic effluents at Thurles, which cause serious impairment
in water quality over a 5 km length (Table 14.2). Sewage-fungus is common here, especially in winter during the sugar beet campaign.

Downstream of Holycross the river flows through a rural hinterland
where the water improves rapidly in quality, a fact reflected in the
excellent salmonid stretches that occur from Two-Ford Br. (station 11)
to just upstream of Caher. From Caher to Clonmel there is moderate
impairment of water quality. Downstream of Caher there is a marked

Table 14.2 R. Suir: water chemistry (Lennox & Toner, 1980) and biological quality (Flanagan & Toner, 1972; Clabby, 1981). Chemistry at stations 1 - 4 for 1978-79 and stations 5 - 19 for 1972-78. Sampling usually carried out between 1000 and 1600. Quality values based on classification system simplified as follows: 5, good; 4, fair; 3, doubtful; 2, poor; 1, bad.

	distance from Station 1 km	dissolved oxygen (% saturation)	BOD (mg l⁻¹)	NH$_3$-N (mg l⁻¹)	NO$_3$-N (mg l⁻¹)	reactive phosphate-P (mg l⁻¹)	Quality rating (Q) 1978	1976	1971
1 Knockanroe Br.	0	84 - 138	1.0 - 3.5	0.02 - 0.05	0.20 - 1.70	0.010 - 0.050	5	5	5
2 Knocknageragh Br.	5	80 - 104	0.6 - 2.5	0.02 - 0.09	0.60 - 3.40	0.010 - 0.100	4 - 5	4	5
3 d/s Templemore, Penane Br.	9	68 - 124	0.8 - 3.4	0.05 - 0.24	1.00 - 3.90	0.020 - 0.100	2 - 3	3 - 4	3 - 4
4 Loughmore Br.	11	81 - 168	0.3 - 2.9	0.03 - 0.10	1.10 - 3.40	0.020 - 0.170	3	3 - 4	3 - 4
5 Rossestown Br.	18	60 - 120*	0.8 - 7.1	tr - 0.19	0.46 - 2.40	0.013 - 0.130	3 - 4	4	4 - 5
6 d/s Thurles Br.	23	41 - 110*	1.3 - 7.0	tr - 0.21	0.43 - 2.80	0.010 - 0.183	3		4 - 5
8 Turtulla Br.	25	37 - 112*	1.4 - 8.9	0.02 - 1.80	0.58 - 2.30	0.013 - 0.330	{ 3 (L.H.) 2 (R.H.)	3 (mid channel)	2 (mid-channel)
9 Cabragh Br.	28	32 - 108*	1.2 - 20.0	tr - 3.00	0.45 - 3.70	0.008 - 0.293	3 - 4	2 - 3	2
10 Holycross Br.	32	44 - 146*	0.8 - 9.3	0.01 - 2.65	0.24 - 2.20	0.003 - 0.200	4	4	4
11 Two Ford Br.	36	40 - 132*	0.9 - 9.1	tr - 0.80	0.26 - 2.20	0.003 - 0.183	5	4	5
12 Ardmayle Br.	45	66 - 126	1.0 - 6.6	tr - 0.16	0.86 - 1.90	0.010 - 0.175	4 - 5	4 - 5	4 - 5
13 Golden Br.	58	69 - 141	1.0 - 4.3	tr - 0.14	0.84 - 1.80	0.010 - 0.070	4	4 - 5	4 - 5
16 Br. at Swiss Cottage	82	25 - 102	1.6 - 8.0	0.04 - 0.37	0.84 - 1.60	0.010 - 0.090	3 - 4	3 - 4	4
17 0.5 km u/s Clonmel	113	79 - 116	1.3 - 3.4	0.01 - 0.11	1.10 - 1.60	0.007 - 0.085	5	3 - 4	5
18 Twomile Br.	119	68 - 108	2.1 - 3.7	0.06 - 3.40	1.10 - 1.60	0.010 - 0.095	4	3	3 - 4
19 Kilsheelan Br.	125	71 - 112	1.3 - 2.3	0.03 - 0.18	1.10 - 1.70	0.020 - 0.080	2 - 3	3 - 4	4 - 5

*lower values for D.O. may have been recorded during diel surveys (see Table 14.3).

'sag' of oxygen under summer conditions and recently fish-kills have
been reported (Lennox & Toner, 1980). The latter appear to be attributable to "once off" discharges under low flow conditions rather than
overall serious impairment.

14.4   MACROPHYTES AND ALGAE

Some indication that a river is eutrophic can be obtained by visual
observation of relative abundances of macrophytes and macroalgae. More
detailed examination often reveals differences in the species composition of the macrophyte and algal communities of enriched and unenriched
reaches (Horkan, 1981). In recent years stretches of the River Suir
have become choked with massive weed and algal growths.

During the summer of 1978 a detailed survey was carried out in the
river near Thurles, as well as a nearby tributary, the Drish River, to
ascertain the main effects of such plant growths. This survey was
extended in 1979 upstream of Templemore and also downstream as far as
Golden; three further tributaries, the Rossestown, Clodiagh and Multeen
Rivers were also included. The more important observations are outlined
below.

Station 1 (Knockanroe Br.: 9 km downstream of the source of the Suir)
is not influenced by any large industrial or domestic effluents. Weed
and algal growth in a riffle area covered a maximum of 20% of the stony-
gravel substratum and the resulting biomass was low at all times. By
the time the river reaches Rossestown (station 5), 18 km further downstream, it has increased in volume due to the addition of some tributaries and is enriched organically by Templemore town sewage. Massive
growths of weeds and algae occur here and the dominants are *Ranunculus*
spp. and *Cladophora*.

At Thurles Br., in addition to some large growths of silt-loving
Potamogeton spp., there were large "streamers" of *Cladophora* and some
patches of sewage-fungus. The trend in this stretch of river of a
shift from *Ranunculus* to *Potamogeton* is presumably a result of organic
enrichment, where the effluent load produces organics which deposit
sediments. Such an effect has been noted by Haslam (1978). Two kilometres further downstream at Turtulla (station 8) the effects of town
sewage are more pronounced on the right-hand side than on the left-hand
side, the latter being diluted by a clean tributary, the Drish River.
As at Thurles Bridge, some silting occurs, most notably on the more
polluted right-hand side, where *Potamogeton* and *Cladophora* dominated.
Cladophora was also common on the cleaner left-hand side, together with
Ranunculus (Horkan, 1980, 1981).

At Cabragh Br. (station 9) the combined effects of town sewage and
industrial effluent, particularly sugar beet wastes, are felt. The
sampling point here is located in a fast-flowing stretch of water

producing a strong current which prevents extensive silt deposition and results in a reduction in *Potamogeton* growth. *Cladophora* is dominant here, but in reduced amounts, possibly because of inhibition through chemical effects and/or by light reduction due to shading and high levels of suspended solids. Westlake (1959) reported similar observations in a polluted English stream.

Further downstream the river enters a recovery zone and by the time it reaches Holycross (station 10), a river distance of 4 km, it shows improved quality. The large growths of *Cladophora* are replaced by the less problematic *Ranunculus*. At Two-Ford Br. (station 11) water quality is good; dominance by one or two genera in large quantities no longer occurs and species diversity is greater than at previous stations.

At Ardmayle Br. (station 12) the river is 40 - 50 m wide and about 0.5 m or more deep and the water takes on the characteristics of a (relatively clean) meandering lowland river. Diversity of plants remains high and mosses become more common (Horkan, 1981). At Golden (station 13) a slight reduction of water quality results from organic enrichment from Cashel entering a tributary, the Black Stream, about 4 km upstream of the sampling point. This enrichment is reduced however, because of dilution by a very clean tributary, the Multeen River, which enters the Suir about 1.5 km upstream of Golden. Nevertheless, the macrophyte biomass at this location was higher than at any other surveyed on the river during 1979, indicating very eutrophic conditions. *Ranunculus* is dominant here in the faster riffles and *Potamogeton* in the slower parts.

Of the four tributaries examined, the smallest is the Rossestown River (station 2A) which is slightly enriched from agricultural sources. Some *Cladophora* spp. growths occur here as well as mosses and *Rorippa* spp. The Drish River (station 3A) enters the River Suir upstream of Turtulla Br. (station 8) and is regarded as being relatively clean but slightly eutrophic. The flora is dominated by a mixture of mosses and *Ranunculus* spp. Long term changes in water quality may occur in this river as a peat-briquette factory (a fuel made from compressed turf) is being built nearby. This gives rise to the possibility of problems from peat silt covering the substratum and possibly interfering with invertebrates and trout spawning. The Clodiagh River (station 7A) sampling station has a loose silty gravel substratum with few stones and is regarded as being a slightly unstable stream bed. *Ranunculus* occurs in patches here with *Potamogeton* being found in the more silted areas. The fourth tributary, the Multeen River (station 8A), is clean and has a very sparse flora consisting mostly of small patches of *Ranunculus* and occasional small stands of *Potamogeton*.

The general impression gained from these observations is that balanced, stable, diverse plant communities occur in the clean stations of the Suir and some of the tributaries examined. In eutrophic areas the community is less stable, has low diversity featuring either *Potamogeton* spp. (usually *P. pectinatus*) or *Cladophora* or both, depending on the degree of siltation of the substratum.

   Praeger (1974) lists a number of macrophytes of interest found along
the Suir, with the following near Clonmel: *Rorippa sylvestris,*
Ceratophyllum demersum, Lemna polyrrhiza.   *Butomus umbellatus* provides
an interesting record near Carrick-on-Shannon.

14.5   MACROINVERTEBRATES

The macroinvertebrate fauna of the River Suir is described below.   The
stations sampled are those listed in Table 14.2; the substrata examined
were in riffle areas and collections made in summer.

Station 1 has a diverse fauna, reflecting the 'excellent' water quality
found there.   The main components of this fauna include Baetidae and
Helmidae, with a number of other forms being relatively common, of
which cased caddis, *Hydropsyche, Polycentropus, Ephemerella*, Ecdynurids,
Leuctra, Gammarus, Chironomidae and other dipteran larvae are the most
important.   At station 2 the quality of the river is 'good'.   Mites are
dominant, with lesser numbers of helmids, chironomids, Trichoptera,
Gammarus, Potamopyrgus and *Ancylus.*

At stations 3 and 4, there is a decline in water quality to a 'doubtful'
condition, reflecting the effects of enrichment from Templemore sewage.
The fauna reflects this condition, showing a reduction in numbers of the
more sensitive forms.   More tolerant Chironomidae (including *Chironomus*
spp.) and haliplids dominate here.   At station 5 the river, although
still eutrophic, shows some improvement in quality.   Baetidae and
helmids are dominant and a wide range of fauna are also found including
Austropotamobius, Simulium, chironomids including a few *Chironomus,*
cased caddis, *Leuctra* and a number of other genera including several
species of molluscs.

Downstream of station 6, the river is in 'poor' condition, with massive
weed and algal growths of *Cladophora* and *Potamogeton* (see section 14.4).
At this location chironomids dominate and Tubificidae are present in
large numbers.   Recorded here also are *Asellus,* Hirudinea (several
species), Sphaeriidae, *Lymnaea, Planorbis* and *Baetis rhodani.*   At
Turtulla and Cabragh Bridges (stations 8, 9) water quality is 'poor -
doubtful' and chironomids (with *Chironomus* spp. included) occur as
dominants.   There are also large numbers of molluscs and leeches found
here and a reduction in the numbers of tubificids.

At station 10 there are definite signs of recovery and here *B. rhodani*
dominates, the chironomids are numerous and other genera observed include
Simulium, Asellus, haliplids, *Lymnaea, Hydropsyche* and some cased caddis
larvae.   The fauna at station 11 is similar to that found at station 10.
Further downstream at station 12, the river deepens and widens and the
water quality is 'good'.   *Gammarus* is dominant here.   There is a diverse

fauna here with good representation of cased and uncased caddis larvae, *Leuctra*, *Ephemerella*, chironomids, *Simulium*, *Austropotamobius* and *Apheilocheirus*.   At station 13, the river shows signs of enrichment in terms of large growths of macrophytes (*Ranunculus* and *Potamogeton*), but the quality remains 'good'.   The fauna dominated by *Gammarus* also has numerous *B. rhodani*, *Simulium*, *Ryacophila*, chironomids, other Diptera, haliplids, *Gyrinus* and mites.

Downstream of Caher at station 16 the river shows 'doubtful' quality, reflecting the effects of urban and industrial sewage effluent. Low dissolved oxygen has been recorded here at times and at least one fish-kill is recorded. The fauna includes many tolerant forms with vast numbers of chironomids, numerous *Gammarus*, mites, Mollusca, *Asellus*, tubificids, lumbricids, *Ephemerella* and *Helobdella*. At station 17 the water quality is improved slightly, being 'doubtful-fair'. Current speeds are lower and fauna which are suited to such conditions such as Corixidae are common; the dominant invertebrate genus is *Valvata*.

Below Clonmel at station 18 and also at Kilsheelan the quality is 'doubtful'. The river is very nutrient-rich, particularly at the latter station, where abundant algal growths occur. The fauna at these locations includes large populations of *Simulium* and Tubificidae. Other invertebrates include: Chironomidae, *Limnaea*, Sphaeriidae, *Physa*, *Planorbis*, *Valvata*, *Potamopyrgus*, *Glossiphonia*, *Baetis rhodani*, *B. scambus*, *Caenis*, *Asellus*, *Gammarus*, mites and Dytiscidae.

14.6   FISHERIES

In common with most large Irish rivers the commercial and angling fisheries of the River Suir are based on salmonids (*Salmo salar*, *S. trutta*, *S. trutta fario*). Eels are numerous, but the commercial fishery is not well developed and returns amount to about 10 tonnes yr^{-1} at present. Other fish found in this catchment include pike, perch, gudgeon, stone loach, minnow and stickleback; none of these are fished commercially. There are cockle and mussel beds in the estuary, but their full commercial potential has not been developed as yet.

The commercial salmonid fishery is confined mainly to drift net catches in the estuary. These catches produce an annual catch currently valued at approximately £0.4 million. The general impression gained is that this is an underestimate, as some illegal fishing is carried out and many anglers do not send in catch returns. Reduced water quality in some areas is unfavourable to game fish at present due to oxygen depletion in summer (Table 14.4, Fig. 14.4) and has resulted in complaints from angling clubs.

Although no detailed statistics are available on fish stocks in the catchment, some idea of the distribution of the more valuable stocks may be gained from Fig. 14.5, where known spawning areas of game fish

Table 14.3 Comparison between four different stations on the Suir showing biomass, dominant macrophyte genus and approximate duration dissolved oxygen in water stayed below 50% saturation during a 24 hour survey, lowest D.O. recorded during 24h. survey and effect on caged fish (see Horkan, 1981). L = left-hand side; R = right-hand side.

location	situation	biomass in July g m^{-2} dry weight	dominant genus	approx. time D.O. <50% (hours)	lowest D.O. (% sat.)	effect on caged fish
Rossestown Br.	open	1978 / 320	Ranunculus	0	87%	none
Thurles Br.	open	570	Cladophora	9.2	41	distressed
Turtulla	open	1979 / L: 380 R: 320	L: Cladophora R: Potamogeton + Cladophora	*15.5	L: 48 R: 27	L: little (some distressed) R: some died, others distressed
Cabragh Br.	part shaded	240	Cladophora	15.8	15	all fish died

*D.O. in mid-channel; insufficient measurements taken on each side to determine duration of D.O. <50% on right- or left-hand sides. However, it is known that D.O. fell to lower levels on the polluted right-hand side.

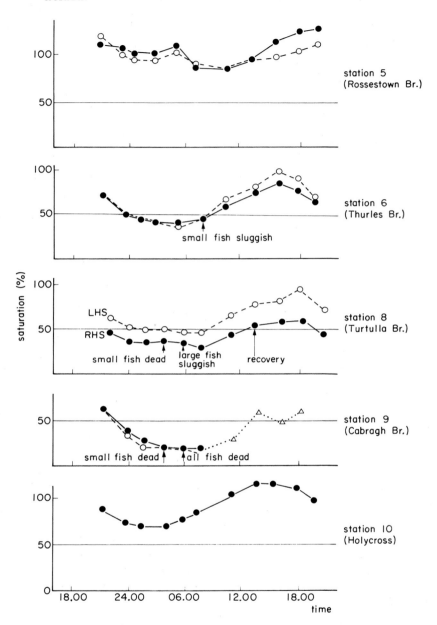

Fig. 14.4 Diel variation in dissolved oxygen and its effect on brown trout kept in cages at different stations. RHS (right-hand side) at Turtulla is influenced by sewage, while LHS (left-hand side) by clean water. D.O. measured beside cages.

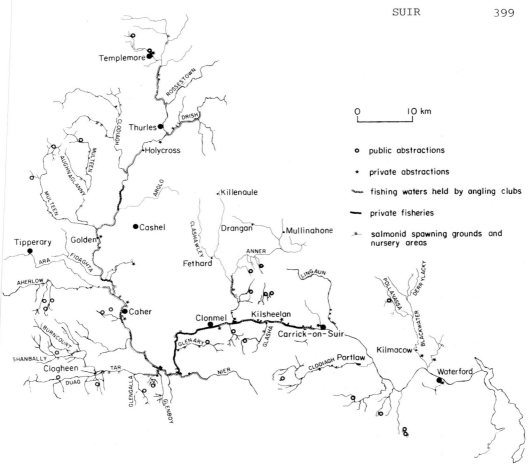

Fig. 14.5 River Suir and main tributaries, showing locations of surface water abstractions, salmonid spawning areas and locations of private and angling club fishing waters (data from An Foras Forbartha, 1980).

are shown. The stretches owned or leased by private individuals or clubs for salmon or trout angling are also shown on this map. The general impression gained is that virtually the whole catchment could be classified as game fish waters at present because of the widespread distribution of salmonid species, although as stated above, some areas have poor water quality at times. Long term water quality management plans are now being developed to ensure that the catchment be retained as a game fishery. A new sewage works has recently been brought into use at Thurles and improvements in water quality are already evident.

REFERENCES

An Foras Forbartha (1980) *Draft Water Quality Management Plan for the River Suir Catchment. Vol. 2.* Summary of Water Resources. 17 pp.

An Foras Forbartha (1980) *Draft Water Management Plan for the River Suir Catchment. Vol. 4. Beneficial Uses and Water Quality Criteria.* 80 pp.

Beaumont P. (1975) Hydrology. In: Whitton B.A. (Ed.) *River Ecology.* 725 pp., pp. 1-38. Blackwell, Oxford.

Bolas P.M. & Lund J.W.G. (1974) Some factors affecting the growth of *Cladophora glomerata* in the Kentish Stour. *J. Soc. Wat. Treat. Exam.* 23, 25-51.

Central Statistics Office (1980) *Census of population of Ireland, 1979. Vol. 1. Population of District Electoral Divisions, Towns and Larger Units of Area.* 95 pp. Stationery Office, Dublin.

Clabby K.J. (1981) *The National Survey of Irish Rivers. A Review of Biological Monitoring, 1971-1979.* 322 pp. Water Resources Division, An Foras Forbartha WR/R 12.

Council of the European Communities (1978) Council directive of 18 July on the quality of fresh waters needing protection or improvement in order to support fish life (78/659/EEC). Official Journal of the European Communities No. L222/7.

Edwards R.W. & Owens M. (1960) The effect of plants on river conditions. I. Summer crops and estimates of new productivity of macrophytes in a chalk stream. *J. Ecol.* <u>48</u>, 151-60.

Haslam S.M. (1978) *River plants.* 396 pp. Cambridge University Press.

Horkan J.P.K. (1980) *Interim Report on Eutrophication and Related Studies of the Thurles Area of the River Suir, May-July, 1980.* 39 pp. An Foras Forbartha, Water Resources Division, WR/R10.

Horkan J.P.K. (1981) *Second Interim Report on Eutrophication and Related Studies of the Thurles Area of the River Suir, July/Sep. 1979.* An Foras Forbartha, Water Resources Division, WR/R13.

Jorga W. & Weise G. (1977) Biomasseentwicklung submerser Makrophyten in langsam fliessenden Gewässern in Beziehung zum Sauerstoffhaushalt. *Int. Rev. Ges. Hydrobiol.* <u>62</u>, 209-34.

Lennox L.J. & Toner P.F. (1980) *The National Survey of Irish Rivers.* A third Report on Water Quality. 104 pp. An Foras Forbartha, Water Resources Division.

Marker A.F.H. (1976) The benthic algae of some streams in southern England. I. Biomass of the epilithon in some small streams. *J. Ecol.* <u>64</u>, 359-73.

Marker A.F.H. & Gunn R.J.M. (1977) The benthic algae of some streams in Southern England. III. Seasonal variation in chlorophyll *a* in the seston. *J. Ecol.* <u>65</u>, 223-34.

Nevill W.E. (1972) *Geology and Ireland.* 263 pp. Allen Figgis & Co. Ltd., Dublin.

Praeger R.L. (1974) *The Botanist in Ireland,* 587 pp. E.P. Publishing Ltd.

Westlake D.F. (1959) The effects of biological communities in polluted streams. In: *Symposium 8, The Effects of Pollution on Living Material,* pp. 25-31. Institute of Biology, London.

Whitton B.A. (1971) Filamentous algae as weeds. In: *3rd Symposium on Aquatic Weeds.* European Weed Research Council, Oxford, 1971.

15 · Po

G. CHIAUDANI & R. MARCHETTI

15.1 INTRODUCTION

The Po basin lies in the continental part of the Italian peninsula, of
which it covers about one-quarter. The catchment is delimited by the
Alps, which fix, from west to north, the boundaries with France, Switzerland and Austria. The southern limit is set by the northern
Appennines, while towards the east the catchment faces the Adriatic.
The basin runs from west to east and corresponds approximately to the
45° N parallel (Fig. 15.1). The most important lakes in Italy,
Maggiore, Como and Garda, lie near the transition between the two main
regions of the basin, the mountain arc and the lowland plain. The
cities of Torino (Turin) and Milano (Milan) lie on the plain, with about
1.2 and 1.7 million inhabitants, respectively. In addition there are
many smaller centres of population.

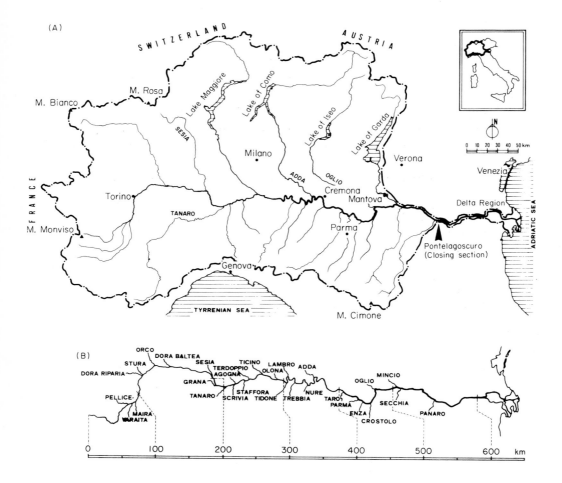

Fig. 15.1 A) Po catchment; B) location of main tributaries.

15.2   GENERAL DATA

DIMENSIONS

Area of total catchment	70500 km^2 (including 600 km^2 of permanent glacier)
Length of river	676 km (Ministero dei Lavori Pubblici, 1968)
Peaks: Monte Bianco	4810 m
Monte Rosa	4633 m
Gran Paradiso	4061 m
Source of Po on Monviso glacier	3841 m
Po plain	24450 km^2
Average height of basin	740 m
Average height of plain	c 100 m
Gradient: Section 1	1785 m in first 67 km
Section 2	0.2%
Section 3	0.01%
Width: maximum in central part	2.5 km (3 km at flood)
downstream	0.8 km (because of embankments)

CLIMATE

Precipitation

average over whole catchment (500 rain stations)   1100 mm
annual contribution from catchment   70-80 x 10^9 m^3

Temperature

winter average: western part of plain	c 0 ^{o}C
near great lakes	2 ^{o}C
spring average: western part of plain	12 ^{o}C
summer average: western part of plain	23 ^{o}C
autumn average: western part of plain	12 ^{o}C

Solar radiation in eastern part of plain:

daily absolute maximum	864 cal cm^{-2} (July, (1966)
average annual	113880 cal cm^{-2} yr^{-1}
= average daily rate	312 cal cm^{-2} d^{-1}

Foggy days: Mantova 71; Parma 65; Milano 38.

PHYSICAL FEATURES OF RIVER

Tributaries

141 second order tributaries of which 95 enter on the left and 46 on on the right
major ones (all with a catchment of at least 6500 km^2)

	length
Adda	313 km
Ticino	248 km
Oglio	280 km
Tanaro	264 km

Hydrological data at Pontelagoscuro gauging station
(90 km from mouth; 8.35 m altitude; data for 1918-1980)

catchment area of this section	70091 km^2 (of which 0.8% glacial, 1.3% lacustrine)
rainfall	1107 mm
runoff	668 mm
loss	439 mm
runoff rainfall ratio	0.60
mean monthly maximum flow	8940 m^3 s^{-1} = 128 l s^{-1} km^{-2}
absolute maximum flow (14.11.51)*	10300 m^3 s^{-1} +
mean flow	1490 m^3 s^{-1} = 21.3 l s^{-1} km^{-2}
mean monthly minimum flow	275 m^3 s^{-1}

Distribution of flows throughout year:

	Jan	Feb	Mar	Apr	May	June	July	Aug	Sep	Oct	Nov	Dec
rainfall (mm)	60	57	80	111	121	100	76	84	104	122	115	77
runoff (mm)	45	43	57	58	73	69	48	36	45	63	73	58
loss (mm)	15	14	23	53	48	31	28	48	59	59	42	19
Q max (m^3 s^{-1})	5010	5810	5380	5570	8850	6990	4150	3750	6900	7290	8940	7730
Q med (m^3 s^{-1})	1180	1220	1480	1570	1920	1870	1260	950	1230	1650	1970	1520
Q min (m^3 s^{-1})	395	418	518	275	312	306	307	293	320	446	490	430
q (l s^{-1} km^{-2})	16.8	17.4	21.1	22.4	27.4	26.7	17.0	13.6	17.5	23.5	28.1	21.6

*The embankment broke before the gauged section; estimated flow was 11580 m^3 s^{-1}. The flood covered 113000 ha (Ministero dei Lavori Pubblici, 1979).

Fig. 15.2 Middle Po, showing nuclear power station.

15.3 PHYSICAL AND GEOGRAPHICAL BACKGROUND

Climate

As would be expected with a catchment which covers such a range in altitude, there are marked differences in both temperature and precipitation
in different parts. The coldest month is January, followed by December,
while the hottest is July, followed by August. The greatest difference
recorded on the plain between minimum and maximum temperature is 57 C
degrees (-18 ^{o}C, 39 ^{o}C, respectively). In 1979 the average temperature
in the central part of the plain (Milano) was 13.5 ^{o}C, with monthly
averages ranging from -4.6 ^{o}C in February to 28.1 ^{o}C in July. Storms,
which are more frequent in summer, increase from west to east, with an
average yearly total of 30 storm days. Relative humidity fluctuates
between 80% (winter average) and 50% (summer average); in the centre of
Milano the annual average is 69%. Because of the mountain chain, strong
winds are relatively infrequent on the plain. In winter there is
typically a complete cloud cover for a period of 10 - 15 days, followed
by 10 days or so of clear sky; in summer the number of totally overcast
days drops to 4 - 5 per month. Fog is frequent in autumn and winter in
the valley, increasing in frequency from September to February.

Geology

While the Alps and the soils derived from them are mainly calcareous,
the Appennines are mainly sandstone. The plain is formed by a continuous alluvial stratum, corresponding to a Quaternary basin with scarce
pre-Quaternary outcrops. The stratum is mostly several hundred metres
thick, but reaches 2500 m in the delta; it is made of clastic materials,
which become finer the further away from the mountain slopes,
alternating with pelitic sediments. Deep water-bearing strata consti-
tude large water reserves.

Hydrography

The initial slope of the river-bed is very steep, dropping off gradually
as the delta is approached (see 15.2). The bed is deepest in the
terminal stretch, reaching 10 - 12 m. Since there is also a 10 m
embankment, the river depth at peak flows is 20 - 22 m. The course of
the river has been influenced by the deposition of alluvial materials
brought down by the tributaries. The alpine tributaries have less
suspended solids than those on the right bank, partly because of the
presence of the lakes.

   Most tributaries can be divided into three main parts (Raffa &
Giuffrida, 1973):

   (i)   Mountain stretch, showing a narrow river-bed, with steep and
         well-defined sides;
   (ii) Intermediate stretch, where ancient and recent alluvial
         materials penetrate into the plain; the bed increases in width

and the sides come to be determined mainly by cultivated
fields;

(iii) A stretch through the plain which is narrow, winding and
generally embanked ; the course of the river is determined
artificially.

A peculiar element of Po hydrography is the delta, which starts some
50 km from the sea and extends over 400 km^2. The mean altitude of this
region is below sea level and there is a marked intrusion of saline
waters into the different branches of the delta. The present delta
structure is due to natural events and human intervention. There are
six branches, with the central one carrying 50 - 60% of the flow. The
area of the delta used to increase by about 50 - 60 ha yr^{-1}, but this
dropped recently to about 35 ha yr^{-1}.

Hydrology

The bed of the Po has been influenced by human activity to such an
extent that there is now little correlation between the nature of the
watershed and the pattern of water flow. The first changes were
brought about by the reduction of the expansion zone due to embankment.
More recent changes which have influenced the natural hydrological
behaviour are lake regulation, construction of reservoirs, land
reclamation, extraction of water for irrigation, and the mining of
inert materials. The key hydrological changes are the concentration of
the downflow flood and the withdrawal of water during periods of low
flow (Raffa, 1973).

Flows show an annual series of changes, of which the following are
the most important features:

(i)   Winter low flow (January, February) due to absence of thaw and
scarce rain.

(ii)   Maximum flow (May, June) caused by thaw.

(iii) Summer low flow (August) much more marked than the winter one.

(iv)   Flood period (October, November). A particularly ruinous
flood took place in November 1951, which covered 113000 ha
(Ministero dei Lavori Pubblici, 1979).

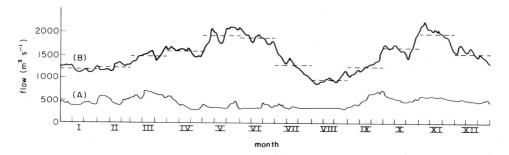

Fig. 15.3 Minimum (A) and mean (B) values for daily flow at
Pontelagoscuro closing station since 1918.

15.4 RIVER WATER USES

Agriculture covers 22894 km^2 of the basin, of which about 15000 km^2 are
irrigated with water direct from the Po or its tributaries. Irrigation
canals fed by Po water were built first at the time of ancient Rome and
there was a marked increase at about the year 1000. Since the end of
the last war, the discharge diverted for this has trebled and exceeds
240 m^3 s^{-1} at the period of peak demand. The direct use of Po water
for drinking purposes is limited to part of the supplies taken for the
two big urban centres of Torino and Ferrara. Five open-cycle thermal
power stations (together producing more than 8500 MW) and two nuclear
power stations (1100 MW) take their cooling water from the river.
Reservoirs, which are sited in the upper part of the basin and were
built originally mainly for hydroelectric purposes, have an overall
storage capacity of about 1.5 x 10^9 m^3 (Ministero dei Lavori Pubblici,
1979).

Only moderate use is made of ships, mainly 1000 - 1300 tonne vessels
carrying oil, chemical products, sand and gravel. This traffic is
expected ultimately to reach a carrying capacity of 12 x 10^6 tonnes
(Ugolini, 1971). Tourist boating and angling are practiced in many
places, but swimming is banned almost everywhere because of poor
hygienic conditions. Commercial fishing is unimportant, except for the
delta region where it is still possible to catch sturgeon. On the
other hand, gravel and sand extraction is intense along the river. The
total is given officially as 7 - 8 x 10^6 m^3 yr^{-1}, but it is probably
more than ten times as much, as demonstrated by the deepening of the
river-bed (2.5 m in 15 years, in certain places).

15.5   ESTIMATED POLLUTION LOAD

A study carried out by the Istituto di Ricerca sulle Acque (1977) made
an estimate of pollution by industry and the human population. Industrial activities have been converted to inhabitant equivalents, using
current conversion equivalents (Cicioni et al., 1977). The 'total
population' (TP) has been obtained by adding up the number of inhabitants (P) and the industrial equivalent (PE). Only towns with a PE
of more than 15000, which constitute 83% of the total basin population,
have been considered. The results were in millions:

$$TP\ (53.8)\ =\ P\ (14.9)\ +\ PE\ (38.9)$$

The river receives an average input per km length equivalent to a TP of
84000. Most of the IE is due to the food industry, followed by the
steel and chemical industries. The distribution of these polluting
sources is very heterogeneous. The highest TP (35% of Po basin) is in
the Milano sub-basin, where most human and industrial wastes reach the
Po via the R. Lambro.

   The Istituto di Ricerca sulle Acque did not consider agricultural
wastes in its study.   Provini et al. (1979) estimated the following
totals for the catchment (all x 10^6): cattle, 3.7; pigs, 3.5; sheep
and goats, 0.34; poultry, 80. Using conversion factors of 10.2, 3, 3.3
and 0.2, respectively (Farnesi et al., 1976), an estimate can be made
for the animal equivalent population of 65 x 10^6 inhabitants. This
raises the TP for the whole Po basin to about 119 x 10^6 inhabitants.
Although these data are merely theoretical, they provide a useful means
of describing the distribution of the potential polluting load bordering
the Po.

15.6   PHYSICO-CHEMICAL FEATURES

15.61   INTRODUCTION

The data in this account are drawn mainly from research carried out in
1971-73 by Istituto di Ricerca sulle Acque along the river (21
stations) and by ENEL (Ente Nazionale per l'Energia Elettrica) in a
central stretch about 370 km from the source. The ENEL research formed
part of a study to evaluate the effects of thermal discharges from a
nuclear power plant (Ioannilli, 1981). Previous surveys include that
by Berbenni (1971) along the whole river. Investigations prior to 1974
have been summarized by Piantino et al. (1977).

15.62   TEMPERATURE

The average annual temperature in the central stretch of the Po is about
14 $^{\circ}$C, the maximum, 29 $^{\circ}$C and the minimum, 3 $^{\circ}$C. Although these data
are based on a relatively short period (1974-77), they can be considered

representative of this stretch (Bertonati & Ioannilli, 1981). Fluctuations in the lower stretch over the period 1961-1977 are shown in Fig.
15.4.

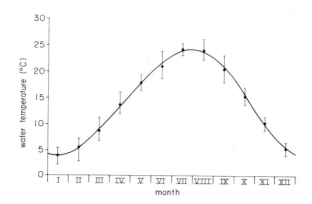

Fig. 15.4 Mean monthly water temperature at Pontelagoscuro closing
station; values shown are the minimum, mean and maximum for the
period 1961 to 1971. (Modified from Raffa, 1973)

15.63 SUSPENDED AND DISSOLVED SOLIDS

The quantity of suspended materials transported by the Po varies
markedly and is influenced mainly by meteorological events. The mean
values for 20 stations range from 45 to 450 mg l^{-1} (105 ^{o}C suspended
solids) with an absolute maximum of 6400 mg l^{-1}; the corresponding
values for volume of settling materials are 0.2, 2.1 and 40 ml l^{-1},
respectively. The concentration of dissolved substances is more
variable, with means ranging from 220 - 370 mg l^{-1} and an absolute
maximum of 560 mg l^{-1}. Starting from the source, the load of solid
materials increases at the rate of 1.209 kg s^{-1} km^{-1}; 86% of this load
is of inorganic materials (Marchetti *et al.*, 1977). In the closing
stretch (Pontelagoscuro), the load is:

particulate	12.55 x 10^6 tonnes yr^{-1}
dissolved	14.14 x 10^6 tonnes yr^{-1}
total	26.69 x 10^6 tonnes yr^{-1}

The settling load, expressed in volume, is 28.4 x 10^6 m^3 yr^{-1}. Using
a value of 52290 km^2 as the area of basin subjected to erosion, this
gives a particulate solid loss of 240 tonnes km^2 yr^{-1} (neglecting any
contribution from sources other than erosion. This calculation based
on 1971-73 data corresponds well with an estimate based on thirty
years' observations of 258 tonnes km^{-2} (Canali & Allodi, 1962).

15.64   ORGANIC LOAD AND DISSOLVED OXYGEN

BOD$_5$ values are not particularly high, with a mean for all stations of 3 mg l^{-1} and COD/BOD ratio of about 5 (Provini & Marchetti, 1977). The BOD$_5$ total load in the last stretch of the basin is about 110600 tonnes yr^{-1} (Marchetti *et al.*, 1977). Using data from many BOD, COD and TOC determinations in the central stretch, Bertonati (1977) and Bertonati & Ioannilli (1981) estimated the oxidation state of organic carbon with the index of 4(TOC - COD) : TOC. This index fluctuates around zero in stable hydrological conditions, but becomes negative in incipient flood phases as a result of resuspension of reduced materials in the sediments.

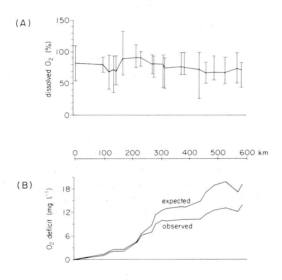

Fig. 15.5 A) Percentage oxygen saturation (minimum, mean, maximum) along the river course for the period 1971-1973, based on monthly sampling. (Modified from Marchetti & Provini, 1977)
B) Expected and observed mean oxygen deficit along the river course. (From Marchetti & Provini, 1977)

Changes in dissolved oxygen along the river are shown in Fig. 15.5. The typical condition is that of appreciable deficit, this deficit reaching a maximum 136280 tonnes yr^{-1} in the lowermost stretch. The oxygen concentrations show marked differences in summer between day and night e.g. 6 mg l^{-1} in Fig. 15.6. A daily variation of about 9 mg l^{-1} has been found in the central stretch of the river (Bertonati, 1977; Bertonati & Ioannilli, 1981). A vertical gradient has been noted under conditions of intense solar radiation, with 19, 16 and 14 mg l^{-1} at the surface, 2.5 m and 5 m depth, respectively.

Using the method of Odum (1956), it has been possible to evaluate

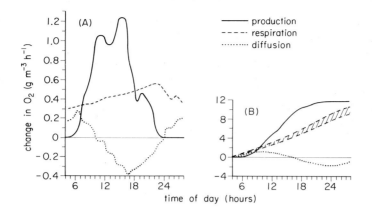

Fig. 15.6 Changes in dissolved oxygen due to photosynthesis, respiration and diffusion measured by Odum method at km 219 on 15.7.74. A) Hourly change; B) cumulative values. Shaded area represents plankton respiration measured by light and dark bottle method. (From Marchetti & Provini, 1977)

gross primary productivity of the community in different parts of the river (2.28 - 15.1 g m^{-3} O$_2$ in July: see 15.8). The oxygen deficit at some stations is probably due mainly to plankton respiration and at others to benthic respiration (Marchetti & Provini, 1977). More recently, measurements of sediment respiration made in marginal areas under low flow conditions have given values of 1.8 - 4.8 g O$_2$ m^{-2} d^{-1} (Bertonati & Ioannilli, 1981).

15.65   TOXIC SUBSTANCES

Studies by Vivoli & Vecchi (1977) and Prati (1977a, b) report data for arsenic, mercury, nickel, lead, copper, zinc, the pesticides α- and γ-BHC, DDT, DDE, heptachlorepoxide, dieldrin, thiodan and total phenols. Copper and lead were detectable and arsenic, nickel and zinc very frequently so (Table 15.1). Mercury was detectable less frequently, with an absolute maximum of 17 mg m^{-3} (Vivoli & Vecchi, 1977). Among the metals listed above, zinc has the highest load in the most downstream stretch, with 2646 tonnes yr^{-1}. In a more recent study (Provini *et al.*, 1980), cadmium was detectable in 26% of samples (max. 0.8 mg m^{-3}) and chromium in all (2 - 107 mg m^{-3}). In the central part of the Po, peak concentrations for all these elements except arsenic and zinc are usually associated with floods (Bertonati & Ioannilli, 1981).

A study of mercury has been made for the fish *Leuciscus cephalus cabeda* taken from nine stations (Pennacchioni & Marchetti, 1973; Marchetti *et al.*, 1975). The concentrations of total and methylmercury expressed as mg per kg fresh weight of dorsal flesh were: 0.144 and 0.14 for the means; 0.068 and 0.035 for the minima; 0.380 and 0.274 for

Table 15.1 Concentrations of heavy metals (mg m^{-3} or µg l^{-1}) at 21 sampling stations on Po together with annual load at the most downstream station (Pontelagoscuro).

	Nickel	Copper	Zinc	Arsenic	Mercury	Lead
number of samples	189	189	169	111	132	189
detection limit					0.2	
positive samples						
(%)	98.9	100	88.2	98.2	43.2	100
absolute minimum	0	3.5	0	0	<0.2	2
absolute maximum	132	137	147	15	17	17.1
load (tonnes yr^{-1})	89	1154	2646	243	65	485

the maxima. Gaggino (1981) has reported a similar study for other heavy metals, but the data given as mg per kg dry weight: Cr, 0.2 - 0.4; Ni, 0.8 - 3.7; Cd, 0.005 - 0.05; Pb, 0.09 - 2.0.

In a study of 150 water samples, Prati (1977a) found that BHC was detectable in 48% and DDT and DDE in 40% of cases. The cyclodienics (heptachlor, dieldrin, thiodan) were found in only 4.4% of cases. The maxima were in ng l^{-1}:

α-BHC	180	DDT	372	thiodan	170
γ-BHC	274	DDE	450	dieldrin	32
				heptachlorepoxide	59

In the most downstream stretch, the total load of pesticides was estimated as 6.7 tonnes yr^{-1} in the period 1971-73. More recently (1977-78), Galassi & Provini (1981) found that the concentrations of some pesticides had decreased which they ascribed to prohibitions or restrictions on use. The concentrations of α-BHC and γ-BHC ranged from 3 - 37 and 2 - 13 ng l^{-1}, respectively. Dieldrin was detectable in only one sample, while aldrin, heptachlor, heptachlorepoxide, endrin, HCB and pp'DDE were below detection limits. In the most downstream stretch, PCB was detectable in 8 out of 18 samples, with a maximum of 42 ng l^{-1} dissolved and 100 ng l^{-1} adsorbed. γ-BHC, HCB, pp'DDE and PCB were reported to be detectable regularly in flesh of *Leuciscus cephalus cabeda* and *Alburnus alburnus alborella* (Galassi *et al.*, 1981b).

Phenols were detectable in 44% of cases (500 samples along whole river) studied by Prati (1977b), with a maximum of 0.146 mg l^{-1}. The load increased on passing downstream and the annual average load in the lowermost stretch was 75 tonnes (Marchetti *et al.*, 1977).

15.66 NITROGEN AND PHOSPHORUS

Except for stations close to the source, the mean ammonium concentration along the whole river is about 0.3 mg l^{-1} N, although the absolute maxima are at times much higher in zones influenced by large urban discharges. Nitrate increases from the mean of 0.3 mg l^{-1} N near the source to 1.7 mg l^{-1} N downstream.

The mean concentration of orthophosphate in the most downstream stretch in 1971-73 was reported to be 0.151 mg l^{-1} P (Istituto di Ricerca sulle Acque, 1977). A more recent study of this stretch led to estimates for the amount of nutrient flowing into the Adriatic: total N, 93091 tonnes yr^{-1}; total P, 11798 tonnes yr^{-1} (Provini et al., 1980).

15.67 OTHER FEATURES

Anionic detergents and ether-extractable oils were studied in 1971-73 (Prati, 1977b). The mean concentrations were 0.05 mg l^{-1} MBAS and 0.5 mg l^{-1}, respectively, but there were remarkable fluctuations. Total radioactivity, also measured in 1971-73, was very low and relatively constant (0.6 - 2.9 pCi l^{-1}) which were due exclusively to radionuclides occurring naturally in the water (Allegrini et al., 1977). Data on other variables (conductivity, turbidity, pH, alkalinity) are included in 15.2.

15.7 BIOLOGY

15.71 INTRODUCTION

Knowledge about the biology of the Po is much less complete than that of physics and chemistry. Certain stretches of the river have been studied intensively, but others are scarcely known. We will try to synthetize here our knowledge, most of which derives from research either carried out by Istituto di Ricerca sulle Acque all along the river or co-ordinated by ENEL (Ioannilli, 1980) on the central part.

15.72 MICROBIOLOGY AND VIROLOGY

Surveys of microbial and viral components have been concerned almost entirely with hygiene. The data available concern the content of total bacteria at 20 $^\circ$ and 37 $^\circ$C (459 samples) of fecal and total coliforms (641), streptococci (217), sulphate-reducing clostridia (218) and Salmonella (49 samples), during surveys in 1971-75 down the whole river (Fara et al., 1977). Nardi (1977) considered the viruses Polio, Coxsackie, Echo, Reo and Adenovirus.

The overall pattern, even if showing large fluctuations, is characterised by the constant presence of all the bacterial group listed above.
From the minimum, recorded at the source, values increase steadily on
passing down to the delta. The number of bacteria per ml shown at 20 ^{0}
ranges from 10^5 to 10^6 and at 37 ^{0} from 10^4 to 10^5. Total coliforms
average around 10^3, fecal coliforms around 10^2, streptococci around 10^2
and clostridia around 10 (all per ml). A survey made in 1971-1974 at
10 stations found *Salmonella* in 98% of samples, with in some cases seven
different strains (species). The most frequent were *S. agona, S. panama,
S. brandeburg, S. paratyphi* and *S. typhimurium*.

The survey (49 samples) was positive in 30% of cases, with a higher
frequency in the central-terminal stretch. Among the 15 viral groups
isolated, strains of *Coxsackie, Echo* and *Adenovirus* were the most
frequent. When Po tributaries were investigated as well (85 samples),
representatives of 42 virus groups were isolated with a marked pre-
ponderance of *Enterovirus* (78.5%), followed by *Echo, Coxsackie* and
Polio (21.4%).

Yeasts have been surveyed only at four stations in the central part
of the Po (1971-75, 209 water samples). There were 80 species and 15
genera. *Saccharomyces* (*S. cerevisiae* and *S. uvarum*) were the dominants,
followed by *Candida, Torulopsis, Cryptococcus* and *Rhodotorula* (Viviani
& Tortorano, 1977). Among these, potential pathogens (*Torulopsis
glabrata, Candida albicans, C. tropicalis, C. guilliermondi* etc)
occurred close to the mouth of very polluted tributaries.

15.73 ALGAE

Researches on Po algae have been limited to a central stretch of the
river, where a quantitative study of the plankton was made from June
1974 to June 1977. Peak populations (10 - 15 x 10^3 cells ml^{-1}) have
been recorded in June-August.   223 species have been identified
(Cironi, 1977), 16 blue-green algae, 128 diatoms, 4 dinoflagellates,
2 Euglenophyla, 67 greens, 6 chrysophytes. Only 20 species were
present during the whole period at densities higher than 100 cells ml^{-1}
(Andreoli & Fricano, 1981). Diatoms formed 50 - 60% of the summer crop
and 80% of the winter one. Abundant species were:

blue-greens	diatoms	greens
Microcystis aeruginosa	*Melosira granulata*	*Scenedesmus quadricauda*
Oscillatoria rubescens	*Cyclotella comta*	*Actinastrum hantzschii*
	C. meneghiniana	*Coelastrum microporum*
	Diatoma vulgare	
	Synedra ulna	
	Nitzschia acicularis	
	N. fruticosa	

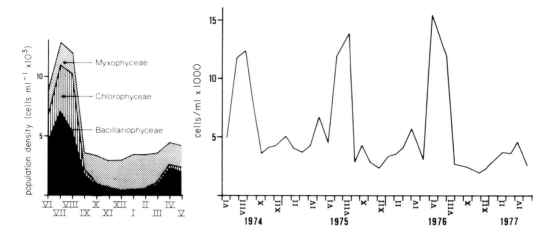

Fig. 15.7 (left)   Mean monthly population density for three main algal groups in the central stretch of the Po for the period June 1974 to May 1977.

Fig. 15.8 (right)   Monthly changes in density of total algal population during the same period.   (Both figures modified from Andreoli & Fricano, 1981)

15.74   MACROPHYTES AND RIPARIAN VEGETATION

Detailed observations about macrophytes are available only for the central stretch of the Po, but some records have been made on upper zone.   Here there are occasional stands of the Helosciadietum, Scirpo-Phragmitetum and Myriophylleto-Nupharetum associations (Corbetta & Zanotti-Censoni, 1977).   In the central stretch of the river, macro-phytic vegetation is well developed, with large banks of *Phragmites australis* by itself or together with *Typha latifolia, T. angustifolia, Sparganium erectum* and *Glyceria fluitans*.   In the backwaters there are dense stands of the floating species, *Lemna minor* and *Salvinia*, and among the submerged species, *Ceratophyllum demersum, Potamogeton crispus, P. lucens, P. perfoliatus, Elodea densa*.

The Scirpo-Phragmitetum association is present in some places on the lower part of the river and, where velocity and depth are favourable, also Myriophilleto-Nupharetum and *Trapa natans*.   In the delta branches influenced by intrusion of sea water, *Phragmites australis* is over-whelmingly the most abundant macrophyte.

Corbetta & Zanotti-Censoni (1977) also studied the riparian vege-tation in 1971-73; that is, the vegetation living on the wet stream margin.   In the upstream stretch, when suitable substrata are available, the Personato-Petasitetum association is present.   On passing downstream there is an increasing representation of nitrophile species, with first *Saponaria officinalis* and then the Artemisietum velotori. In the middle stretch the vegetation tends to be more homogeneous, with the Polygono-Chenopodietum association and the *Cyperus glomeratus* sub-association dominant. In the lower stretch the phytosociological picture is not changed much, except for the frequency of the variant *Amaranthus* in the *Cyperus glomeratus* sub-association, which continues almost to the delta.

In the upper part of the river (from 2020 to 490 m altitude), woodland consists of *Alnus glutinosa, Fraxinus excelsior* and different species of *Salix*. Further downstream woodlands become extremely scarce, since the areas adjacent to the river are nearly all cultivated. *Salix alba* is quite common along the river sides, but natural stands are very reduced. Even islands in the river are usually cultivated. The only example of Po climax forest vegetation on one island in the lower stretch (about 550 km from the source), with specimens of *Quercus pedunculata, Populus alba, P. nigra, Ulmus campestris* and *Acer campestre* (Corbetta & Zanotti-Censoni, 1977).

15.75 PROTOZOA, NEMATODES AND ROTIFERS

Protozoa, especially those in the bottom sediments, were studied in 1978 at four stations from the middle zone to downstream (Madoni, 1980). Both number of species (16, 20, 22, 13 at the four stations) and population density (46, 41, 105, 36 per cm^2 of sediment) were low. Among Ciliophora the prevailing species were *Carchesium polypinum, Cyclidium glaucoma, Mesodinium pulex*, while among Sarcodinea the prevailing one was *Actinophrys sol*.

Information about nematodes is available for the upper (Zullini, 1969) and middle (Zullini, 1974a, b) parts of the river. *Tobrilus longus* and *T. helveticus* are the most frequent in the former and *Paroigolaimella bernensis* in the latter. 58 species belonging to 18 families and settled on different substrata (sand, mud, vegetation) have been recognised. In relation to this species abundance, the number of individuals is modest, probably because of the periodical washing effect of floods. The community with the greatest number of species and individuals lies in a narrow littoral strip of land considered as an "ecotone" (soil-water-air).

Rotifers were studied in 1975 (ENEL, 1980) along the central stretch of the Po. 51 species were found, with *Brachionus calcyformis* and *Polyarthra vulgaris major* the most frequent. Subsequent surveys have raised the total to 65, belonging to 12 sub-families. *Brachionus, Polyarthra* and *Synchaeta* have remained the dominants, especially in August.

15.76 MACRO-INVERTEBRATES

Turbellaria

Five species have been recorded from submerged and emergent macro-phytes with a population density ranging from 5 to 5500 per m^2; *Dugesia tigrina* is the most common. Temporal changes are well charac-terised, with maxima in autumn and minima in summer (Cironi, 1979).

Oligochaeta

Oligochaetes have been studied both on the river bottom (19 species: Paoletti Di Chiara, 1981) and in association with macrophytes (37 species: Samburgar, 1981). On bare surfaces, more or less subject to the influence of currents, the association is composed of *Tubifex, Limnodrilus hoffmeisteri, L. udekemianus, L. claparedeianus* and *L. profundicola*. The dominant is *L. hoffmeisteri,* which constitutes typically about 75% of the total population. The population structure is, however, broadly influenced by sediment size (Fig. 15.9). On bare surfaces, densities range from hundreds of thousands per m^2 to a few individuals, this being related to flood events. After floods the number of species increases to a maximum within about two months, but the number of individuals more slowly (Cironi & Ruffo, 1981). Oligo-chaetes associated with *Typha* and *Phragmites* are more abundant and mostly represented by Tubificidae and Naididae.

Hirudinea

The 10 species recorded include *Thereruyzon tessulatum* (Müll.), a species new to the Italian fauna (Minelli & Mannucci Minelli, 1981). 7 species have been found on bare surfaces, among which *Erpobdella testacea* and *Helobdella stagnalis* are the most frequent. As with the oligochaetes, population densities are higher on macrophytes. The highest values have been recorded in June, with about 9000 individuals per m^2 of *Bratracobdella paludosa*, 3000 of *Helobdella stagnalis*, 2000 of *Erpobdella testacea*.

Cladocera

These were studied during 1974-1977 by Rossaro (1981). The plankton of the middle part of the Po included 33 species, mostly Chidoridae and Daphnidae. Densities ranged from a maximum of about 1000 per m^3, usually in August, to a minimum between October and April. The most abundant species were *Moina branchiata, Bosmina coregoni, Alona rectan-gula*.

Copepoda

There are 17 species of planktonic copepod (Rossaro, 1976), one Diaptomidae, one Arpacticidae and all others, Cyclopidae (Rossaro & Cotta Ramusino, 1976).

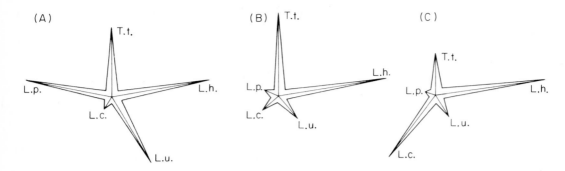

Fig. 15.9 Relative density of various species of oligochaete associated with different sediment sizes: A) <0.074 mm; B) 0.074 –
4.76 mm; C) >4.76 mm. (After Paoletti di Chiara, 1981)

   L.c. *Limnodrilus claparedeianus,* L.h. *L. hoffmeisteri,*
   L.p. *L. profundicola,* L.u. *L. hudekemianus,*
   T.t. *Tubifex tubifex*

   The temporal changes in Copepoda density were much more variable than
those for the Cladocera, with summer maxima (August) over 1000 individuals per m^3 in 1974, reducing to less than 100 in summer 1975 and
1976, and with winter minima under 10 per m^3. The most abundant species
were *Eucyclops serrulatus* and *Acanthocyclops robustus*. *Cyclops vicinus,*
a cold stenothermal species, and *Thermacyclops crassus,* a warm steno-
thermal species, were both found at the same time.

Isopoda

   Only one isopod, *Asellus aquaticus*, is present in the part of the
Po studied. This species is abundant in summer months on sediments
rich in decomposing vegetation.

Amphipoda

   Four species have been found in the middle Po, constituting an
Amphipoda community typical for all slow current stream of the Po
Valley. This community, which shows a maximum density in spring and
a minimum in autumn (Cironi, 1979), is characterised by the dominance
of *Echinogammarus* (*E*. *veneris* and *E*. *stammeri*) and the absence of
Gammarus, the typical genus of French, German and Austrian waters.
Innifargus elegans and *Synurella ambulans* are also present, but in low
numbers. These two species, characteristic of East European waters,
are at present spreading towards west in the rivers of Po Valley, but
are rare in Central Italy and absent in the south (Cironi & Ruffo,
1981). *Echinogammarus veneris* is, on the contrary, a typical Mediterranean species, and probably of marine origin. The two *Echinogammarus*
species, which are rarely present at the same time in other Italian

rivers, co-exist in the middle part of the Po.

Decapoda

Research on the middle Po have shown the presence of two species, *Palaemonetes antennaris* and *Austropotamobius pallipes fulcisianus*. Both are widespread in Italian freshwaters (ENEL, 1980).

Ephemeroptera

There are 15 species in the stretch studied. These belong to the families Siphlonuridae, Baetidae, Heptageniidae, Ephemerellidae, Caenidae, Leptophlebiidae and Ephemeridae. Querena (1980) has compared the population structure of bare sediments (with 6 species) and macrophytes (with 12 species). *Ecdyonurus aurantiacus, Heptagenia coerulans* and *Ephemera vulgata,* species associated with macrophytes, were always absent on bare substrata. The numerically prevailing species is *Cloëon dipterum* (reaching 200 m^{-2}), which is found only on macrophytes, followed by *Caenis* species and *Siphlonurus lacustris*.

Odonata

22 species have been found in the central stretch of the Po, of which 9 belong to Zygoptera and 13 to Anisoptera (Galletti & Ravizza, 1977; Galletti, 1980). The most common is *Ischnura elegans*. The mean density is under 100 individuals per m^2, with a maximum of 550 in spring and minima in autumn (Cironi, 1979; Cironi & Ruffo, 1981).

Hemiptera

Hemiptera are represented, in Po central reach, by 14 species, associated with macrophytes, raising 80 individuals per m^2 as a maximum density (Cironi & Ruffo, 1981). *Plea minutissima* is the prevailing species, present as imago all the year round with a maximum density during autumn. *Ilyocoris cimicoides, Micronecta meridionalis, Sigara dorsalis* and *S. italica* are well represented too. *Nepa cinerea* and *Ranatra linearis* are rare, and *Notonecta glauca* is very rare (Ravizza, 1976, 1981). The presence of *Mesovelia fucata* is to be pointed out because this species, considered to be typical of sub-mountainous running waters, was found associated with *Phragmites* in stagnant waters.

Trichoptera

Information about Trichoptera from the upper zone of the river, close to the source (Moretti *et al.*, 1964) and the central stretch (G. P. Moretti, pers. comm.). There are 29 species classified but this is probably less than the real total, especially in the upper zone where only 16 species were recorded. These were: 5 *Rhyacophila* spp. (*galaerosa, intermedia, kelnerae, tristis, vulgaris*), 2 *Wormaldia* (*copiosa, occipitalis*) and others typically rheophilous and cold steno-thermal (e.g. *Philopotamus ludificatus, Drusus discolor, Halesus rubricollis, Odontocerum albicorne).*

The Po central zone Trichoptera populations are clearly different. The 13 species classified are typical of lentic waters. Among these are *Hydroptila sparsa, Agraylea sexmaculata, Allotrichia pallicornis* and *Hydropsyche dissimulata.* The last one has been shown to be present in remarkable quantities both in the river and in the guts of *Leuciscus cephalus* and *Barbus barbus.* It prefers turbid waters and can be found also in polluted and oxygen depleted environments, where are often covered by ciliates.

Coleoptera

Dytiscidae are the family most represented in the middle Po, with the most prevalent species being *Guignotus fuscillus, Noterus clavicornis, Laccophilus hyalinus* and *L. minutus.* Among the Halipida, *Halipus ruficollis* is the most frequent (Ravizza, 1980). 47 species have been identified and densities range from the abundance values 100 to 1000 individuals per m^2, values which may be considered to be quite low (Cironi & Ruffo, 1981). The Coleoptera population shows spring minima and autumn maxima; it colonizes *Typha* rather than *Phragmites* vegetation.

Diptera

Diptera, together with Oligochaeta, are the group with most species in the middle Po. 109 taxa have been identified, belonging to 13 families, with many species still not classified (Ferrarese, 1980). Chironomidae clearly prevail, with 4 species of Diamesinae, 1 of Prodiamesinae, 34 of Orthocladiidae, 28 of Chironominae, 6 of Tanytarsinae, 5 of Tanypodinae. The most abundant species are: *Cricotopus (Isocladius) sylvestris, Rheosmittia spinocornis* (recorded for the first time in continental Europe), *Chironomus riparius, Chironomus* cf. *plumosus, Polypedilum cultellatum, Cryptochironomus* cf. *defectus.* The chironomid species composition is typical of a beta-mesosaprobic environment (B. Rossaro, pers. comm.).

Mollusca

22 species have been recorded with population densities from 10 to over 1000 m^{-2} and represented chiefly by *Acroluxus lacustris* (Cironi & Ruffo, 1981). The groups with most species are Sphaeriidae (5 species), Limniids (3 species) and Planorbidae (6 species).

15.77   CYCLOSTOMA AND FISH

Recent research on the Po ichthyofauna includes extensive investigations of the whole river course (Gandolfi & Le Moli, 1977a, b) and intensive investigations in the middle zone (Leone, 1981 ; Vitali, 1977, Vitali & Braghieri, 1979, 1981 ; Ioannilli, 1981 ; Aisa & Gattaponi, 1981) and

in the delta system (Gandolfi, 1973; Gandolfi *et al.*, 1978, 1979;
Giannini *et al.*, 1979).

44 species of fish have been identified (excluding the delta):

27 freshwater, present already in the last century
 7 freshwater, appearing at the beginning of this century
 5 migrant anadromous
 1 migrant catadromous
 4 euryaline

The complete list and their distribution along the river is given in
Table 15.2. At the beginning of this century *Salmo gairdneri,
Salvelinus fontinalis, Micropterus salmoides, Lepomis gibbosus,
Ictalurus melas, Gambusia affinis holbrooki* (from North America) joined
the original population. *Silurus glanis*, coming from Yugoslavia,
appeared more recently (Gandolfi & Giannini, 1979).

Along with the increase in species number, there has been a change
in the population structure, resulting from the interaction of factors
such as restocking, dams and pollution. It is not possible to quantify
the changes because of the lack of reference data. However, it is
possible to document a reduction in the distribution area of some
species. In the medium-upper zone *Thymallus thymallus, Cottus gobio,*
and the Cyclostoma *Lampetra planeri zanadreai* have all disappeared;
Cobitis taenia bilineata, Sabanejewia larvata (Cobitida) and *Padogobius
martens* (Gobida) have decreased greatly. Sturgeon migration is
restricted to the lower half of the river and few individuals are able
to overcome the power plant dam. This dam has affected even more
drastically the reproductive migration of *Alosa fallax nilotica* (Fig.
15.10).

In the medium lower stretch, *Anguilla anguilla, Barbus barbus
plebejus, Tinca tinca* and *Esox lucius* are all on the decrease. On the
other hand, other species are increasing in abundance and areal distribution. These include the recently introduced species, *Micropterus
salmoides* and *Lepomis gibbosus*, and species subject to intensive restocking (*Salmo trutta fario* etc.) and *Alburnus alburnus alborella*
(Gandolfi & Le Moli, 1977a). *Carassius auratus* is caught occasionally,
but how this exotic species was introduced to the Po is unknown. *Salmo
gairdneri*, which was spreading until a few years ago, is now decreasing
in density and area.

In the central stretch (Vitali & Braghieri, 1981), the most important
species in the annual catch are *Alburnus alburnus alborella* (60 - 70%),
Rutilus rubilio (11 - 16%), *Scardinius erythroptalmus* (4 - 11%),
Barbus barbus plebejus (2 - 4%) and *Leuciscus cephalus cabeda* (2 - 3%).
Records in terms of weight are shown in Fig. 15.11.

CHIAUDANI & MARCHETTI

Table 15.2 A. Fish species recently recorded in four stretches of the Po. I source - km 103; II km 103 - km 282; III km 282 - km 492; IV km 492.
Abundance: 1 rare; 2 occasional; 3 common; 4 very common (modified from Gandolfi & Le Moli, 1977a).
B. 19th century records for the Po taken from Festa (1892) and Pavesi (1896) by Gandolfi & Le Moli (1977a). + = present.
(Most of the binomials used by these authors differ from the ones in present use and listed here)

| | | A. 1970s stretch | | | | B. | |
		I	II	III	IV	1892	1896
Petromyzonidae	*Petromyzon marinus*			1	1	+	+
	Lampetra planeri zanandreai	2	2	2	1	+	+
Acipenseridae	*Acipenser sturio*		1	2	2	+	+
	A. naccarii			2	2	+	+
	Huso huso				?	?	
Clupeidae	*Alosa fallax nilotica*		?	2	3	+	+
Salmonidae	*Salmo trutta* m. *fario*	3	2	1	1	+	+
	S. trutta marmoratus	2	1			?	+
	S. gairdneri	4	3	2	1		
	Salvelinus fontinalis	?					
Thymallidae	*Thymallus thymallus*	2	?	1			
Esocidae	*Esox lucius*	3	3	3	3		+
Cyprinidae	*Rutilus rubilio*	3	4	4	4	+	+
	R. pigus	3	3	2	1		+
	Leuciscus cephalus cabeda	3	4	4	4		
	L. souffia muticellus	4	3	2	1		
	Phoxinus phoxinus	3	2	1	1	+	+
	Tinca tinca	3	3	3	3	+	
	Scardinius erythrophthalmus	3	4	4	4	+	+
	Alburnus alburnus alborella	3	4	4	4	+	+
	Chondrostoma soetta	3	4	4	4	+	+
	C. toxostoma	3	4	4	3	+	+
	Gobio gobio	3	3	3	2	+	+
	Barbus barbus plebejus	4	4	4	4	+	+
	B. meridionalis	2	1	1		+	+
	Carassius carassius		1	3	3		
	C. auratus		1	1	1		
	Cyprinus carpio	3	4	4	4	+	+
Cobitidae	*Cobitis taenia bilineata*	3	3	2	1	+	+
	Sabanejewia larvata	1	1	1			
Ictaluridae	*Ictalurus melas*	2	3	4	4	+	+
Siluridae	*Silurus glanis*			1	1		
Anguillidae	*Anguilla anguilla*	3	4	4	4	+	+
Gadidae	*Lota lota*		?	1	?		+
Gasterosteidae	*Gasterosteus aculeatus*			1	1		
Poeciliidae	*Cambusia affinis holbrooki*			1	3		
Mugilidae	*Liza ramada*			2	3		
	L. saliens				?		
Percidae	*Perca fluviatilis*	1	2	3	3	+	+
Centrarchidae	*Lepomis gibbosus*	3	4	4	4		
	Micropterus salmoides	2	3	4	4		
Gobiidae	*Padogobius martensi*	2	3	3	1	+	+
Cottidae	*Cottus gobio*	2	1	?		+	+
Pleuronectidae	*Platichthys flesus luscus*				2		+

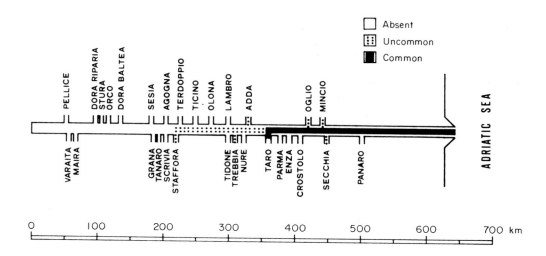

Fig. 15.10 Longitudinal distribution of the diadromous species *Alosa fallax nilotica* down the Po (modified from Gandolfi & Le Moli, 1977a).

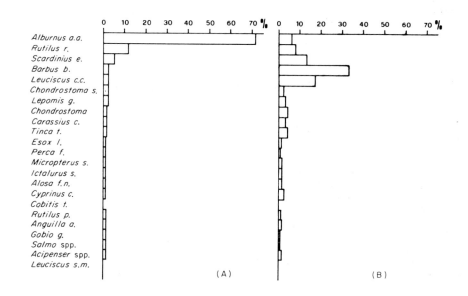

Fig. 15.11 Species composition of fish catches in the central stretch of the Po during 1976-77 as A) % number, and B) % yield (modified from Vitali, 1977; Vitali & Braghieri, 1981).

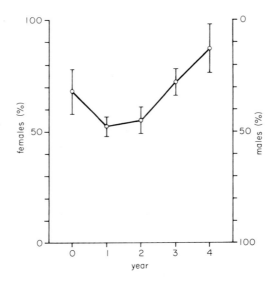

Fig. 15.12 Age and sex distribution of *Alburnus albidus alborella* population in the central reach of the Po, showing mean values and 95% confidence limits (from Vitali & Braghieri, 1981).

The sex ratio showed a dominance of females (60 - 75%) for all the common species. For *Alburnus alburnus* the distribution of sex ratio for age classes showed an increase in the differences between sexes with age (Fig. 15.12).

Research between 1974 and 1977 (Vitali, 1977; Aisa & Gattaponi, 1981) showed that 41 - 63% of fish suffered from diseases of internal and external organs, chiefly of an inflammatory nature. Moreover, 65% of 878 individuals examined, were infected with parasites. These include Protozoa (*Mixobolus, Trichodina*), Trematoda (*Asymphylodora tincae, Cariophyllaeus* sp.), Nematoda (Ascaridea), Acanthocephala, Hirudinea and Crustacea (*Lamproglena pulchella* and others).

Studies at 11 sampling stations in the delta region showed 47 fish species, some of freshwater, some euryaline and others typical of sea coastal waters (Gandolfi, 1973). More recent research (Gandolfi *et al.,* 1978; Giannini *et al.,* 1979) has investigated annual fluctuations in species. A gradual impoverishment in fish populations is taking place here, related to the reduction in flow and greater sedimentation.

15.78 BIRDS

The birds of stream margins of the middle Po have been characterized qualitatively in two ways, species phenology and the choice of habitats. Of the 70 species observed, 24 are nesting (8 resident and 16 present in summer), 36 appear more or less regularly during migrations or in the winter season and 10 are present only occasionally.

Among the nesting birds, three heron species are particularly important, both in terms of biomass and for their close trophic relations with the aquatic food web; these are the night heron *Nycticorax nycticorax*, the little egret *Egretta garzetta*, and the purple heron *Ardea purpurea*. Their colonial nests can be found in the residual riparian woods and their trophic territory corresponds to the river and its tributaries. Two gulls (the common tern, *Sterna hirundo* and the little tern, *Sterna albifrons*) nest during their reproductive on the ground of sandy and almost bare islands. They catch fish chiefly in the main stretch of the river. The habitat of the common moorhen (*Gallinula chloropus*) and coot (*Fulica atra*) is represented usually by scrub at the sides and shallow waters with floating and emergent vegetation. In winter the resident coots are joined by other birds and aggregate in large groups in zones of deep water. A community of flycatchers (Muscicapidae) and thrushes (Turdinae) populates the reed and scrub belts in spring and summer.

The middle Po habitats are potentially, at least, among the most important resting stations of Northern Italy for the migratory aquatic avifauna. However, recent changes in the physiognomy of the river and the various anthropogenic influences have strongly reduced their suitability. Ducks (Anatidae), shore birds (Caradriidae), loons (Gaviidae), grebes (Podicipedidae) and cormorants (Phalacrocoracidae) now all occur in lower numbers and less regularly.

Species diversity is richest during the period of autumn migration and least in winter. The number of individuals and size of biomass are, on the average, however, relatively high in winter (Toso & Tosi, 1981).

15.8 GENERAL COMMENTS

It will be clear from the above account that most attention has been given to the central stretch of the Po and that most of the biological data are either descriptive or quantitative accounts of populations. To frame the Po under a "metabolic" point of view, it is possible only to refer to data obtained by the Odum (1956) oxygen method. This study (Marchetti & Provini, 1977) dealt with six stations sampled over the same 24-hour period in July 1974. These data show a pattern of passing down the river of first autotrophy (P/R ratio higher than 1), then heterotrophy in the central stretch (P/R declining to 0.29) and finally rising again in the last station (Fig. 15.13).

Although such a study does not permit a natural heterotrophic condition to be distinguished from one due to organic pollution (Mann,

1975), it seems almost certain that waste load brought in by the R. Lambro is chiefly responsible. This contains the wastes from 31% of the inhabitants and 36.4% of the industrial equivalents of the whole basin. There was a considerable decrease in gross production below the entry of the Lambro, being five times less than in the upper part of the Po. It seems possible that inhibitory substances present in Lambro water may have influenced this result. Algal growth inhibition has been verified on several occasions by *Selenastrum capricornutum* bioassays, with growth being less than expected on the basis of nutrient concentration (Chiaudani & Vighi, 1977).

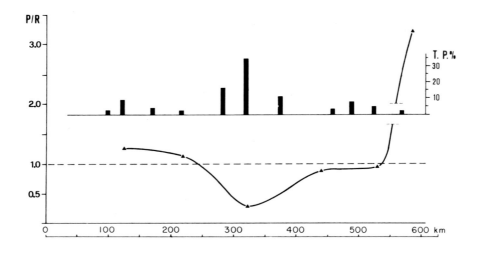

Fig. 15.13 Longitudinal distribution down the Po of pollution loads expressed as % total.

In addition to the subsequent increase in P/R on passing further downstream, an increase in gross production was also observed, at levels typical of recovery zones (Knöpp, 1960; Hammer, 1965; Woods, 1965). This is related to reduced organic pollution and probably also improved hydrological conditions.

Although the pattern described above is based on one survey, it underlines the existence in the Po as well as other lotic environments, of a "spectrum of metabolic states" (Wetzel, 1975) involving auto- and heterotrophy. The spectrum depends on the role played by, on the one hand, autochthonous primary production, and, on the other by alloch-thonous materials supporting heterotrophic metabolism (Hynes, 1969, 1970; Mann, 1969). For a complete knowledge of the Po, it would be interesting to superimpose the structural imagine of the biological communities on this "spectrum of metabolic states". Such comparison is possible for the central Po and the results suggest that the expected relationship does not exist. The temporal trends (during

1975-77) of the diversity index values calculated by Cironi (1979)
for the macro-invertebrates on *Typha* and *Phragmites* in the central
Po lie between 3 and 4 (Fig. 15.14), values typical of moderately
polluted waters. The redundancy values, providing an index of species
dominance, remained nearly constant during the period of observation.
The results of the 1978 studies co-ordinated by Ghetti (1980) are also
partly in contrast with the conclusions drawn from the trend of P/R
values. The use of biotic indices (Woodiwiss, 1964; Verneau & Tuffery,
1968; Woodiwiss, 1978) all indicate good conditions in the central Po
and deterioration in the final stretch (Ghetti, 1980: see Table 15.3).
This trend is confirmed by decreasing diversity index values on passing
downstream (Table 15.4) and also by other quality indices considered by
Ghetti's co-ordinated working group (Verneaux & Faessel, 1976;
Gardenier & Tolkamp, 1976).

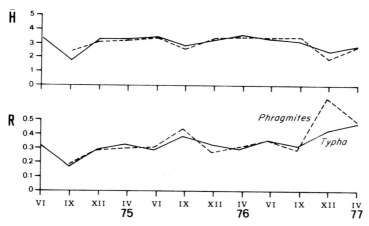

Fig. 15.14 Changes in diversity index (H) and redundance (R) for
macroinvertebrates on *Phragmites* and *Typha* in the central stretch of
the Po during the period June 1974 to April 1977 (modified from
Cironi, 1977; Cironi & Ruffo, 1981).

    The lack of correspondence between the two approaches can be
ascribed mostly to the shortage of metabolic data and the unsuitability
of structural analyses to describe the situation in a river. If the
complete faunistic list for the central Po is considered, it suggests
a community consisting chiefly of common species typical of polluted
environments, Oligochaetes with *Limnodrilus* dominant and chironomids
of the *Chironomus tummi* group; Plecoptera are virtually absent. The
studies co-ordinated by Ghetti (1980) and carried out by Schmitz *et
al.* (1980) and Tittizer & Wiemers (1980) on four zones of the Po
showed the dominance of β-mesosaprobic organisms, the rarity of α-
mesosaprobic and oligosaprobic ones and the absence of polysaprobic
ones. These data are in better accord with the pattern of metabolic
data.

Table 15.3. Biotic index values calculated by various authors for four stations on the Po (Ghetti, 1980).

Index	Authors	km from source			
		387	445	535	585
Biotic Index	Bonazzi & Ghetti (1980)	9	8	6	5
(Woodiwiss, 1964)	De Pauw *et al.* (1980)	9	9	9	5
	Vanhooren & Dehavay (1980)	8	6	6	5
	Woodiwiss & Rees (1980)	8	6	6	4
Biotic Index	Bonazzi & Ghetti (1980)	9	6	5	5
(Verneau &	Debout & Tachet (1980)	8	6	5	5
Tuffery, 1968)	De Pauw *et al.* (1980)	6	6	5	4
	Mouthon & Faessel (1980)	10	9	5	6
	Woodiwiss & Rees (1980)	7	6	5	4
Extended Biotic	Woodiwiss & Rees (1980)	8	6	6	4
Index	De Pauw *et al.* (1980)	9	10	9	5
(Woodiwiss, 1978)					

Table 15.4. Diversity index values calculated by various authors for four stations on the Po (Ghetti, 1980).

Diversity Index	Authors	km from source			
		387	445	535	585
Margalef (1951)	Vanhooren & Dehavay (1980)	2.1	2.6	1.9	1.5
	Woodiwiss & Rees (1980)	2.4	1.4	1.1	0.5
Shannon & Weaver	Madoni (1980)	2.8	4.2	3.7	2.8
(1963)	Woodiwiss & Rees (1980)	1.8	1.2	0.4	0.8
Wilhm & Dorris (1968)	Woodiwiss & Rees (1980)	2.5	1.8	0.6	1.2

Many other elements concur to prove the importance of pollution not only for the central stretch, but for nearly the whole river. Among these are the changes in fish populations, the presence of toxic compounds in fish flesh, the inhibitory effects of waters on algal growth and the various chemical and microbiological indices. The Po, the most important Italian river is, except for the uppermost stretch, unfit for bathing and as a source of drinking water and some restrictions have to be placed on its use for irrigation. It is also responsible for the extreme eutrophication phenomena which have occurred in recent years in the coastal waters which receive the Po loading (Chiaudani *et al.*, 1980).

The Po is a river whose ecosystem needs to be studied much more fully in order to characterize its structural and functional aspects. Yet more than enough is known to demonstrate the extent to which the river has been compromised and to press for immediate intervention.

REFERENCES

Aisa E. & Gattaponi P. (1981) Indagine idrobiologica sugli effetti degli scarichi termici di centrali termoelettriche nel medio Po. Patologia e parassitologia della fauna ittica de medio Po. *Riv. Idrobiol.* 20, 305-36. (The ecosystem of the middle reaches of the River Po. Pathological processes and conditions and parasitism in the ichthyological fauna of the mid-stretch of the river.*)

Allegrini M., Lanzola E. & Marchetti R. (1977) Indagine sulla qualità delle acque del fiume Po. Cap. 8: Radioattività delle acque. *Quaderni Istituto di Ricerca sulle Acque* 32, 495-501. (Survey on the water quality of the Po river. Water radioactivity measurements.)

Andreoli C. & Fricano G. (1981) Variazioni stagionali (giugno 1974-maggio 1977) del fitoplancton nel tratto medio del Po interessato dalla costruenda Centrale termonucleare di Caorso (Piacenza). *Riv. Idrobiol.* 20(1), 139-52. (The ecosystem of the middle reaches of the River Po. Seasonal variations (from June 1974 to May 1977) of the phytoplankton in the section of the River Po on which the thermo-nuclear plant of Caorso (Piacenza) will be built.*)

Berbenni P. (1971) Igiene e inquinamento del Po. Atti 2[o] Congresso Nazionale del Po. Mantova 9-10 ottobre 71. Vol. 1, 189-215. (Sanitary conditions and pollution of the Po river.)

Bertonati M. (1977) Indagine idrobiologica per la valutazione degli effetti degli scarichi termici di centrali termoelettriche sull' ecosistema del medio Po. Cap. 2: Qualità dell'acqua, and Cap. 3: Qualità dei sedimenti. ENEL-DCO, III Rapporto annuale. (Hydro-biological survey on the effects of thermal discharges on the Eco-system of the middle reaches of the Po river. Ch. 2: Water quality, Ch. 3: Sediment quality.)

Bertonati M. & Ioannili E. (1981) Indagine idrobiologica sugli effetti degli scarichi termici di centrali termoelettriche del medio Po. Considerazioni sulle caratteristiche dell'acqua e dei sedimenti nel medio Po nel triennio 1974-1977. *Riv. Idrobiol.* 20(1), 123-38. (Hydrobiological survey of the effects of thermal discharges on the middle Po. Remarks about water and sediment characteristics of the middle Po river during the triennium 1974-1977.)

*Literal translation of title of volume: "Hydrobiological survey of the effects of thermal discharges on the middle Po."

Bonazzi G. & Ghetti P.F. (1980) see: Ghetti (1980) pp. 241-69.

Canali L. & Allodi G. (1962) Contribution a l'étude sur le transport solide in suspension dans le cours d'eau et sur la dégradation du sol dand le basin du Po. Symposium AIHS, Bari.

Chiaudani G., Marchetti R. & Vighi M. (1980) Eutrophication in Emilia-Romagna coastal waters (North Adriatic Sea, Italy): a case history. *Prog. Wat. Tech.* 12, 185-92.

Chiaudani G. & Vighi M. (1977) Indagine sulla qualità delle acque del fiume Po. Cap. 9: Caratteristiche trofiche e fenomeni fitotossici. *Quaderni Istituto di Ricerca sulle Acque* 32, 503-21. (Survey on the water quality of the Po river. Ch. 9: Trophic and toxic condition.)

Cicioni G., Marchetti R. & Spaziani F.M. (1977) Indagine sulla qualità delle acque del fiume Po. Cap. 3: Fonti potenziali di inquinamento. *Quaderni Istituto di Ricerca sulle Acque* 32, 69-114. (Survey on the water quality of the Po river. Ch. 3: Potential sources of pollution.)

Cironi R. (1977) Indagine idrobiologica per la valutazione degli scarichi termici delle centrali termoelettriche sull'ecosistema del medio Po. Cap. 5: Organismi inferiori del complesso trofico. ENEL-DCO, III Rapporto annuale. (Hydrobiological survey on the effects of thermic discharges on the ecosystem of the middle reaches of the River Po. Ch. 5: Phytoplankton, zooplankton and benthos.)

Cironi R. (1979) Struttura delle comunità macrobentoniche in zone a macrofite nel medio Po. XLVII Conv. Unione Zool. Italiana, Bergamo, settembre 1979. (The macroinvertebrate community associated with macrophytes in the middle Po.)

Cironi R. & Ruffo S. (1981) Indagine idrobiologica sugli effetti degli scarichi termici di centrali termoelettriche nel medio Po. Il popolamento a macroinvertebrati bentonici del medio Po, in relazione alla qualità dell'acqua. *Riv. Idrobiol.* 20, 47-82. (The ecosystem of the middle reaches of the River Po. Some studies on the macro-invertebrate community of the Po near Caorso.*)

Corbetta F. & Zanotti-Censoni A.L. (1977) Indagine sulla qualità delle acque del fiume Po. Cap. 17: Cenosi macrofitiche. *Quaderni Istituto di Ricerca sulle Acque* 32, 679-722. (Survey on the water quality of the River Po. Ch. 17: Macrophyte distribution.)

Debout O. & Tachet H. (1980) see: Ghetti (1980) pp. 162-71.

De Pauw N., Verreth J. & Talloen M. (1980) see: Ghetti (1980) pp. 34-93.

Fara G.M., Pagno A., Nardi G., Bracchi U. & Legnani L. (1977) Indagine sulla qualità delle acque del fiume Po. Cap. 14: Aspetti microbiologici. *Quaderni Istituto di Ricerca sulle Acque* 32, 619-37. (Survey of the water quality of the River Po. Ch. 14: Microbiological characteristics.)

Farnesi B., Puddu A. & Spaziani F.M. (1976) Coefficienti di popolazione equivalente delle attività economiche. *Quaderni Istituto di Ricerca sulle Acque* 33, 106 pp. (Industry and population equivalents as a means to estimate pollution loads.)

*See p.429.

Ferrarese U. (1981) Indagine idrobiologica sugli effetti degli scarichi termici di centrali termoelettriche nel medio Po. Nota sui Ditteri del medio corso del fiume Po a Caorso (Piacenza). *Riv. Idrobiol.* <u>20</u>(1), 245-54. (The ecosystem of the middle reaches of the River Po. Diptera of the middle stream of the river at Caorso (Piacenza.*)

Festa E. (1892) I pesci del Piemonte. *Boll. Musei Zool. Anat. comp. Torino* <u>7</u>, 1-125. (The fish of Piemonte Region.)

Gaggino G.F. (1982) Contenuti di alcuni metalli pesanti nella fauna ittica del fiume Po. *Inquinamento* <u>24</u>(6), 25. (Heavy metal content of some fish in the River Po.)

Galassi S. & Provini A. (1981) Chlorinated pesticides and PCB contents of the two main tributaries into the Adriatic sea. *Sci. tot. Environ.* <u>17</u>, 51-57.

Galassi S., Gandolfi G. & Pacchetti G. (1981) Chlorinated hydrocarbons in fish from the River Po (Italy). *Sci. tot. Environ.* <u>20</u>, 231-40.

Galletti P. (1981) Indagine idrobiologica sugli effetti degli scarichi termici di centrali termoelettriche nel medio Po. Indaginin idrobiologiche sul medio Po a Caorso: *Odonata*. *Riv. Idrobiol.* <u>20</u>, 205-16. (The ecosystem of the middle reaches of the River Po. Hydrobiological survey of the middle Po at Caorso: Odonata.*)

Galletti P. & Ravizza C. (1977) Note sull'entomofauna acquatica del corso medio-inferiore del Po: *Odonata*. *Rend. Ist. Lomb. Sci. Lett.* <u>111</u>, 89-100. (Notes on *Odonata* of middle part of the Po river.)

Gandolfi G. (1973) Primi dati sul popolamento ittico nelle acque interne del delta padano. *L'Ateneo Parmense* <u>9</u>, 409-17. (The fish population of the Delta of the River Po: preliminary observations.)

Gandolfi G. & Le Moli F. (1977a) Indagine sulla qualità delle acque del fiume Po. Cap. 18: Distribuzione della fauna ittica nel Po. *Quaderni Istituto di Ricerca sulle Acque* <u>32</u>, 723-45. (Survey on the water quality of the River Po. Ch. 18: Fish distribution in the River Po.)

Gandolfi G. & Le Moli F. (1977b) A preliminary report on fish distribution in the Po river. *Boll. Zool.* <u>44</u>, 149-54.

Gandolfi G., Giannini M. & Torricelli P. (1978) Osservazioni sulla biocenosi ittica della Sacca del Canarin (Delta del fiume Po). *Boll. Zool.* <u>45</u> (Suppl.), 24. (Observations on the fish population in the Delta of the River Po (Sacca del Canarin).)

Gandolfi G. & Giannini M. (1979) La presenza di *Silurus glanis* nel fiume Po. *Natura,* Milano <u>70</u>, 3-6. (On the presence of *Silurus glanis* in the River Po.)

Gardeniers J.J.P., Tolkamps H.H. & Biejer J.A.J. (1980) see: Ghetti (1980) pp. 312-33.

Ghetti P.F. (1980) Biological Water Assessment Methods. 3rd Technical Seminar on Torrente Parma, Torrente Stirone, Fiume Po. Vol. 1. Publ. Commission of the European Communities.

Ghetti P.F. & Bonazzi G. (1980) Biological Water Assessment Methods. 3rd Technical Seminar on Torrente Parma, Torrente Stirone, Fiume Po. Vol. 2. Publ. Commission of the European Communities.

Giandotti M. (1941) Le caratteristiche fisiche del Po nei rapporti con la sua navigabilità. *Annali del Ministero dei Lavori Pubblici* 6. (Physical characteristics of the River Po in relation to its navigability.)

Giannini M., Vitali R. & Gandolfi G. (1979) Studio quantitativo sul popolamento ittico di un ambiente salmastro nel delta del fiume Po (Sacca del Canarin). *Atti Soc. Tosc. Sci. Nat. Mem.*, Ser. B. 86 (Suppl.), 100-103. (Quantitative studies on fish populations in brackish water of the Delta of the River Po (Sacca del Canarin).)

Hammer L. (1965) Photosynthese und Primärproduktion im Rio Negro. *Int. Rev. ges. Hydrobiol.* 50, 335-39.

Hynes H.B.N. (1969) The enrichment of streams. In: *Eutrophication, Consequences, Correctives.* Proc. Symp. nat. Acad. Sci., Washington. pp. 188-96.

Ioannilli E. (1981) Indagine idrobiologica sugli effetti degli scarichi termici di centrali termoelettriche nel medio Po. Progetto sperimentale per la valutazione degli effetti degli scarichi termici e trattamento statistico dei dati. *Riv. Idrobiol.* 20(1), 83-122. (The ecosystem of the middle reaches of the River Po. Experimental design to assess the impact of thermal discharge and statistical treatment of data.)

IRSA (Istituto di Ricerca sulle Acque, Consiglio Nazionale delle Ricerca) (1977) Indagine sulla qualità delle acque del fiume Po. *Quaderno* 32, 821 pp. (Survey of the water quality of the River Po.)

Knöpp H. (1960) Untersuchungen über das Sauerstoff-Produktions-Potential von Flussplankton. *Schweiz. Z. Hydrol.* 22, 152-66.

Leone V. (1981) Indagine idrobiologica sugli effetti degli scarichi termici di centrali termoelettriche nel medio P. Quadro faunistico del medio Po: i Vertebrati. *Riv. Idrobiol.* 20(1), 83-92. (The ecosystem of the middle reaches of the River Po. Faunistic picture of the middle river: Vertebrata.)

Madoni P. (1980) see: Ghetti (1980) pp. 295-312.

Mann K.H. (1969) The dynamic of aquatic ecosystems. In: Cragg J.B. (Ed.) *Advances in Ecological Research* 6, 1-81.

Mann K.H. (1975) Patterns of energy flow. In: Whitton B.A. (Ed.) *River Ecology.* 725 pp., pp. 248-63. Blackwell, Oxford.

Marchetti R., Garibaldi L. & Zambon G. (1977) Indagine sulla qualità delle acque del fiume Po. Cap. 7: La qualità delle acque del Po, quadro di insieme. *Quaderni Istituto di Ricerca sulle Acque* 32, 345-493. (Survey of the water quality of the River Po. Ch. 7: The overall picture.)

Marchetti R., Pennacchioni A., Ottolenghi L. & Gaggino G.F. (1973) Indagine sul mercurio totale e metile in acque interne italiane e in specie ittiche dulcicole. *Acqua Aria* 32, 41-52. (Inorganic and organic mercury concentration in water and fish of different Italian freshwater environments.)

Marchetti R. & Provini A. (1977) Indagine sulla qualità delle acque del fiume Po. Cap. 10: Ossigeno e carico organico. *Quaderni Istituto di Ricerca sulle Acque* 32, 523-49. (Survey of the water quality of the River Po. Ch. 10: Oxygen balance and its relationship to organic loads.)

Margalef R. (1951) Diversidas de species en las communidades naturales. *P. Ins. Biol. Appl.* 9, 5-27. (Species diversity in natural communities.)

Minelli A. & Mannucci Minelli M.P. (1981) Indagine idrobiologica
    sugli effetti degli scarichi termici di centrali termoelettriche nel
    medio Po. Irudinei del medio corso del Po a Caorso (PC). *Riv.*
    Idrobiol. 20(1), 187-94. (The ecosystem of the middle reaches of the
    River Po. Leeches of the middle river at Caorso.)
Ministero dei Lavori Pubblici (1979) *Per il Po e i suoi affluenti.*
    Pubbl. Magistrato per il Po, Parma. 148 pp. (River Po and its
    tributaries.)
Moretti G.P., Vigano' A. & Taticchi Vigano' M.I. (1964) Some informa-
    tion on the orobiontic fauna of Trichoptera of the Italian Western
    Alps above 2000 m. *Proc. Internat. Symp. Trichoptera*, Lunz, pp.
    87-92. (Some information on the orobiontic fauna of Trichoptera of
    the Italian Western Alps above 2000 m.)
Mouthon J. & Faessel B. (1980) see: Ghetti (1980) pp. 174-205.
Nardi G. (1977) Indagine sulla qualità delle acque del fiume Po. Cap.
    15: Aspetti virologici. *Quaderni Istituto di Ricerca sulle Acque*
    32, 639-59. (Survey on the water quality of the River Po. Ch. 15:
    Viral component.)
Odum H.T. (1956) Primary production of flowing waters. *Limnol.*
    Oceanogr. 1, 102-17.
Paoletti di Chiara A. (1981) Indagine idrobiologica sugli effetti
    degli scarichi termici di centrali termoelettriche nel medio Po.
    Gli Oligocheti del benthos del medio Po a Caorso (PC). *Riv.*
    Idrobiol. 20, 173-78. (The ecosystem of the middle reaches of the
    River Po. Benthic Oligochaeta of the middle river near Caorso.)
Pavesi P. (1896) La distribuzione dei pesci in Lombardia. Fusi, Pavia.
    (Fish distribution in Lombardy.)
Pennacchioni A. & Marchetti R. (1973) Mercury contamination in some
    Italian waters, organisms and sediments. In: Proc. Int. Symposium,
    Problems of the Contamination of Man and Environment by Mercury and
    Cadmium, Luxembourg, July 1973.
Piantino P., Vanini G.C. & Verde L. (1977) Indagine sulla qualità delle
    acque del fiume Po. Cap. 2: Precedenti bibliografie sulla qualità
    delle acque. *Quaderni Istituto di Ricerca sulle Acque* 32, 21-67.
    (Survey on the water quality of the River Po. Ch. 2: Annotated
    bibliography on the water quality of the River Po.)
Prati L. (1977a) Indagine sulla qualità delle acque del fiume Po.
    Cap. 12: Pesticidi. *Quaderni Istituto di Ricerca sulle Acque* 32,
    581-603. (Survey of the quality of the River Po. Ch. 12: Pesti-
    cides.)
Prati L. (1977b) Indagine sulla qualità delle acque del fiume Po.
    Cap. 13: Detergenti anionici e fenoli. *Quaderni Istituto di Ricerca*
    sulle Acque 32, 605-18. (Survey of the quality of the River Po.
    Ch. 13: Anionic detergents and phenols.)
Provini A., Mosello R., Pettine M., Puddu A., Rolle E. & Spaziani F.M.
    (1979) Metodi e problemi per la valutazione dei carichi di
    nutrienti. In: *Atti del "Convegno sulla Eutrofizzazione in Italia".*
    C.N.R., Roma 3-4 October 1978. (Methods and problems for nutrients
    load calculation.)

Provini A., Gaggino G.F. & Galassi S. (1980) Po e Adige: valutazione statistica della frequenza di campionamento in un programma di monitoraggio. *Ing. Ambientale* 9, 379-90. (Po and Adige: statistical evaluation of sampling frequency in a monitoring program.)

Querena E. (1981) Indagine idrobiologica sugli effetti degli scarichi termici di centrali termoelettriche nel medio Po. Gli Efemerotteri nel medio Po a Caorso (PC). *Riv. Idrobiol.* 20(1), 195-204. (The ecosystem of the middle reaches of the River Po. Ephemeroptera of the middle river at Caorso.*)

Raffa U. (1972) Particolarità del regime dei corsi d'acqua padani in relazione ai fenomeni di inquinamento. *Quaderni Assoc. Idrotecnica Italiana* (sez. Padana) 1, 15-36. (Hydrological particularities of the rivers of the Po valley in relation to pollution.)

Raffa U. (1973) L'evoluzione recente del regime idrico e termico del Po. 2º Convegno sullo stato di avanzamento della radioecologia in Italia. Parma 24-25 May 1973. (Recent hydrological and thermal changes in the River Po.)

Raffa U. & Giuffrida G. (1973) Alimentazione delle falde attraverso d'alveo dei corsi d'acqua. Atti 2º Colloquio Internazionale sulle acque sotterranee. Palermo, 28-30 April. (Artificial recharge of groundwater by river water infiltration.)

Ravizza C. (1976) Note sull'entomofauna acquatica del corso medio-inferiore del Po: *Hemiptera heteroptera*. *Rend. Ist. Lomb. Sci. Lett.* 110, 55-66. (Note on the aquatic entomological fauna of the mid-lower stretch of the Po: *Hemiptera heteroptera.)*

Ravizza C. (1981) Indagine idrobiologica sugli effetti degli scarichi termici di centrali termoelettriche nel medio Po. Note sugli Emitteri ed i Coleotteri popolanti le acque del medio Po, fra Caorso e Isola Serafini. *Riv. Idrobiol.* 20, 217-230. (The ecosystem of the middle reaches of the River Po. Notes on the aquatic Hemiptera and Coleoptera inhabiting the river between Caorso and Isola Serafini.*)

Rossaro B. (1976) Lo zooplancton del Po a Caorso. *Rend. Ist. Lomb. Sci. Lett.* 110, 35-54. (Zooplankton of the Po at Caorso.)

Rossaro B. (1978) Composizione tassonomica e fenologica delle *Orthocladiinae* (Dipt. Chironomidae) nel Po a Caorso (PC) determinata mediante analisi delle exuvie. *Riv. Idrobiol.* 17, 277-300. (The taxonomic composition and phenology of the Orthocladiinae, Dipt. Chir. in River Po near Caorso, as determined by analysis of pupal exuviae.)

Rossaro B. (1981) Indagine idrobiologica sugli effetti degli scarichi termici di centrali termoelettriche nel medio Po. Il carico biologico del medio Po presso Caorso (PC): la sua composizione specifica e le relazioni con le proprieta dell'acqua esaminate mediante l'analisi delle componenti principali. *Riv. Idrobiol.* 20(1), 152-166. (The ecosystem of the middle reaches of the River Po. The biological load in the middle Po near Caorso: specific composition and its relation with water characteristics, examined by principal components analysis.)

Rossaro B. & Cotta-Ramusino M. (1976) I Copepodi del Po a Isola Serafini. *Riv. Idrobiol.* <u>15</u>, 187-203. (Copepoda of the River Po near Isola Serafini.)

Rossetti M. & Raffa U. (1973) Evoluzione idrologica e idrografica della regione del Delta del Po. *Ann. Univ. Ferrara*, Sez. 1 (Ecologia) <u>1</u>, 5-28. (On the evolution of Po hydrography and hydrology in the Delta.)

Sambugar B. (1981) Indagine idrobiologica sugli effetti degli scarichi termici di centrali termoelettriche nel medio Po. Gli Oligocheti raccolti tra i banchi di macrofite nel medio Po a Caorso (PC). *Riv. Idrobiol.* <u>20</u>, 179-186. (The ecosystem of the middle reaches of the River Po. The Oligochaeta sampled from the bank macrophytes of the middle Po at Caorso.[*])

Shannon C.E. & Weaver W. (1963) *The Mathematical Theory of Communication.* Univ. Illinois Press, Urbana.

Schmitz W., Buck H. & Vobis H. (1980) see Ghetti (1980) pp. 383-415.

Tittizer T. & Wiemers W. (1980) see: Ghetti (1980) pp. 335-81.

Tosi G. & Toso S. (1981) Indagine idrobiologica sugli effetti degli scarichi termici di centrali termoelettriche nel medio Po. Caratteristiche strutturali e dinamiche della popolazione ornitica del medio Po. *Riv. Idrobiol.* <u>20</u>(1), 337-82. (The ecosystem of the middle reaches of the River Po. The bird population along the Po near Caorso.[*])

Ugolini P. (1971) La navigazione interna nella Valle Padana. Atti 2° Congresso Nazionale del Po. Mantova, 9-10 ottobre 71, Vol. 1, 107-23. (Navigation through the Po river system.)

Vanhooren G. & Dehavay P. (1980) see: Ghetti (1980) pp. 96-139

Verneaux J. & Tuffery G. (1968) Méthode zoologique pratique de détermination de la qualité biologique des eaux courantes. Indices Biotiques. *Annls scient. Univ. Besançon Zool.* <u>3</u>, 79-90.

Verneaux J. & Faessel B. (1976) Quoted in Mouthon & Faessel (1978).

Vitali R. (1977) Indagine idrobiologica per la valutazione degli effetti degli scarichi termici di centrali termoelettriche sull' ecosistema del medio Po. Cap. 6: Popolazione ittica. ENEL-DCO, III Rapporto annuale. (Hydrobiological survey of the effects of thermal discharge on the ecosystem of the middle reaches of the Po. Ch. 6: Fish stock assessment.)

Vitali R. & Braghieri L. (1979) Parametri dinamici di popolazioni di barbo (*Barbus barbus plebejus*) e cavedano (*Leuciscus cephalus cabeda*) nel medio Po. XLVII Conv. Unione Zool. Italiana. Bergamo, September 1979. (Dynamic characteristics of fish populations (*Barbus barbus plebejus*) and (*Leucisus cephalus cabeda*) in the middle stream of Po river.)

Vitali R. & Braghieri L. (1981) Indagine idrobiologica sugli effetti degli scarichi termici di centrali termoelettriche nel medio Po. Caratteristiche strutturali e dinamiche del popolamento ittico del medio Po nella zona di Caorso. *Riv. Idrobiol.* 20(1), 269-304. (The ecosystem of the middle reaches of the River Po. Structural and dynamic characteristics of the fish community of the middle stream of Po river at Caorso.[*])

Viviani M.A. & Tortorano A.M. (1977) Indagine sulla qualità delle acque del fiume Po. Cap. 16: Aspetti micologici. *Quaderni Istituto di Ricerca sulle Acque* 32, 661-78. (Survey of the water quality of the River Po. Ch. 16: Survey of yeasts.)

Vivoli G. & Vecchi G. (1977) Indagine sulla qualità delle acque del Fiume Po. Cap. 11: Metalli tossici. *Quaderni Istituto di Ricerca sulle Acque* 32, 551-78. (Survey of the water quality of the River Po. Ch. 11: Heavy metals.)

Wetzel R.G. (1975) Primary Production. In: Whitton B.A. (Ed.) *River Ecology*. 725 pp., pp. 230-47. Blackwell, Oxford.

Wilhm J.L. & Dorris T.C. (1968) Biological parameters for water quality criteria. *Bioscience* 18, 477-81.

Woodiwiss F.S. (1964) The biological system of stream classification used by the Trent River Boards. *Chem. Ind.,* 443-47.

Woodiwiss F.S. (1978) Comparability study of biological-ecological assessment methods. 2nd Technical Seminar on the River Trent and tributaries. Commission of European Communities, Brussels.

Woodiwiss F.S. & Rees S. (1980) see: Ghetti (1980) pp. 417-40.

Woods W. (1965) Primary production measurements in the Upper Ohio River. *Spec. Publ., Pymatuning Hab. Ecol. Univ. Pittsburg* 3, 66-78.

Zullini A. (1969) Osservazioni sui Nematodi di un biotopo fluviale (Po a Trino Vercellese). *Rend. Accad. naz. Lincei,* sez. VIII, 47, 109-15. (Notes on aquatic Nematoda in the River Po (Trino Vercellese).)

Zullini A. (1974) The nematological population of the Po river. *Boll. Zool.* 41, 183-210.

Zullini A. (1975a) Nematodi dello Psammon del Po. *Rend. Ist. Lomb. Sci. Lett.* 109, 79-83. (Nematoda as a component of psammon of the River Po.)

Zullini A. (1975b) Nematodi ripicoli del Po. *Rend. Ist. Lomb. Sci. Lett.* 109, 130-42. (Riparian nematodes of the Po.)

16 · Lower Rhine-Meuse

G. VAN URK

16.1 INTRODUCTION

The geomorphology of the central part of the Netherlands is largely determined by the rivers Rhine and Meuse.

Originally, the Rhine meandered freely in its lower reaches and continually formed new courses. Its present partition into three branches - Waal, Nederrijn and IJssel - was effected by means of two canals, the Pannerdens Kanaal, dug in 1707, and the Bijlands Kanaal, which dates back to 1780 ; both are described in detail by Van de Ven (1976). Further downstream the river ran through a large peat swamp, where it also flowed through different courses. For embankment and land reclamation purposes the river needed to be restricted in its freedom. The present configuration was not reached until 1870 when the Nieuwe Merwede was completed, which leads most of the Waal water into the Hollands Diep.

The Hollands Diep also receives the water from the Meuse by way of the Bergsche Maas, which was completed in 1904 (Figs. 16.1, 16.3). Until very recently, no port or industry of any significance has grown up along the Hollands Diep and Haringvliet. The port of Rotterdam is situated further north, at a former mouth of the Meuse, and was connected to the sea by the Nieuwe Waterweg in 1872. Therefore, large parts of the Rhine-Meuse estuary long remained fairly unspoiled. Its freshwater tidal area was unique in Europe, in spite of the pollution of the rivers Rhine and Meuse.

In 1970 a dam with sluice was completed across the mouth of the Haringvliet, as part of the so-called "Delta Works", which are to protect the south-western part of the Netherlands against flooding from the sea. The closure has reduced considerably tidal movements in the Haringvliet and Hollands Diep (Deltadienst of Rijkswaterstaat, 1973) and robbed the area of most of its unique natural features. Before the closure a number of studies had been carried out in the freshwater tidal area, which enables me to include a general description of this particular habitat. In order to understand the hydrography and hydrochemistry of the main part of this area, the Netherlands stretch of the Meuse will be described briefly, as will the present state of one of the branches of the Rhine.

16.2   GENERAL DATA
(Meuse in the Netherlands)

DIMENSIONS
    Area of catchments
        whole Meuse (including other countries) 33000 km^2
        below Eysden (Belgian-Dutch border)   12000 km^2 (part of catchment
                                                                in Germany)
        major tributaries: Roer (Rur)   2350 km^2 (majority in Germany)
                           Niers   1350 km^2 (majority in Germany)
                           Dieze   2800 km^2
    Length of river
        Eysden to most downstream weir (Lith)   195 km
        Lith to confluence with branch of Rhine
            (Nieuwe Merwede)   60 km
        final stretch to the sea (estuary of
            Haringvliet)   50 km
    Drop in height of river in Netherlands   45 m

CLIMATE

Precipitation
    In Dutch part of the catchment about 700 mm yr^{-1}; 200 mm yr^{-1} carried
off by the tributaries of the Meuse.

LAND USE
    agriculture   4000 km^2
    woodlands   900 km^2
    urban   700 km^2

POPULATION
    Meuse catchment in the Netherlands: 3 x 10^6 people = 250 people km^{-2}

PHYSICAL FEATURES OF RIVER
    Discharge (means for the period 1951-1960: m^3 s^{-1}

	Borgharen	Lith		Borgharen	Lith
January	504	609	August	108	167
February	456	538	September	126	185
March	356	468	October	159	232
April	227	306	November	257	332
May	149	208	December	377	465
June	116	169			
July	83	127	year	243	317

These mean values do not reflect the large fluctuations in discharges
that occur because the Meuse is a rain-fed river. During dry periods
(e.g. summer 1976) the discharge at Borgharen may be less than 10 m^3 s^{-1}.
A treaty between Belgium and the Netherlands should guarantee a minimum
discharge of 25 m^3 s^{-1} by building reservoirs upstream.

Tributaries

	discharge $(m^3 s^{-1})$	point at km entry to Meuse
Roer (= Rur)	20	79
Niers	8	157
Dieze (confluence of Aa and Dommel)	18	220

Major Weirs

	km sign	water level above weir (+NAP)
Borgharen	15	43.90
Linne	71	20.50
Roermond	81	16.75
Belfeld	100	14.00
Sambeek	146	10.75
Grave	175	7.50
Lith	201	4.60

Regulation changed the character of the river completely except for the section between Borgharen and Linne.

Abstraction from Meuse

Important abstraction points for Rotterdam and the Hague are near km sign 250. Both cities use Meuse water for their drinking water supply.

Substratum

Gravel, sand, silt, clay.

Temperature (monthly mean values at Borgharen)

	1976	1977		1976	1977
January	6.6	5.1	August	19.7	20.9
February	4.5	6.6	September	16.3	19.3
March	7.3	10.5	October	14.3	16.0
April	11.2	10.2	November	9.3	11.3
May	15.4	15.8	December	5.3	7.5
June	19.4	20.0			
July	22.0	21.8	year	12.6	13.8

These temperatures are of course not natural, but influenced by regulation and cooling water discharges.

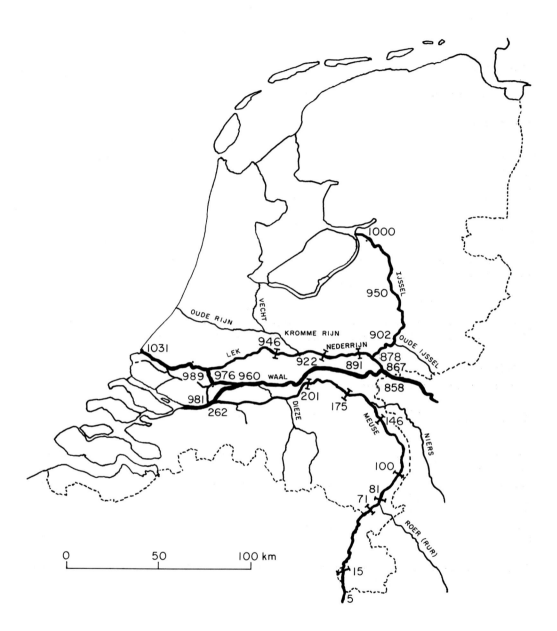

Fig. 16.1 The Rhine-Meuse system in the Netherlands. The Kromme Rijn, Oude Rijn and Vecht do not form a part of the main system nowadays. Canals are not shown. The figures indicate km signs.

16.3   THE MEUSE IN THE NETHERLANDS

16.31   HYDROGRAPHY

The main data on the hydrography of the Netherlands stretch of the Meuse are summarized in 16.2. The regulation of the river was carried out in the years 1919-1926, except for the river section between km 17 and km 63, which forms the boundary between Belgium and the Netherlands. Here a lateral canal was constructed instead. As a result, this section of the river has retained some of its original character (16.35). More data on the hydrography of the Meuse basin are found in Van der Made (1972) and Chapter 1. (The km values in the present chapter refer to distance from where the Meuse first meets the Netherlands border.)

16.32   POLLUTION

16.321 *Sewage discharges*

A survey of the discharges of biodegradable organic wastes in 1970 was presented in a report by a working group on the sanitation of the Meuse (Werkgroep Sanering Maas, 1971). Altogether some 3.4×10^6 population equivalents were discharged into the river, directly or indirectly, the greater part between km 12 and km 108. The greatest source, the river Roer, discharged some 10^6 population equivalents into the Meuse at km 79. As a result of a sanitation programme, sewage discharges have been cut down considerably (Dijkzeul, 1981).

16.322 *Industrial waste water discharges*

In 1970, industry discharged large amounts of ammonia, cyanides and heavy metals into the Meuse. The river's high cadmium and zinc contents - about 4 µg l^{-1} and 250 µg l^{-1}, respectively - are partly due to former mining activities, but metallurgical works have also made substantial contributions. Mercury used to be discharged by a chlor-alkali plant near km 74 and chromium by a plant near km 109. In recent years these discharges have been reduced significantly.

16.323 *Water quality*

Due to long residence times in the river sections between weirs, as well as the aerating effect of the weirs, the average oxygen content of the Meuse water is nearly always above 6 mg l^{-1} (Fig. 16.2; Koolen, 1973). The NH_4^+-content has a maximum of more than 5 mg l^{-1} in the unregulated section of the river between km 17 and km 52, but decreases rapidly downstream, at least in summer, due to nitrification. The phosphate-P content is 0.3 - 0.7 mg l^{-1}, the "natural" content being about 0.07 mg l^{-1} (Zuurdeeg, 1980). Heavy metals, except nickel, are for the greater part adsorbed on the suspended matter and therefore

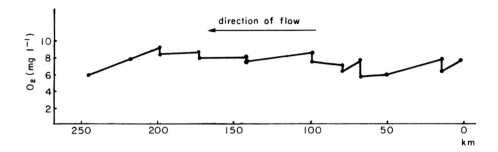

Fig. 16.2 Mean oxygen content of Meuse in the Netherlands during summer 1970. The km distances are those indicated on signs by the river. (Data from Rijksinstituut voor Zuivering van Afvalwater)

accumulate in the sediments. At high river flow the sediments may be resuspended in the water; consequently, exceptionally high metal contents may be measured, but normally the metal contents of the downstream reaches are not very high.

Table 16.1 Trade metal concentrations in sediments from the Meuse and the Rhine, expressed as $\mu g\ g^{-1}$ at 50% <16 μm. (Data derived from: Salomons & Mook, 1977; Rijksinstituut voor Zuivering van Afvalwater, 1980; unpublished reports)

	Cr	Ni	Cu	Zn	Cd	Pb
river sediments deposited 15th-16th century	77	33	21	93	0.5	31
Meuse sediment about 1960	216	44	160	1516	28	382
Rhine sediment about 1900	140	35	42	380	1.4	82
Rhine sediment about 1960	640	54	295	2420	14	530
Rhine sediment 1977	830	75	290	1660	37	390

So, despite considerable waste water discharges in its middle course, the water quality of the Meuse, where it enters its tidal section (km 201), has always been relatively good, or at least far better than that of the Rhine in the freshwater tidal area. The same applies to the content of heavy metals - except cadmium and zinc- in Meuse sediments as compared with Rhine sediments (Table 16.1).

16.33   PLANKTON

The composition of the plankton of the Meuse, analysed since 1966 (Rijksinstituut voor Zuivering van Afvalwater, 1970), has not changed in recent years. In the river sections between the weirs algal blooms may develop. These blooms do not occur in regular seasonal patterns; their formation depends on river flow (Table 16.2).

Table 16.2   Algal blooms in the Meuse at km 201 in five successive years.

year	month	biomass (μg l^{-1} chlorophyll *a*)	dominant species
1974	April	78	*Synedra acus*
	May-October	65-107	*Melosira granulata*
1975	June	78	*Diatoma elongatum*
1976	April	53	*Asterionella formosa*
	May	80	*Diatoma elongatum*
1977	May-October	56-103	*Melosira granulata*
1978	July	49	*Stephanodiscus hantzschii*
	August	80	*Melosira granulata*

16.34   MACROINVERTEBRATES

16.341   *Before regulation*

Shortly before work on the regulation of the Meuse was started, a working group was formed to study the ecological impact of regulation. In the summer of 1918 the group went on a rowing expedition down the river in order to collect the material, and published a preliminary report the following year (Romijn *et al.*, 1919). Typical rheophilous species were still found at that time, such as the gyrinid beetle *Orectochilus villosus*, which was abundant on the stones of the jetties (see 16.521) and the water-bug *Aphelocheirus aestivalis*. On the whole, the identification of Ephemeroptera, Trichoptera and Chironomidae was not up to species level, so that no good picture of these groups can be derived from this report. Imagines of the damsel fly

Calopteryx splendens were reported to be abundant in the vegetation at
the riverside, but it is not explicitly stated if the larvae occurred
in the river itself. The closely related *C. virgo* was found only along
the brooks that flow into the Meuse. Larvae of *Ophiogomphus serpentinus*
were recorded in the river itself. Imagines of the water-beetle
Helophorus arvernicus were also abundant, but again nothing was said
about the occurrence of the larvae. Among the molluscs, *Unio crassus*
var. *batavus* was characteristic of the Meuse.

    Although the report announces subsequent studies, the difficulties
encountered in the organization of more expeditions may have been too
great; as it is, no reports on the Meuse prior to its regulation have
appeared since then.

16.342 *After regulation*

Recent data on the macroinvertebrates of the Meuse are scarce; they
point to a rather diverse macroinvertebrate fauna which resembles that
of stagnant eutrophic waters in the Netherlands, so from the viewpoint
of river ecology there is little reason to go into this any further.
The crayfish *Orconectes limosus* seems to have invaded the Netherlands
via the Meuse basin and is likely to replace the indigenous species
Astacus astacus in most habitats (Geelen, 1975).

16.35 FISH

Although the barbel *Barbus barbus* is uncommon in the Netherlands, it
still occurs in the Meuse. It has survived the regulation, especially
in the section between km 15 and km 71, as mass fish kills due to excess
discharges of ammonia or cyanides by a local industry have demonstrated.
Another common fish in the Meuse is the beaked carp *Chondrostoma nasus*.
Locally numerous in the Meuse, but also occurring elsewhere, are the
dace *Leuciscus leuciscus* and the chub *Leuciscus cephalus*. For the
rest, the fish stock of the Meuse resembles that of other - stagnant -
waters in the Netherlands (De Groot & Muyres, 1980).

16.4   THE FRESHWATER TIDAL AREA OF THE BIESBOSCH

16.41   HISTORY

The western part of the Netherlands between the river branches (Vecht,
Oude Rijn, Lek, Merwede and former branches of the Meuse situated south
of the Merwede; Figs. 16.1, 16.3) originally consisted of peat swamps
lying above mean river level. In the Middle Ages the swamps were
embanked and cultivated as pastures. Drainage caused the peat soil to
set. As a consequence, the rivers rose to levels higher than pre-

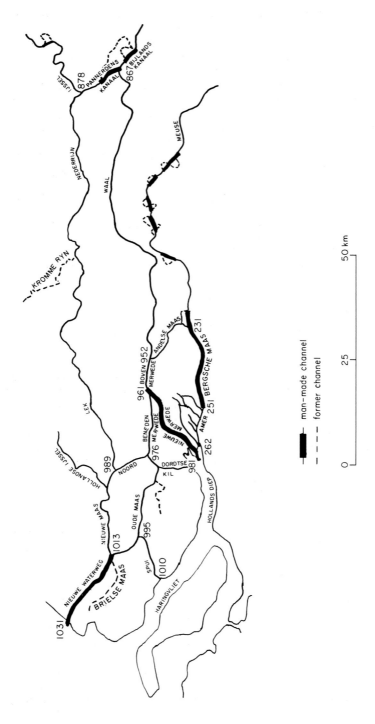

Fig. 16.3 Map showing man-made channels in the main system of the Rhine-Meuse, together with former river channels.

viously and overflowed if a dike was breached. Flooded areas were
usually reclaimed by repairing the dikes.

In 1421 the so-called "Sint Elizabethsvloed" inundated the Groote
Waard, a very large polder area between the Merwede and a former course
of the Meuse, which was never fully re-embanked again due to administrative disorganization. The rivers Waal and Meuse flowed into this
newly formed sheet of water. The sand transported by these rivers was
deposited on the original peat soils, forming sandflats which developed
a vegetation of their own, which in turn facilitated the deposition of
silt by a reduction of the currents. Many such sand- and mud-flats
were embanked and transformed into small polders suitable for farming.
Eventually, the sand- and mud-flats and polders hampered the rivers in
their flow, increasing the risk of inundation upstream at high river
discharges. Therefore, at the end of the 19th century two new channels
were dug, the Nieuwe Merwede and the Bergsche Maas, which were to carry
off the water of the Waal and the Meuse, respectively. With the completion of these channels, the present configuration of the area was
achieved. The area was called the Biesbosch, which stands for "wood
of rushes". The development of the Biesbosch is illustrated in Fig.
16.4. Unless stated otherwise, the following paragraphs deal with the
situation which existed before 1970 (16.1).

16.42 HYDROGRAPHY AND HYDROCHEMISTRY

The isohaline of 300 mg Cl^- l^{-1}, which may be considered as the boundary between fresh and brackish water (Den Hartog, 1968), normally
lies in the Hollands Diep, a few kilometres west of the Biesbosch
(Peelen, 1967). Mean high and low water levels on the Nieuwe Merwede/
Waal and on the Amer/Bergsche Maas/Meuse are given in Fig. 16.5; data
on the ebb and flood currents in the rivers are summarized in Table
16.3 (data from Rijkswaterstaat). The currents in the tidal channels
of the Biesbosch were about the same as in the Amer. The same applies
to mean high and low water levels. At ebb-tide many of the smaller
creeks fell dry.

The creeks north of the Nieuwe Merwede were fed by water from the
Rhine. The channels and creeks south of the Nieuwe Merwede were not
fed directly by the Nieuwe Merwede thanks to a dike along the south
bank of the river, but by the Amer, which carries off water from the
Meuse. From the Cl^- contents it was concluded that only the western
part received some water from the Rhine which entered at flood tide
from the Hollands Diep, as the flow of the Nieuwe Merwede is much
greater than that of the Meuse (Parma, 1968). This resulted in short
residence times of the water in the northern part of the Biesbosch, so
that the water quality here was nearly as bad as in the Rhine itself.
In the Meuse sector, on the contrary, the oxygen content was near
saturation, except in the small creeks where oxygen consumption by
sediments was high. The NH_4^+ content was lower and the NO_2^- and NO_3^-
contents were higher than in the Meuse (Parma, 1968).

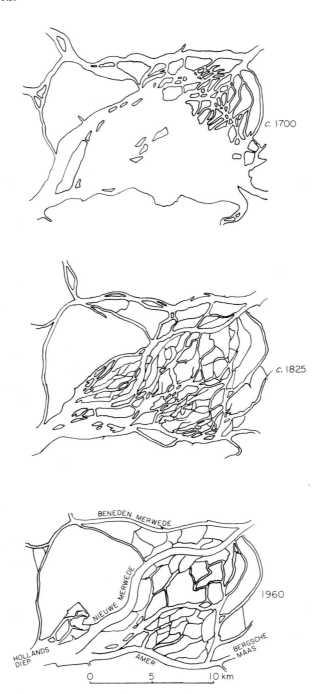

Fig. 16.4 Development of the Biesbosch area from 1700 to 1960. (Based on historical maps)

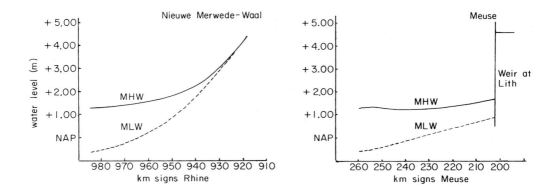

Fig. 16.5 Mean high water levels (MHW) and mean low water levels (MLW) on the Nieuwe Merwede-Waal (Rhine) and the Amer-Bergse Maas-Maas (Meuse) in 1969. (Data from Rijkswaterstaat)

Table 16.3 Velocity (m s^{-1}) and direction of the maximum ebb and flood currents on the rivers in the tidal area.

		Rhine flow (m^3 s^{-1})					
		980		2300		9400	
		ebb	flood	ebb	flood	ebb	flood
Nieuwe Merwede	979	0.65 ←	0.55 →	0.70 ←	0.40 →	1.15 ←	0.35 →
Boven Merwede	960	0.75 ←	0.40 →	0.90 ←	0.10 ←	1.85 ←	1.55 ←
Amer	261	0.75 ←	0.85 →	0.80 ←	0.85 →	0.85 ←	0.70 →

16.43 PLANKTON

A typical plankton community developed only in the Meuse sector. It is however difficult to obtain a full picture of the plankton composition: in the absence of complete mixing in the channel system, different populations may have developed at different places (Ringelberg, 1968) and due to changing current velocity the plankton composition changed within a tidal cycle. Maximal concentrations of phytoplankton were found, when the flood current was maximal. At high water, part

of the phytoplankton sank to the bottom and was only partly resuspended
in the ebb current. Different species reacted in different ways to
changes in the current velocity (Dresscher *et al.*, 1968). At low water,
many planktonic diatoms could be found on the sand- and mud-flats
(Zonneveld, 1960, Appendix 15).

The characteristic species of the Meuse sector of the Biesbosch were
the diatom *Actinocyclus normannii* and the copepod *Eurytemora affinis*
(Dresscher, 1968; Leentvaar, 1968). These species also occur in great
numbers in the freshwater tidal areas of other rivers, such as the
Elbe (Nöthlich, 1972) and the Weser (Behre, 1961). In both the Elbe
and the Biesbosch, the greatest numbers of *E. affinis* were found in
spring, while *Actinocyclus normannii* reached its maximum in the summer
months. It seems clear that there is no trophic relationship between
these species, but that each species is adapted to the current conditions in the freshwater tidal area. Otherwise, there was a great
resemblance to the phytoplankton of the Meuse, in which *Melosira
granulata* is the most abundant species (Table 16.2). Since the closure
of the Haringvliet in 1970, *Actinocyclus normannii* has disappeared
(Peelen, 1975).

In the zooplankton of the small creeks, many primarily benthic
organisms (e.g. *Chydorus sphaericus, Alona* sp.) were found (Leentvaar,
1968).

16.44   BENTHIC PLANTS

16.441 *Phanerogams*

The early stages of development of the vegetation on an intertidal
sandflat are shown in Fig. 16.6. All species but *Scirpus triqueter*
are also common in non-tidal habitats in the Netherlands. However,
the morphology of the plants in the tidal area often differs from those
in the non-tidal habitats e.g. in the Biesbosch, *Caltha palustris*
forms typical "spins" of new roots and is considered as a separate
variety, *Caltha palustris* var. *araneosa* (Van Steenis, 1971). Elsewhere, *Scirpus triqueter* is also found in non-tidal habitats e.g. in
the Upper Rhine area, but is never as common as in the freshwater
tidal areas. Some of the species mentioned in Fig. 16.6 also grow
in brackish water e.g. *S. maritimus*.

Where the soil level was above 0.80 m + NAP (Table 16.4), willows
were usually planted. In the natural situation no trees other than
willows would have grown; once the tidal influence had diminished to
a level where it no longer dominated, willow coppice would probably
have developed, but before this stage was reached in the Biesbosch,
the land was embanked and cultivated as farmland. In the intertidal
willow coppices the composition of the herb layers was strongly
influenced by the way the willow coppices were exploited (Zonneveld,
1960).

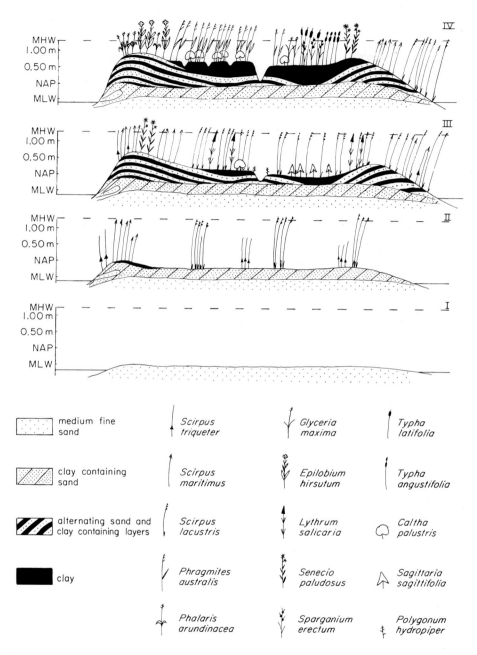

Fig. 16.6 Development of vegetation on an intertidal sandflat in the freshwater tidal area. In stage IV on the left a cultured reed marsh with ditches is shown and on the right the more natural vegetation developing without any artificial drainage. (Modified from Zonneveld, 1960)

Table 16.4 Summary of data on the Biesbosch area. (For references, see text)

	subtidal zone	intertidal zone		
	open water	sand- and mud-flats	rush and reed marshes	willow coppices
area	12 km²	together 32 km²		20 km²
level with regard to NAP	below - 0.70 m	- 0.70 m to NAP	- 0.50 m to + 0.80 m	above + 0.80 m
water movement	current to c 1.0 m s⁻¹	decreasing exposure to water movements ⟶		
frequency of flooding	100%	100%	100%	may be less than 100%
duration of flooding in a tidal cycle		c 12 - 7 h	c 11 - 4 h	less than 4 h
main phanerogam	locally *Potamogeton* sp.		*Scirpus* sp. *Phragmites australis*	*Salix* sp.
characteristic algae	planktonic *Actinocyclus normannii*	locally *Vaucheria compacta*	at banks of creeks *Vaucheria compacta*	*Vaucheria* sp.
macro-invertebrates more or less restricted to the habitat	Unionidae	*Trocheta bykowskii*?	at banks of creeks *Pseudamnicola confusa*	*Orchestia cavimana*
macro-invertebrates with wide distribution in the area	Chironomidae sp., Tubificidae sp., *Pisidium* sp., *Sphaerium corneum*, *S. solidum*			

Fig. 16.7 Tidal creek with
natural levee. (Photo by
O.M. van Hoorn)

   The study of Zonneveld was confined to the Meuse sector of the
Biesbosch, but the vegetation of the Rhine sector was almost the same
(Verhey, 1961a)´, which shows that it is the frequency and duration of
flooding that are the key factors in the development of the vegetation,
not the composition of the water.

16.442 *Cryptogams*

Blue-green algae and diatoms occurred on intertidal sand- and mud-flats
near MLW (mean low water). Among diatoms, the planktonic species
Melosira granulata and *Actinocyclus normannii* were common, but benthic
species like *Rhoicosphenia curvata* and *Cocconeis placentula* were also
found in large numbers (Zonneveld, 1960, Appendix 15). At a somewhat
higher level in the intertidal zone *Vaucheria compacta* formed extensive
growths (Simons, 1974, 1975). This species was also found on the banks
of creeks in the rush and reed marshes, together with *Callitriche
stagnalis*. Where sedimentation was not too intensive, *Rhizoclonium
riparium* was abundant on all kinds of substratum, together with
Ulothrix tenerrima, Microcoleus vaginatus or *Vaucheria* sp. (Nienhuis,

1975, 1980). *Cladophora okamurai* was found on hard substrata. This species seems to be confined to freshwater tidal areas, at least in the Netherlands (Den Hartog, 1967).

On the soil of the willow coppices locally rich *Vaucheria* growths were encountered, the main species being *V. frigida, V. bursata* and *V. canalicularis* (Simons, 1975). On the trunks of the willows, the bryophytes up to MHW (mean high water), *Marchantia polymorpha, Fissidens bryoides, F. taxifolius, Mnium* sp. and *Chiloscyphus pallescens*, occurred in a vertical zonation (Barkman, 1953).

16.45 MACROINVERTEBRATES

16.451 *Tidal channels*

A tidal channel is like a major river, as far as its width, depth and current velocity is concerned. Only the current is not unidirectional, but alternating. Tubificidae, Chironomidae and, in particular, bivalve molluscs occurred in the main tidal channels below MLW. In addition to ubiquitous species like *Sphaerium corneum*, species with a fluviatile distribution were found, including *S. rivicola, S. solidum, Unio crassus* var. *batavus, U. tumidus, Anodonta anatina* and a number of species of *Pisidium* (Wolff, 1968, 1970; Kuiper & Wolff, 1970). These fluviatile species were absent in channels which were directly fed by water from the Rhine.

Gammarids were represented in the Biesbosch by *Gammarus zaddachi* ; between 1958 and 1960 this species was very common in the Meuse sector of the Biesbosch, whereas it occurred only sparsely in the Rhine sector. After a sudden break-down of the populations of *G. zaddachi* in 1960, some specimens of *G. pulex* were occasionally found in the tidal creeks of the Biesbosch (Den Hartog, 1964). Among Tubificidae, *Potamothrix moldaviensis* was the most numerous species, followed by *Limnodrilus claparedeianus* and *L. hoffmeisteri* (Verdonschot, 1980).

16.452 *Intertidal zone*

Aquatic species can only colonize the intertidal sand- and mud-flats if the substratum retains enough water during ebb-tide, as has been demonstrated by Kuiper & Wolff (1970) for *Pisidium* sp. Unionidae are however not found above MLW (Wolff, 1968). Hardly anything is known about the distribution of Tubificidae and Chironomidae on the mudflats. As a rule, in the intertidal zone the same species of Tubificidae occur as in the sediments below MLW, as for example in the Elbe (Pfannkuche, 1977). *Sphaerium corneum* was found in small numbers on the mud-flats (Wolff, 1970). Sand does not retain much water and consequently few aquatic species can maintain themselves on intertidal sand-flats. Only a few individuals of *Gammarus zaddachi* and *Sphaerium solidum* were found in this habitat and in many places the sand-flats

looked almost barren (Wolff, 1973). Den Hartog & Van Rossum (1957)
consider that the lower part of the intertidal zone was the favourite
habitat of the leech *Trocheta bykowskii*, which fed mainly on Tubificidae.
However, it was also found higher up in the intertidal zone (Heyligers,
1961). None of the species, occurring on intertidal sand- and mud-
flats are completely confined to this habitat and many niches seem to
be unoccupied there (Wolff, 1973).

Higher up the intertidal zone the small snail *Pseudamnicola confusa*
was found. This is an intertidal species, which occurred most commonly along the banks of small tidal creeks in the willow coppices and
reed marshes, characterized by a vegetation of *Vaucheria compacta*,
Callitriche stagnalis and *Polygonum hydropiper*. It lived only in
places where it was sheltered against the currents and also occurred in
brackish water (Butot, 1960; Den Hartog, 1960). Aquatic organisms were
still dominant in the soil of the lower willow coppices; at 0.40 m
below MHW the soil fauna consisted mainly of Tubificidae, *Pisidium* sp.
and *Potamopyrgus jenkinsi*. Near MHW the species present were mainly
terrestrial (Heyligers, 1961). At this level there occurred also the
semi-terrestrial amphipod *Orchestia cavimana* (Den Hartog, 1963).
Although this species is not confined to the freshwater tidal area,
it seems more numerous there than in any other habitat. It also occurs
regularly along the major Dutch lakes (Den Hartog & Tulp, 1960) and
along the Meuse. The condition for its occurrence seems to be that the
water level either remains the same, as in the lakes and the regulated
Meuse, or that it fluctuates in a predictable way, as in a tidal area
O. cavimana has not been found along the branches of the Rhine, where
changes in water level occur irregularly and where there may be a
great distance between the high and low water line. In the soil of
the willow coppices there were also larvae of Chironomidae, Ceratopogonidae and Dryopidae, but these could not be identified to species
(Heyligers, 1961).

Among the smaller animals, species generally regarded as rare occurred, e.g. the ostracod *Candona sarsi* (Heyligers, 1961). Boaden (1976)
presents some qualitative data on the meiofauna of the sand-flats.

16.46 FISH

Characteristic species of the lower courses of large rivers are shad
Alosa alosa and twait *Alosa fallax*. Before 1970 twait occurred in
rather large numbers in the Biesbosch. Shad had already virtually
disappeared, mainly because of overfishing (Verhey, 1961b). The
houting *Coregonus oxyrhynchus*, previously also numerous, had already
disappeared by about 1910. A history of fishing in the lower river
reaches is presented by Verhey (1961b).

Fig. 16.8 A) End of creek in tidal willow coppice. B) Intertidal sand-flat at ebb-tide. (Photos by O.M. van Hoorn)

16.47 CONCLUDING REMARKS

Unfortunately, there were no studies of production and energy flow in the Biesbosch. The ecosystem was supplied with enormous amounts of

nutrients from the rivers and probably was a highly productive one.
Man exploited the area, harvesting large amounts of rushes, reed and
osiers. Among the species living in the Biesbosch, the most conspicuous were a number of euryhaline ones, particularly among the benthic
algae (Den Hartog, 1968). Here again, it appears that the transition
from the river community into the brackish water community was a gradual one, rather than distinctly marked. It should be remembered that
at storm floods brackish water penetrated into the Biesbosch (Peelen,
1967). The transition from the freshwater tidal area to the brackish
part of the Rhine-Meuse estuary is discussed at length by Wolff (1973).
On the other hand, no clear picture can be given of the transition from
the non-tidal river to the freshwater tidal area. Nearly all freshwater species occurring in the Biesbosch are also found either in the
Meuse or the Rhine, but the reverse need not be true. In a natural river
ecosystem, aquatic insects such as Ephemeroptera and Trichoptera constitute a major part of the macroinvertebrate fauna. Surveys of the
macroinvertebrate fauna of a branch of the Rhine in 1973-1977 showed
that these groups were absent from the river at that time (Van Urk,
1978). The regulation of the Meuse has certainly affected the distribution of rheophilous species of Ephemeroptera and Trichoptera in this
river. There were very few records of these groups from the Biesbosch,
whereas the area suffered far less from human activities than the
rivers. This scarcity might be a natural feature of the freshwater tidal
area, as the habitat offers special problems to successful colonization.
But other factors may still play a role. With the closure of the
Haringvliet, the habitat of the freshwater tidal area was not entirely
lost. Along the Oude Maas there is a river stretch that has a great
resemblance with the Biesbosch. Water flows through the Oude Maas
either from the Waal or from a mixture of the Waal and the Meuse,
depending on the flow of the Rhine. The area is much smaller than the
Biesbosch and is therefore vulnerable, as it is situated near the port
and industry of Rotterdam. Nevertheless, it would be of great value in
order to study yet unsolved ecological problems of the freshwater
tidal habitat, once the water quality of the Rhine has improved sufficiently.

16.5 PRESENT STATE OF THE LOWER RHINE

16.51 POLLUTION MONITORING IN THE LOWER RHINE

The Rhine is the major source of freshwater in the Netherlands. In the
country's water management system the Rhine water affects water quality
both in the Dutch polders and in coastal waters, including the Wadden
Sea (Wolff, 1978). The Rhine used to be the source of drinking water
for Amsterdam, Rotterdam and The Hague, but the latter two cities are
now supplied by the Meuse. Monitoring the water quality of the Rhine
is therefore of great importance and a regular chemical monitoring
programme was established in 1953. Particular attention was given at
first to the oxygen balance, but interest is now shifting to heavy

metals and organic micropollutants. Attention is being paid especially
to the geochemical behaviour of heavy metals in estuarine areas (e.g.
Salomons & Mook, 1977; Salomons et al., 1981). The behaviour of
organic micropollutants is followed only incidentally (Duinker &
Hillebrand, 1979). Yet effects of organic micropollutants have been
demonstrated more clearly than those of heavy metals: the insecticides
telodrin and dieldrin, originating from an industry near Rotterdam,
have caused a reduction of the numbers of sandwich terns in the Wadden
Sea, and polychlorinated biphenyls from the Rhine are held responsible
for the decline of harbour seals. From their experiments with rainbow
trout, Poels et al. (1980) conclude that the Rhine water is chronically
toxic to salmonids.

Hydrobiological research in the Rhine itself constitutes only a minor
part of the pollution monitoring programme. It started with the analysis of the phytoplankton composition (Wibaut-Isebree Moens, 1956), at a
later stage supplemented with analyses of periphyton on artificial
substrata, but these groups have not shown much change in recent years
(Rijksinstituut voor Zuivering van Afvalwater, unpublished data).
Research into the macroinvertebrate fauna in one of the branches of
the Rhine in the Netherlands was started in 1973.

16.52 MACROINVERTEBRATE FAUNA OF THE IJSSEL

16.521 *Description of the IJssel*

The flow of the IJssel can be kept high by means of weirs in the
Nederrijn (Deltadienst of Rijkswaterstaat, 1973). In the period of
1975-1980 the mean flow was 343 m^3 s^{-1}; the lowest flow was 139 m^3 s^{-1}
in July 1976 and the highest flow 1165 m^3 s^{-1} in February 1980. The
current velocity at km 885 is 0.5 - 1.1 m s^{-1} and decreases to 0.3 -
0.7 m s^{-1} at km 996. The river bed consists of gravel, sand and clay,
median grain size decreasing downstream like in the Nederrijn/Lek
(Terwindt et al., 1963).

Jetties were built on the river in the 19th century, in order to
protect its banks from erosion (Fig. 16.9). Current velocity increased,
as a result of the constriction of the river channel by these jetties
(Jansen et al., 1979). In some stretches the banks have recently been
consolidated with metal slag. In the lower course, a vegetation of
Nuphar luteum, Scirpus sp., *Typha* sp. and *Phragmites australis* has developed at the banks. In the upper reaches only isolated plants of
Potamogeton nodosus can be found. Bryophytes and filamentous algae,
growing on the upper side of the stones at jetties and on the metal
slag, form a natural "trap" to the suspended matter in the river water.
Much silt is deposited and forms a substratum for many midge larvae
and worms. The bottom of the stones, however, is almost clean and here
species which require a hard substratum, such as *Ancylus fluviatilis*,
can settle.

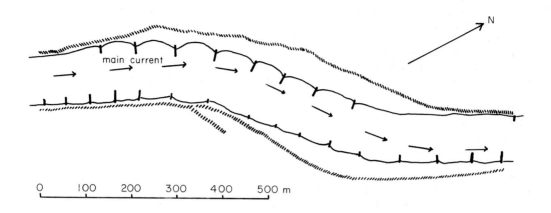

Fig. 16.9 Section of the IJssel at km 916, showing the position of the jetties (from a river map of Rijkswaterstaat).

16.522 *Macroinvertebrates on stones*

The composition of the macroinvertebrate fauna on the stones from the jetties in the period 1975-1981 is presented in Table 16.5. Bryozoa and sponges were not included in this survey. The most conspicuous feature of the fauna was the appearance of a population of *Hydropsyche contubernalis* and the mass development of Chironomidae since 1978. Among the latter, *Dicrotendipes* gr. *nervosus* and *Cricotopus* sp. are the most numerous (Van Urk, 1981). At the most downstream sites *Glypto-tendipes* dominates; it is also very common in the reed-beds. Less abundant on the stones are *Parachironomus* gr. *arcuatus*, *Xenochironomus xenolabis*, *Endochironomus albipennis* and *Nanocladius* sp. Tanytarsini are found in very low numbers, as are Tanypodinae (names according to Moller Pillot, 1979). These changes in numbers can hardly be considered merely as natural fluctuations. It has to be assumed that water quality factors play a role, in particular the oxygen content and perhaps also the contents of cholinesterase inhibi-tors such as organophosphorus insecticides (Table 16.6). Besch *et al.* (1977) reported that concentrations as low as 1 µg l^{-1} of the carbamate insecticide Fenethcarb, which is also a cholinesterase inhibitor, interfered with netspinning of *Hydropsyche angustipennis*. Chironomidae are equally sensitive to cholinesterase inhibitors (Karnak & Collins, 1974). Other water quality factors may of course also play a role. It is interesting to compare this mass occurrence of *H. contubernalis* with that reported for the middle Rhine in 1975-1977 (see p.290).

Among the species found, *H. contubernalis* is the only one that is

Table 16.5 Composition of the macroinvertebrate fauna on jetties of the River IJssel in autumn.

	1975	1976	1977	1978	1979	1980	1981
Hydra sp.	3	-	-	136	111	383	58
Dugesia polychroa	-	29	22	86	136	89	260
D. tigrina	-	-	-	-	7	84	639
Nais bretscheri	-	-	-	5	-	30	5
N. elinguis	-	-	-	21	-	75	45
N. simplex	-	-	-	74	45	67	2
Stylaria lacustris	48	15	219	609	631	371	799
Glossiphonia complanata	16	14	12	19	20	23	30
G. heteroclita	9	17	35	14	34	9	38
Erpobdella octoculata	35	41	244	281	126	181	104
Acroloxus lacustris	31	1	26	53	59	73	68
Ancylus fluviatilis	30	-	104	376	74	349	122
Bithynia tentaculata	50	3	43	51	709	645	1626
Lymnaea peregra	43	13	71	28	47	62	67
Physa fontinalis	7	5	39	10	40	50	8
Dreissena polymorpha	114	4	3	18	52	91	204
Asellus aquaticus	142	345	4370	3168	2326	2373	1621
Hydropsyche contubernalis	-	-	-	10	253	2104	598
Cyrnus trimaculatus	-	-	-	-	-	-	107
Ecnomus tenellus	-	-	-	-	-	-	35
Chironomidae	19	56	254	4680	7282	10198	15287
others	33	2	19	39	95	313	234
total number of animals	580	545	5461	9678	12047	17570	21957

Table 16.6 Water quality variables of the Rhine at km 862 (German-Dutch border). Mean values for April-September.

		1973	1974	1975	1976	1977	1978	1979	1980	1981
discharge	$m^3\ s^{-1}$	1818	1592	2207	1128	2035	2663	2125	2674	2288
suspended solids	$mg\ l^{-1}$	44	43	39	63	39	56	38	27	34
dissolved oxygen	$mg\ l^{-1}$	4.1	3.9	6.1	5.4	6.3	6.8	6.9	7.1	7.1
total organic C	$mg\ l^{-1}$		14.3	11.4	15.1	10.0	10.1	10.2	8.5	9.8
NH_4^+-N	$mg\ l^{-1}$	1.9	1.9	0.6	1.1	0.6	0.5	0.5	0.4	0.4
chlorophyll a	$\mu g\ l^{-1}$					69	51	47	56	85
cholinesterase inhibitors (as para-oxon)	$\mu g\ l^{-1}$	2.0	1.6	6.5	11.6	4.6	1.3	1.2	0.6	0.9

characteristic of the potamon. At some sites it is particularly abundant, densities approaching one individual per cm^2 of stone surface.
It was found along the whole river from km 885 to km 1000. The factors
determining the zonation of Hydropsychidae in a stream have been discussed by Hildrew & Edington (1979), who consider temperature as the
main factor. In the IJssel, temperatures show a wide range, with
summer maxima exceeding 20 °C.

16.523 *Macroinvertebrates in sediments*

Constriction of the river channel by jetties and the increased current
velocities have influenced the sedimentation pattern. The river bed
of the IJssel consists of hard packed sand and gravel and here hardly
any macroinvertebrates are found. A regular sedimentation of the suspended matter occurs only near the mouth of the river, especially near
the banks where the current diminishes. A mainly tubificid macroinvertebrate population here reaches densities of some thousands of
individuals per m^2. There is no distinct seasonal trend in their
numbers. The dominant species are *Limnodrilus hoffmeisteri* and *L.
claparedeianus*; their proportions vary from place to place and from
year to year (Table 16.7).

Table 16.7 Percentage composition of Tubificidae in samples of
sediment at a depth of about 2 m from the IJssel at km 1000; a) near
the left bank; and b) near the right bank.

	1976		1978		1979		1980	
	a	b	a	b	a	b	a	b
Limnodrilus claparedeianus	51	41	25	16	15	51	6	42
L. hoffmeisteri	22	22	68	73	72	30	89	56
Tubifex tubifex	12	18	3	3	3	<1	5	2
Potamothrix moldaviensis	7	–	<1	2	1	7	–	–
other species	8	19	4	6	10	11	–	<1

Dominance of *L. claparedeianus* has seldom been reported elsewhere,
the dominant species in this kind of habitat nearly always being *L.
hoffmeisteri*. It is not clear what biotic or abiotic factors favour
L. claparedeianus here. Relations between different species of Tubificidae may be very complex (Pfannkuche, 1977) and there is no simple
explanation for the changes in relative proportions of the various
species. A large proportion of *L. hoffmeisteri* has often been considered to be indicative of pollution by organic wastes, but it seems
doubtful whether this also holds in this habitat.

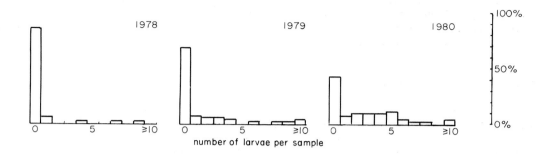

Fig. 16.10 Frequency distribution of numbers of Chironomidae in grab samples (surface area 0.03 m^2) from the sediments of the IJssel at km 1000.

In the IJssel a quite different association of Tubificidae is found among vegetation near the banks. The main species are *Psammoryctides albicola*, *P. barbatus* and *Limnodrilus udekemianus* (Van Urk, 1978). *Tubifex tubifex*, *Potamothrix hammoniensis* and *P. moldaviensis* are found in all habitats in low numbers. Chironomidae occur at much lower densities than Tubificidae, but appear to increase slightly (Fig. 16.10). The species (or larval groups) that are found regularly, are: *Procladius* s.1, *Dicrotendipes* gr. *nervosus*, *Cryptochironomus* gr. *defectus*, *Parachironomus* gr. *arcuatus* and *P.* gr. *longiforceps*, *Microchironomus tener*, *Chironomus* f.l. *fluviatilis* and *reductus*, *Glyptotendipes* sp., *Cricotopus* sp. Their densities are too low to detect any periodicity or trends for individual species. Comparison of species composition with that of other rivers is difficult for the same reason. Besides Chironomidae, occasional specimens of Unionidae, in particular of *Anodonta anatina* and *Unio pictorum*, were found in grab samples in 1978-1980. As mentioned above (16.451), Wolff (1968) was unable to find any Unionidae in the Rhine branches. This absence was attributed to the pollution of the river water and it seems reasonable to conclude that their present occurrence has been permitted by recent improvements in water quality, such as the increase in oxygen content. Among the Pisidiidae only *Pisidium henslowanum* and *Sphaerium corneum* are found, both in relatively low numbers. The fluviatile species *S. rivicola* and *S. solidum* have so far not been encountered in the IJssel.

16.524 *Prospects*

Monitoring the macroinvertebrate fauna of the IJssel has revealed some signs of restoration of the river's ecosystem. However, it is far from being unpolluted, and even further from being a natural river. Many species, such as the burrowing Ephemeroptera that were reported once to have been abundant, may have disappeared forever, since there may be no suitable refuge left to permit re-introduction. Nevertheless, biological surveillance of the Rhine is not only valuable from a viewpoint of pollution monitoring, but may also contribute to knowledge of the autecology of species whose occurrence is restricted to the lower reaches of large rivers, which form a very vulnerable habitat in a densely populated area like Western Europe.

Acknowledgements

I thank Miss S.M. Wiersma and A. Espeldoorn for carrying out part of the work on the IJssel reported here, L.W.G. Higler and I. Wolters for identification of the Trichoptera, and F. Repko for checking the identification of Chironomidae.

REFERENCES

Barkman J.J. (1953) Over de Mosvegetaties van onze getijgrienden. *Buxbaumia* 7, 42-49. (On the moss vegetation of our tidal willow coppices.)

Behre K. (1961) Die Algenbesiedlung der Unterweser unter Berücksichtigung ihrer Zuflüsse. *Veröff. Inst. Meeresforsch Bremerhaven* 7, 71-263. (The algae of the Lower Weser and its tributaries.)

Besch W.K., Schreiber I. & Herbst D. (1977) Der Hydropsyche- Toxizitätstest, erprobt an Fenethcarb. *Schweiz. Z. Hydrol.* 39, 69-85. (The *Hydropsyche* toxicity test, tried out with Fenethcarb.)

Boaden P.J.S. (1976) Soft meiofauna of sand from the delta of the Rhine, Meuse and Scheldt. *Neth. J. Sea Res.* 10, 461-71.

Butot L.J.M. (1960 *Pseudamnicola confusa* (Frauenfeld 1863) algemeen in de Biesbosch (Gastropoda, Prosobranchia). *Basteria* 24, 60-65. (*Pseudamnicola confusa*, a common mollusc in the Biesbosch.)

De Groot A.T. & Muyres W.J.M. (1980) Visserijkundige waarnemingen vispassages 1975 tot en met 1979. *Visserij* 33, 446-61. (Observations on fish migration 1975-1979.)

Deltadienst of Rijkswaterstaat (1973) The Realization and Function of the Northern basin of the Delta project. *Rijkswaterstaat Communications* 14, 73 pp. Government Publishing Offices, Gravenhage, Netherlands.

Den Hartog C. (1960) Verspreiding van het slakje *Pseudamnicola confusa* in het Deltagebied van Rijn en Maas. *Basteria* 24, 66-74. (Distribution of *Pseudamnicola confusa* in the Deltaic region of Rhine and Meuse.)

Den Hartog C. (1963) The amphipods of the deltaic region of the rivers Rhine, Meuse and Scheldt in relation to the hydrography of the area. Part II: The Talitridae. *Neth. J. Sea Res.* 2, 40-67.

Den Hartog C. (1964) The amphipods of the deltaic region of the rivers Rhine, Meuse and Scheldt in relation to the hydrography of the area. Part III: The Gammaridae. *Neth. J. Sea Res.* 2, 407-57.

Den Hartog C. (1967) Brackish water as an environment for algae. *Blumea* 15, 31-43.

Den Hartog C. (1968) Een karakteristiek van het zoetwatergetijdenge-bied en de plaats van de Biebosch in dit geheel. *Meded. Hydrobiol. Ver.* 2, 173-82. (A delineation of the freshwater tidal area and the position of the Biesbosch in that area.)

Den Hartog C. & Tulp A.S. (1960) Hydrobiologische waarnemingen in Friesland. *De Levende Natuur* 63, 133-40. (Hydrobiological obser-vations in Friesland.)

Den Hartog C. & Van Rossum E. (1957) De bloedzuiger *Trocheta bykowskii* in de benedenrivieren. *De Levende Natuur* 60, 228-33. (The leech *Trocheta bykowskii* in the lower river reaches.)

Dijkzeul A.H.J. (1981) De waterkwaliteit van de Maas in Nederland in de periode 1953-1980. H_2O 14, 628-34. (The water quality of the Meuse in the Netherlands in the period 1953-1980.)

Dresscher T.G.N. (1968) De soortencombinaties van de in het water zwevende plantaardige micro-organismen van enkele plaatsen in de Brabantse Biesbosch en de omgeving. *Meded. Hydrobiol. Ver.* 2, 154-66. (The phytoplankton of the Brabantse Biesbosch and adjacent rivers.)

Dresscher T.G.N., Bakker C. & Geelen J.F.M. (1968) Planktonconcentra-ties en getijbeweging in het Gat van de Turfzak. *Meded. Hydrobiol. Ver.* 2, 192-206. (Plankton concentrations and tidal movements in a small tidal creek.)

Duinker J.C. & Hillebrand M.T.J. (1979) Behaviour of PCB, pentachloro-benzene, hexachlorobenzene, α-HCH, γ-HCH, β-HCH, dieldrin, endrin and p,p'-DDD in the Rhine-Meuse estuary and the adjacent coastal area. *Neth. J. Sea Res.* 13, 256-81.

Geelen J.F.M. (1975) *Orconectes limosus* (Raf.) and *Astacus astacus* L. (Crustacea, Decapoda) in the Netherlands. *Hydrobiol. Bull.* 9, 109-13.

Heyligers P.C. (1961) De bodemfauna van de grienden. In: Verhey C.J. (Ed.) *De Biesbosch, land van het levende water*, 255 pp., pp. 85-117. Thieme, Zutphen, Netherlands. (The soil fauna of the willow coppices.)

Hildrew A.G. & Edington J.M. (1979) Factors facilitating the co-exis-tence of Hydropsychid caddis larvae (Trichoptera) in the same river system. *J. Anim. Ecol.* 48, 557-76.

Jansen P.P., Van Bendegon L., Van den Berg J., De Vries M. & Zanen A. (1979) *Principles of River Engineering. The Non-tidal Alluvial River*, 509 pp. Pitman, London.

Karnak R.E. & Collins W.J. (1974) The susceptibility to selected insecticides and acetylcholinesterase activity in a laboratory colony of midge larvae, *Chironomus tentans* (Diptera:Chironomidae). *Bull. environ Contam. Toxicol.* 12, 62-69.

Koolen J.L. (1973) De kwaliteit van het Maaswater in Nederland. H_2O 6, 3-14.

Kuiper J.G.J. & Wolff W.J. (1970) The Mollusca of the estuarine region of the rivers Rhine, Meuse and Scheldt in relation to the hydrography of the area. III. The genus *Pisidium*. *Basteria* 34, 1-40.

Leentvaar P. (1968) Het zooplankton van de Brabantse Biesbosch en enige eigenschappen van de planktongemeenschap. *Meded. Hydrobiol. Ver.* 2, 167-72. (The zooplankton of the Brabantse Biesbosch with notes on the plankton association of the area.)

Moller Pillot H.K.M. (1979) De larven der Nederlandse Chironomidae (Diptera). *Nederl. Faun. Meded.* 1, 1-276. (The larvae of the Dutch Chironomidae (Diptera).)

Nienhuis P.H. (1975) Biosystematics and ecology of *Rhizoclonium riparium* (Roth) Harv. (Chlorophyceae, Cladophorales) in the estuarine area of the rivers Rhine, Meuse and Scheldt, 240 pp. Ph.D. Thesis, University of Groningen.

Nienhuis P.H. (1980) The epilithic algal vegetation of the SW Netherlands. *Nova Hedwigia* 33, 1-94.

Nöthlich I. (1972) Trophische Struktur und Bioaktivität der Planktongesellschaft im unteren limnischen Bereich des Elbe-Aestuars. *Arch. Hydrobiol., Suppl.* 43, 33-117. Trophic structure and bioactivity of the plankton communities in the lower reach of the Elbe estuary.)

Parma S. (1968) Hydrografie van de Biesbosch. *Meded. Hydrobiol. Ver.* 2, 95-145. (Hydrography of the Biesbosch.)

Peelen R. (1967) Isohalines in the delta area of the rivers Rhine, Meuse and Scheldt. *Neth. J. Sea Res.* 3, 575-97.

Peelen R. (1975) Changes in the plankton of the estuarine area of the Haringvliet-Hollands Diep-Biesbosch in the S.W. Netherlands caused by the dams through Volkerak and Haringvliet. *Hydrobiol. Bull.* 8, 190-200.

Pfannkuche O. (1977) Ökologische und systematische Untersuchungen an naidomorphen Oligochaeten brackiger und limnischer Biotope, 138 pp. Ph.D. Thesis, University of Hamburg. (Ecological and taxonomic investigations of naidomorph Oligochaeta of brackish and freshwater habitats.)

Poels C.L.M., Van der Gaag M.A. & Van de Kerkhoff J.F.J. (1980) An investigation into the long-term effects of Rhine water on rainbow trout. *Water Res.* 14, 1029-35.

Rijksinstituut voor Zuivering van Afvalwater (1971) *Hydrobiologisch onderzoek van de Maas*, 63 pp. Mededeling No. 9, Staatsuitgeverij's Gravenhage, Netherlands. (Hydrobiological research of the Meuse.)

Rijksinstituut voor Zuivering van Afvalwater (1980) *De waterkwaliteit van de Rijn in Nederland in de periode 1972-1979*, 71 pp. Nota No. 80-032, Lelystad, Netherlands. (Water quality of the Rhine in the Netherlands 1972-1979.)

Ringelberg J. (1968) De relatieve abundantie van de planktonische raderdieren in de Biesbosch en omgeving. *Meded. Hydrobiol. Ver.* 2, 183-91. (The relative abundance of planktonic Rotifers in the Biesbosch and adjacent rivers.)

Romijn G., Uyttenbogaart D.L. & Heimans J. (1919) Verslag van het biologisch onderzoek van de Maas en hare oevers. In: *Natuurhistorisch Genootschap in Limburg Jaarboek 1918*, 177 pp., pp. 93-145. Maastricht, Netherlands. (Report on the biological exploration of the Meuse and its banks.)

Salomons W. & Mook W.G. (1977) Trace metal concentrations in estuarine sediments: mobilization, mixing or precipitation. *Neth. J. Sea Res.* 11, 119-29.

Salomons W., Mook W.G. & Eysink W. (1981) *Biogeochemical and Hydro-Dynamic Processes affecting Heavy Metals in Rivers, Lakes and Estuaries*, 68 pp. Delft Hydraulics Laboratory Publication No. 253

Simons J. (1974) *Vaucheria compacta*: a euryhaline estuarine algal species. *Acta bot. neerl.* 23, 613-26.

Simons J. (1975) *Vaucheria* species from estuarine areas in the Netherlands. *Neth. J. Sea Res.* 9, 1-23.

Terwindt J.H.J., De Jong J.D. & Van der Wilk E. (1963) Sediment movement and sediment properties in the tidal area of the Lower Rhine (Rotterdam Waterway). *Verh. Kon. Ned. Geol. Mijnbouw Genootsch* 21, 243-58.

Van der Made J.W. (1972) Hydrografie van het Maas-Bekken. H_2O 5, 356-62. (Hydrography of the Meuse basin.)

Van de Ven G.P. (1976) *Aan de wieg van Rijkswaterstaat Wordings-geschiedenis van het Pannerdens Kanaal*. Gelderse Historische Reeks VIII, 439 pp. Walburg Pers, Zutphen. (The roots of Rijkswaterstaat History of the "Pannerdens Kanaal".)

Van Steenis C.G.G.J. (1971) De zoetwatergetijde-dotter van de Biesbosch en de Oude Maas, *Caltha palustris* L. var. *araneosa* var. nov. *Gorteria* 5, 213-19. (The freshwater tidal marsh marigold of the Biesbosch and the Oude Maas.)

Van Urk G. (1978) The macrobenthos of the River IJssel. *Hydrobiol. Bull.* 12, 21-29.

Van Urk G. (1981) Veranderingen in de macro-invertebraten-fauna van de IJssel. H_2O 14, 494-99. (Changes in the macroinvertebrate fauna of the IJssel.)

Verdonschot P.F.M. (1980) Aquatische Oligochaeta III Het Deltagebied. *Delta Instituut voor Hydrobiologisch Onderzoek Rapp Versl 1980-9*, 115 pp. (Aquatic Oligochaeta III The Delta area.)

Verhey C.J. (1961a) Vegetatie en fauna van de Sliedrechtse Biesbosch. *De Levende Natuur* 62, 138-44. (Vegetation and fauna of the Sliedrechtse Biesbosch.)

Verhey C.J. (1961b) De vissen en de visvangst. In: Verhey C.J. (Ed.) *De Biesbosch, land van het levende water*, 255 pp., pp. 139-64. Thieme, Zutphen, Netherlands. (The fish and fisheries.)

Werkgroep Sanering Maas (1971) *De verontreiniging van de Maas, aanbevelingen tot sanering*. Rapport, 68 pp. 's Gravenhage, Netherlands. (The pollution of the Meuse, recommendations with respect to sanitation.)

Wibaut- Isebree Moens N.L. (1956) *Rivierenonderzoek 1954-1955*. Report in 6 pts., 37 pp. and annexes, mimeographed. Amsterdam. (Investigations in the rivers 1954-1955.)

Wolff W.J. (1968) The Mollusca of the estuarine region of the rivers
    Rhine, Meuse and Scheldt in relation to the hydrography of the area.
    I The Unionidae. *Basteria* 32, 13-47.
Wolff W.J. (1970) The Mollusca of the estuarine region of the rivers
    Rhine, Meuse and Scheldt in relation to the hydrography of the area.
    IV The genus *Sphaerium*. *Basteria* 34, 75-90.
Wolff W.J. (1973) The estuary as a habitat. An analysis of data on
    the soft-bottom macrofauna of the estuarine area of the rivers Rhine,
    Meuse and Scheldt. *Zool. Verh.* 26, 1-242. Brill, Leiden.
Wolff W.J. (1978) The degradation of ecosystems in the Rhine. In:
    Holdgate M.W. & Woodman M.J. (Eds.) *The Breakdown and Restoration
    of Ecosystems*, 496 pp., pp. 169-88. Plenum Press, New York.
Zonneveld I.S. (1960) De Brabantse Biesbosch A study of soil and
    vegetation of a freshwater tidal delta. Ph.D. Thesis, Agricultural
    University of Wageningen. *Versl. Landbouwk Onderz* 65.20, 3 vols.,
    Pudoc Wageningen.
Zuurdeeg B.W. (1980) De natuurlijke chemische samenstelling van Maas-
    water. H_2O 13, 2-7. (The natural chemical composition of River
    Meuse water.)

17 · Glåma

O. M. SKULBERG & A. LILLEHAMMER

17.1 INTRODUCTION

The Glåma is chosen as a representative of the rivers of Fennoscandia, a region of Archaean crystalline rocks in North Europe. Fennoscandia, which received its name from the Finnish geologist Ramsay (1898), has a geological structure and natural environment quite different from that of the neighbouring countries. This chapter is based partly on research carried out in recent years by the authors, but to a larger extent on publications and survey reports.

Fig. 17.1 River Glåma at Orvos - a streaming silver-band in the landscape.

17.2   GENERAL DATA

DIMENSIONS

   Area of catchment   41767 km^2

LAND USE

    agriculture   5%

    forest   34%

    marshes and peatland   5%

    freshwater   4%

POPULATION

    catchment above junction of Glåma and R. Vorma   93000 people

    catchment of R. Gubrandsdalslågen and R. Vorma   205000

    catchment below meeting of the two major tributaries
and Lake Øyeren   100000

    catchment below Lake Øyeren   120000

518000 = approx
12 people km^{-2}

PHYSICAL (AND OTHER) FEATURES OF RIVER

   Discharge etc

location	catchment (km^2)	mean annual discharge $(m^3\ s^{-1})$	forest area	agricultural area	population in catchment
Glåma at Funnefoss	20670	320	7822 km^2	580 km^2	93000
Glåma at Fredrikstad	41767	706	13600 km^2	1934 km^2	518000

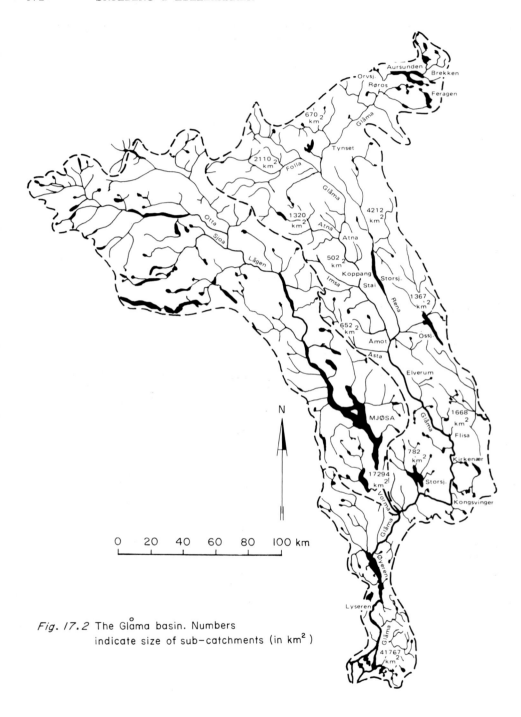

Fig. 17.2 The Glåma basin. Numbers
indicate size of sub-catchments (in km²)

Discharge ($m^3 s^{-1}$) at Solbergfoss, downstream of the Lake Øyeren.
(Data from Glommens og Laagens Brukseierforening)

month	year				mean for
	1958	1969	1970	1980	1955-1964
January	355	406	367	414	395
February	332	379	335	386	380
March	300	309	295	267	354
April	308	475	326	427	419
May	673	1289	1042	948	1094
June	1178	1103	909	1212	1182
July	1132	405	789	993	1060
August	789	401	572	607	817
September	504	379	582	653	784
October	880	463	674	891	747
November	579	388	546	492	581
December	382	383	474	448	408
Mean for year	618	647	576	645	685

Selected waterfalls along the Glåma

name	distance from Lake Rien (km)	approximate vertical fall (m)	mean discharge for 1946-1975 ($m^3 s^{-1}$)
Kuråsfoss	34	48.2	20
Røstefoss	51	11.0	-
Brufoss	72	6.0	-
Erlifoss	77	8.0	50
Barkaldfoss	149	4.0	102
Braskereidfoss	323	7.6	247
Funnefoss	434	11.0	305
Rånåsfoss	457	14.5	628
Bingsfoss	463	3.0	650
Solbergfoss	506	22.6	656
Kykkelsrud	516	25.5	658
Vamma	520	27.5	660
Sarpsfoss	577	20.1	669

17.3   PHYSIOGRAPHIC CONDITIONS

17.31 THE RIVER SYSTEM

The name Glåma includes both the main river in the valley of Østerdalen, and the river below the confluence with the R. Vorma, the main tributary. It extends down to the estuary in the archipelago of Hvaler on the eastern side of the outer Oslofjord (Fig. 17.2). The name is interpreted linguistically to identify the brightness and clarity of its water. River Glåma means the shining river - reflecting much light. And the Glåma is still a streaming silver band in the landscape of East Norway, close to the border between Norway and Sweden (Fig. 17.1). It is the largest water course in Scandinavia with respect to discharge and its catchment comprises 13% of Norway's total land area (see 17.2).

A multitude of small lakes in the mountain area of Røros form the headwaters of the Glåma, but the river is not usually called by this name until downstream of Lake Rien (762 m altitude). From L. Rien the Glåma flows southwards to L. Aursunden. This lake (catchment area 830 km^2, surface area 44 km^2, depth 60 m) is regulated between the 684 and 690 contour lines. The mean discharge of the Glåma at the outlet of L. Aursunden is 20 m^3 s^{-1} (Skulberg, 1967a).

The Glåma proper is divisible into four sections (Fig. 17.3) based on the distribution of waterfalls (Norwegian *foss*); the first 178 km down to Barkaldfoss; the 185 km from Barkaldfoss to Braskereidfoss;

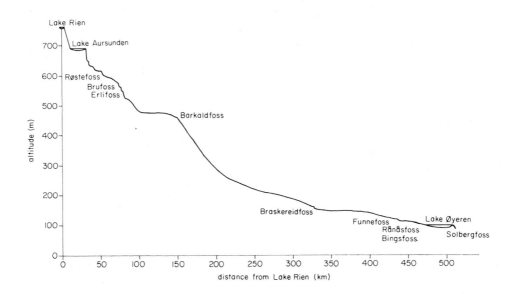

Fig. 17.3 Longitudinal profile of the Glåma (Lake Rien to Solbergfoss), showing major waterfalls.

the 170 km from Braskereidfoss to the outlet of Lake Øyeren; the last
65 km from L. Øyeren to the sea. Information on selected waterfalls is
included in 17.2. In the valley of Østerdalen, the Glåma receives a
number of tributaries, including the Folla, the Atna and the Rena, the
largest of the three. The Rena flows through L. Storsjøen (catchment area 2270 km^2, surface area 48.8 km^2, depth 309 m). Further south
the Glåma broadens its valley. The Flisa and the Opstad are among the
larger tributaries here. The latter drains L. Storsjøen (catchment
area 774 km^2, surface area 42.5 km^2, depth 17 m) in the Odalen district.

The Glåma's catchment area at Funnefoss, before the confluence with
the Vorma, is 20670 km^2, while the catchment of the Vorma-Lågen water-
course is 17294 km^2. After joining with this major tributary - and
almost doubling its size, the Glåma becomes a mighty river. The region
here has a landscape of low hills and small plains. The river is some
250 to 400 m broad and relatively shallow, with a slope of round about
0.39 m km^{-1}. Where the river enters L. Øyeren at Fetsund, it has a
mean discharge of about 650 m^3 s^{-1}.

L. Øyern (catchment area 39964 km^2, surface area 85.2 km^2, depth
70 m) is a typical fjord lake (Strøm, 1935). Due to the large volumes
of water entering the lake, the water mass has a short renewal time.
Based on the assumption that the throughflow occurs only in the epilim-
nion during stagnation periods, the theoretical retention time of L.
Øyeren will vary between 8-17 days, when discharge in the Glåma lies
in the 500-1000 m^3 s^{-1} range.

From L. Øyeren (101 m altitude) down to the estuary there is a series
of four step-like waterfalls. Tributaries are all relatively small and
enter mainly on the east bank; they include the Lekumelva and the Rak-
kestadelva. The Glåma is slow-flowing and almost lake-like in the area
upstream of the waterfall Sarpsfossen. Here the river is divided in
two branches, with much higher current speeds in the eastern, larger
one, which includes the waterfall Sarpsfossen. The western branch has
lake-like extensions (L. Mingevatn, L. Vestvatn, L. Visterflo). The
two branches meet again in the estuary at Greåker.

The estuary of the Glåma, with a salinity gradient 10-20 o/oo, is
of the halocline type. Depending on river discharge and the force of
the tide, the water can reach up to Sarpsfossen and L. Visterflo.
The tidal range is very low, approximately 0.13 m.

17.32 GEOLOGICAL CHARACTERISTICS

The bedrock of the drainage basin dates from the Archaean era (Holtedahl,
1953a; 1960). Granites, gneisses and other hard crystalline rocks are
dominant (Fig. 17.4). Towards the west - in the catchment of the
Gudbrandsdalslågen - the geology is partly characterised by deposits
from Lower Palaeozoic times in the Caledonian mountain range. The
central northern part consists mainly of Eocambrian sediments (thick

Fig. 17.4 East Norway—solid rocks

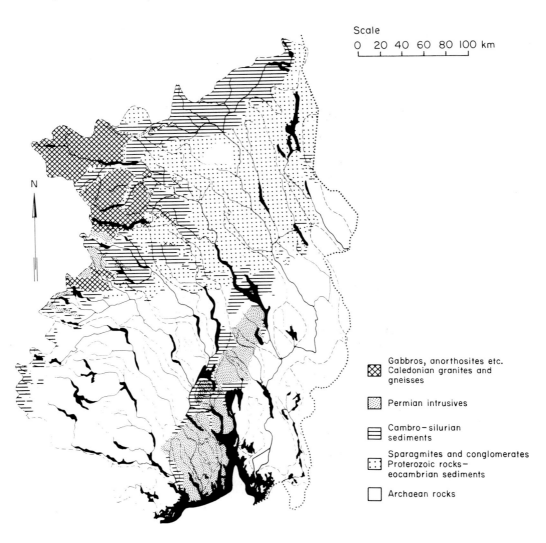

Scale

0   20 40 60 80 100 km

Gabbros, anorthosites etc.
Caledonian granites and
gneisses

Permian intrusives

Cambro-silurian
sediments

Sparagmites and conglomerates
Proterozoic rocks—
eocambrian sediments

Archaean rocks

sandstones, conglomerates etc.) known as sparagmite. Towards the
south, rocks of gneiss and granite predominate, both types showing wide
variation with respect to mineral composition and granular size. This
extremely hard bedrock - belonging to the Fennoscandian Shield - is
about 1800 million years old (dating by radioactive methods).

During the Quaternary period the area was affected by glaciation

Fig. 17.5 Superficial deposits

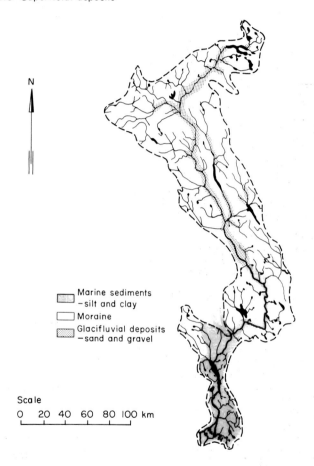

N

Marine sediments
– silt and clay
Moraine
Glacifluvial deposits
–sand and gravel

Scale
0   20 40   60   80 100 km

and glacial deposits are extensively present (Holtedahl, 1953b; Fig. 17.5). Deglaciation and postglacial uplift have been reasonably well elucidated by Scandinavian geologists (John, 1979). The glacial streams from the melting ice masses were heavily loaded with erosion materials and they left important deposits in the landscape.

Above the upper marine limit (in the valley of Østerdalen 190 - 200 m

altitude) terraces, deltas and overflow gorges are numerous. They are
remnants of the ice-dammed lakes between the mountains and the glaciers
(Holmsen, 1915; Holtedahl, 1953b); moraines are particularly common.
Glacial gravel and river deposits of sand and gravel are present in the
valleys.

   In areas, which were situated below sea level during the period of
deglaciation, conditions are different. Deposits - partly morainic and
partly glacifluvial - form series of conspicuous transverse ridges.
Glacifluvial deltas are frequent. The fine material of glacial streams
(clay and silt) was carried away to be deposited at a distance, partly
as varved clay in fresh or brackish water. The delta in the northern
part of Lake Øyeren is still the most considerable area of sedimentation in Norway. Extensive parts of the catchment area below the marine
limit consist mainly of clay (Bjørlykke, 1940). The late Weichselian
deglaciation in the southern part of the Glåma drainage basin has
recently been the subject of careful study (Sørensen, 1979).

17.33 TOPOGRAPHY

The Glåma system drains the central and eastern side of the Scandinavian mountains - the Scandes. The highest places in the catchment area
exceed 2000 m altitude (e.g. Høgronden, Rondeslottet). More than half
lies above 500 m and about a quarter above 1000 m. The inland region
is dominated by valleys with moderate hills and more or less rounded
mountains (locally named *voler*). In the southern part of the drainage
basin a dissected lowland landscape of varying height rises to about
200 m above sea level (Fig. 17.6). Rounded hills (locally named *åser)*
of Archaean rocks and relatively low altitude (Precambrian peneplain)
are characteristic land forms (Fig. 17.7).

17.34 CLIMATE AND HYDROLOGY

The openness of Norway to oceanic influences and the topography of the
country combine to produce outstanding differences in physical features
and climate. This is demonstrated in the drainage basin of the Glåma,
which is relatively narrow, but extends for more than five hundred
kilometres from south to north and an altitude range of about 2000 m.
The region belongs to the Boreal Forest climatic zone, with the mean
temperature in the coldest month below -3 $^{\circ}$C and a snow cover every
year (Birkeland, 1935). The length of the vegetative season, defined
as the number of days between the date in spring when the daily mean
temperature reaches +3 $^{\circ}$C and the date in autumn when the daily mean
again falls to this figure, is a good indication of seasonal variation
of temperature. This period corresponds roughly to the growing season
for grass (Lauscher *et al.*, 1955). The vegetative season at Røros
(673 m altitude) is 7 May - 29 September and at Lillestrøm (104 m
altitude) 11 April - 25 October. (These dates are averages for the
period 1901-1930.)

Fig. 17.6 The Glåma at Braskereidfoss - low relief is predominant. (Photo by Widerøes Flyveselskap A/S)

Fig. 17.7 The Glåma at Bingsfoss during storm flood. Lowland landscape with hills of Archaean rocks. (Photo by Widerøes Flyveselskap A/S)

The monthly means of temperature and precipitation for selected
stations are shown in Fig. 17.8. The Scandes mountain range forms an
effective barrier against the influence of the ocean, and marked conti-
nentality is therefore found in the central and northern part of the
catchment area. Maritimity is experienced to some degree towards the
south, with more wind activity and greater cloud cover.

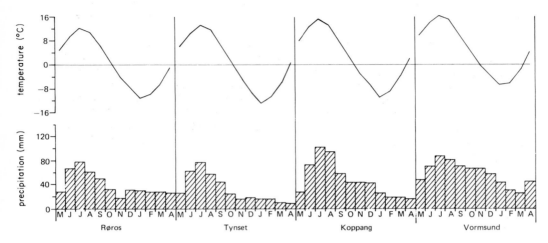

Fig. 17.8 Mean temperature and precipitation for individual months
during the period 1931-1960.

Ice conditions in the lakes give further important information about
the climate, since the key factor in the formation of ice is temperature. The average dates for freezing and break-up of the ice are given
in Table 17.1 for selected lakes. The lakes of the area are typically
cold dimictic-temperate lakes, with circulation in spring and autumn.
Thermal stratification is inverse in winter and direct in summer.

Table 17.1 Duration of the ice cover. Average dates for freezing
and break-up of ice.*

lake	period of observations	date of freezing	date of breaking
Aursunden	1923 - 1950	November 20	June 5
Savalen	1945 - 1950	November 17	May 23
Atnasjøen	1917 - 1950	November 18	May 5
Storsjøen (Rendalen)	1902 - 1950	February 7	May 2
Mjøsa	1863 - 1980	December 10	April 25
Øyeren**	1930 - 1960	December 10	April 20

*Data from Norges vassdrags- og elektrisitetsvesen.
**Local information.

The hydrology of the Glåma can be characterized by the annual variation in discharge (Otnes & Ræstad, 1978). Owing to the precipitation
regime and the storage of snow during winter, rivers of east Norway
have a maximum discharge in spring and a minimum discharge in late
winter and during the summer (Tollan, 1972). Two discharge maxima are
usually experienced during the spring and early summer. "Home flood"
is due to the melting of snow in the lowlands, and "mountain flood"
when melting from snow and glaciers in the highlands starts later on.

The Glåma, however, shows modifications to this general drainage
pattern (see Fig. 17.9). This is due partly to peculiarities of the
catchment and partly to extensive management of water flow for hydro-
electric power production. The spring flood is an unfailing phenomenon,
usually with two maxima. The first corresponds with the flood from the
valley of Østerdalen, the second one with the flood through the valley
of Gudbrandsdalen. This is caused by several circumstances, among
which is the delay to the water of the R. Gudbrandsdalslågen due to its
passage through L. Mjøsa.

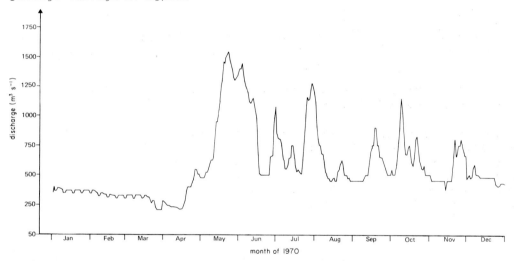

Fig. 17.9 Changes in discharge of the Glåma at Sarpsborg hydrometric station throughout 1970.

The hydrological data included in Section 17.2 are based on
measurements at official hydrometric stations. At Solbergfoss (downstream of L. Øyeren) the spring flood on average peaks on June 1. The
earliest and latest dates are 6th May (1959) and 7th July (1927),
respectively. The average discharge in the period 1911-1950 was 686 m^3
s^{-1}. This corresponds with a discharge per unit area of 17.1 l s^{-1}
km^{-2} and a mean precipitation of 540 mm for the whole Glåma catchment.
Coincidence of floods in the two major valleys can give rise to catastrophic high waters (Tollan, 1967). This has taken place at intervals,
with the years 1740, 1743, 1749, 1754, 1760, 1771, 1773, 1789, 1795,
1850, 1860, 1910, 1916, 1927, 1934, 1966 and 1967 being outstanding.

The largest discharge was experienced during 1789 and 1860, when the peak was 4700 m^3 s^{-1} and 4200 m^3 s^{-1}, respectively (at Solbergfoss, downstream of L. Øyeren). The level of L. Øyeren rose by 11 m and the surface of the lake increased from 35 to 54 km in length (Blakstad, 1939).

17.35 AREA AND SETTLEMENT

The natural environment accounts for the character of the human settlement (Arnesen, 1970; Holtan, 1973). The population is concentrated mainly towards the coast. The inland pattern of distribution is linear, following the valleys that dissect the mountains. Of a total of 518000 inhabitants, approximately 50% live in towns, 10% in town-like agglomerations and 40% are dispersed. Their distribution is evident from data in 17.2. The economy of the Glåma basin is centred on agriculture, forestry and manufacturing industries (Sund, 1960). Although now somewhat less important than urban industries, agriculture has been the dominant factor in forming the pattern of settlement and economic activities in the area.

The Glåma is important as an energy resource. Hydroelectric plants both with a low head and a large flow of water, and plants with a high head and a small flow are used. The first category, the lowland type, is confined to waterfalls in the lower part of the basin. The production of hydroelectricity from the Glåma amounts to 9000 GWh (9 x 10^{12} Wh), approximately 12% of the water-power used at present in Norway.

17.4 CHEMICAL AND BIOLOGICAL QUALITY OF WATER

17.41 WATER CHEMISTRY

The quality of run-off water depends mainly on the chemistry of the precipitation, the bedrock, the plant cover and human activity in the catchment area. The distance from the sea coast has an important influence on precipitation. The prevailing south-western air currents bring moisture laden air of oceanic origin towards Norway and there is a gradient in chloride concentrations from west to east across the Scandes - from approximately 30 to 0.3 mg Cl l^{-1}. The Glåma basin belongs to the part of the gradient with low concentrations.

Precipitation over southern Scandinavia contains large amounts of H^+, SO_4^{2-} and NO_3^- that originate as air pollutants in the industrialized areas of Europe. The inland waters in regions with Archaean rocks and thin cover of unconsolidated glacial till and soil have extremely low buffer capacities. Increasing acidity results in changes in chemical and biological conditions. Interference with fish reproduction and spawning, leading to decline in fish populations, are particularly well-documented (Wright et al., 1976; Drabløs & Tollan, 1980).

The bedrock has a marked influence on the composition of inland waters (Kjensmo, 1966). The drainage water from the hard-schistose and siliceous rocks of the Scandes is very poor in electrolytes and its conductivity is sometimes below 10 μS cm^{-1}. The softer, frequently somewhat calcareous rocks are richer in electrolytes and conductivity lies in the range 100 - 200 μS cm^{-1}. Areas of this type are, however, sparse and restricted in the Glåma basin (Holtan, 1975).

Examples of hydrochemical data are given in Tables 17.2 and 17.3. The highest conductivity values occur during periods of low flow. The percentage of ground water in the Glåma is normally high, and consequently the water will contain a relatively high content of dissolved inorganic substances under such conditions. Daily measurements of hydrographical parameters reveal a complex relationship between meteorological conditions and the composition of river water (Skulberg, 1968). The range for conductivity and turbidity during a 24-hour period can approach that of the yearly range. Important factors influencing the quality of Glåma water are discharge, surface water run-off from cultivated fields, pollution from sewage and transport of seston (Arnesen, 1969).

17.42 WATER FERTILITY

For water management and pollution control it is important to understand changes in algal vegetation of waters receiving nutrients in excess of the natural supply. The use of the algal growth potential test (AGP) for establishing qualitative and quantitative responses to the major plant nutrients has rendered possible effective regional comparisons of water fertility (Skulberg, 1967b; Källqvist, 1973; Kotai et al., 1978). The inland waters of the Glåma basin belong to several categories with regard to water fertility, ranging from a distinct oligotrophic-dystrophic type to an extreme eutrophic type. Based on land use, the predominant type of watershed may be classified into three categories; bioassay determinations of the fertility of various water show marked differences with respect to their potential for supporting algal growth (Table 17.4).

The size of algal crops in the predominantly oligotrophic waters of the Glåma has in most cases been shown to depend on the amount of available phosphorus. In some instances, however, enrichment experiments with essential nutrients have indicated that availability of nitrate as well as orthophosphate was important for the total yield. In such balanced conditions both nitrogen and phosphorus may act as limiting nutrients, either simultaneously or alternately.

Table 17.2 Aqueous chemistry at selected stations in the Glåma. Data of April 1967.

variable	unit	Aursunden	Elverum	Funne-foss	Solberg-foss
temperature	$^\circ$C	1.8	0.2	3.3	2.5
colour	mg l^{-1} Pt	8	40	112	55
turbidity	F.T.U.	0.3	1.1	5.4	8.0
conductivity 20°C	μS cm^{-1}	35	42	29	45
acidity	pH	6.9	6.9	6.5	6.8
magnesium	mg l^{-1}	0.8	1.0	0.8	1.4
calcium	mg l^{-1}	4.4	5.2	2.9	4.3
manganese	mg l^{-1}	0.005	0.016	0.031	0.11
iron	mg l^{-1}	0.02	0.13	0.31	3.3
nitrate	μg l^{-1} N	65	50	265	330
nitrogen (total)	μg l^{-1} N	295	220	525	650
reactive phos-phate	μg l^{-1} P	2	8	3	5
phosphorus (total)	μg l^{-1} P	5	10	15	83
chloride	mg l^{-1}	6.5	1.0	1.3	2.0
sulphate	mg l^{-1} S	0.7	2.0	1.5	2.6
oxygen demand (KMnO$_4$)	mg l^{-1} O$_2$	1.4	6.5	7.9	5.0

Table 17.3 Water chemistry at monthly intervals during 1977 in the Glåma at Solbergfoss.

variable date	colour mg l^{-1} Pt	turbi-dity F.T.U.	conduc-tivity μS cm^{-1} 20°C	pH	nitrogen (total) μg l^{-1} N	phospho-rus (total) μg l^{-1} P
January 9	26.0	0.96	39.5	6.9	500	12
February 19	21.5	0.60	41.1	7.1	420	8
March 13	26.0	0.80	40.9	6.9	470	11
April 17	197.0	6.8	47.0	7.0	640	17
May 22	104.5	3.8	33.2	7.0	490	13
June 12	95.5	2.9	34.2	6.9	400	12
July 18	85.5	1.4	31.3	6.8	270	10
August 5	37.5	0.76	35.0	7.3	230	12
August 25	40.5	0.77	33.8	6.7	270	8
September 25	46.0	1.1	38.4	7.2	370	7
October 23	32.3	1.5	40.9	7.2	390	6
November 18	197	5.8	40.3	6.8	380	15
December 18	92	3.0	37.6	7.2	400	11.5

Table 17.4 Examples of algal growth potential (AGP) and associated water chemistry.

predominant watershed	locality	AGP cells l^{-1} x 10^6	phospho-rus (total) $\mu g\ l^{-1}$ P	nitrogen (total) $\mu g\ l^{-1}$ N	conduc-tivity $\mu S\ cm^{-1}$ 20°C
forest and	Savalen	2	3	50	49
mountain	Ossjøen	3	8	205	16
areas	Storsjøen (Rendalen)	4	6	155	26
	Storsjøen (Odalen)	4.5	8	210	25
rural and	Otta	7	4	130	16
agricultural	Mjøsa	18	10	400	39
areas	Øyeren	20	9	265	35
	Nitelva(Glåma)	145	28	600	48
urban and	Svartelva	156	55	1200	118
industrial	Rømua	965	170	570	115
areas	Nitelva(Kjeller)	1600	174	1590	93

17.5 PLANT AND ANIMAL LIFE

17.51 CLASSIFICATION OF THE NATURAL GEOGRAPHICAL REGION

By use of descriptions of the geology, climate, plant and animal life in the Glåma basin, it is possible to classify the area into a system of natural, geographical regions (Udvardy, 1975). A subdivision of Norden into vegetation zones has been made, based mainly on floristical, phytosociological and ecological principles (Nordic Council of Ministers, 1977). According to this the Glåma basin belongs largely to the boreo-nemoral and boreal zones of the palaearctic realm. The alpine zone is represented in areas above the tree line (600 - 800 m altitude) in the mountain range of the Scandes.

17.52 BENTHIC VEGETATION

17.521 Attached algae

A vegetation rich in species and sometimes abundant occurs in the Østerdalen part of Glåma (Skulberg & Kotai, 1977). More than two hundred algal species have been reported from attached vegetation, about eighty of which may be considered common. These consist primarily of several genera of green algae and diatoms, but species of blue-green and red algae may also be abundant (Table 17.5). Species which frequently

Table 17.5 Common species of benthic algae in the upper Glåma.
++++ abundant

BLUE GREEN ALGAE

Lyngbya sp.
Nostoc cf. *verrucosum*
Oscillatoria cf. *bornetii*
Oscillatoria sp.
Phormidium autumnale
Scytonema sp.
Stigonema mamillosum

GREEN ALGAE

Bulbochaete sp.
Closterium spp.
Cosmarium undulatum
Cosmarium spp.
Draparnaldia glomerata
Microspora amoena +++
Mougeotia spp.
Oedogonium sp.
Scenedesmus obliquus
Scenedesmus spp.
Spirogyra spp.
Staurastrum alternans
S. apiculatum
Staurastrum spp.
Stigeoclonium tenue
Ulothrix zonata +++
Ulothrix spp.
Vaucheria sp.
Zygnema spp. ++++

CHRYSOPHYTES

Hydrurus foetidus ++++

DIATOMS

Achnanthes spp.
Amphora ovalis
Ceratoneis arcus +++
Cymbella cf. *affinis*
Cymbella spp.
Diatoma elongatum
Didymosphenia geminata ++++
Eunotia arcus
Fragilaria construens
Fragilaria sp.
Gomphonema olivaceum
Gomphonema spp.
Meridion circulare
Navicula spp.
Nitzschia palea
Nitzschia spp.
Pinnularia cf. *brevicostata*
P. viridis
Pinnularia sp.
Synedra ulna
Tabellaria fenestrata
T. flocculosa +++

RED ALGAE

Batrachospermum sp.
"*Chantransia chalybea* var.
   leibleinii"
Lemanea fluviatilis ++++

form mass developments are *Zygnema* spp., *Ulothrix zonata*, *Microspora amoena*, *Tabellaria flocculosa*, *Ceratoneis arcus*, *Didymosphenia geminata*, *Hydrurus foetidus* and *Lemanea fluviatilis* (Skulberg & Kotai, 1978). Measurements of conspicuous benthic communities have shown values for mass in the range 50 - 500 g organic dry weight per m^2 of river bottom.

The seasonal variation in the development of attached algae is pronounced, with light climate, temperature and rhythm of discharge as the key factors. Four periods may be distinguished:

November to February: dark part of winter, low flow;
March to April: light part of winter, low flow;
May to mid-July: summer, high flow;
Mid-July to November: summer and autumn, low flow.

During the light part of winter a varied flora of diatoms, *Hydrurus foetidus* and *Ulothrix zonata* are well developed. This is torn away from the substratum by the powerful turbulence of storm flows. During the period of high summer flows, there is a luxuriant vegetation of *Zygnema* spp., *Microspora amoena* and *Lemanea fluviatilis*. The dark part of winter shows a decline in algal crops, but cold-water species may persist for longer or shorter periods.

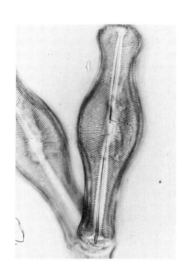

Among the diatoms of quantitative importance *Didymosphenia geminata* (Fig. 17.10) deserves mention. This species may flourish in late summer and, where growth continues undisturbed, the grey and brown colonies reach several centimeters in diameter. In some places this diatom forms a dense cover on boulder surfaces directly facing the current. In rapids which are not ice-covered it can grow throughout the year.

In the lower stretches of the Glåma (downstream of its confluence with the Vorma) current speeds are lower and substrata of sand, clay and mud predominate. The vegetation of attached algae both reflects these conditions and shows a distinct eutrophic trend. Species of *Vaucheria* (e.g. *walzi*) develop typically as cushion-like growths close to the soft substratum. Abundant epipelic algae include *Oscillatoria brevis*, *O. limosa*, *Phormidium uncinatum*, *Fragilaria capucina*, *Nitzschia acicularis* and *Diatoma elongatum*. Various Chlorococcales and green flagellates occur sporadically.

The attached algal vegetation of the Glåma may be classified according to the chief floral component (Israelson, 1949), with an upstream *Zygnema*-type (relatively oligotrophic) and a downstream *Vaucheria*-type (relatively eutrophic). Vegetation of the *Vaucheria*-type has however increased during recent years in some parts of the Østerdalen stretch. This is the result of changes due to pollution and regulation for hydroelectric purposes.

17.522 Bryophytes

Bryophytes have not been the subject of special study. Species identified in connection with algal studies include: *Brachythecium rivulare, Ceratodon purpureus, Dichelyma falcatum, Drepanocladus exannulatus,*

Fontinalis antipyretica, F. dalecarlica, Hygrohypnum ochraceum,
Philonotis fontana, Pohlia albicans, Schistidium alpicola.

17.523 Higher Plants

Although the river may be characterized generally as oligotrophic,
higher plant vegetation is abundant in some areas, especially on finer
sediments (Rørslett, 1972; Arnesen, 1970). Some indication of the
relative abundance of the various species is given in Table 17.7.

The most common long-shoot plants are species of *Potamogeton* and
Sparganium (Table 17.6). *Myriophyllum alterniflorum* is typical of the
more oligotrophic parts of the water course. Isöetids are usually predominant in shallow parts of the lower reaches. The zone between high
and low water is colonized by *e.g. Eleocharis acicularis, Juncus fili-*
formis, Subularia aquatica and *Ranunculus reptans.* A helophyte vegetation can grow abundantly on fine-grained river beds: prominent species are *Equisetum fluviatile, Carex aquatilis* and *Carex rostrata.*

Vegetation dominated by *Phragmites australis* and/or *Schoenoplectus*
lacustris occurs only occasionally, except in areas of slow-moving
water upstream of the estuary (broads and lake-like extensions of the
river). Here they are the predominant species, with the associated
plants being ones indicative of highly eutrophic conditions (*e.g.*
Ceratophyllum demersum, Myriophyllum verticillatum). The brackish-
water area of the estuary has a vegetation of freshwater and marine
species (Rørslett, 1974). Among those which are abundant include
Potamogeton perfoliatus and two species of *Ruppia* (Rørslett, 1975).

17.53 SESTON - ESPECIALLY PHYTOPLANKTON

The term "seston", as used by Kolkwitz (1912), means the particulate
mass of various living and non-living substances in water. The composition of the suspended matter in the Glåma is very complex, depending
on the geology, climate and biology of the catchment area, and on the
biology of the river itself. Man's activities greatly influence the
amount of suspended material. Observations on the seston, and especially the phytoplankton, in the lake-river system Lake Mjøsa - Lake
Øyeren - R. Glåma have been carried out extensively by the Norwegian
Institute for Water Research (Lindstrøm et al., 1973). Several methods
have been used for the measurement and identification of the seston
(Baalsrud & Henriksen, 1964; Skulberg, 1978).

The phytoplankton of the Glåma is dominated mainly by diatoms and
chrysophytes. A comparison of fluctuations in seston and diatoms in
the water column upstream of Sarpsfossen during one year (1970) with
fluctuations in discharge are shown in Fig. 17.11. Seston concentrations were highest just as the spring flood started to develop.
Erosion in the lowland parts of the catchment gives rise to increased

Table 17.6 Examples of vascular plants in the upper, lower and
estuarine Glåma.   ++++, abundant; +++, frequent; ++, occasional;
+, rare.

species	upper	lower	estuary
Equisetum fluviatile	+++	++++	
Isoetes echinospora	++	+++	
Sparganium angustifolium	++	+	
S. ramosum	+	++	++
S. simplex	+	++	
Potamogeton alpinus	++		
P. gramineus	++	+	
P. perfoliatus	++	+++	++
Alisma plantago-aquatica	++	+++	
Sagittaria sagittifolia	+	+++	
Phragmites australis	++	+++	++++
Phalaris arundinacea	++	+++	++
Scirpus lacustris	++	+++	
S. tabernaemontanii			++++
Eleocharis acicularis	+++	++++	
E. palustris	+++	+++	
E. uniglumis	+	+	+++
Carex aquatilis	++	+++	
C. gracilis	+++	++++	
C. rostrata	++++	+++	
Acorus calamus		++	
Juncus filiformis	+++	++++	
Ceratophyllum demersum		+++	++
Ranunculus peltatus	++	+++	
R. reptans	+++	++++	
R. trichophyllus	++		
Subularia aquatica	+++	++++	
Crassula aquatica	++	+++	
Callitriche spp.	++	++++	
Elatine hydropiper	++	+++	
E. triandra	+	+++	
Peplis portula	+	+	
Myriophyllum alterniflorum	+++	++	
M. verticillatum		++	++
Hippuris vulgaris	++	+	
Cicuta virosa	++	++	
Limosella aquatica	+++	+++	
Bidens tripartita	++	+++	

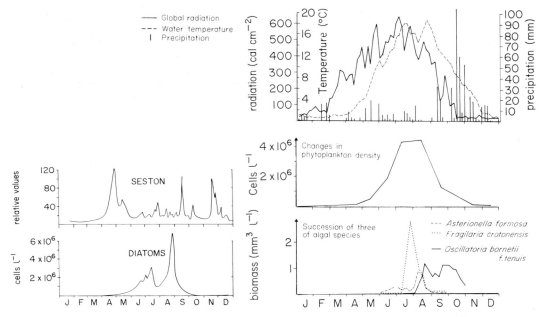

Fig. 17.11 Changes in relative amounts of seston and population density of diatoms at Sarpsborg throughout 1970. (Compare with Fig. 17.9 which shows discharge at the same site and through the same year)

Fig. 17.12 Comparison of climatic conditions (radiation and precipitation), water temperature, phytoplankton population density and volume of selected species in the Glåma at Solbergfoss throughout 1976.

amounts of clay and silt in the river water during this period. Two species of diatoms dominated the plankton community. *Asterionella formosa* was largely responsible for the early summer burst (June-July) and during the late summer peak (August-September) *Fragilaria crotonensis* exceeded *Asterionella* in abundance.

In the lowland part of the river, the species composition (Table 17.7) reflects the increased fertility of the water resulting from agricultural drainage and sewage effluents. Nutrient enrichment leads both to an increase in the blue-green algal crop and in its species composition. During August and September 1976 the first bloom of *Oscillatoria bornetii* f. *tenuis* took place in the water system L. Mjøsa-R. Glåma. The population density of this species increased during August, although diatoms were still the dominant. By the end of September the density of blue-green alga was considerable (L. Mjøsa, approximately 2 g m^{-3}: Holtan, 1979) and it dominated the planktonic vegetation during late autumn (October-November). The development of this population resulted in an obnoxious and abnormal taste and odour

of the water in the lake-river system. Investigations made it clear
that the alga was involved, directly or indirectly, in the establishment
of conditions leading to taste and odour problems in the waterworks
using L. Mjøsa-R. Glåma as raw water supply. Among the substances
associated with the *Oscillatoria* bloom was geosmin (trans-1,10 dimethyl-
trans-9-decalol) giving the water a very distinct and offending flavour.

Table 17.7 Common planktonic algae in the lower Glåma.
++++ abundant

BLUE-GREEN ALGAE

Anabaena flos-aquae ++++
Gomphosphaeria lacustris
Merismopedia tenuissima
Oscillatoria bornetii f.
    tenuis ++++

GREEN ALGAE

Ankistrodesmus falcatus
Chlamydomonas spp.
Closterium pronum
Crucigenia tetrapedia
Dictyosphaerium pulchellum ++++
Gemellicystis neglecta
Scenedesmus spp.

CHRYSOPHYTES

Chrysococcus spp.
Dinobryon borgei
D. divergens
D. sociale
Kephyrion spirale
Mallomonas elongata ++++
M. reginae ++++
Synura spp. ++++

DIATOMS

Asterionella formosa ++++
Diatoma elongatum ++++
D. vulgare
Fragilaria crotonensis ++++
Cyclotella stelligera
C. comta
Melosira granulata var.
    angustissima
M. islandica ssp. *helvetica*
M. italica
Rhizosolenia eriensis ++++
R. longiseta ++++
Stephanodiscus hantzschii
Synedra acus
S. ulna
Tabellaria fenestrata var.
    asterionelloides ++++
T. flocculosa
T. fenestrata

17.54 BENTHIC INVERTEBRATES

Studies of the benthos of stony substrata at eleven stations from the
alpine zone down to the Lake Øyeren area have shown that insect nymphs
and larva completely dominate the fauna. Chironomids are most numerous,
while Trichoptera have the largest biomass. There are, however, marked
longitudinal differences in the faunistic composition. Plecoptera are
best represented in the upper stretches, whereas Ephemeroptera and
Trichoptera are most abundant further downstream, where pollution from

sewage and the influence of flow regulation increase the benthic algal crops. At the station above L. Øyeren, where pollution is prominent, Chironomids dominate entirely (Fig. 17.13). Reservoirs influence the faunistic composition by increasing the abundance of filter feeding Trichoptera, *e.g.* below the Barkaldfoss and Rånåsfoss reservoirs (Lillehammer & Saltveit, 1980).

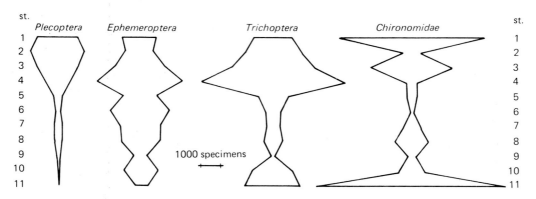

Fig. 17.13 Longitudinal changes in main insect groups on passing down the Glåma (from stations 1 to 11).

Nymphs of the stoneflies *Capnia pygmaea* and *C. atra* occur in especially large numbers in the Østerdalen part of the Glåma, forming the basis for local ice fishing for grayling and whitefish each April. The nymphs crawl in profusion onto the river ice ready for emergence. They are then collected by fishermen, who use them as bait (Jensen, 1968; Lillehammer, 1971).

Blood-sucking blackflies are a considerable nuisance for humans and livestock in the area of Sarpsborg in the lower Glåma. A notorious example is the large numbers of *Simulium truncatum* in a tributary connected to a dam spillway with lacustrine conditions above. Development during the spring and summer floods benefits from the rich supply of organic seston in the river water. Raastad (1974) estimated that more than a billion flies emerge from the rapids in this tributary during swarming.

17.55 FISH

The freshwater fish recorded from the Glåma water-course comprise 28 species (Huitfeldt-Kaas, 1918; Halvorsen, 1968; Hansen, 1978; Pethon, 1974, 1978, 1980). The fauna is especially rich in the lower river reaches - *e.g.* the L. Øyeren area with 24 species and the river below L. Øyeren with 27 species (Table 17.8).

Table 17.8 Freshwater fish in different parts of the Glåma.
Records: x, frequent; (x), occasional.

species	upper	middle sub-section one	middle sub-section two	L. Øyeren	lower	estuary
sea lamprey					(x)	x
lamprey		x	x	x	x	
char	x					
trout	x	x	x	x	x	x
whitefish	x	x	x	x	x	x
cisco				x	(x)	(x)
grayling	x	x	x	x	x	
salmon					(x)	x
smelt			x	x	x	x
pike	x	x	x	x	x	(x)
roach		x	x	x	x	x
dace			x	x	x	x
chub			x	x	x	(x)
ide			x	x	x	x
rudd					x	(x)
minnow	x	x	x	x	x	
asp			x	x	x	(x)
bleak				x	x	(x)
bream			x	x	x	x
white bream				x	x	
crucian carp				x	x	
ten-spined stickleback				x	x	x
three-spined stickleback					x	x
burbot	x	x	x	x	x	(x)
perch	x	x	x	x	x	x
pikeperch				x	x	(x)
ruffe			x	x	x	x
bullhead	x	x	x	x	x	

The water-course is favourable for grayling and trout over an extensive stretch. From the alpine zone down to the Østerdalen stretch of
the Glåma, grayling, bullhead and minnow are the most common species in
riffles, with smaller populations of trout and burbot (Borgstrøm *et al.*,
1976). Further downstream, pike, perch, whitefish and some cyprinids are
common, and productive fishing grounds are found in L. Øyeren and elsewhere. The Atlantic salmon (*Salmon salar*) occurs in the Glåma below
Sarpsfossen and in the tributary, R. Ågårdselv. The population is small

in recent years (e.g. Halvorsen, 1968), but at the beginning of this
century catches could be of some economic importance (an annual average
of 3500 kg during 1878-1882: Kiær, 1915). Twelve freshwater species are
present in the estuary, with ide, roach, perch, dace, bream and ruffle
being those most commonly observed. In this area it is possible to
fish freshwater fish in the outstreaming, surface water, and saltwater
fish in the inward-streaming bottom (marine) water.

17.6 THE LAKE-RIVER SYSTEM

The Glåma can be divided on faunal criteria into three main parts:

<u>Upper</u>   above the tree line, i.e. the alpine zone.
<u>Middle</u> from the tree line down to the L. Øyeren area.
<u>Lower</u>   downstream of L. Øyeren area, via estuary to the sea.

<u>Upper</u>. Studies of Norwegian streams have shown that the benthic fauna
decreases in species above the tree line, and most of the species
present are less abundant than downstream (Lillehammer, 1974; Lillehammer et al., 1980). The reduction is marked in most of the benthic
groups and filter-feeding Trichoptera larvae are rarely recorded above
the tree-line. Only nine species of fish have been recorded in the
upper Glåma (Table 17.8).

<u>Middle</u>. This part can be divided into two subsections, above and down-
stream of the central area of the Østerdalen valley. In the former,
Plecoptera are abundant and grayling and bullhead are the most common
species of fish; only two species of cyprinid have been recorded,
minnow and roach. The benthic fauna of the lower subsection is dominated by Ephemeroptera and Chironomidae. Plecoptera are sparse and
the number of cyprinid species increases gradually on passing downstream.

<u>Lower</u>. Two species of fish not recorded in the other parts of the
Glåma are present below L. Øyeren. These are the Atlantic salmon and
the rudd. In the estuary, besides the twelve species of freshwater
fish (see above), there are eighteen marine species (Pethon, 1980).
Insects are sparsely represented in the benthos, except for chironomids.

    The changes in plankton composition are demonstrated downstream from
L. Mjøsa through the Vorma and Glåma, L. Øyeren and the estuary and out
to the sea of the Hvaler archipelago. Diatoms are in general predominant, but blue-green algal blooms can occur (17.53). In L. Øyeren
flagellates are relatively more frequent (Skulberg, 1972). There is a
distinct similarity between the plankton of L. Mjøsa and that of the
Vorma and Glåma down to the inlet of L. Øyeren, with *Asterionella
formosa*, *Fragilaria crotonensis*, *Diatoma elongatum*, *Tabellaria fenestrata* var. *asterionelloides* and *Rhizosolenia* spp. the dominants. These
species also characterize to a large extent the plankton down to the
estuary (Lindstrøm et al., 1973). This shows that the populations

developing in an upstream lake can influence the composition of plankton in the whole water-course downstream.

REFERENCES

Arnesen R. (1969) *En Undersøkelse av Glåma i Østfold. Kjemiske og Bakteriologiske Forhold.* 0-217. Norsk Institutt for Vannforskning, Oslo. (A survey of the Glåma in Østfold. Chemical and bacteriological situation.)

Arnesen R. (1970) *En Undersøkelse av Glåma i Østfold. Sammenfattende del.* 0-217. Norsk Institutt for Vannforskning, Oslo. (A survey of the Glåma in Østfold. Summary.)

Birkeland B.J. (1935) *Mittel und Extreme der Lufttemperatur.* Geofysiske Publikasjoner, Vol. XIV, No. 1, 155 pp. Det Norske Videnskaps-Akademi, Oslo.

Bjørlykke K.O. (1940) *Utsyn over Norges jord og jordsmonn.* Nor. Geol. Unders. 156, 235 pp. Aschehoug, Oslo. (Overview of Norwegian soil conditions.)

Blakstad W. (1939) *Tømmerfløtingen i Nedre Glomma gjennom de siste hundre år.* 326 pp. Thorvald Nielsens Boktrykkeri, Porsgrunn. (Timber floating in the lower Glåma over the last hundred years.)

Baalsrud K. & Henriksen A. (1964) Measurement of suspended matter in stream water. *J. Am. Wat. Wks Assoc.* 56, 1194-200.

Borgstrøm R., Brittain J.E. & Lillehammer A. (1976) Østerdalsskjønnet. Glåma mellom Auma og Høyegga. Virkning på fisket. *Rapp. Lab. Ferskv. Økol. Innlandsfiske* 25, 1-6. (Effect of stream regulation on fisheries.)

Drabløs D. & Tollan A. (Eds) (1980) *Ecological impact of acid precipitation.* Proceedings of an international conference, Sandefjord, Norway, March 11-14 1980. 383 pp. SNSF-project, ÅS-NLH.

Halvorsen O. (1968) Resultater av månedlige garnfangster i Glomma ved Sarpsborg. *Fauna* 21, 166-74. (Result of monthly net catches in the Glåma at Sarpsborg.)

Hansen L.-P. (1978) Forekomst og fordeling av noen fiskearter i Nordre Øyeren. *Fauna* 31, 175-83. (Occurrence and distribution of some fishes in northern Øyeren.)

Holmsen G. (1915) Brædæmte sjøer i Nordre Østerdalen. *Nor. Geol. Unders.* 72. (Glacier - dammed lakes in northern Østerdalen.)

Holtan H. (1973) *Glåma i Hedmark. Undersøkelser i Tidsrommet 1966-1972.* 0-138/70. Norsk Institutt for Vannforskning, Oslo. (Glåma in Hedmark. Surveys in the period 1966-1972.)

Holtan H. (1975) *Gudbrandsdalsvassdraget, Mjøsa, Vorma. Resipient-Undersøkelser i Forbindelse med Planlagte Vassdragsreguleringer 1974-1975.* 0-151/73. Norsk Institutt for Vannforskning, Oslo. (Gudbrandsdal river-system, Mjøsa, Vorma. Recipient surveys in connection with planned stream regulations 1974-1975.)

Holtan H. (1979) The Lake Mjøsa story. *Arch. Hydrobiol. Beih. Ergebn. Limnol.* 13, 242-58.

Holtedahl O. (1953a,b) *Norges Geologi.* Bind I, pp. 1-583. Bind II, pp. 585-1118. Aschehoug, Oslo.

496   SKULBERG & LILLEHAMMER

Holtedahl O. (1960) *Geology of Norway.* Nor. Geol. Unders. 208. 580 pp. Aschehoug, Oslo.

Huitfeldt-Kaas H. (1918) *Ferskvandsfiskenes Utbredelse og Indvandring i Norge.* 106 pp. Kristiania (Centraltrykkeriet). (Freshwater fish in Norway and their immigration.)

Israelson G. (1949) On some attached Zygnemales and their significance in classifying streams. *Bot. Not.* 4, 313-58.

Jensen K.W. (1968) Abbor. In: Jensen K.W. (Ed.) *Sportsfiskerens Leksikon I*, pp. 1-11. Gyldendal, Oslo. (Perch.)

John B. (Ed.) (1979) *The winters of the world, earth under the Ice Ages.* 256 pp. David & Charles, London.

Källqvist T. (1973) Algal assay procedure (bottle test) at the Norwegian Institute for Water Research. *Nordic Symposium on Algal Assays in Water Pollution Research, Oslo 1972*, pp. 5-17. Nordforsk Publication 2, Helsinki.

Kiær A.Th. (1915) *Smaalenenes Amt 1814-1914.* 543 pp. E. Sem, Fredrikshald.

Kjensmo J. (1966) Electrolytes in Norwegian lakes. *Schweiz. Z. Hydrol.* 28, 29-42.

Kolkwitz R. (1912) Plankton und Seston. *Ber. deutsch. Bot. Gesell.* XXX, 334-46.

Kotai J., Krogh T. & Skulberg O. (1978) The fertility of some Norwegian inland waters assayed by algal cultures. *Mitt. int. Verein. theor. angew. Limnol.* 21, 413-36.

Lauscher A., Lauscher F. & Printz H. (1955) *Die Phänologie Norwegens. Teil I. Allgemeine Übersicht.* Skriftet utgitt av Det Norske Videnskaps-Akademi i Oslo. I. Mat.-Naturv. Klasse, 1955. No. 1, 99 pp. Jacob Dybwad, Oslo.

Lillehammer A. (1971) Døgnfluer. Øyestikkere. Steinfluer. *Norges Dyr* 4, 217-230. Cappelen, Oslo. (Norwegian fauna - ephemeras, dragon-flies and stoneflies.)

Lillehammer A. (1974) Norwegian stoneflies, II. Distribution and relationship to the environment. *Norsk ent. Tidsskr.* 21, 195-250

Lillehammer A., Borgstrøm R. & Skulberg O. (1980) Miljøpå-virkninger i norske vassdrag. 1. Økosystembeskrivelse. *Fauna* 33, 109-116. (Environmental impacts on Norwegian water-courses. Ecosystem description.)

Lillehammer A., Brittain J. & Borgstrøm R. (in prep.) Bunnfauna og fiskeregistreringer på strykstrekninger i Glåma. (Bottom fauna and fishery registrations in the fast-flowing stretches of Glåma.)

Lillehammer A. & Saltveit S.J. (1980) Stream regulation in Norway. In: *The Ecology of Regulated Streams*, pp. 201-213. Plenum Publishing Corporation. ...s in tne rast-riowing stretches of Glama.)

Lindstrøm E.A., Skulberg R. & Skulberg O.M. (1973) Observations on planktonic diatoms in the lake-river system Lake Mjøsa - Lake Øyeren - River Glåma, Norway. *Norwegian J. Bot.* 20(2/3), 183-95.

Nordic Council of Ministers (1977) *Naturgeografisk Regionindelning av Norden*, 130 pp. NU B 1977:34. (Nature geographical classification of regions in the Nordic countries.)

Otnes J. & Ræstad E. (1978) *Hydrologi i Praksis*, 314 pp. Ingeniør-forlaget, Oslo. (Hydrology in Practice.)

Pethon P. (1974)   Fiskeundersøkelsene 1973. In: Pethon P. (Ed.)
Øraundersøkelsene årsrapport 1973, pp. 139-52. Zoologisk museum,
Oslo. (Fisheries survey for 1973.)

Pethon P. (1978)   Age, growth and maturation of natural hybrids between
roach (*Rutilus rutilus* L.) and bream (*Abramis brama* L.) in Lake
Øyeren, SE Norway. *Acta Hydrobiol.* 20, 281-95

Pethon P. (1980)   Variations in the fish community of the Øra Estuary,
SE Norway, with emphasis on the freshwater fishes. *Fauna norv. Ser.
A* I, 5-14

Ramsay W. (1898)   Über die geologische Entwicklung der Halbinsel Kola
in der Quartärzeit. *Fennia* 16(1), 4. Helsingfors

Rørslett B. (1972)   *Resipientforholdene i Romeriksvassdragene Nitelva,
Leira og Rømua. II. Botaniske undersøkelser.* 0-55/68. Norsk
Institutt for Vannforskning. (Recipient conditions in the Romerike
rivers, Nitelva, Leira and Rømua.)

Rørslett B. (1974)   *Hydrobotaniske forhold i Øra-omradet ved Fredrikstad.* 0-50/73. Norsk Institutt for Vannforskning. (Hydrobotanical
conditions in Øra region at Fredrikstad.)

Rørslett B. (1975)   *Potamogeton perfoliatus* i Øra, et brakkvannsområde
ved Fredrikstad. *Blyttia* 33, 69-82. (*Potamogeton perfoliatus* in
the Øra, a brackish water area at Fredrikstad.)

Raastad J.E. (1974)   Outbreaks of blood-sucking black flies (Simullidae)
in Norway. In: *Proc. Third International Congress for Parasitology*
2, 918-919. München

Skulberg O.M. (1967a)   *Beskrivelse og Undersøkelser av Vannforekomster.
Glåma.* 0-110/65. Norsk Institutt for Vannforskning. (Descriptions
and surveys of water in the Glåma.)

Skulberg O.M. (1967b)   Algal cultures as a means to assess the fertilizing influence of pollution. *J. Wat. Pollut. Control Fed.*, 113-27.
Washington D.C.

Skulberg O.M. (1968)   Some experimental investigations of self-purification processes. *Grundförbättring* 21, 25-37

Skulberg O.M. (1972)   *Resipientforholdene i Romeriksvassdragene Nitelva,
Leira og Rømua.* Hovedrapport I, 0-55/68. Norsk Institutt for
Vannforskning, Oslo. (Recipient conditions of the Romerike water
courses, Nitelva, Leira and Rømua.)

Skulberg O.M. (1978)   Sestonobservasjoner ved vassdragsundersøkelser.
Fauna 31, 48-54. (Seston observations in connection with the investigation of water.)

Skulberg O.M. & Kotai J. (1977)   *Utredning om Begroingsforhold og
Vannkvalitet for Østerdalsskjønnet. Vassdragsstrekningen fra Stai
til Samløp med Rena. Undersøkelsene i Vegetasjonsperioden 1977.*
Norsk Institutt for Vannforskning, Oslo. (Survey of benthic vegeta-
tion and water quality in Østerdal. Stretch from Stai to Rena.
Investigations during the growth period of 1977.)

Skulberg O.M. & Kotai J. (1978)   Miljøfaktorer og algeutvikling i
strømmende vann. Noen observasjoner av innvirkningene av vassdragsreguleringer på begroingsforhold i Glåma i Østerdalen. In: *Norsk
Institutt for Vannforskning, Årbok 1977*, pp. 63-73. Norsk Institutt
for Vannforskning, Oslo. (Environmental factors influencing algal
development in flowing waters. Some observations on the implication
of water regulation on the state of growth in the Glåma in Østerdal.

Strøm K. (1935)   Sammensatte botnsjøer og litt om geomorfologiske
   sjøtyper i Norge. *N. Geogr. T.* 5, 334-341. (Glacier-eroded lakes
   and geomorphological lake types in Norway.)

Sund T. (1960) Norway. In: Sømme A. (Ed.) *A Geography of Norden*,
   363 pp. J.W. Cappelens Forlag, Oslo.

Sørensen R. (1979) Lake Weichselian deglaciation in the Oslofjord
   area, south Norway. *Boreas* 8, 241-246.

Tollan A. (1967) Når vannet stiger. Om flom og flomberegninger.
   Elektroteknisk Tidsskrift 32, 1-6. (When water rises: about floods
   and flood estimation.)

Tollan A. (1972) Variability of seasonal and annual runoff in Norway.
   Nordic Hydrology 3, 72-79.

Udvardy M. (1975) *A Classification of the Biogeographical Provinces of
   the World*. IUCN Occasional Paper No. 18, 48 pp. International
   Union for Conservation of Nature and Natural Resources, Morges,
   Switzerland.

Wright R.F., Dale T., Gjessing E.T., Hendrey G.R., Henriksen A.,
   Johannessen M. & Muniz I.P. (1976) Impact of acid precipitation on
   freshwater ecosystems in Norway. *Wat. Air Soil Pollut.* 6, 483-99.

18 · Dunajec

B. KAWECKA & B. SZCĘSNY

18.1   INTRODUCTION

The Dunajec* has the most beautiful landscape of any mountain river in
Poland and is the most popular with tourists. This landscape includes
the extraordinarily picturesque mountains around the springs of the
tributaries in the Tatra Mountains and the famous gorge in the Pieniny
Mountains, where it is traditional to float on rafts down the river.
These rides, which have gone on for more than a century, offer the parti-
cipants unforegettable and exciting moments. It is not surprising,
therefore, that the river has been the subject of interest for naturalists
over the past century and a half. Detailed hydrobiological studies on
the Dunajec were however initiated on a wide-scale by Karol Starmach
only in the 1960s. Although some aspects are now understood quite well,
information about others is of only a preliminary and inventory nature.
This is because of the numerous methodological difficulties which have
been encountered and the lack of specialists for many taxonomic groups.

Although the river is only 251 km long and the direct distance from
source to its entry to the Vistula (Wisła) is 120 km, it flows through
a wide range of geographical and climatic conditions and geological
formations. As a consequence there are marked changes in character on
passing downstream. In the upper part of the catchment there are high
mountain streams with gradients exceeding $400^o/oo$ separated by young
post-glacial lakes. In contrast the downstream part of the river has
almost a lowland character. Between these there are all the intermediate forms associated with the natural development of a river. The
variety of biotopes and the history of the region result in communities
of aquatic organisms with diverse ecologies and which frequently
include endemic post-glacial relic species. The upper section of the
Dunajec and most parts of the tributaries are still of an entirely
natural character.

There are three national parks within the 70% of the catchment which
lies within Poland. These are: the Tatra National Park, the Pieniny
National Park and the Gorce National Park. Because of its striking
landscape and scientific interest, the Dunajec was included in "Project
AQUA" (Luther & Rzóska, 1971). The river is also of great economic
importance for Poland. In its mid-section there are two on-stream
reservoirs with dams, which are also used for recreational purposes
and production of hydro-electric energy. Other reservoirs are planned
or under construction, including ones on tributaries.

Plans for construction of reservoirs totalling 2.32×10^8 m^3 capacity close to the boundary of the Pieniny National Park encountered
strong opposition not only from scientists but much of society, because
they endangered the unique nature of the Pieniny Mountains.

---

*English transliteration = Dunaiyetz

18.2   GENERAL DATA

DIMENSIONS

Area of catchment	6798 km^2
Length of river	251 km
Highest points	2663 m Gerlach Mt in Tatras; 1310 m Turbacz Mt in Beskids
Altitude at source	1540 m Chochłowski Stream springs in Wolowiec Mt.

CLIMATE

Climate associated with vertical zones in Polish Northern Carpathians. Data calculated from 10 year period (from Hess, 1965)

zones	Upper limit (m)		Mean temperature (°C)			Precipitation mean ann. (mm)	Snow cover occurrence mean ann. days	depth mean max. (cm)	Cloudiness mean ann. days sunny	cloudy
	Tatras	Beskids	annual	Jan.	July					
Cold	2665		- 4.0	- 12.0	4.0	1625	290	> 200	45	140
Temperate cold (climatic snow line)	2200		- 2.0	- 10.0	6.0	1750	250	> 200	35	160
Very cool	1850		0.0	- 8.5	8.2	1800	215	> 200	40	165
Cool (upper forest limit)	1550		2.0	- 6.0	10.5	1600	180	> 120	45	145
Temperate cool	1150	1100	4.0	- 5.5	13.0	1400	140	> 60	40	160
Temperate warm		750	6.0	- 4.0	15.5	1000	105	> 25	55	175
		280	6.2	- 8.5	16.5	500	105		20	110

POPULATION

1.4 x 10^6 people

PHYSICAL FEATURES OF RIVER

<u>Flow</u> in m^3 s^{-1} (Data from Informacje G.U.S., 1979)

km	absolute minimum	mean of minima	mean	mean of maxima	absolute maximum
18	9.2	15.6	84.0	1210	3500
150	3.0	6.8	30.1	448	1240

VEGETATION ZONES IN POLISH NORTH CARPATHIANS

zone	main associations	mean upper limit (m) Tatras granite	limestone	Beskids
Supra-alpine (subnival)	Distichetum subnivale	2663		
Alpine zone meadows	Trifido-Distichetum Seslerion Tatrae	2300	2250	
Mountain pine zone, *Pinus mughus*	Pinetum mughi carpaticum	1850	1800	
Upper mountain forest zone	Piceetum tatricum	1650	1590	1310
Lower mountain forest zone	Fagetum carpathicum		1250	1150
Foothills	Tilio-Carpinetum, Pino-Quercetum			550

18.3   THE CATCHMENT

18.31   GEOGRAPHY

The Dunajec is the right-bank mountain tributary of the River Vistula. Almost all its catchment lies within the Northern Carpathians. The catchment includes the highest massif of the Carpathian range, the Tatra Mountains, and some minor ranges of the West Beskid Mountains (18.2). The source of the river is regarded as the head-spring of the Chochołowski stream. This stream joins with the Kościeliski stream at the base of the Tatra Mountains to form the Czarny Dunajec. This in turn joins with the Biały Dunajec to form the Dunajec proper (Fig. 18.2). The Poprad, the largest tributary and over half of whose length lies in Czechoslovakia, drains the highest granite parts of the Tatra Mountains, the High Tatras.

   The lowest part of the drainage basin of the Dunajec lies outside the Carpathians and constitutes part of the Sandomierska Basin. The Dunajec discharges to the Vistula at an altitude of only 170 m.

Fig. 18.1   The Dunajec near Krościenko in the Beskid Mountains.

18.32   PHYSIOGRAPHY

The highest part of the catchment is of an alpine character (18.2).

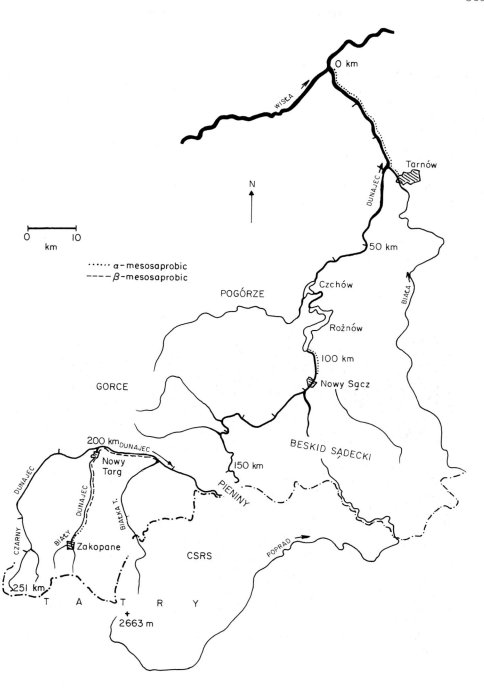

Fig. 18.2 The Dunajec and its tributaries.

It is an area of bare, sheer granite rocks with glacial action evident in the form of eroded valleys and numerous cirque lakes - the High Tatras. The unitary gradient of the streams is often considerable and may reach 600°/oo. Falls are a common phenomenon. The West Tatras lacks post-glacial lakes, in spite of similar post-glacial activity in the past. Instead, there are karst phenomena associated with an accumulation of limestones.

The prevailing character of the catchment as a whole is of mountains of medium height, with softly inclined slopes and afforested tops. The lowest part of the catchment has a typical lowland and agricultural character.

18.33 METEOROLOGY

The catchment of the Dunajec is characterized by a great variety of climatic type, with six zones distinguished (see 18.2). The difference between the mean annual air temperature in the lowest and highest parts of the catchment area is 12 $^\circ$C (+8 $^\circ$ to -4 $^\circ$C), a decrease of 0.5 $^\circ$C per 100 m of altitude. The snow-line lies at 2300 m in the Tatra Mountains.

Thermal inversion is a relatively frequent phenomenon in the valleys. The highest precipitation and the greatest number of cloudy days occur between 1550 and 1850 m. There is a relatively high precipitation and a large number of cloudy days in July (Hess, 1965). The climate of the Pieniny Mts (Romer, 1949) is slightly warmer and drier than that of the adjacent mountain ranges, as may be expected from their lower elevation.

18.34 GEOLOGY

The bed-rock of the majority of the catchment is formed by the Carpathian Flysch or the thick series of sediments of the Magurian sandstones. The highest part of the catchment, lying within the limits of the Tatra Mts, is built of crystalline rocks or granite. Around this granite trunk there are various sediment formations from Permian to cretaceous. Calcareous rocks form the east part of the High Tatra (the Bielskie Tatra), some parts of the West Tatra and the Pieniny Mts.

From the foot of the Tatra Mts down to the Pieniny Mts, the Dunajec flows over glacial formations (loams, sands) derived mostly from the Tatra range. The river then falls into the gorges of the Pieniny and Beskid Mts. Below Nowy Sącz the river flows over alluvial sediments of its own origin combined with diluvial ones from the glaciation of the Cracovian Stage (= Mindel). Information on soils of the catchment are given by Pasternak (1968).

18.35   LAND USE: AGRICULTURE AND FORESTRY

Over half the catchment area is under agricultural utilization, the
greatest part of which is arable land, though there are also meadows
and pastures. Forests cover 35% of the area, but most lie within the
mountain ranges - the Tatra, Gorce and Pieniny Mts, where the national
parks are situated, and the Beskid Sądecki Mts.

In the Carpathian part of the drainage basin the natural plant
associations form characteristic zones within the vertical range. Six
such zones can be distinguished whose vertical ranges correspond with
the climatic zones in the particular mountain chains (see 18.2). Three
zones, alpine meadow, supra-alpine and dwarf mountain pine, occur only
in the Tatra Mts.

The relatively large area lacking forest cover in the upper catchment, the steep slopes and the frequent occurrence of driving rains in
summer, all combine to favour soil erosion and the linear erosion of
streams. Soil wash-off has been estimated to be about 0.1 to 0.5 mm
yr^{-1}, but can sometimes reach 1.0 mm yr^{-1}. Soil landslips are frequent
(Figula, 1956). Soil wash-off results in a considerable increase in
suspended materials transported by the Dunajec during periods of flood.

There is little peat within the catchment. A peat bog above Nowy
Targ is 11.47 km^2, less than 1% of the Czarny Dunajec catchment.

18.36   MAN'S ACTIVITY

Among the population of 1.4 million people only 25% live inside towns.
The populations of the four largest towns are: Zakopane, 30000; Nowy
Targ, 30000; Nowy Sącz, over 61000; Tarnów, 100000. The towns are
centres of industry. These include food and shoe industries at Nowy
Targ, and chemical industry at Nowy Sacz and Tarnów. Many mountain
villages are used for health or recreational purposes. Lodging
facilities are well developed and there is a high tourist capacity.
Numerous villages and industrial plants discharge their sewages to the
Dunajec and to its tributaries.

The middle section of the Dunajec is separated by dams associated
with two reservoirs, the Rożnów and the Czchów. The former, which was
built in 1942, had originally a capacity of 2.29 x 10^8 m^3 (Wróbel,
1965), but due to silting this is now less than 1.7 x 10^8 m^3. The
Czchów acts as a retarding reservoir for the Rożnów and has a capacity
of only 0.12 x 10^8 m^3. Other reservoirs planned for the Dunajec and
its tributaries will have a total capacity of about 7.6 x 10^8 m^3 by
the end of this century.

The dam of the Rożnów reservoir is 32 m high. As the outlet is
relatively low, it lies below the thermocline in summer.

18.37 MANAGEMENT

The main organizations are as follows:

Polish	English
Urzędy Wojewódzkie	Voivodeship Authorities
Ośrodek Badań i Kontroli Środowiska	Centre for Monitoring the Environment
Okręgowa Dyrekcja Gospodarki Wodnej	District Office of Water Management
Zakłady Energetyczne Okręgu Południowego w Katowicach	Energy Plants of the Southern District of Katowice
Polski Związek Wędkarski	Polish Angling Union

The Dunajec belongs territorially to Voivodeship Authorities located in Nowy Sącz and Tarnów. They both determine the regulations for use of water and release of pollution and give any permission needed. They also request the chemical and biological investigations to be made by the Centre for Monitoring the Environment in Kraków. They own the results, which are official and published in the Atlas of Pollution of Polish Rivers (Atlas zanieczyszczenia rzek Polskich).

The District Office of Water Management in Kraków is engaged in river management, preservation of plant facilities and construction of buildings associated with water treatment. The reservoirs are subject to the Energy Plants of the Southern District of Katowice.

The fish in both the river and reservoirs belong to the Polish Angling Union, who administer all aspects of stocking and angling and make claims in the case of fish mortalities due to pollution.

18.4 ENVIRONMENT

18.41 SUBSTRATUM

For most of its length, the river bed consists mostly of loosely arranged boulders and smaller rocks. Sheets of rock occur only in short sections of the Tatrian tributaries and in gorges. The bed of the most downstream section of the river is covered by pebbles and sand, which are found also in marginal pools along the whole river, often together with much silt.

Upstream of the reservoirs, the banks of the river are quite low and flat, formed by natural stony deposits often overgrown with willow brakes. Downstream of the reservoirs, the banks are managed and covered with large angular stones.

18.42 FLOW

The Dunajec shows marked variation in flow. In the middle section the
ratio of maximum to minimum flow exceeds 400, in the Tatrian affluents
it reaches 1230 and in Biała Tarnowska it can even reach 3300
(Ziemońska, 1965). The highest flows occur in summer, especially in
July, and are connected with driving rains. Spates develop rapidly,
but have a short duration time as a result of the small retention
capacity of the drainage basin (Figuła, 1965). Relatively high flows
also occur in spring due to snow melt. During periods of flood the
water level of the unregulated part of the river can reach 5.7 m above
the mean annual height. Minimum flows occur in winter, when some
sections of Tatrian streams freeze up entirely. A few streams also
have sections which are dry from late autumn through the winter,
although they will of course have a snow cover for much of the time.

18.43 TEMPERATURE AND LIGHT CLIMATE

The thermal characteristics of the Dunajec from Nowy Targ to the mouth
were reported in detail by Gołek (1961). The highest monthly mean
occurred in July, August and June (Fig. 18.3). The highest temperatures recorded anywhere were in the Pieniny section, although the
monthly means here were still less than further downstream. This is
due to the influence of a warm local climate and perhaps also the high
turbulence of the water as it passes among the rocky rubble there.
The reservoirs Roznów and Czchów disturb the natural thermal profile
of the river, generally increasing the temperature of water in the
downstream section.

The temperature regimes of the spring tributaries have not yet been
studied in detail. A number of measurements have however been made in
the streams Chochołowski, Kościeliski and Czarny Dunajec during the
course of chemical (Oleksynowa & Komornicki, 1965) or hydrobiological
studies. The water temperature of the lower Chochołowski stream
(890 m altitude) ranged from 1.0 to 14.0 $^{\circ}$C, according to season; in
the lower Kościeliski stream (930 m altitude) the range was 2.0 to
11.0 $^{\circ}$C. The highest temperatures in the upper Czarny Dunajec did not
exceed 13.0 $^{\circ}$C.

Many of the streams and narrower stretches of river are shaded by
evergreen forest. A fringe of deciduous trees is common in the section
above the reservoirs, with alder further upstream tending to give way
to willows further downstream. The section downstream of the reservoir
is relatively open, with shading likely to have little impact on the
ecology of the river.

18.44 WATER CHEMISTRY

Details of the chemistry of waters of the Dunajec and tributaries are

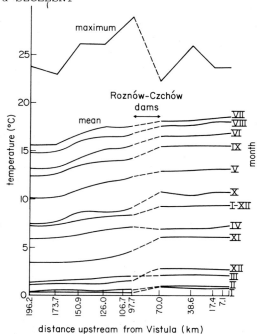

Fig. 18.3 Thermal profile of the Dunajec from 196.2 km upstream of the mouth to the mouth, showing mean monthly temperatures for 1948–1958 (except for 38.6 km) and the maximum recorded during the same period (after Gołek, 1961). Data for 38.6 km upstream based on data for 1901–1958.

given in Table 18.1. The water is mostly of calcium-carbonate type with high values for calcium and total alkalinity along the whole length of the river. The inflow of waste waters from the nitrogen industry in Tarnów via the Biała Tarnowska river considerably enriches the waters of the Dunajec in combined nitrogen (over 4 mg l^{-1} N).

The physical and chemical conditions of the water change markedly during floods caused by heavy rain. The water becomes muddy and appears yellow-brown to an observer on the banks. The river transports a great amount of suspended material and rolls cobbles and boulders over the bottom. The total amount of the materials in suspension or dragged by the Dunajec in its middle section is estimated to be 486000 t yr^{-1}; this amount is discharged to the Roznów reservoir, where 90% of it is deposited, filling the reservoir at a rate of 0.84% of its capacity per year (Brański, 1975).

18.45 POLLUTION

The Dunajec is considered to be a relatively clean river in spite of

Table 18.1 Chemical composition (mg l⁻¹) of River Dunajec.

river	Kościeliski Potok	Chochołowski Potok	Czarny Dunajec	Biały Dunajec	Dunajec	Dunajec
distance from mouth	km 2.5	km 236.0	km 205.0	km 6.3	km 197	km 185.0
conductivity (μS cm⁻¹)		177.9 - 202.8	176.9 - 292.8	174.1 - 317.2	233.3 - 361.9	138.9 - 351.9
pH	7.1 - 8.1	7.0 - 7.8	- 8.0	- 8.0	8.0 - 8.1	8.0 - 8.1
total alkalinity (meq. l⁻¹)	1.24 - 1.72	1.1 - 1.44	3.0 - 3.2	2.8 - 3.1	2.7 - 3.4	2.6 - 3.3
Na	0.6 - 3.0	- 3.7	- 6.2	- 11.8	- 12.2	- 10.8
K		- 2.0	- 2.3	- 3.2	- 3.8	- 3.8
Mg	4.6 - 9.9	3.6 - 13.0	8.7 - 12.3	10.4 - 10.7	11.4 - 14.6	9.7 - 10.4
Ca	21.8 - 33.3	19.6 - 39.4	39.2 - 51.3	39.2 - 51.3	42.4 - 52.9	43.9 - 54.5
Fe	0 - 0.03	0 - 0.05	0.15 - 0.4	0.2 - 0.3	0.2 - 0.45	0.2 - 0.3
Cl	0 - 0.35	0 - 0.35	6.2 - 5.9	8.9 - 14.4	11.9 - 16.5	9.9 - 14.4
SO4-S	5.6 - 13.5	4.6 - 15.6	5.3 - 6.0	3.4 - 10.2	3.2 - 9.4	7.6 - 7.7
NH3-N		0.09 - 0.24	0.10 - 0.28	0.10 - 0.55	0.45 - 0.61	0.20 - 0.34
NO3-N		0.70 - 1.06	1.10 - 1.50	1.38 - 1.50	0.96 - 1.62	1.00 - 1.18
PO4-P	0 - 0.01	0 - 0.005	0.006 - 0.02	0.16 - 0.19	0.13 - 0.16	0.04 - 0.06
O2	10.5 - 11.4	9.8 - 11.4	12.7 - 13.0	11.6 - 12.5	11.0 - 12.7	12.0 - 13.5
BOD5-O2		1.7 - 2.0	2.6 - 3.7	4.3 - 5.1	4.2 - 6.3	3.0 - 3.3
oxidability -O2		1.7 - 2.4	0.6 - 2.9	1.9 - 5.2	2.4 - 6.7	1.4 - 5.9

river	Dunajec	Dunajec	Dunajec	Dunajec	Dunajec	Dunajec
distance from mouth	km 120.0	km 106.0	km 34.0	km 28.0	km 22.0	km 0.5
conductivity (μS cm⁻¹)	230.8 - 351.9	288.5 - 377.5	230.8 - 363.9	346.2 - 477.3	346.3 - 548.8	243.0 - 429.5
pH	8.0 - 8.1	7.8 - 8.0	7.9 - 8.0	7.8 - 7.9	7.6 - 7.7	7.8 - 7.9
total alkalinity (meq. l⁻¹)	2.8 - 3.1	3.5 - 3.8	3.1 - 3.6	3.2 - 3.8	3.2 - 5.2	3.1 - 4.1
Na	- 7.1	- 15.2	- 14.2	- 24.7	- 20.2	- 17.5
K	- 2.2	- 4.4	- 4.4	- 4.9	- 4.8	- 4.6
Mg	10.7 - 10.8	11.7 - 11.8	10.8 - 14.6	11.7 - 12.7	12.7 - 18.5	11.7 -
Ca	43.7 - 59.7	55.0 - 67.7	50.1 - 60.9	50.1 - 67.3	59.9 - 92.9	53.4 - 74.2
Fe	0.2 - 0.32	0.32 - 0.5	0.17 - 0.35	0.3 - 0.45	0.4 - 0.5	0.3 - 0.4
Cl	9.9 - 13.4	14.9 - 18.5	12.9 - 16.5	21.8 - 27.8	19.9 - 23.7	17.9 - 20.6
SO4-S	8.6 - 11.4	9.3 - 12.8	6.7 - 16.0	11.8 - 17.2	13.9 - 21.9	11.4 - 15.9
NH3-N	0.240 - 0.250	0.40 - 0.55	0.125 - 0.26	2.400 - 3.48	1.850 - 1.90	0.98 - 2.00
NO3-N	1.300 - 1.480	1.46 - 2.24	1.44 - 1.68	2.040 - 3.28	2.100 - 2.48	2.50 - 3.28
PO4-P	0.006 - 0.05	0.16 - 0.19	0.01 - 0.05	0.006 - 0.06	0.016 - 0.04	0.016 - 0.06
O2	13.4 - 13.7	11.4 - 12.7	9.8 - 11.4	9.6 - 10.4	9.6 - 10.2	10.0 - 11.9
BOD5-O2	2.7 - 2.7	7.4 - 7.5	2.4 - 4.0	3.0 - 6.2	2.0 - 6.0	3.0 - 4.0
oxidability -O2	1.8 - 2.3	2.6 - 4.7	2.7 - 8.2	2.5 - 5.1	1.8 - 4.1	2.0 - 5.2

the fact that, together with its tributaries, it receives annually 55 x 10^6 m^3 of sewage purified only moderately and 50 x 10^6 m^3 of sewage with no treatment (Informacje G.U.S., 1979). Both minor and some more important tributaries are heavily polluted, in some cases for considerable stretches. This pollution is accompanied by such phenomena as a marked oxygen deficit, complete elimination of sensitive invertebrates (Ephemeroptera, Plecoptera, Trichoptera) and the mass development of the sewage-bacterium *Sphaerotilus natans* and fungi. However no degradation of environmental conditions has been observed in the Dunajec river above the mouth of the Biała Tarnowska river i.e. neither Nowy Targ nor Nowy Sącz lead to a lack of oxygen or elimination of sensitive invertebrates (Dratnal *et al.*, 1979).

18.5 PLANTS AND ANIMALS OF THE UPPER AND MIDDLE (UNREGULATED) RIVER

18.51 PLANTS

Algal communities are rich in species. Wasylik (1971) recorded 459 taxa from the upper section of the river, while Chudyba (1965) recorded 214 taxa in the middle and lower section. Overall, diatoms constitute the bulk of the species (Table 18.2), though population densities are relatively low in the highest Tatra streams. As altitude decreases their densities increase (Kawecka, 1980). Other groups of algae can form species-rich aggregations, such as locally the blue-green algae in both calcareous and non-calcareous upper sections of Tatra streams and the green algae in the mid-section of the river.

Table 18.2 Composition of algal flora in upper, middle and lower Dunajec, expressed as percentage of total flora. (Based on data in Chudyba (1965) and Wasylik (1971).)

phylum	% flora in sections of river		
	upper	middle	lower
Cyanophyta	6.9	11.8	8.0
Rhodophyta	0.4	0.7	
Bacillariophyta	76.9	81.2	83.1
Chrysophyta	0.7	0.2	
Chlorophyta	15.1	6.1	8.9

The golden-brown alga *Hydrurus foetidus* occurs in masses in streams of the Tatra Mts, especially in spring. The bryophyte flora of the upper part of the river and tributaries is rich. Mosses are also obvious in the middle section, near the Pieniny Mts, but much less abundant than further upstream. They presumably also occur downstream of the dams, but no surveys have been made. Rooted submerged macrophytes are apparently absent from the Dunajec, but again there have been no detailed surveys to establish this with certainty.

18.52   MACROINVERTEBRATES

The list of invertebrates in the Dunajec and its spring tributaries is still far from complete. Key publications are those of Dratnal and Szczęsny (1965), Dratnal *et al*. (1979), Kamler (1967), Kownacka (1971), Kownacka & Kownacki (1965), Kownacki (1971, 1977), Sowa (1975a, b), Wojtas (1964). Diptera, and especially Chironomidae, are the most numerous group. They have the greatest number of species (over 100) and, in most samples, the greatest number of specimens. However they are out-numbered in polluted sections by the Oligochaeta, on the stony bed by Naidae (mainly *Nais*) and on the fine-grained bed by Tubificidae.

Among the other groups, Ephemeroptera, Plecoptera and Trichoptera are numerous. In spite of the fact that Chironomidae and Naidae out-number other groups of invertebrates, they seldom constitute the bulk of the biomass. It seems probable that Trichoptera and Ephemeroptera are the dominant groups in this respect.

A classification has been made (Dratnal *et al*., 1979) of macro-invertebrates in the Pieniny section of the Dunajec into functional feeding groups, based on the categories of Cummins (1974): predators, collectors, filter-feeders, grazers, scrapers, shredders. A summary of the quantitative data is given in Table 18.3. The participation of shredders is extremely low, because their source of food, pieces of vascular plant tissue, is low in that section of the river. The proportionally high participation of grazers and collectors indicates that epiphytic microphytes are the main food source.

Table 18.3 Composition of functional feeding groups of macro-invertebrates in the Pieniny section of the Dunajec, expressed as percentage of the total. Data calculated from mean annual number of specimens (= 32395 m^{-2}).

predators	1.15%
collectors	87.50%
grazers and scrapers	9.70%
shredders	0.04%
filter-feeders	1.45%

18.53 FISH

The fish fauna of the Dunajec drainage basin has been a subject of interest to fish biologists from Nowicki (1880, 1882, 1883) to Kołder (1967), Solewski (1963, 1964, 1965) and Bieniarz & Epler (1972); since 1976 studies have been carried out by the Department of Water Biology in Kraków. Different fish communities occur in different sections, though angling has a marked influence on the relative composition of the species. In the mountain sections of the Czarny Dunajec and Biały Dunajec there are only two species, *Cottus poecilopus* and *Salmo trutta* f. *fario*. Downstream, in the vicinity of Nowy Targ, where the river loses the character of a mountain stream, the number of species increases markedly. In addition to the above, there occur also *Thymallus thymallus*, *Phoxinus phoxinus*, *Barbus meridionalis*, *Leuciscus cephalus*, *Cottus gobio* and *Chondrostoma nasus*. Downstream of Nowy Targ, near Knurów, further species appear: *Barbus barbus*, *Rutilus rutilus*, *Abramis brama* and *Hucho hucho*. This last was introduced to the Dunajec from the catchment area of the Danube (Dunaj). With the exception of *Cottus poecilopus* all these species occur down to Nowy Sącz. The relative proportion of the various species changes gradually in favour of fish typical of submontane rivers - *Barbus barbus*, *Chondrostoma nasus* and *Leuciscus cephalus*.

Below Nowy Sącz the Dunajec loses its submontane character, probably due to the heavy pollution. Salmonids, *Phoxinus phoxinus* and *Gobio gobio* are all lacking. Instead, the following species appear: *Leuciscus leuciscus*, *Alburnus alburnus*, *Perca fluviatilis*, *Acerina cernua*, *Alburnoides bipunctatus*, *Lucioperca lucioperca* and *Esox lucius* (Starmach, in preparation).

18.54 ZONATION

It will be clear from the above account that both individual species and communities as a whole show marked changes on passing down the river. Such zonation is evident not only in the main sequence of the spring streams of the Dunajec, but also in its other tributaries. Wayslik (1971) distinguished three zones of microphytes in the spring tributaries (Table 18.4), based on community similarities calculated according to the method of Czekanowski and Kulczyński (Pawłowski, 1959). Their limits were indistinct and species composition changed gradually.

In the first zone *Chamaesiphon polonicus* was particularly abundant. *Homoeothrix varians* (probably identical with *H. janthina*: Starmach, 1966) and *Hydrurus foetidus* also formed rich aggregations. These algae were accompanied by numerous diatoms, forming the association Diatometo hiemalis-Meridionetum with the characteristic species *Diatoma hiemale*, together with its variety *mesodon,* and *Meridion circulare*.

In the second zone a community of intermediate character occurred.

Table 18.4 Zonal distribution of algal communities in the upper
Dunajec, showing the most abundant species. In the case of groups
other than diatoms these are the species which form macroscopic
conglomerations, covering more than 10% of bottom area. The diatoms
are those species with more than 90 cells in an area of 20 mm^2
(modified from Wasylik, 1971).

species	zone of Wasylik	1	2	3
	altitude (m)	1500–1000	1000–750	750–600
Chamaesiphon fuscus		+		
Meridion circulare		+		
Diatoma hiemale		+	+	
D. hiemale var. *mesodon*		+	+	
Chamaesiphon polonicus		+	+	
Phormidium favosum		+	+	
Hydrurus foetidus		+	+	
Ulothrix zonata		+	+	
Homoeothrix cf. *varians*		+	+	+
Ceratoneis arcus		+	+	+
Cymbella ventricosa		+	+	+
Synedra ulna		+	+	+
Cocconeis placentula		+		+
Gomphonema intricatum var. pumilum			+	
Achnanthes pyrenaica			+	+
Diatoma vulgare			+	+
D. vulgare var. *ehrenbergii*			+	+
D. vulgare var. *productum*			+	
Cymbella helvetica			+	+
C. affinis				+
Achnanthes minutissima				+
Navicula cryptocephala				+
N. cryptocephala var. intermedia				+
N. gracilis				+
N. radiosa				+
Stigeoclonium tenue				+
Cladophora glomerata				+

Hydrurus foetidus, Homoeothrix cf. *varians* and *Chamaesiphon polonicus*
were still quite abundant, but less so than in the first zone.
Phormidium favosum and *Ulothrix zonata* were relatively frequent. The
diatoms formed the sub-association vulgaretosum of the association
Diatometo-hiemale-Meridionetum, in which, in addition to *Diatoma
hiemale* and var. *mesodon, Diatoma vulgare* with varieties *ehrenbergii*
and *productum* were numerous.

In the third zone *Cladophora glomerata* and *Stigeoclonium tenue* were
frequent. The diatoms formed the association Diatometo vulgaris-
Cymbelletum with the characteristic species: *Diatoma vulgare, D.*
vulgare var. *ehrenbergii, Cymbella affinis, C. helvetica, Navicula*
gracilis. This community resembles that in the middle section of the
Dunajec, where *Cladophora glomerata* together with diatoms *Diatoma*
vulgare (including var. *ehrenbergii* and var. *productum), Cymbella*
affinis, C. ventricosa, Navicula gracilis, N. radiosa, N. viridula,
Nitzschia dissipata and *Rhoicosphenia curvata* were the dominant species
(Chudyba, 1965).

Three general zones of microphytes and invertebrates were also
recognized in the streams of the High Tatras by Kawecka *et al.* (1971).
These authors distinguished the main zones of microphytes on the basis
of the most numerous species in limited sections of streams (Kawecka,
1971), and of invertebrates by means of so-called domination indices
(Kownacki, 1971). The most numerous microphytes and invertebrates of
the highest values of domination index determined the zones (Table
18.5). The three zones were called (I) the high montane streams, (II)
streams of the wooded hills and (III) streams and rivers of the foot of
the Tatra Mts. Two sub-zones determined mainly by the dominant chironomids were additionally distinguished in both zones I and II. Zone I
is missing entirely in the West Tatras (Kownacki, pers. comm.).

The zones distinguished by Kownacka (1971) and Kownacki (1971)
depended heavily on representation of various species of chironomid.
This was a consequence of the method used for statistical analysis of
the data, which depends on the most numerous and frequent species in
the samples collected. Chironomids usually exceed 50% of specimens in
samples from Tatra streams.

The classification of rivers and streams using chironomids as indicator species is perhaps well grounded if one takes into account their
numerical predominance in aquatic habitats. So far, however, the
ecologies of individual species are poorly understood. Such a classification is also made difficult by their small size, poor morphological
differentiation, and the taxonomic problems associated larvae and
with imaginal forms. For instance, *Orthocladius rivicola* and *O.*
thienemanni were both given as characteristic for zone III because
their larvae are indistinguishable. *Cricotopus* group *algarum* comprises at least nine species which are not distinguishable at the
larval stage (Dratnal *et al.*, 1979).

Another attempt to distinguish general zones of distribution along
the whole river was made by Sowa (1975a) exclusively on the basis of
mayflies. The author analysed their distribution in the main aquatic
systems of the Polish Northern Carpathians. He distinguished five
zones in running waters by grouping stations of a similar species
composition within any one stream, and then applying the percentage
coefficient of similarity of Jaccard. In addition to these five zones
the author distinguished zone 0, spring outflows where mayflies do

Table 18.5 Zonal distribution of benthic communities in streams of the High Tatras. (After Kawecka et al., 1971)

zone	sub-zone	upper limit (m)	dominant taxa	
			algae	macroinvertebrates
I high-mountain streams	1	2100	crust of blue-green algae Chamaesiphon polonicus	Diamesa steinbocki
	2	2000		Diamesa gr. latitarsis Simuliidae (Prosimulium sp.)
II streams of wooded hills	3	1550	Hydrurus foetidus Homoeothrix janthina Diatoma hiemale D. hiemale var. mesodon	Eukiefferiella minor Parorthocladius nudipennis Baetis alpinus Rhithrogena loyolaea
	4	1500		Parorthocladius nudipennis Orthocladius rivicola Baetis alpinus Rhithrogena loyolaea
III streams and rivers at foot of the Tatras	5	1000	Ulothrix zonata Diatoma vulgare D. vulgare var. ehrenbergii Cymbella affinis Synedra ulna	Orthocladius rivicola + O. thienemanii Cricotopus gr. algarum Simuliidae

Table 18.6 Altitudinal distribution of Trichoptera in upper Dunajec (B. Szczęsny, unpublished).

species	presence at various altitudes (m)														
	1680	1600	1500	1400	1300	1200	1100	1000	900	800	700	600	500	400	300
Drusus monticola	+	+	+	+	+										
Rhyacophila philopotamoides	+	+		+	+										
R. glareosa	+	+	+	+	+		+	+							
Allogamus uncatus			+	+	+	+	+	+							
Acrophylax zerberus	+	+	+	+	+	+	+	+	+						
Drusus discolor	+	+		+	+		+	+	+	+					
Chaetopteryx polonica		+	+	+	+			+							
Lithax niger				+	+										
Apatania carpathica				+	+		+								
Psilopteryx psorosa psorosa				+	+	+	+	+							
Rhyacophila tristis				+				+	+	+	+				
Drusus carpathicus					+										
Melampophylax nepos					+	+	+	+							
Acrophylax vernalis					+		+	+	+						
Rhyacophila vulgaris					+		+	+	+						
Halesus rubricollis					+		+	+	+						
Rhyacophila polonica					+		+	+	+						
R. fasciata					+		+			+					
Allogamus starmachi						+	+								
Drusus annulatus							+	+							
Rhyacophila obliterata							+	+		+					
Drusus biguttatus							+	+	+	+					
Potamophylax cingulatus							+	+	+	+					
Allogamus auricollis							+	+	+	+	+				
Ecclisopteryx madida								+	+	+					
Glossosoma conformis								+	+	+					
Potamophylax latipennis									+	+	+				
Rhyacophila mocsaryi									+	+	+				
Ecclisopteryx dalecarlica									+	+	+				
Glossosoma intermedia									+	+	+	+			
Halesus interpunctatus									+	+					
Annitella obscurata									+	+					
Silo pallipes									+		+				
Sericostoma sp.									+	+	+	+			
Rhyacophila nubila									+	+	+	+	+	+	+
Polycentropus flavomaculatus									+	+	+	+	+	+	+
Hydropsyche pellucidula									+	+	+	+	+	+	+
Glossosoma boltoni										+		+			
Psychomyia pusilla										+	+	+	+	+	
Hydropsyche bulbifera										+					
Chaetopteryx fusca										+					
Annitella thuringica										+					
Goera pilosa										+					
Lasiocephala basalis											+	+			
Agapetus delicatulus												+			
Brachycentrus subnubilus												+			
Cheumatopsyche lepida												+	+		
Hydroptila forcipata												+	+	+	
Oligoplectrum maculatum												+	+	+	

Table 18.7 Altitudinal distribution of Ephemeroptera in upper Dunajec. (From Sowa (1980) and M. Olechowska (unpublished).)

species	\	\	\	\	\	presence at various altitudes (m)	\	\	\	\	\	\	\	\
	1680	1600	1500	1400	1300	1200	1100	1000	900	800	700	600	500	400	300
Rhithrogena loyolaea	+	+	+	+	+	+	+	+							
Baetis alpinus	+	+	+	+	+	+	+	+	+	+	+	+	+		
Ameletus inopinatus						+	+	+	+						
Rhithrogena iridina							+	+	+						
R. hybrida							+	+	+						
Ecdyonurus venosus							+	+	+	+	+	+	+	+	+
Baetis vernus								+	+	+	+	+	+	+	+
B. muticus								+	+	+	+	+	+	+	+
Rhithrogena ferruginea								+	+	+	+	+	+	+	+
Baetis melanonyx								+	+	+	+	+	+	+	+
Ephemerella krieghoffi									+	+	+				
Habroleptoides modesta									+	+	+	+	+	+	+
Baetis rhodani										+	+	+	+	+	+
Rhithrogena hercynia										+	+	+	+	+	+
Epeorus sylvicola										+	+	+	+	+	+
Baetis sinaicus										+	+	+	+	+	+
Caenis beskidensis										+	+	+	+	+	+
Ecdyonurus lateralis											+	+	+	+	+
Ephemerella major											+	+	+	+	+
Baetis scambus											+	+	+	+	+
B. lutheri										+	+	+	+	+	+
Siphlonurus lacustris												+	+	+	+
Ecdyonurus dispar												+	+	+	+
Centroptilum luteolum												+	+	+	+
Ephemera danica												+	+	+	+
Habrophlebia fusca												+	+	+	+
Baetis fuscatus												+	+	+	+
Ephemerella ignita												+	+	+	+
Oligoneuriella rhenana												+	+	+	+
Rhithrogena semicolorata												+	+	+	+
R. germanica												+	+	+	+
R. diaphana													+	+	+
Ecdyonurus insignis													+	+	+
E. torrentis													+	+	+
Heptagenia sulphurea													+	+	+
Habrophlebia lauta													+	+	+
Baetis vardarensis													+	+	+
Centroptilum pennulatum													+	+	+
Procloeon bifidum													+	+	+
Cloeon cognatum													+	+	+
Potamanthus luteus														+	+
Caenis horaria														+	+

not occur at all.   The upper limits of the vertical range of zones in
the Tatra tributaries of the Dunajec are shown in Table 18.6.

Table 18.8   Upper limit of zones populated with determined communities of Ephemeroptera (see text) in Tatra streams (after Sowa (1980)
and Olechowska (1982)).

zone	High Tatras	West Tatras
0	2100	1700
1	1450	1680
2	1050	1150
3	680	700
4	–	560

Some of the zones determined by Kawecka *et al*. (1971) and Sowa (1980)
coincide.   Zones I-II of the former authors coincide with zones 0-I of
the latter.   The group best suited to indicate the character of the
environmental factor will be   one whose representatives are narrowly
adapted to the particular factor.   At the same time the scale of adap-
tation of the whole group to this factor ought to be as wide as
possible.   An example of the different character of distribution of
two groups of invertebrates important in the Dunajec is provided by a
comparison of caddis-flies and mayflies (Tables 18.6 and 18.7).
Among caddis-flies is a regular exchange of species with decrease in
altitude.   With mayflies this exchange is only slight and instead the
number of species increases with decrease in altitude.

Without going into the problem as to which factors determine the
distribution of Trichoptera and Ephemeroptera down the river, it is
evident that with decreasing altitude and stream gradient, the sections
inhabited by the particular species or groups of species become longer.
This needs to be borne in mind when choosing sample sites.

18.55   SEASONAL CHANGES

The development of plant and animal communities takes place throughout
the year in the Dunajec and most of its tributaries.   Ice rarely covers
the whole surface of a river or stream and the water is frozen to the
bottom only near the banks.   Although data on algae in the Dunajec in
winter are lacking, changes have been followed over a whole year in
streams of the High Tatras (Kawecka, 1980).   In general, it is possible

to distinguish the spring period (May, June) when communities are subjected to heavy destruction caused by the rise in water resulting from
melting snow.   In summer (July, August) there occurs the natural re-
construction of communities, although hampered by frequent rain.   In
autumn (September, October) and winter (mid-November to mid-May in the
Tatras) algal communities change relatively little.   The greatest diversity of species is recorded in spring and summer and the least in
winter.   Most diatoms show a tendency to increase in abundance in
winter and spring; green algae develop well in summer and blue-green
algae all the year though with a tendency to increase in biomass in
autumn; *Hydrurus foetidus* is conspicuous in winter and spring. Wasylik
(1971) reported that in the Czarny Dunajec the maximum algal crop occurs
usually in September, October and November, and the minimum one in May
or June.   In the course of studies carried out from spring to winter,
Chudyba (1965) did not notice distinct differences in the occurrence
of algae in the middle and lower sections of the Dunajec.

A full year's quantitative studies on invertebrates in the Pieniny
section of the Dunajec (Dratnal *et al.*, 1979) showed that the greatest
density occurs in June (Fig. 18.4).   This results partly from the mass
development of *Nais* (Fig. 18.5) associated with the growth of an algal-
layer on stones.   The small size of the bodies of *Nais*, together with
its method of reproduction by fragmentation, enables the species to
live even in very rapid currents.

Emergent insects are characterized by two maxima, in January–March
and in June–September.   The period of decrease in April and May occurs

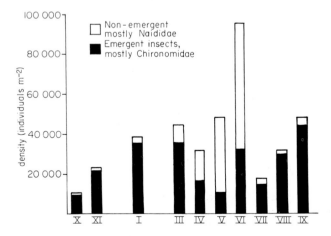

Fig. 18.4 Seasonal changes in density of invertebrates in the
Dunajec at stations 165 - 170 km upstream from the mouth.

after the emergence of winter generations and before the increase in numbers of the young larvae of the summer generations. The beginning of the second insect maximum coincides with the June maximum number

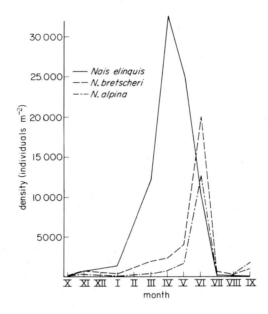

Fig. 18.5 Seasonal changes in density of three Nais species in the Dunajec at stations 165 - 170 km upstream from the mouth.

for *Nais*; hence the greatest number of invertebrates occurs in June in this section of the river. Summer floods reduce pronouncedly the number of invertebrates, especially the worms. This decrease in number of worms depends to a considerable extent on the destruction of the algal layer. Under extreme conditions it may approach 100%.

The seasonal success of species within the genus *Nais* (Fig. 18.5) is probably connected with conditions in the river after the ice cover has melted. *Nais elinguis*, a species preferring habitats rich in the organic matter, first dominates; it is then succeeded by *N. bretscheri* and *N. alpina*.

18.56   PARTICULAR HABITATS

The most peculiar aquatic communities are those in the sections of Tatra streams which become dry or freeze at the end of autumn or in winter. Here there are rich aggregations of blue-green algae, which appear to be very resistant to such environmental conditions. *Chamaesiphon polonicus* is the dominant among these species. Its cell envelopes perhaps provide some protection from the effects of drying (Kann, 1978). Among the bottom fauna *Melampophylax nepos, Allogamus starmachi* and the chironomid *Diamesa starmachi* live exclusively in this habitat. *Allogamus starmachi* is endemic to the Tatra. All three species survive the period of freezing in the form of eggs laid in late autumn, which begin to hatch only during the spring flow of waters from snow-melt. It seems likely that development from eggs is possible only after freezing, but no experimental studies have been made.

18.6   PLANTS AND ANIMALS OF THE LOWER
(REGULATED) RIVER

The section of the Dunajec from the present dams to its entry into the Vistula was regulated before the Second World War and as a consequence its natural ecological conditions were changed. Further changes occurred during the war due to the damming and in the post-war period due to discharges from plants of the nitrogen industry in Tarnów through the Biała Tarnowska river.

In the algal and animal communities inhabiting this section of the river there is a decrease in number of species combined with mass development of a few of them. This is manifest in the abundant development of *Sphaerotilus natans*, especially in the mouth part of the river (Chudyba, 1965). The algal community lacks golden-browns and reds (Table 18.5). Among the green algae, *Cladophora glomerata*, which is very abundant in the middle river, disappears and *Stigeoclonium tenue* is abundant instead. This species is very resistant to organic (Palmer, 1969) and heavy metal pollution (Harding & Whitton, 1976). In comparison with the middle river, the number of species of diatom

decreases and *Diatoma vulgare* disappears. *Nitzschia palea*, another
species known to be resistant to many types of pollution, also increases
(Kawecka, 1981). *Cyclotella meneghiniana, Melosira granulata* and
Fragilaria crotonensis are abundant. The reservoirs provide an inoculum for all three species although the outflow water does lie below the
thermocline in summer (18.36); these species can apparently continue to
grow well in the plankton of the main river (Bucka, 1965).

Among the bottom fauna the most obvious example of a decrease is in
the Plecoptera, which are completely absent; the groups next more
seriously affected are the Ephemeroptera, Trichoptera and even the
Diptera. Ephemeroptera and Trichoptera are eliminated not so much by
changes in the environment caused by flow regulation as by pollution.
In the polluted section a mollusc (*Ancylus fluviatilis*), a bloodsucker
(*Erpobdella octoculata*), worms (*Nais elinguis*, Tubificidae) and midges
(*Cricotopus* spp., *Microcricotopus bicolor, Limnochironomus* ex. gr.
nervosus) develop in such masses that there is an almost three-fold
increase in macroinvertebrate density as compared with upstream sites.

Below the dam reservoirs the following species of fish prevail:
*Alburnus alburnus, Leuciscus leuciscus, L. cephalus, Barbus barbus,
Rutilus rutilus, Esox lucius, Perca fluviatilis.* The lowest part of
the Dunajec was examined extensively by the Fisheries Division of the
Institute of Applied Zoology, University of Agriculture, Kraków, during
1973-1976. Special attention was paid to the influence of domestic and
chemical effluents from Tarnów. The study included cage experiments,
aquarium biotests with rainbow trout and electro-fishing. The results
revealed that this section of the river is drastically affected by
these effluents. The number of fish species, amounting to about 20
upstream of these effluents was drastically reduced to 6 in a zone
6 - 20 km downstream of them. Much of the zone was in fact totally
fishless and most fish that were caught were unsuitable for consumption
because of their abnormal "chemical" odour (A. Łysak, pers. comm.).

18.7   CONCLUDING COMMENTS

It will be evident that there are many gaps in our present knowledge
of the hydrobiology of the Dunajec. In fact, many of the results so
far obtained can be considered only as preliminary. Nevertheless they
do indicate the value of studies on this river. The extensive stretches
of stream and main river which are in more or less natural condition
make the studies of much more than local interest. Unfortunately,
there are also sections where the pronouncedly destructive activity of
man (regulation, pollution) is marked. Polish hydrobiologists are
under an obligation to study the river biocoenosis in all its aspects
to determine changes connected with the gradual eutrophication of the
environment, and to participate actively in the protection of the
river.

The authors are very grateful to Professor K. Starmach and Professor
S. Wróbel for critical comment on the manuscript. Thanks are due also
to Professor A. Łysak and Professor J. Starmach for aid with the section on fish and to Dr B. A. Whitton for valuable comments.

REFERENCES

Bieniarz K. & Epler P. (1972) Ichtiofauna niektórych rzek Polski Połud-
niowej. *Acta Hydrobiol.* 14, 419-444. (English abstract and summary.
Ichthyofauna of certain rivers in Southern Poland.)

Brański J. (1975) Assessment of the Vistula catchment area denudation
on the basis of bed load transport measurements. *Prace Inst. Gosp.
Wodn.* 6, 5-58.

Bucka H. (1965) The phytoplankton of the Roznów and Czchów reservoirs.
Limnological investigations, in the Tatra Mountains and Dunajec River
basin. *Kom. Zagospod. Ziem Górskich PAN* 11, 235-63.

Chudyba D. (1965) Benthic algae in the river Dunajec. Limnological
investigations in the Tatra Mountains and Dunajec River Basin. *Kom.
Zagospod. Ziem Górskich PAN* 11, 153-59.

Cummins K.W. (1974) Structure and function of stream ecosystems.
Bioscience 24, 631-41.

Dratnal E., Sowa R. & Szczęsny B. (1979) Zgrupowania bezkregowców
bentosowych Dunajca na odcinku Harklowa - Sromowce Niżne. *Ochrona
Przyrody* 42, 183-215. (English summary. Benthic invertebrate
communities in the Dunajec River between Harklowa and Sromowce Nizne.)

Dratnal E. & Szczesny B. (1965) Benthic fauna of the Dunajec river.
Limnological investigations in the Tatra Mts and Dunajec River Basin.
Kom. Zagospod. *Ziem Gorskich PAN* 11, 161-214.

Figuła K. (1956) Monografia górnego Dunajca. *Pr. i Stud. Kom. Gosp.
Wodnej* 1, 327-57. (Monograph of the upper part of the Dunajec River.)

Gołek J. (1961) Termika rzek Polskich. *Prace Inst. Hydrol.-Meteorol.* 62,
5-79. (Ru. and Fr. summaries. Thermal features of Polish rivers)

Harding J.P.C. & Whitton B.A. (1976) Resistance to zinc of *Stigeoclonium tenue* in the field and the laboratory. *Br. phycol. J.* 11,
417-26.

Hess M. (1965) Piętra klimatyczne w Polskich Karpatach Zachodnich.
Prace Naukowe Uniwersytetu Jagiellońskiego, CXV, Prace Geograficzne
Seszyt 11, *Prace Instytutu Geograficznego, Zeszyt* 33, 5-267.
(English summary. Vertical climatic zones in the Polish Western
Carpathians.)

Informacje G.U.S. (1979) Zasoby wodne i gospodarka wodno-ścienkowa
Wisly i jej dorzecza w 1978 r. Warszawa. (Water resources and
management of the Upper Wistula basin .... This information is
available only at the Central Statistical Office.)

Kamler E. (1967) Distribution of Plecoptera and Ephemeroptera in
relation to altitude above mean sea level and current speed in Mountain waters. *Pol. Arch. Hydrobiol.* 14(27), 2: 29-42.

Kann E. (1978) Systematik und Ökologie der Algen österreichischen
  Bergbächen. *Arch. Hydrobiol., Suppl.* 53, 405-643.
Kawecka B. (1971) Zonal distribution of alga communities in streams of
  the Polish High Tatra Mountains. *Acta Hydrobiol.* 13, 393-414.
Kawecka B. (1980) Sessile algae in European mountain streams. 1. The
  ecological characteristics of communities. *Acta Hydrobiol.* 22,
  361-420.
Kawecka B. (1981) Sessile algae in European mountain streams. 2. Taxo-
  nomy and autecology. *Acta Hydrobiol.* 23, 17-46.
Kawecka B., Kownacka M. & Kownacki A. (1971) General characteristics
  of the biocenosis in streams of the Polish High Tatras. *Acta
  Hydrobiol.* 13, 465-76.
Kołder W. (1967) Rybactwo w dorzeczu Dunajca. *Gospodarka Rybna* 12,
  18-19. (Fisheries of the Dunajec basin.)
Kownacka M. (1971) The bottom fauna of the stream Sucha Woda (High
  Tatra Mountains) in the annual cycle. *Acta Hydrobiol.* 13, 425-38.
Kownacka M. & Kownacki A. (1965) The bottom fauna of the river Białka
  and of its Tatra tributaries the Rybi Potok and Potok Roztoka. *Kom.
  Zagospod. Ziem Górskich PAN* 11, 130-52.
Kownacka M. & Kownacki A. (1968) Wplyw pokrywy lodowej na faunę denną
  potoków tatrzańskich. *Acta Hydrobiol.* 10, 90-102. (English abstract
  and summary. The influence of ice cover on bottom fauna in Tatra
  streams.)
Kownacki A. (1971) Taxocens of Chironomidae in streams of the Polish
  High Tatra Mountains. *Acta Hydrobiol.* 13, 439-64.
Kownacki A. (1977) Biocenosis of a high mountain stream under the
  influence of tourism. 4. The bottom fauna of the stream Rybi Potok
  of the High Tatra Mountains. *Acta Hydrobiol.* 19, 293-312.
Luther H. & Rzóska J. (1971) Project AQUA. *A source Book of Inland
  Waters Proposed for Conservation. IBP Handbook No. 21.* 239 pp.
  Blackwell, Oxford.
Nowicki M. (1880) *Ryby i Wody Galicji pod wzgledem rybactwa krajowego,
  Kraków (u Korneckiego)*, 96 pp. (Fishes and Waters of Galicia,
  Poland.)
Nowicki M. (1882) Przeglad fauny i rozsiedlenia ryb w wodach Galicji.
  Gazeta Rolnicz, Kraków 39, 465-66. (List of fauna and fish distri-
  bution in the waters of Galicia.)
Nowicki M. (1883) Przeglad rozsiedlenia ryb w Wodach Galicji według
  dorzeczy krain rybnych. (Mapa) Wien. (Review of fish distribution
  in Galicia water according to fish zones.)
Olechowska M. (1982) Zonation of mayflies (Ephemeroptera) in several
  streams of the Tatra Mountains and the Podhale region. *Acta
  Hydrobiol.* 24.
Oleksynowa K. & Komornicki T. (1965) The chemical composition of water
  in the Polish Tatra Mountains and the problems of its variation in
  time. *Kom. Zagospod. Ziem Górskich PAN* 11, 91-111.
Palmer C.M. (1969) A composite rating of algae tolerating organic
  pollution. *J. Phycol.* 5(1), 78-82.
Pasternak K. (1968) Charakterystyka podłoza zlewni rzeki Dunajec.
  Acta Hydrobiol. 10, 299-317. (Abstract and summary in English.
  Characteristics of the substratum of the River Dunajec catchment
  basin.)

Pawłowski B. (1959) Skład i budowa zbiorowisk roślinnych oraz metody
    ich badania. In: Szafer W. (Ed.) *Szata roslinna Polski* 1. pp.
    229-74. (Composition and structure of plant communities and methods
    of investigation.)

Romer E. (1949) Regiony klimatyczne Polski. *Pr. Wroc. Tow. Nauk. B.*
    16, 452-72. (Climatic regions of Poland.)

Solewski W. (1963) Lipień (*Thymallus thymallus* L.) potoku Rogoźnik.
    Acta Hydrobiol. 5, 229-43. (English summary. The grayling -
    Thymallus thymallus L. - in the Rogoźnik stream.)

Solewski W. (1964) Pstrag potokowy (*Salmo trutta morpha fario* L.)
    niektórych rzek karpackich Polski. *Acta Hydrobiol.* 6, 227-53.
    (English summary. *Salmo trutta* f. *fario* L. in some Polish Carpathian
    rivers.)

Solewski W. (1965) Rybostan potoku Białka Tatrzańska ze szczególnym
    uwzględnieniem charakterystyki pstraga potokowego (*Salmo trutta
    morpha fario*). *Acta Hydrobiol.* 7, 197-224. (The ichthyofauna of
    the Białka Tatrzańska stream with particular reference to the brown
    trout (*Salmo trutta* f. *fario* L.)

Sowa R. (1975a) Ecology and biogeography of mayflies (Ephemeroptera)
    of running waters in the Polish part of the Carpathians. 1. Distri-
    bution and quantitative analysis. *Acta Hydrobiol.* 17, 223-97

Sowa R. (1975b) Ecology and biogeography of mayflies (Ephemeroptera)
    of running waters in the Polish part of the Carpathians. 2. Life
    cycles. *Acta Hydrobiol.* 17, 319-53.

Sowa R. (1980) La zoogeografie, l'ecologie et la protection des
    Ephemeroptères en Pologne et leur utilisation en tant qu'indicateurs
    de la puretée des eaux courantes. In: Flannagan J. F. & Marshall K.E.
    (Eds) *Advances in Ephemeroptera Biology*, pp. 141-54. Freshwater
    Inst., Winnipeg, Canada. Plenum Press, New York & London.

Starmach J. (in preparation) Stan ichtiofauny górnego dorzecza Dunajca
    w obecnych warunkach środowiska. (State of the ichthyofauna of the
    upper part of the Dunajec basin under present environmental condi-
    tions.)

Starmach K. (1966) *Cyanophyta-Sinice, Glaucophyta-Glaukofity*, 807 pp.
    Pańtswowe Wydawnictwo Naukowe, Warszawa. (Flora Słodkowodna
    Polski.)

Wasylik K. (1971) Zbiorowiska glonów Czarnego Dunajca i niektórych
    jego doplywów. *Fragm. Flor. et Geobot.* 15, 257-354. (English
    summary. Algal communities in the Czarny Dunajec River (Southern
    Poland) and in some of its effluents.)

Wojtas F. (1964) Widelnice (Plecoptera) Tatr i Podhala, pp. 1-29.
    Uniwersytet Łodzki. (German summary. Stoneflies (Plecoptera) of
    the Tatra Mountains and Podhale region.)

Wróbel S. (1965) The bottom deposits in dam reservoirs at Roznów and
    Czchow. *Komitet Zagospod. Ziem Górskich PAN* 11, 289-94.

Ziemońska Z. (1965) Hydrographic characteristics of the Dunajec River
    basin. Along the Dunajec River. In: Starmach K. (Ed.) *XVI Limnolo-
    gorum Conventus in Polonia MCMLXV,* pp. 5-10. Polish Academy of
    Science Hydrobiological Committee, Kraków.

19 · Llobregat

N. F. PRAT, M. A. PUIG, G. GONZALEZ, M.F. TORT & M. ESTRADA

19.1 INTRODUCTION: GEOGRAPHY OF THE CATCHMENT

The Llobregat, a typical Mediterranean river, runs across Catalonia in
N-E. Spain (Fig. 19.1). The river flows south-east from its source, a
spectacular spring in the Cadi Mountains of the Pre-Pyrenees, to the
Mediterranean sea. The main channel is 145 km long and drops 1360 m
from origin to mouth (19.2). Its slope diminishes very steeply in the
first few kilometres.

The drainage basin of the Llobregat has an area of 4948 km^2 and comprises several climatic and geological zones. The hydrological regime
of the river reflects the Mediterranean climate of its basin. Characteristic features are the variation in discharge (19.2 and Table 19.1)
with frequent floods in spring and autumn. The recent construction of
a dam at La Baells has allowed some degree of flow regulation.

Geographically the Llobregat can be divided into zones of different
physical and geological characteristics (Fig. 19.3).

(i) The head streams are located in the Cadi Mountains and traverse
deep valleys carved by the erosion processes in the calcareous blocks.

(ii) The largest zone of the river basin occupies part of the
Catalonian central plain, between the Cadi Mountains and the junction
of the Cardoner and Llobregat. The head streams of the Anoia watershed
are also situated in this area. Detritic materials cemented by carbonates and layers of gypsum and salts form the river bed in this plain.

(iii) After the confluence with the Cardoner, the main river crosses
the pre-littoral chain near the Montserrat Mountains. The river runs
through a canyon excavated on carbonate conglomerates. Further to the
west the Anoia also traverses the pre-littoral chain. The Llobregat
and Anoia join in the pre-littoral plain, just in the vicinity of the
littoral chain. The pre-littoral plain is recent in geological time

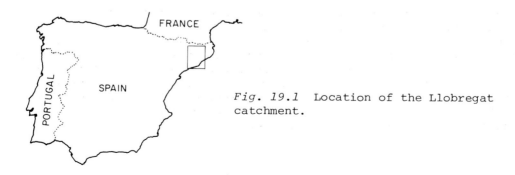

Fig. 19.1 Location of the Llobregat
catchment.

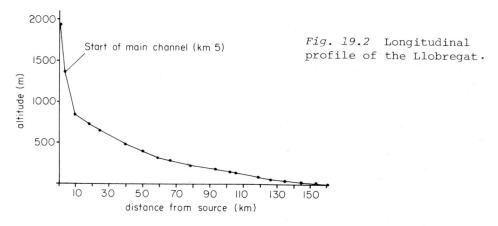

Fig. 19.2 Longitudinal profile of the Llobregat.

(Miocene) and contains detritic materials and outcrops of gypsum and limestone.

(iv) The littoral chain is composed of granite and paleozoic meta-morphic shales, but these materials are often covered locally by cal- , careous soils. After crossing the littoral chain the Llobregat flows into the Mediterranean through a typical delta plain.

The Llobregat receives many tributaries (Fig. 19.4). From the left side, the Marles and the Gabarresa, which also arise in the Cadi Moun-tains, join the main river in the Catalonian central plain. The heavily polluted waters of Riera de Rubi reach the Llobregat in the left lower part of its course. From the right, the Cardoner meets the Llobregat in the Catalonian central plain and the Anoia in the limits between pre-littoral plain and the littoral chain. The Cardoner, which rises also in the Cadi Mountains, crosses several salt deposits and carries waters of very high salinity. The Anoia is a typical stream of limestone country with strong organic and industrial pollution in its lower reaches.

Along the course of Llobregat, numerous weirs retain water for urban and industrial use and irrigation. As a result, in summer only a small volume of water reaches the Mediterranean through the river channel. The principal factors influencing the hydrological, physical and chemi-cal features of the water are flow modification by abstraction, agri-cultural, lumbering and mining activities and urban and industrial wastes. The river may be regarded as strongly polluted in the lower stretch and this is even more so in some tributaries, where human settlements and industries are more abundant.

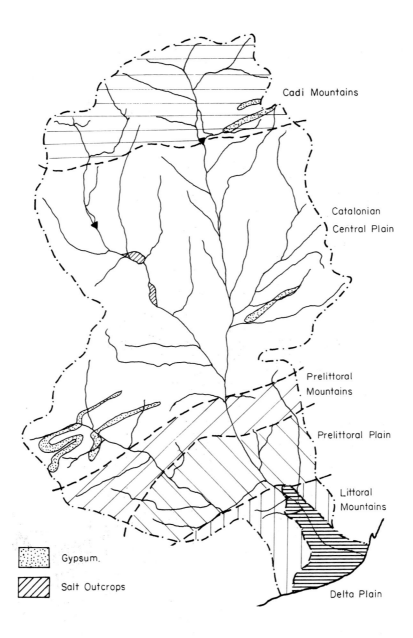

Fig. 19.3 Geographical units in the Llobregat basin. (For details of geology see 19.2)

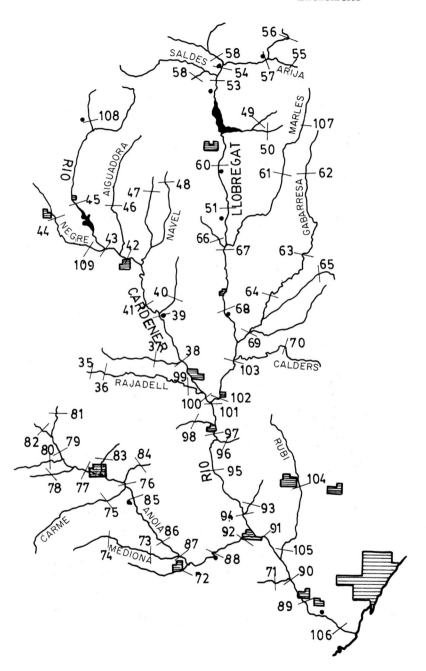

Fig. 19.4 Llobregat and main tributaries, showing location of sampling stations. The principal towns are indicated in hatched polygons whose areas are proportional to the human population.
A, Sant Ponç reservoir; B, La Baells reservoir.

19.2   GENERAL DATA

DIMENSIONS
   Area of catchment: 4938 km^2
   Length of river   145 km
   Highest point in catchment   2531 m
   Source of river   1360 m

CLIMATE
   (Data from Panareda, 1979)
   The river runs across three climatic zones from spring to mouth.
   (i)   The Pre-Pyrenean zone, with high rainfall in spring and relatively
abundant rainfall in summer. There are differences between the eastern
and western basin areas with mean annual precipitation of 905 and 703 mm
respectively.
   (ii)   Central zone, including part of the central plain and the mountains
of pre-littoral chain and the pre-littoral zone, with dry summers and abundant rainfall in spring and autumn. Rainfall is abundant in the top of some
mountains (800 mm yr^{-1}), but scarce in valleys (550 mm yr^{-1}). The mean
temperatures are low to very low in winter.
   (iii)   Littoral zone, with low annual rainfall (550 mm yr^{-1}) most of
which occurs in spring and autumn. Moderate temperatures.

GEOLOGIC
   Related to geographical zones (Riba, 1979)
   (i)   Cadi mountains. Mesozoic calcareous rocks with deep canyons.
   (ii)   Central plain. Eocene and Oligocene detritic materials, with
calcareous cement and gypsum and saline outcrops.
   (iii)   Pre-littoral chain. Calcareous conglomerates of Eocene origin.
In the central part of this chain these form the very characteristic
Montserrar Mountains. Very altered metamorphic materials of Palaeozoic
origin, often covered by Quaternary materials are also present.
   (iv)   Pre-littoral plain. Detritic materials of Miocene origin with
some layers of gypsum and limestone.
   (v)   Littoral chain. Granite and Palaeozoic shales covered by Mesozoic
limestones.
   (iv)   Delta plain. Quaternary sediments.

POPULATION
   About 782000 people inhabit the watershed, but if the city of Barcelona
(located near the river mouth) is considered, this number rises to 25 x 10^6.
The mean density in the drainage area (Barcelona excluded) is 116.25 km^2
(Linati *et al*., 1977).

LAND USE
   In the higher parts forestry and cattle rising activities are important.
There are also some industries (textile, papermills) and mines. Alteration
to waters is limited in extension.
   In the intermediate zone (central plain and pre-littoral chain) agriculture and livestock are important, together with textile industries.
Mining (sodium and potassium salts) has a strong influence on water
quality. Many weirs provide water to textile industries for electricity
power generation, and to villages for drinking water supply (Manresa).
In this zone Anoia is grossly polluted by industrial activities, especially
wastes from paper-mills.

In the lower zone many industries with toxic wastes are situated in Riera de Rubi. Untreated wastes are discharged also in the Llobregat. Abstraction of water for irrigation and urban water supply can leave the end part of the river channel with a very reduced flow during the dry period of the year.

PHYSICAL FEATURES OF RIVER

Flow at selected points (See also Table 19.1)

village	sampling station no.	range of monthly mean discharge ($m^3 s^{-1}$)	annual discharge (Hm^3)	maximum recorded ($m^3 s^{-1}$)
Castellbell	101	13.4 - 26.8	576	1502
Martorell	91	14.6 - 29.4	680	2200
Prat del Llobregat	106	7.2 - 68	707	3800

Seasonal changes in flow
   Minimum and maximum recorded discharge fluctuate between 2 and 3800 $m^3 s^{-1}$, according to fluctuations of the Mediterranean climate. Storms in spring and fall may cause catastrophic floods.
   Snowfall is only important in the higher parts of the basin and its melt does not represent an important input to the discharge of the river. On the other hand the head of the watershed (Cadi Mountains) is a karstic zone and acts as a regulator of flow. A further regulation is provided by La Baells reservoir.

On-stream reservoirs and major weirs
   Reservoirs La Baells (Llobregat) 125 Hm^3 (1976)
              Sant Pono (Cardoner   247 Hm^3 (1957)
Many weirs along the river, very frequent in central plain.

Abstraction
   350000 $m^3 d^{-1}$ abstracted for Barcelona

Substratum (related also to geographical zones)
   (i) Cadi Mountains. Abundant rocks of various sizes and sheets of rock. Sedimentation zones scarce.
   (ii) Central plain. Pebbles, with cobbles and in some zones, boulders. Riffles and pools developed. Weirs are frequent and accumulate large amount of silt.
   (iii) Pre-littoral chain. The Anoia and the Llobregat flow through canyons. The substratum is formed by boulders and cobbles, but pools are also common.
   (iv) Pre-littoral plain. Sand and pebbles. Sand and gravel are extracted for building purposes, creating artificial ponds near the river banks. These ponds are in some cases filled with urban or industrial waste.
   (v) Delta plain. Lotic areas are scarce. Cobbles, pebbles and sand. Silted bottoms in large pools are frequent. Sand and gravel extraction has been carried out intensively. In the final part the river runs as a channel.

Temperature
   Thermal constancy near the sources and regulated temperature below La Baells reservoir. For explanation, see Fig. 19.5.

19.3   WATER QUALITY

19.31   TEMPERATURE

Reservoirs play an important role in temperature regulation of the
rivers along which they are sited; they lower the river temperature in
summer and increase it in winter, especially when the water flows from
the hypolimnion (cf Ward, 1976a, b; Short & Ward, 1980).   Such an effect
is evident in the middle and lower stretches of the Llobregat due to the
presence of the La Baells dam (Fig. 19.4).   In the study summarized in
Fig. 19.5, differences in the water temperature between stations upstream
and downstream of the reservoir (stations 53 and 60) were evident at all
sampling periods.   Except in summer, there was very little change in the
temperature during passage of the river downstream of La Baells.   In
summer, at the tail of the reservoir the temperature was high (23 °C)
due to the reduction of flow caused by water abstraction.   The temperature was lower (15 °C) below the dam, but rose steadily as the water
passed downstream because of the effect of the high air temperatures
and low flow.   The maximum temperature downstream of the dam was recorded in autumn (19 °C), coinciding with the vertical mixing of the
reservoir water.

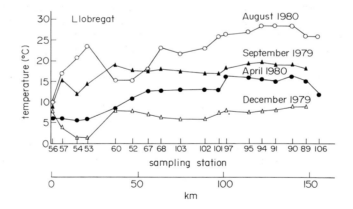

Fig. 19.5 Temperature profiles along the river in four sampling
periods. The numbers on the abcissa correspond to the stations on
Fig. 19.4.

19.32   CHEMISTRY

Due to the composition of the geological substratum, the concentrations of carbonates and bicarbonates are very high in the whole

Llobregat basin, and sulphate is also high in some tributaries. Alkalinity is always higher than 2 meq l^{-1} and exceeds 4 meq l^{-1} in some limestone bed streams of the Anoia basin. As a result of mining activities
in salt beds (Queralt, 1974), the water is loaded heavily with sodium
and potassium chlorides in middle and lower parts of the Cardoner and
Llobregat.

A summary of surveys of water quality in the Llobregat (*La contaminación de los cauces públicos*, Ed. Laia, 1977) includes chapters by
Linati *et al.* (1977), who studied sources of pollution in the surface
waters and Vilaro (1977) who dealt with groundwater problems. Episodic
but dangerous incidents of heavy metal contamination were reported by
Guardiola *et al.* (1977).

In order to provide a more comprehensive framework for biological
and ecological studies, the Department d'Ecologia of the Universitat de
Barcelona has carried out an extensive survey on the Llobregat basin.
Part of this work is described by Prat *et al.* (1982). In this survey
samples were taken on four occasions (September 1979, December 1979,
April 1980, August 1980) for physicochemical analysis of the water and
to study the benthic fauna at 75 stations from the Llobregat drainage
area (Fig. 19.4). The analytical methods employed were those given in
Margalef *et al.* (1976). Certain geographical variables were determined
for all sampling stations.

From the 27 geographical and physicochemical variables considered,
19 were selected as the most representative to perform a multivariate
statistical analysis: altitude, slope, distance from origin, drainage
area, conductivity, temperature, dry matter, alkalinity, dissolved
oxygen, chloride, silicate, sulphate, nitrite, nitrate, phosphate,
calcium, magnesium, sodium, potassium. The technique chosen was principal component analysis. (This and other related methods have been
applied only rarely to rivers - see, for example, Angelier *et al.* (1978)
- but can be very useful in summarizing information and pointing out
trends of variation contained in large data sets.) The analysis was
based on the correlation matrix among the selected variables, without
previous logarithmic transformation of the data and following the
methods described in Estrada (1975) and Margalef *et al.* (1976). 248
samples were included.

The two first principal components explain,   respectively, 30 and
17% of the variance of the original data (Fig. 19.6). As can be seen
in this figure, the first component is correlated positively with
altitude and slope and negatively with distance from the origin,
conductivity, chloride, sodium, potassium and magnesium. Thus the
major source of variability of the original data is related to the
increase in conductivity due to the ion content that increases along
the river (Figs. 19.7, 19.9, 19.10). (The distribution of the mean
values of the score of the first component for all the samples taken
at each point, reflects this situation and presents decreasing values
from the upper to the lower parts of the basin (Fig. 19.7). A strong

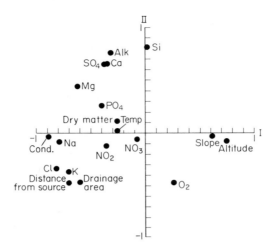

Fig. 19.6 Position of the extremes of the vectors corresponding to the 19 variables in the space of the first two axes of the principal component analysis.

drop in the scores occurs at stations 68 of the Llobregat and 42 of the Cardoner, corresponding to the passage across salt deposits and the quantitatively important input to the river of chloride, sodium and potassium (Figs. 19.9, 19.10). Between sampling points 102 and 101 there is a rise in conductivity and the score for the first component drops in the Llobregat as a result of the entry of the Cardoner which carries the higher salt concentration (Fig. 19.9).

The strongly negative scores of the first component in the Anoia and Riera de Rubi (Fig. 19.7) are not related to mining activities, but reflect elevated ionic concentrations due to high solubility of the rocks in their upper basin areas and to heavy industrial and urban pollution in their lower stretches, where chloride concentrations may reach 1 g l^{-1}.

The second principal component (Fig. 19.6) shows positive correlations with alkalinity, silicate, sulphate, calcium and magnesium, and negative correlations with distance from the source, drainage area and oxygen concentration.

The mean scores of the second component at each sampling point (Fig. 19.8) appear to discriminate between the Anoia basin and the other basins. The values are negative with a decreasing trend down the Llobregat course and positive in the Anoia and the polluted Riera de Rubi. The source of variation expressed by the second component is

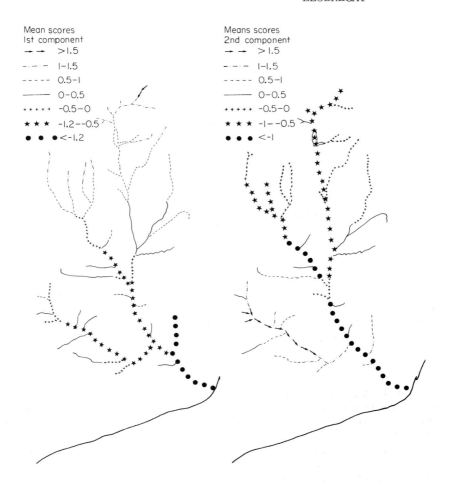

Fig. 19.7 Mean scores of the first principal component in the Llobregat catchment. Note the range along the river.

Fig. 19.8 Mean scores of the second axis principal component in the Llobregat basin. Note the differences between the Anoia and Riera de Rubi (positive values) and the other river basins (negative values).

due to the concurrence, in the Anoia basin, of high alkalinity and elevated concentrations of calcium, magnesium and sulphate of geological origin, with high silicate and phosphate concentrations due to the inflow of urban and industrial wastes. The Riera de Rubi has lower alkalinity and concentrations of sulphate, calcium and magnesium, but more severe pollution problems.

In the middle and lower stretches of the Llobregat and Cardoner, silicate concentrations are sometimes reduced by phytoplankton and periphyton growth. Alkalinity, sulphate, calcium and magnesium do not vary in a significant way along these rivers and oxygen depletion is rare ; in some cases oversaturation may exist during daylight hours due to the activity of the primary producers. The negative scores of the Llobregat and Cardoner for the second component reflect these features.

19.33   POLLUTION PROBLEMS

The Llobregat basin is densely populated (19.2). There are virtually no sewage treatment plants and the run-off from villages and industries goes directly to the river channel, which thus receives numerous sources of pollution. In the lower parts of the Llobregat and specially in the Anoia basin there is intense organic pollution leading to high BOD and COD values which have increased progressively in recent years (Canto *et al.*, 1975).

Monthly data collected by river authorities (D.G.O.H., 1978) at stations coinciding with our sampling points show the increase in BOD and COD on passing down the river (Table 19.1). Variability in the values was related principally to river discharge (Canto *et al.*, 1975) and as a result, marked fluctuations in BOD and COD occurred during the year. The maximum values were recorded in summer, when the river flow was minimum.

The saline load due to substratum composition and mining activities has been already considered (19.32). Cases of pollution by heavy metals such as chromium have occasionally forced the water supply to one of the drinking water treatment plants for Barcelona to be discontinued (Guardiola *et al.*, 1977).

As a result of organic and other type of contamination, plant nutrients such as phosphate show also a progressive increase along the Llobregat course (Fig. 19.11). This trend persists throughout the year but is specially evident in summer, coinciding with the lower water flow.

19.4   THE LLOBREGAT AND ITS LIFE

19.41   INTRODUCTION

Very few biological studies have been carried out on any Spanish river (Meynell, 1973; Prat *et al.*, 1979; Prat, 1981; Viedma & Jalon, 1980). In the case of the Llobregat, there are the publications of Oliver & Bernis (1967), Puig *et al.* (1981) and Tomas & Sabater (in press).

The data on submerged macrophytes and periphyton growths on stones

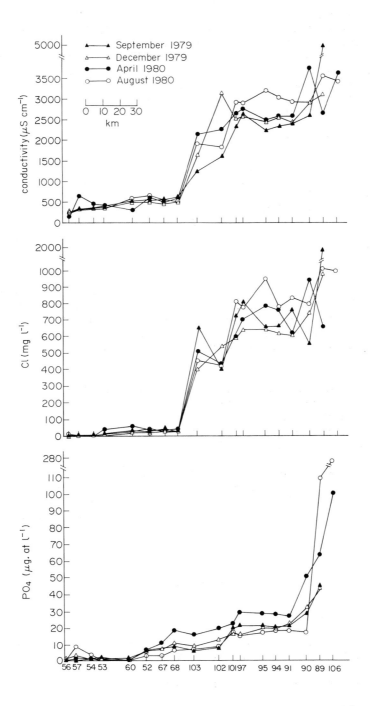

Figs. 19.9, 19.10, 19.11 Profiles of conductivity, chloride and phosphate along the Llobregat.

A

C

B

Fig. 19.12 Llobregat. A) Station 53: immediately upstream of La Baells reservoir showing Roman Bridge. B) Station 102: Pont de Vilomira, above the confluence of the Llobregat with the Cardener. C) Station 91: Martorell, downstream of the entry of the Anoia.

Table 19.1 Geographical features, discharge, COD (chemical oxygen demand) and BOD (biological oxygen demand) along the Llobregat. Data from D.G.O.H. (1978), with mean, standard deviation (in some cases) and range for the three variables.

sampling station	nearest village	altitude (m)	catchment (km^2)	distance from springs (km)	discharge annual mean (m^3 s^{-1})	COD annual mean (mg l^{-1})	BOD annual mean (mg l^{-1})
53	Figols	620	352	22.3	10.1 ± 6.8 (5.4 - 12.5)	5.32 (1.8 - 16)	2.46 (0.5 - 5.3)
68	Balsareny	229	1093	61	14.4 ± 4.7 (5.3 - 22.7)	3.37 (2.6 - 4.6)	2.87 (0.5 - 6.2)
102	Vilomara	176	1885	87	16.3 ± 5.5 (5.9 - 25.8)	4.51 (3 - 7.8)	3.55 (1.4 - 6.1)
97	Castellbell	155	3293	99	25 ± 10.9 (1.3 - 38.5)	5.03 (3 - 9.3)	3.91 (1 - 7.2)
91	Martorell	44	4561	131	43.6 ± 59 (3.9 - 176)	32.12 (6.8 - 276)	20.58 (4.4 - 163)
89	Sant Joan Despi	17	4848	149.4	46.3 ± 60.4 (8.5 - 180)	11.85 (7.9 - 23)	9.12 (5.6 - 23.7)

are sparse. Some general information on the algae is given in Margalef (1951), while Tomas and Sabater have surveyed the diatom flora. In the zone above the La Baells reservoir (see 19.36) the most abundant species are *Achnanthes minutissima, Cymbella minuta, Diatoma hiemale* var. *mesodon* and, in winter, *Meridion circulare*. In a zone from stations 60 to 94 the above species are gradually replaced by species of *Navicula* and *Gomphonema*. The last zone of the Llobregat, from station 91 downstream, only pollution-tolerant species are found. These include *Nitzschia frustulum, N. palea, Navicula pellulosa, N. atomus, N. permitis* and various centric diatoms. In the lower zone of the watershed where chloride concentrations are high, there is a proliferation of *Cladophora* (see Margalef, 1951).

Cladophora is very abundant on stones from station 57 down to the lower river. After station 68 which coincides with the high salt inputs (Fig. 19.10) *Potamogeton pectinatus* appears in addition to the abundant *Cladophora*. Between stations 68 and 102 clumps of this species are only infrequent, but below the entrance of the Cardoner it can cover up to 50% of the river bed at some stations. In the Cardoner, in areas of scarce flow, *P. pectinatus* forms a dense carpet covering all the bottom of the river. Animals associated with *Cladophora* and *P. pectinatus* may be very abundant and some macroinvertebrates are found only on these plants.

Although fishing is a popular sporting activity, the fish fauna has not been surveyed extensively in the Llobregat. As in other Spanish rivers, fish diversity is low: only four species can be found with some abundance. Rainbow trout (*Salmo gairdneri*), which has been introduced from fish-farms, is abundant in inpolluted stretches of the upper courses. *Barbus* spp. and especially *Leuciscus cephalus* are frequent in the middle and lower river; carp (*Cyprinus carpio*) can be found in the lower river where pollution is not so strong.

The more detailed information on macroinvertebrates reported below is based on an extensive survey carried out by Departament d'Ecologia of the Universitat de Barcelona between September 1979 and August 1980, the same period as that covered by the physicochemical survey.

19.42   STUDY METHODS AND TAXONOMIC COMPOSITION

Samples for qualitative study were collected during the four sampling periods at all the stations in Fig. 19.4. The sampling strategy was based on the stratification guide-lines suggested by Resh (1979). All attached material was scraped from stones taken, where possible, from areas subject to fast or moderate current speeds in the central part of the stream. The organisms (mostly larval and pupal insect stages) retained by a net of 250 μm mesh size were fixed with 10% formalin and transported to the laboratory for identification and enumeration under a stereoscopic microscope (with a magnification of 10x). This account is based on all the material obtained from the 18 Llobregat stations,

with the exception of the chironomids, which have been studied only in
samples from the first sampling period (September 1979). More detailed
studies can be found in Prat *et al.* (1983), Gonzalez *et al.* (in press),
Millet & Prat (in press) and Prat (in press).

More than 120 species or groups of species have been found in these
18 stations. As with for other rivers (Illies & Botosaneanu, 1963;
Hynes, 1970), insects form the major part of the fauna, with 7 species
of Plectoptera, 15 Ephemeroptera, 14 Trichoptera, 9 Coleoptera and 65
Diptera. From the Diptera, 45 were Chironomidae, 2 Blepharoceridae and
10 Simuliidae. Among the other groups, the most common were the Turbellaria (1), Hirudinea (2), Mollusca (6) and Ostracoda. However, while
some species were very frequent, others could be found only in one or
very few sampling points.

19.43 LONGITUDINAL DISTRIBUTION

Fig. 19.13 shows the longitudinal distribution of the most important
macroinvertebrates along the Llobregat course. As can be seen, several
species showed a very wide distribution along the river. This group
included the chironomids *Rheocricotopus chalybeatus, Cricotopus* gr.
tremulus and *Polypedilum* gr. *convictum*, the caddis *Hydropsyche exocellata* and *Hydroptila* sp ; the mayfly *Caenis moesta* and the molluscs
Physa acuta and *Ancylus fluviatilis*. On the other hand, some rheo-
philic and stenothermic species were found only at the most upstream
station (56); these comprised the blepharococerid *Liponeura cinerascens*,
the chironomid *Diamesa* sp., the stoneflies *Isoperla* sp. and *Protonemura risi
spinulosa*, the caddis *Tinodes dives* and *Micrasema morosum* and the flat-
worm *Polycelis felina*.

Many species were found only between stations 68 (280 m altitude)
and 57 (840 m) or between stations 57 and 53 (500 m), upstream of La
Baells reservoir. These included mayflies (*Baetis rhodani, Rhithrogena
diaphana*), stoneflies (*Dinocras cephalotes*) and the blackflies *Simulium
monticola* and *S. rheophilum*.

No stonefly larvae were found downstream of station 60, which can be
attributed to their cold stenothermy (Berthélemy, 1966); they were very
abundant at stations 56 and 54, but absent from the intermediate station
57, probably due to pollution by wastes of nearby town and a papermill.
The Simuliidae were also restricted to the upper part of the river (Fig.
19.13).

Some species, such as the mayfly *Baetis meridionalis* and caddisflies
Hydropsyche contubernalis and *Psychomyia pusilla*, and the chironomids
Cardiocladus fuscus and *Orthocladius* sp. 2 were found exclusively in
intermediate zones of the river. Their distribution downstream was
probably limited by pollution.

Only very few species were present in the heavily contaminated lower
reaches of the river. Among them were the chironomids *Cricotopus* gr.

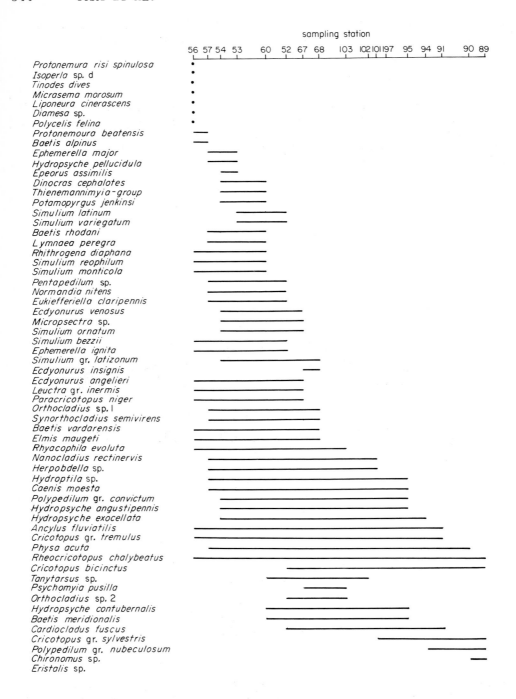

Fig. 19.13 Longitudinal distribution of macroinvertebrates along the Llobregat. Numbers correspond to stations in Fig. 19.4.

sylvestris, Polypedilum gr. *nubeculosum* and *Chironomus* sp. and another Diptera, *Eristalis* sp. Tubificids were also abundant here.

19.44 SEASONAL VARIATION

The relative importance (based on total number of individuals collected in each station) of the main macroinvertebrate groups was relatively
constant for the different sampling periods (Fig. 19.14). The highest
temporal differences were shown by the chironomids, which were very
abundant in spring in polluted zones. Blackflies, which were present in

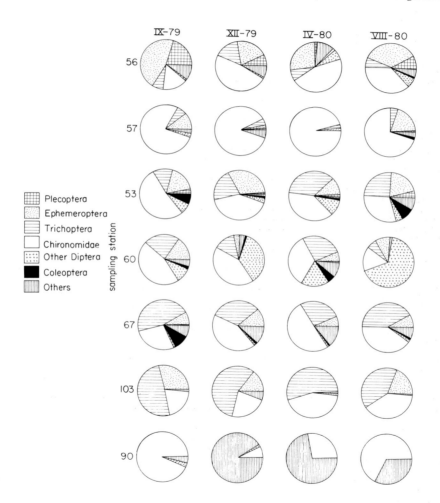

Fig. 19.14 Percentage composition (based on number of animals) of
the macroinvertebrate fauna at seven stations of the Llobregat during
four different sampling periods. Only the most important groups are
figured.

masses in both summer and winter at station 60, just below the La Baells
reservoir, were scarce in autumn and spring at the same site. Plankton
richness and low temperatures seem to be the two key factors bringing
about this abundance; the decrease in numbers could be related to the
life cycles of the species present. Among the caddisflies, *Psychomyia
pusilla* proliferated on stones in spring and summer and was scarce in
autumn and winter, as had been reported previously in the river Ter
(Prat, 1981). *Hydroptila* larvae were also more abundant in spring and
summer, the behaviour in this case being due to its association with
Cladophora.

19.45   CHANGES IN RELATIVE ABUNDANCE ALONG THE RIVER

The relative abundance of many species (based on percentage of all
individuals taken at each station in September 1979) showed interesting
changes along the Llobregat. The genus *Baetis* presented a gradual
pattern of species replacement (Fig. 19.15) from the upper to the lower
parts of the river course. The sequence was: *B. alpinus, B. rhodani,
B. vardarensis* and *B. meridionalis* and is consistent with known ecological characteristics of these species. For example, *B. alpinus* has
been found in Andorra above 1000 m (Puig, 1981) and *B. meridionalis* is
common in the lower stretches of many rivers (Müller-Liebenau, 1969).
Its absence from the most downstream sampling sites on the Llobregat can
be attributed to pollution. Among the caddis, *Rhyacophila evoluta*
could not be found downstream of station 68 and was more abundant in
the upper river. Conversely *Hydropsyche exocellata* was common in the

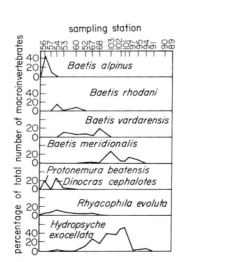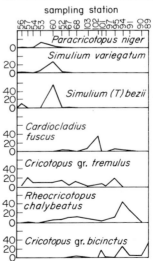

Fig. 19.15 Distribution of the most common species of Ephemeroptera,
Plecoptera and Trichoptera along the Llobregat.

Fig. 19.16 Distribution of the most common Diptera along the
Llobregat.

middle river, especially between stations 67 and 101, where it was the most abundant invertebrate.

In the case of midges (Chironomidae), the lack of detailed data for three of the sampling periods and the difficulties of species determination (only possible by examination of mature pupae with male genitalia) made it difficult to interpretate their abundance patterns. In general, they prefer polluted zones (compare, for instance, their numbers at the polluted station 57 with those at neighbouring stations in Fig. 19.14). However examples were found in each section of river where particular species reached their greatest abundance. *Paracricotopus niger* was most abundant upstream La Baells reservoir (Fig. 19.14), *Cardiocladus fuscus* larvae in the central part of the river and *Cricotopus sylvestris* and *Polypedilum* gr. *nubeculosum* in lower polluted stretch. The abundance of some midge species but related to their substratum preferences *Rheocricotopus chalybeatus* was especially common on stones with *Cladophora* or on *Potamogeton* clumps, and its abundance increased in the middle and lower river (Fig. 19.16). Classification difficulties made it impossible to distinguish patterns of distribution within the *Cricotopus* gr. *tremulus* species complex, which involves two or three species (*Cricotopus annulator, C. triannulatus* and another unidentified species). *C. bicinctus* is an example of species where pollution was the key factor; as can be seen in Fig. 19.16, it was abundant both in the lower Llobregat and also in the polluted station 57 in the upper river.

19.46 ZONATION OF COMMUNITIES

The establishment of macroinvertebrate communities along the course of rivers has been subject of much discussion. River zonation has been considered as a succession of more or less well defined discrete units (Illies & Botosaneanu, 1963) or as the result of continuous change in faunistic composition throughout the river (Hynes, 1970; Vannote *et al.*, 1980). Sharp faunistic changes can be found as the result of sudden variations in water quality due, for example, to pollution sources and in rivers less affected by contamination a gradual replacement of species can be found without sharp changes (Hawkes, 1975).

Analysis of common species between neighbouring stations following the method of Illies & Botosaneanu (1963) has been applied to the data of September 1979 (Fig. 19.17). If the longitudinal distribution of species and their relative importance is considered in the Llobregat, five faunistic units can be distinguished (Table 19.2).

The first zone includes only most upstream station (56), which has a species composition very similar to that of Pyrenaean streams (see for example Prat *et al.*, 1979). In this area water is very cold, current speeds are high and dissolved oxygen is near saturation.

The limit between the second and third zone is set by the La Baells

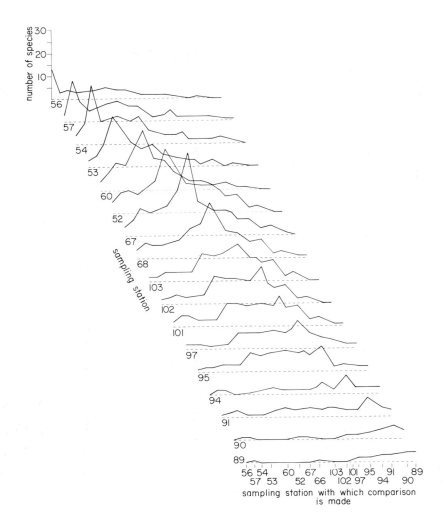

Fig. 19.17 Number of species at each sampling station on the Llobregat in September 1979 present also at other stations at the same time. Each graph corresponds to the station indicated in the left side of the abcissa.

reservoir. Upstream of the dam (the second zone) the Llobregat is a typical mountain stream, with cool water, a relatively steep slope, large rocks in the river bed and a poorly developed riffle-pool struc- ture. The faunistical similarity among the three stations of the second zone (57, 54, 53) is somewhat confused because of local pollu- tion (station 57) or water abstraction (station 53). Station 60, down- stream of La Baells reservoir, is influenced strongly by the hydrolo- gical regime of reservoir and the fauna is intermediate between second

Table 19.2 Zonation along the Llobregat: sampling stations, physicochemical characteristics (with mean and range) and important indicator species.

	FIRST ZONE (epirhithron)	SECOND ZONE (upper metarhithron) polluted	SECOND ZONE (upper metarhithron) clean	BELOW RESERVOIR	THIRD ZONE (lower metarhithron)	FOURTH ZONE (hyporhithron)	FIFTH ZONE (potamon)
sampling station	56	57	54, 53	60	52, 67, 68	103, 102, 101, 97, 95	94, 91, 90, 89
slope (range) (%)	16	8.6	1.11 - 1.4	1.1	0.46 - 0.88	0.25- 0.52	0.12- 0.35
temperature (°C)	8.5 (7 - 10)	10.8 (4 - 17)	10.5 (1.5 - 23.5)	12.7 (8.5 - 19)	14.2 (6.5 - 23)	16.5 (7.5 - 28.5)	17.6 (8.2 - 28.5)
conductivity (μS cm^{-1})	228 (159 - 270)	418 (330 - 667)	396 (329 - 474)	493 (323 - 605)	566 (498 - 672)	2369 (1250 - 3200)	3082 (2400 - 5000)
phosphorus (μg - at l^{-1})	0.67 (0.01 - 1.45)	3.14 (0.8 - 9.7)	1.8 (0.58- 4.2)	1.87 (0.8 - 1.76)	8.69 (3 - 19.2)	17.95 (7 - 30.6)	40.5 (19.1 - 112.6)
oxygen (mg l^{-1})	9.31 (8.61 - 10.8)	11.5 (8.4 - 16.4)	9.0 (4.1 - 16.4)	10.8 (6.41 - 8.5)	9.1 (6.5 - 12.9)	9.3 (<0.1 - 16.4)	8.1 (2.5 - 21.0)
INDICATOR SPECIES exclusive	Protonemura risi; Tinodes dives; Micrasema morosum; Liponeura cinerascens; Diamesa spp.; Polycelis felina		Epeorus assimilis; Paracricotopus niger	Simulium bezzii; Simulium ornatum	*Ecdyonurus insignis*	Baetis meridionalis; Hydropsyche contubernalis; Cardiocladius fuscus; Psychomyia pusilla	Cricotopus gr. sylvestris; Polypedilum gr. nubeculosum; Chironomus spp.; Eristalis spp.
more abundant in this zone	Baetis alpinus; Protonemura beatensis	Cricotopus bicinctus	Baetis rhodani; Rhithrogena diaphana; Dinocras cephalotes; Hydropsyche pellucidula; Ephemerella major	Baetis vardarensis	Hydropsyche exocellata		

and third zone. As has been indicated previously the high abundance of blackflies is a very characteristic feature here at some seasons.

The third zone extends between stations 52 and 68, where many species have their downstream limit due to sharp changes in water quality and geographical characteristics. High faunistic variety characterises this zone; the maximum number of species per sample was recorded at station 67.

Dominance of *Hydropsyche exocellata* is the most striking feature of the fourth river zone, between stations 103 and 97. High salinity and conductivity, warm summer temperatures and the succession of weirs and riffle-pool areas are the most characteristic physicochemical features. The presence of abundant filter-feeding species on stones is associated with the development of plankton behind weirs and pools due to nutrient inputs of sewage from villages and farms.

Downstream of station 95 (fifth zone) the benthic fauna is very poor. The river increases in width and depth and true lotic zones are infrequent. The concentration of salts is very high and there is a high contribution of urban and industrial pollution. Dissolved oxygen concentrations are low in some periods of the year. Only a very few species resist these conditions, some of which are abundant but some of which are very scarce.

Acknowledgements

We thank Prof. R. Margalef, J.D. Ros and R. Julià for their helpful criticisms and for reviewing the text, Anna M. Domingo for the drawings and V. Rull, X. Millet, C. Marrasé, T. Vegas, Nuria Prat, T. Cassasayas and J. Carbonell for sampling assistance. Financial support has been provided by Universitat de Barcelona and "Servei del Medi Ambient, Diputació Provincial de Barcelona".

REFERENCES

Angelier E., Bordes J.M., Luchetta J.C. & Rochard M. (1978) Analyse statistique des paramètres physico-chimiques de la rivière Lot. *Annls Limnol.* 14, 29-58.

Berthélemy C. (1966) Recherches écologiques et biogéographiques sur les Plécoptères et Coléoptères d'eau courante (Hydraena et Elminthidae) des Pyrénées. *Annls Limnol.* 2, 227-458.

Canto J., Guardiola J. & Salvatella N. (1975) Evolución de la polución del agua en el rio Llobregat. *Agua* 91, 15-24.

D.G.O.H. (1978) *Análisis de la calidad de aguas. 1978-1977.* Publicaciones del M.O.P.U., Centro de Estudios Hidrográficos. No. 129. Ministerio de Obras Públicas, Madrid.

Estrada M. (1975) Statistical considerations of some limnological parameters in Spain reservoirs. *Verh. int. Verein. theor. angew. Limnol.* <u>19</u>, 1849-59.

Guardiola J., Salvatella N. & Canto J. (1977) Polución del agua por Cromo en el rio Llobregat. In: *La contaminacion de cauces públicos*, 251 pp., pp. 117-137. Laia, Barcelona.

Gonzalez G., Puig M.A., Millet J. & Prat N. (in press) Patterns of macroinvertebrate distribution in the Llobregat river basin (NE Spain). *Verh. int. Verein. theor. angew. Limnol.* <u>22</u>

Hawkes H.A. (1975) River zonation and classification. In: Whitton B.A. (Ed.) *River Ecology*, 725 pp., pp. 313-374. Blackwell, Oxford.

Hynes H.B.N. (1970) *The Ecology of Running Waters*, 555 pp. Liverpool University Press, England; University of Toronto Press, Canada.

Illies J. & Botosaneanu L. (1963) Problèmes et méthodes de la classification et de la zonation écologique des eaux courantes, considerées surtout du point de vue faunistique. *Mitt. int. verein. theor. angew. Limnol.* <u>12</u>, 57 pp.

Linati J.A., Aparicio I., Guardiola J. & Verges J.C. (1977) El problema de la contaminación en la cuenco del rio Llobregat. In: *La contaminación de los Cauces Publicos*, 251 pp., pp. 75-92. Laia, Barcelona.

Margalef R. (1951) Regiones limnológicas de Catalunya y ensayo de sistematización de las asociaciones de algas. *Collectanea Botánica* <u>3</u>, 43-67.

Margalef R., Planas D., Armengol J., Vidal A., Prat N., Guiset A., Toja J. & Estrada M. (1976) *Limnologia de los Embalses Espanoles*. Publicaciones de M.O.P.U. Centro de Estudios Hidrográficos. No. 123, 454 pp. Ministerio de Obras Públicas, Madrid.

Meynell P.J. (1973) A hydrobiological survey of a small Spanish river grossly polluted by oil refinery and petrochemical wastes. *Freshwat. Biol.* <u>3</u>, 503-20.

Millet J. & Prat N. (in press) Las comunidades de macroinvertebrados a lo largo del rio Llobregat. Actas del Segundo Congreso Espanol de Limnologia. Murcia, April 1983. (Spanish = Macroinvertebrate communities along the River Llobregat.)

Müller-Liebenau I. (1969) Revision de europäischen Arten der Gattung *Baetis* Leech (Insecta, Ephemeroptera). *Gewässer Abwässer* 48/49, 1-214.

Oliver B. & Bernis J. (1967) Contribución al estudio biológico del rio Llobregat y sus afluentes. *Doc. Inv. Hidrológica* <u>2</u>, 27-71.

Panareda J.M. (1979) El clima y les aigües. In: *Geografia Fisica dels Paisos Catalans*, pp. 69-102. Ketres, Barcelona.

Prat N. (1981) The influence of reservoir discharge on benthic fauna in River Ter (NE Spain). In: Moretti G.P. (Ed.) *Proceedings of the 3rd International Symposium on Trichoptera*, pp. 295-303. Series Entomologicae Vol. 20. W. Junk, The Hague, The Netherlands.

Prat N. (in press) Los rios. In: *Curso Fundamentos de Limnologia*. Murcia, April 1983. Publicaciones de la Universidad de Murcia.

Prat N., Bautista M.I., Gonzalez G. & Puig M.A. (1979) Els cursos d'aigua. In: Folch R. (Ed.) *El Patrimoni Natural d'Andorra*, pp. 261-309. Ketres, Barcelona.

Prat N., Gonzalez G. & Puig M.A. (1983) Predicció i control de la qualitat de les aigües dels rius Besós i Llobregat. II. El poblament faunistic i la seva relació amb la qualitat de les aigües. Monografies, 9. 164 pp. Servei del Medi Ambient. Diputació de Barcelona. (Catalan = Prediction and control of water quality in the Rivers Besós and Llobregat. II. Faunistic composition in relation to water quality)

Prat N., Gonzalez G., Puig M.A. & Millet J. (in press) Chironomid longitudinal distribution and macroinvertebrate diversity along the river Llobregat (NE Spain). Memoirs of the American Entomological Society.

Prat N., Puig M.A., Gonzalez G. & Tort M.J. (1982) Predicció i control de la qualitat de les aigües dels rius Besós i Llobregat. I. Característiques fisicio-químiques. Monografies 6, 206 pp. Servei del Medi Ambient. Diputació Provincial de Barcelona. (Catalan = Prediction and control of water quality in the Rivers Besós and Llobregat. I. Physico-chemical features)

Puig M.A. (1981) Contribució al coneixement dels Plecópters i Efemerópters d'Andorra. *Bull. Ins. Cat. Hist. Nat.* (45 Sec. Zool. 3). (Cat. = Contribution to knowledge of Plecoptera and Ephemeoptera ..)

Puig M.A., Bautista I., Tort M.J. & Prat N. (1981). Les larves de trichoptères de la rivière Llobregat (Catalogne, Espagne). Distribution longitudinale et relation avec la qualité de l'eau. In: Moretti G.P. (Ed.) *Proceedings of the 3rd International Symposium on Trichoptera*, pp. 305-311. Series Entomologicae, vol. 20. Junk, The Hague, The Netherlands.

Queralt A. (1974) La contaminación de las aguas. *C.A.U.* 25, 83-110.

Resh V.H. (1979) Sampling variability and life history features: basic considerations in the design of aquatic insect studies. *J. Fish. Res. Bd Canada* 36, 290-311.

Riba O. (1979) El relleu dels Paisos Catalans. In: *Geografia Fisica dels Paisos Catalans*, pp. 7-66. Ketres, Barcelona.

Short R.A. & Ward J.V. (1980) Leaf litter processing in a regulated rocky mountain stream. *Can. J. Fish. aquat. Sci.* 37, 130-7.

Tomàs X. & Sabater S. (in press) The diatom associations of the Llobregat River and their relations to water quality. *Verh. int. Verein. theor. angew. Limnol.* 22.

Viedma M.G. de & Jalon D.G. de (1980) Descriptions of four larvae of *Rhyacophila (Parahyacophila)* from the Lozoya river, central Spain, and key to the species of the Iberian peninsula (Trichoptera: Rhyacophilidae). *Aquatic Insects* 2, 1-12.

Vilaro F. (1977) La contaminación de las aguas subterráneas. In: *La contaminación de los cauces públicos*, 251 pp., pp. 139-159. Laia, Barcelona.

Ward J.W. (1976a) Effects of thermal constancy and seasonal temperature displacement on community structure of stream invertebrates. *ERDA Symposium Series*, pp. 302-7.

Ward J.W. (1976b) Effect of flow patterns below large dams on stream benthos. A review. In: Osborn J.F. & Allman C.H. (Eds) *Instream flows*, vol. II, pp. 235-53. *American Fisheries Society*.

20 · Rönneå

A. ALMESTRAND & A. ALMESTRAND

20.1 INTRODUCTION

Sweden, a long, narrow country with north-south extension, exhibits many types of river, due to large differences in climate, morphology and geology. There are slow-moving lowland rivers almost hidden in abundant vegetation and mighty, rushing Norrland rivers (Sw. älv) with huge hydroelectric plants.

This chapter describes a river in Scania (Sw. Skåne) which runs mainly through farming land, but whose tributaries drain forest-covered districts. The hydrology of this and other Scanian rivers has been much changed by drainage and the lowering of lake water-levels; farmers strove to improve and extend their fields in this way long ago.

Skåne is the southern-most province of Sweden, with 1/40 (11000 km^2) of the country, but about 1/8 (1.2 x 10^6) of the population. It is the most densely populated province; in Malmöhus county there are about 150 inhabitants per km^2 and in Kristianstad county, 43 per km^2. The climate is mild due to the long coast-line. Precipitation varies from 800 mm per year in the north to <500 mm in the south-west. Skåne is poor in lakes except in the north-east; the biggest is Ivösjön (54.2 km^2, 50 m depth). The rivers follow in the western part the line of the landscape, from S-E. to N-W. Typical examples are Rönneå, Vegeå and Kävlingeån.

Northern Skåne is a monotonous Archaean rock plain rich in mires and sloping gradually towards the S-W. The Archaean rock ridges (Sw. åsar) south of the plain have come up as horsts and present a striking feature, although they rise to only about 200 m above sea-level. They belong to the system of diagonal parallel axes which can be drawn over Skåne from N-W. to S-E. Some axes constitute depressions, others ridges, but all are reflected in both bedrock and surface morphology. The typical plains of Skåne are in the N-W., N-E. and south; thanks to these Skåne has become "the granary of Sweden".

The geology of Skåne differs from the rest of the country (Lundegårdh et al., 1964). A coloured map is like a mosaic. More than half is connected geologically with Middle Europe via Denmark. Along the dividing lines mentioned above, movements in the earth's crust occurred over a long period and probably did not end until the Tertiary. In the depressions "young" rocks (sandstones, slates, limestones) from the Cambrian to Cretaceous layers have been preserved; in other places erosion has sometimes reached down to Archaean rocks. The final surface forms were built up during the Quaternary period. The most important moraine soil in the Rönneå district derives from the N-E. ice (Sw. nordostisen). Ice river deposits gave rise to eskers, gravel ridges and plateaus.

Less than 50% of Skåne is cultivated and this is continuing to decrease. Farming is not profitable and cattle-breeding in particular is on the decline. Meadows where cows once grazed are now planted with spruce. Skåne is however highly industrialized, with food production predominant. Industry employs c. 35% of the working population.

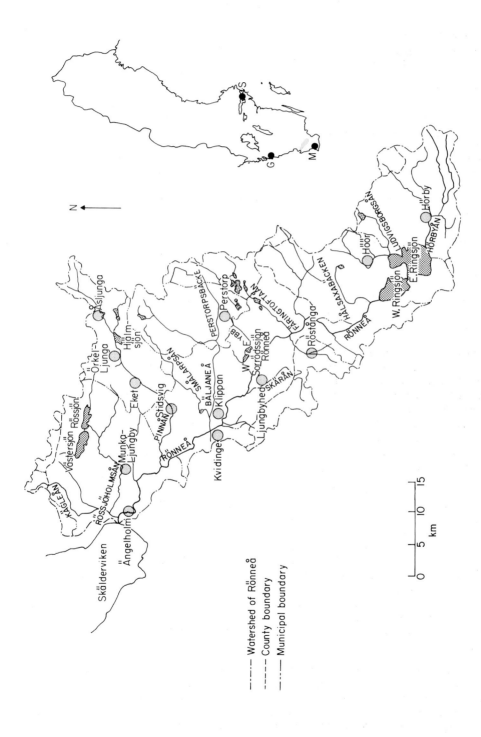

Fig. 20.1 Survey of Rönneå catchment area. Inset: Map of Sweden; Rönneå river system is dotted;
S, Stockholm; G, Göteborg; M, Malmö.

20.2   GENERAL DATA

DIMENSIONS

Area of catchment	1890 km^2 (including 66.5 km^2 of lake)
Ybbarpsån	89 km^2
Bäljaneå	228 km^2
Pinnån	206 km^2
Rössjöholmsån & Kälgeån	270 km^2
Highest point in catchment	160 m
Source	c. 160 m: streams draining the ridge Linerödsåsen
Length: total	120 km
downstream of Ringsjön	83 km
Drop in height below Ringsjön	54 m = 0.54 m km^{-1}

CLIMATE

Precipitation	700 mm (approx.)
Evaporation (mean)	350 mm
Temperature (mean annual)	7.8 $^{\circ}$C

GEOLOGY

(i) Northern side of river	Archaean
Southern side of river	younger
close to Ringsjön	Silurian
Ängelholm	Rhaet-Lias

(ii) In upper part, within the triangle Ringsjön, Lippan and Hässleholm, about fifty basalt domes some 50 m high - probably remnants of Tertiary volcanoes.

(iii) soils: Moraines with and without clay, sea clay, sand and gravel.

LAND USE

Agricultural fields 570 km^2; grazing 45 km^2; forest 1200 km^2.

POPULATION

Farming 13%; industry 40%; trade 17%; transport 6%; public services 24%.

PHYSICAL FEATURES OF RIVER

Discharge   (Data obtained before 1970; $m^3 s^{-1}$)

	at Ringsjön	2 km upstream of Klippan	Klippan, 2 km downstream of Bäljaneå	at Skälderviken	Tributary Bäljaneå at Klippan
maximum flood	36	77	137	200	41
normal flood	18	48	75	100	21
mean flood	4.0	10.1	13.7	21.0	3.3
normal low flood	0.9	2.2	2.5	4.5	0.25
minimum low flood	0.4	0.16	0.75	2.0	0.07

total discharge into bay at Skälderviken $6 \times 10^8 m^3 yr^{-1}$

$$10\text{-}11 \ l \ s^{-1} \ km^{-2}$$

SWEDISH WORDS

Examples of terms frequently encountered in descriptions of Swedish rivers:

å	river or small river
älv	large river
bäck	stream (beck)
sjö	lake
ås	ridge
slätt	plain
mosse	bog with peat
samhälle	small town
mölla	mill

Word endings

A terminal 'n' after the name of a river means 'the'. The present chapter mostly prefixes Rönneå with 'the', but adds the terminal 'n' for other rivers and omits 'the'.

20.3 PHYSIOGRAPHY AND HYDROLOGY

The Rönneå has the second largest catchment in Skåne. (The largest
river, Helgeån, has its source in another province, Småland.) The
Rönneå has its sources in the area east of (lake) Ringsjön and thus
drains the central part of the province. The small source streams run
mainly through mixed forests, and pollution is negligible, since the
region is thinly populated. At Hörby the river leaves the Archaean
ridge, passes through cultivated land and receives sewage effluent.
The stream is now called Hörbyån.

The main river begins below Ringsjön. The bed is about 5 m wide
here, and the river runs in a N-W direction, at first through hilly
forest country, then through an open landscape covered with fields and
meadows. It runs parallel to the Archaean rock horst Söderåsen. The
N-E. side of the ridge is steep and the small pure streams flow very
rapidly - sometimes in narrow ravines - down to the Rönneå. Of the
western tributaries, Bälganeå, rising in the southern-most part of the
ridge, is the largest. Unlike the others it moves rather slowly and
drains agricultural land. The greatest number and largest tributaries
however, come from the north-east. These also have the most water.

In its downstream reaches the Rönneå becomes more and more markedly
a lowland river. It empties into the sandy bay of Skälderviken 5 km
north of Ängelholm. The influence of high sea water levels is obser-
vable up to Tranarps Bridge, about 28 km from the mouth. The Rönneå
runs mainly through soils of moraine and glacial river gravel. Upstream the soils are clayey, and downstream, clay-poor. The hilly
landscape of the glacial riverain drifts consists of small ridges and
cavities sometimes filled by lakes. It is often very beautiful and
excellent recreation ground. The Counties therefore want these areas
to be protected against damage by housing schemes and deforestation.
In the downstream part of the main valley, up to Ljungbyhed, a glacial
sea bay has deposited varved clay and sand. From Ängelholm to the
mouth the river passes through a post-glacial sand area. The Ängelholm plain, built up of the glacial sea sediments, is very flat. It
rises slowly from the sandy dune coast up to 40 m above sea-level at
Ljungbyhed (c. 35 km). The Rönneå has often cut out 10 - 20 m deep
ravines in the plain.

The rapid changes between frost and mild weather in winter result
in frequent changes in type of precipitation. The amount of discharge
therefore varies, although it remains generally high throughout this
season. Usually there is sufficient snow to give rise to an obvious
contribution from snow-melt in March-April. In summer flows are
usually low. If rainfall is insignificant in autumn, the low flows
continue, and their total can extend for a very long time, e.g. until
December in 1975. On the other hand, the rainfall can be very high
in autumn and winter and then there is extensive flooding, especially
where the side of the river is flat. Flooded fields can cause the
farmers serious problems. On a stretch of about 20 km between Höja

and Bäljaneå, as much as 750 ha have been inundated. As a consequence
of these floods alluvial clay is sedimented in the valley. The continuous regulation of discharge at power plants does, however, lessen
some of the effects of the floods.

For the main river the following discharges are characteristic:

Level	below Ringsjön ($m^3 s^{-1}$)	Skälder-viken ($m^3 s^{-1}$)
maximum flood	36	200
normal flood	18	100
mean	4.0	21.0
normal low	0.9	4.5
minimum low	0.4	2.0

20.4 RINGSJÖN AND ITS TRIBUTARIES

The Ringsjön is by far the largest lake within the catchment of the
Rönneå and will be described in some detail. It consists of two units,
E. Ringsjön with Sätoftasjön and W. Ringsjön (Ryding, 1983b).

	area (km^2)	max. depth (m)	vol. ($\times 10^6 m^3$)	mean depth (m)	alti-tude (m)	drain-age (km^2)
E. Ringsjön	20.8	16	104	5	53.7	325
Sätoftasjön	4.5	18	10.4	2.3	53.7	105
W. Ringsjön	15.4	6	52	3.4	53.7	347

History. The land around the Ringsjön became populated early. Archaeo-
logical investigations show that an extensive settlement occurred along
its shores and on swampy islands and also around the prehistoric lakes
Agerödsmosse and Rönneholms mosse to the west (Bunte, 1976; Eskeröd,
1976). The abundance of fish, birds and good hunting grounds first
attracted the nomads; subsequently there came the megalithic farmers.
When the Ringsjön was lowered in the 1880s, a number of Stone Age sites
around the lake were laid bare. The Rönneå was, in prehistoric time,
navigable for most of its course. The falls at Djupadal, Forsmöllan
and Klippan were probably no hindrance to seamen wishing to reach the
Ringsjön.

Geology and soils. Ringsjön is on the border between Archaean gneiss,
Rhaet-lias sediments (mainly sandstones) and Cambrosilurian shales.
The surrounding soils are to the north composed of Archaean rock moraine,
to the south of clayey moraine. Clays occur in many places at the
shore line. The agricultural ground round Ringsjön is extensive (60 -
70%) except north of the lake, where only 30 - 40% of the land consists
of fields.

Tributaries. Three larger tributaries empty into Ringsjön. Hörbyån, considered to be the source of the Rönneå, flows into E. Ringsjön, Ludvigsborgsån and Höörsån into Sätoftasjön. All have their sources in and flow through mixed woods and arable land. The water is therefore more or less polluted when it reaches Ringsjön. Until a few years ago Höörsån received the sewage effluent of the urban district of Höör (c. 5000 population equivalents). Hörbyån still receives sewage effluent from the town of Hörby. Even after biological and chemical purification of this effluent the bacterial content of the river is still so high that bathing in the Ringsjön is sometimes prohibited near the entry of Hörbyån.

Water quality. The quality of Ringsjön water has been studied more or less continuously since 1940 and obvious changes have taken place during this period. The maximum pH has increased from 8 to above 9 in summer and both alkalinity and conductivity are generally higher. The surface water is very often supersaturated with oxygen in summer, while in the hypolimnion the oxygen content is reduced. When the lake is covered in ice, the water at the bottom can be deficient in oxygen. The situation with respect to dissolved oxygen is better in the shallow W. Ringsjön than in the other parts of the lake. In 1978 the average total P concentration was about 0.15 mg l^{-1} for the whole lake, thus indicating a clear eutrophic-hypereutrophic state. Phosphorus seems to be supplied also from the sediments. The P budget is now the subject of a comprehensive study, with the aim to elucidate the eutrophication process and to propose methods for reducing blooms (Ryding, 1983a).

Vegetation. Ringsjön provides a suitable environment for fringing vegetation. It is relatively shallow, the shores slope gradually and, especially with W. Ringsjön, the sediments are calcareous. Grazing still occurs around the lake and consequently the shore vegetation is nowhere intact. The level of Ringsjön was lowered in 1883 by 1.5 m, reducing the area of the lake by 3.4 km^2. New reed beds have developed, but the area overgrown is still small in relation to the total area of the lake. Phragmites australis is the most abundant component (Fig. 20.2). The floating leaf vegetation is insignificant, whereas the isoetid vegetation forms dense mats. This is more or less true for all three parts of the lake. In the 1950s the larger part of the lake bottom in W. Ringsjön was covered with submerged plants, the most abundant of which were charophytes, Myriophyllum alterniflorum, Naias flexilis, Ceratophyllum demersum and Elodea canadensis (Almestrand & Lundh, 1951). Nowadays, the submerged macrophytes have diminished, probably as a result of reduced illumination owing to the much increased phytoplankton crops. A notable finding as late as 1950 was a locality for Isoetes lacustris (Lundh, 1951b), a species belonging normally to non-humic oligotrophic lakes and thus perhaps an indicator of a former more oligotrophic condition in Ringsjön.

Plankton. The earliest detailed investigation of the Ringsjön plankton was made in the 1940s (Lundh, 1951a). In April 1949 Asterionella formosa was dominant and Melosira granulata frequent in E. Ringsjön;

Fig. 20.2 W. Ringsjön at the outflow of Rönneå, showing the broad, dense reeds of *Phragmites* along the shore-line. The background shows a famous beech wood on the southern shore. May 1980.

in September *Ceratium hirundinella* was dominant and *Aphanizomenon flos-aquae* and *Melosira granulata* var. *angustissima* very frequent. In W. Ringsjön *Dinobryon sociale* dominated in April 1949, while *Asterionella, Melosira granulata* and *Synedra acus* var. *angustissima* were frequent. In September *Fragilaria crotonensis* was the dominant and *Melosira granulata* var. *angustissima* the next most frequent. Ringsjön could thus, by this time, be considered a more or less typical lowland lake when compared to other Scanian lakes (Lundh, 1951b).

Some sparse data from older literature (Trybom, 1893) may indicate that Ringsjön contained more oligotrophic species before the lowering of the water level in 1883. Lundh-Almestrand (1954) concluded that the relatively eutrophic species, *Melosira granulata* had increased since the end of the nineteenth century. It is, however, hardly likely that Ringsjön has ever been an extremely oligotrophic lake because of the geological conditions and its position in an ancient agricultural area. Changes in the species composition and abundance of the plankton have occurred very rapidly since the 1950s. During the middle part of the sixties large crops of *Aphanizomenon flos-aquae* were noted. Other blue-green algae species have appeared in tremendous quantities (water bloom) since the end of the 1970s, especially *Microcystis* species (Cronberg, 1983).

The accelerating eutrophication in recent decades is due especially to the increasing load of more or less purified sewage effluent and rapid increases in pollution from farm fertilizers. 75% of N and P comes from farming (Ryding, 1983a). Problems are also created by

recreational settlements, camping, boating and waterski-ing. Ringsjön
was one of the Scanian lakes that was "colonized" by summer guests
particularly early, already by the turn of the century. Everything so
far indicates that excess phosphorus is the key factor and the sewage
treatment plants of Höör and Hörby include chemical treatment to reduce
the phosphorus content. Nevertheless, it cannot be said that the lake
has recovered yet - rather the reverse.

It is of great importance to try to stop the eutrophication process
as far as technology and economy will allow. The detrimental changes
in Ringsjön influence not only aesthetic and recreational quality but
also fishing and grazing: cattle have died after drinking lake water.
Of still greater importance is the fact that the lake serves as a
source of water for a large region including the towns of Lund, Eslöv,
Landskrona and Helsingborg. The water is treated by microstraining,
chemical precipitation and filtration.

20.5 THE MAIN RIVER

20.51 CHANGES ON PASSING DOWNSTREAM

The total fall from Ringsjön to the sea is 53.7 m, 70% of which is
concentrated in a few rapids. Much of the river has, therefore, slow
current speeds and in parts an almost lenitic character. The main
Rönneå comes from W. Ringsjön, where a dam regulates the water level
of the lake. The river then flows through a flat landscape between
two large bogs, in one of which peat-cutting is still going on; part
is used also as a site for the deposition of sludge from Eslöv treatment plant.

At Hasslebro summerhouses have been built along the riverside,
making access to the water impossible for anyone but the owner. A
recent law now prohibits any more private houses so close to the
water's edge. From Hasslebro the river passes over to a lotic section
(Stockamöllan) and the current speeds are high. Where the river forms
rapids, even small ones, water power has been used in the past for
mills (Sw. möllor). Most now no longer operate, but remnants are
still preserved. The old mills have become parking places for nature
lovers. At Stockamöllan a well-known truck industry has replaced the
old mill.

The next part of the valley offers a particularly beautiful natural
landscape. The slopes are covered with forests of various types. The
valley is narrow but sometimes there is a strip of swampy meadow along
the bank which is flooded in rainy seasons. Canoeists love this quiet
landscape without houses and other tourists, where there is an opportunity to observe the rich undisturbed wild life.

Upstream of the mill-pond at Djupadal, the river Snällerödsbäcken

enters the Rönneå. It drains, among other waters, the small lake Dags-
torpssjön, which is surrounded by forests of beech and planted spruce
in a hilly terrain. The landscape is appreciated as recreation ground
and it is possible to swim in the lake, which is said to have the
"purest water" in Skåne. The plankton, which can be classed as meso-
trophic, does not seem to have changed much during the last thirty
years.

Apart from the vicinity of the outflow of Ringsjön, the aquatic
macrophytic vegetation downstream to Djupadal is relatively sparse.
This may be due partly to the high turbidity of the water during the
growing season as a result of phytoplankton originating from Ringsjön.
In the mill-pond at Djupadal, however, there are stands of *Glyceria
maxima*, and downstream from the bridge high stands of *Phalaris arundinacea* and *Phragmites australis*. On stones at the bottom of the river
there are abundant growths of *Fontinalis antipyretica*, *Hildenbrandia*,
Lemanea and *Cladophora*. The volcanic tufa found at this site is
perhaps responsible for the rich vegetation.

Leaving Djupadal, where the bed is very narrow, the river changes
direction towards the north-west. It is fringed by marshy meadows and
surrounded by deciduous forests. The valley is deep and the river
moves slowly in a meandering course. At Ö. Forestad the river touches
a big esker, Färingtoftaåsen, and receives from the east the stream
Klingstorpabäcken (Färingtoftaån). This arises in a lake-rich area of
gravel ridges and cavities. The largest lake is Store Damm, which like
the other lakes in this area has peaty shores. These lakes are known
for their rich bird life. Aquatic birds, such as the whooper, both
rest and breed here. The surface is partly covered with water lilies
mixed with *Equisetum fluviatile* and other species. From the west
comes Bälganeå, which receives the sewage effluent of Röstånga
municipality, and also drains a big rural district. The water of
this river has the highest conductivity (500 - 550 μS cm^{-1}) of the
whole Ronneå catchment.

Further downstream, at Riseberga, the Rönneå becomes wider and the
landscape flattens. The riverside meadows broaden, becoming extensive
flooded areas during the rainy seasons. At the old inn Spången the
river is 25 m wide and filled completely with *Glyceria maxima*
and *Scirpus lacustris*.   Within Ljungbyhed the current is rapid, but
later slows down and at the castle Herrevadskloster the river and its
surroundings are said to make a "pastoral idyll". Here the river,
fringed by alders, winds along through grazing grounds with scattered
oaks, giving shelter for the cattle. At the castle, once a monastery
in the Middle Ages, Ybbarpsån flows into the Rönneå. The monks used
to use its water power.

A couple of kilometres downstream the river is transformed into a
long and narrow lake, the reservoir of Forsmöllan, which is owned by
Sydkraft AB and used for hydro-electricity. The fall is 5 m. The
reservoir and the rapids are popular places for excursions and anglers.

Between Forsmöllan and the next fall, situated about 2 km downstream at Klippan, the current is very fast and the river falls 8 m. In the municipality of Klippan is the oldest industry of Skåne, the paper-mill Klippans Finpappersbruk AB.

The Rönneå is then dammed again, forming a long and narrow lake; this is the Stackarp reservoir (Fig. 20.3), which also has a hydro-electric plant (6 m fall). The reservoir is partly surrounded by marshy ground overgrown with *Glyceria maxima*. The shores are intensively grazed. Submerged macrophytes include *Elodea canadensis*, *Potamogeton crispus* and filamentous algae (especially *Vaucheria*, *Microspora*, *Spirogyra* and *Oedogonium*). The water is strongly turbid with clay particles and during summer also phytoplankton.

Fig. 20.3 Stackarp reservoir downstream from Klippan paper-mill. The clayey shores are trampled heavily by cattle. Saprobic vegeta-tion can sometimes be observed here. August 1979.

From Stackarp the Rönneå flows through a fertile open agricultural district towards Ängelholm, sometimes like a broad ditch between fields. Near Stackarp the river is bordered with alders. The marshy land adjacent is mostly used for grazing and is floristically rich. It is flooded at high water levels. The river winds through Ängelholm in a wide and deep bed fringed with a park-like vegetation. Along the remaining stretch down to the mouth there are forests of alder, beech and pine. The bed of the river is more or less overgrown with reeds, which provide a favourable habitat for birds. At its mouth the Rönneå divides into two branches forming islands. Brackish water can be pushed upstream several kilometres.

20.52   USAGE OF THE WATER AND WATER QUALITY

The Rönneå passes through districts with various cultural traditions and its water has been exploited for different purposes. Since time immemorial the river has surely been used for fishing, transport and water power. Today, however, its principal importance lies in its utilization as a receiving body for polluted water. A tragic fate!

Only two densely populated areas, Ljungbyhed (military airport) and Ängelholm, put treated sewage effluent directly into the river. These include about 26000 people, since small densely populated areas in the vicinity are included. The town of Ängelholm has, among other industries, a tannery which discharges chrome-rich waste water.

The reservoir at Stackarp also receives waste effluent, in this case from Klippans Pappersbruk, which produces annually about 30000 tonnes of paper. Sufficient dilution of the waste water is made possible because of a law stipulating that Ringsjön is regulated to secure a minimum discharge of 2.5 m^3 s^{-1} at Forsmöllan. In addition, the factory has had to reduce fibre discharge by providing the paper machines with fibre regainers. Finally the waste water is treated by sedimentation prior to its release to the Rönneå.

The water of the main river is influenced also by the quality of the tributaries. Table 20.1 presents some results for 1980, based on monthly samples. In general differences between minimum and maximum were extremely large in 1980, because of the high precipitation in June and October-December; total precipitation was, in spite of a dry spring, 40% higher than average. Therefore, the analyses provide a good indication of seasonal variations in water quality.

Nowhere in the main river has the water anything like a natural character. The water discharged from Ringsjön is already greatly affected by eutrophication processes in the lake. During May-October the water leaving Ringsjön is extremely turbid due to the phytoplankton, and with pH values over 9. The turbidity (up to 17 FTU-formazin turbidity units) is visible down to the mouth. It is not aesthetically attractive. At the time of year when the plankton is not so dense, there is an increase in turbidity in the lower river due to erosion and surface run-off from fields. Thus, the lower stretch of the Rönneå is turbid throughout the year. The same holds true for water colour. The Ringsjön water is generally weakly coloured (equivalent c. 40 - 60 mg l^{-1} Pt; in 1980, 15 - 80 mg l^{-1}), but the colour increases more and more downstream on account of the supply of humic water from the tributaries, especially Baljaneå and Pinnån (Table 20.1). In the rainy autumn of 1980 the water was intensively brown at the mouth (200 mg l^{-1} Pt) upstream of Ängelholm. The Rönneå water coming from Ringsjön has relatively high concentrations of total phosphorus (up to 0.3 mg l^{-1} P) and total combined nitrogen (4.4 mg l^{-1} N). While phosphorus usually decreases slightly downstream, nitrogen increases. Thanks to the present effective purification of the sewage effluents,

Table 20.1 Chemistry of water in Rönneå and tributaries: data show the range found during monthly samplings during 1980. Analyses from tributaries refer to a station just above the entry to the main river.

site		pH	conductivity $\mu S\ cm^{-1}$	alkalinity $meq\ l^{-1}$	turbidity FTU	colour (as mg l^{-1} Pt)	O_2 mg l^{-1}	BOD_7 mg l^{-1}	NH_4-N $\mu g\ l^{-1}$	NO_3-N $\mu g\ l^{-1}$	total-N $\mu g\ l^{-1}$	total-P $\mu g\ l^{-1}$
Main river												
Rönneå, below	min.	7.40	261	0.26	1.8	15	8.10	2.4	6	1	710	76
W. Ringsjön	max.	9.05	384	1.76	22	80	20.90	9.0	1110	1400	4400	294
Rönneå, above	min.	7.15	235	0.57	3.8	50	7.15	2.8	126	1200	3200	55
Skälderviken	max.	7.65	371	1.02	17	160	13.10	7.9	1590	3000	6500	210
Tributaries												
Bålganeå	min.	7.30	365	0.43	2.8	40	8.15	2.2	29	1600	2800	37
	max.	7.65	469	2.29	4.6	60	11.90	4.6	1050	3500	7500	106
Ybbarpsån	min.	6.60	171	0.15	2.5	70	7.40	1.6	34	160	1400	32
	max.	7.50	344	0.81	9.5	180	11.70	6.7	470	1200	2600	140
Bäljaneå	min.	6.85	140	0.22	2.7	50	6.90	2.4	343	590	1900	44
	max.	7.35	329	1.00	18	400	13.50	9.5	3175	1600	5700	486
Pinnån	min.	6.70	140	0.18	3.0	60	7.50	3.6	485	1200	2700	80
	max.	7.15	266	0.30	7.6	250	13.15	9.4	2280	2790	6100	254
Rössjöholmsån-Kägleån	min.	7.10	186	0.52	2.0	50	9.10	1.4	22	1200	1900	47
	max.	8.10	239	0.89	7.0	200	13.75	3.1	416	2000	3200	101

there are seldom problems due to reduced oxygen concentrations. The
waste water and surface run-off from Ängelholm have led to increased
heavy metal contents in the sediments, particularly in chromium and
lead.

As mentioned in 20.4, Ringsjön is used as a source of water for
several densely populated areas.   Klippans Pappersbruk takes in river
water for industrial usage. Rönneå water is also important for the
irrigation of fields, which is increasing, and for watering cattle.
There are three large hydro-electric power plants on the main river,
as described above. Most of the numerous water mills are no longer
functioning.

Ringsjön is vital for the fishing trade in the province, while
amateur angling is rewarding and contributes one/third of the total
catch. The conditions for summer anglers have worsened in recent
years because of the enormous plankton crops. Jigging for fish in
winter is more profitable. Valuable species in Ringsjön include
eel, pike, perch and bream. Since the eel cannot migrate up the Rönneå,
a yearly input is needed. The one-time lucrative fishing for whitefish has lessened appreciably, probably another effect of the eutrophication. At present there is serious discussion about the possibility of improving the water quality by selective fishing for bream.
A decrease in the present bream population ought to result in an
increase in the zooplankton density and hence decrease in the phytoplankton. Economically unimportant fish have been used as fish food for
large-scale basin-culture of rainbow trout in the neighbourhood. This
was situated close to the Höör waste water treatment plant, which provided heat energy to warm the basins (now used for algal cultures). The
saprobic conditions which occurred at one time in Hörbyån and Höörsån
have disappeared once improvements in waste treatment started in 1975
and 1978, respectively. The environment for fish will probably improve
in the long run. Electric fishing in 1970-1972 for trout revealed
only a small population. E. and W. Ringsjön have been popular bathing
places for a long time. However, at present the water-bloom acts as a
deterrent. In addition, bathing along the Hörby shore of E. Ringsjön
is often prohibited.

In the main river Rönneå, perch and pike can be caught, and, near
the mouth, also trout. The migration of salmon is prevented at
Stackarp. Around the upper reaches of the Rönneå there are a great
number of put-and-take waters. In recent years the Rönneå has become
the favourite canoe-river in Skåne and the number of canoes is
increasing more and more. A positive consequence of the enormous
interest in recreational activities in and near the river is that the
authorities have sharpened their watchfulness. The river is subject
to continuous checking of public activities, and any deterioration
in water quality should be registered rapidly. The countryside of
Skåne can be damaged easily if visited by too many people, such as by
trampling, discarding litter or disturbing of bird life.

20.6   THE LARGEST TRIBUTARIES

20.61   YBBARPSÅN

This river drains a series of shallow dammed-up, pond-like lakes in a
dead ice landscape south of Perstorp. The lakes serve as regulating
reservoirs for Perstorp AB, a chemical industry and one of the most
important in the world for production of laminates, plastics, formal-
dehyde and polyalcohols. Ybbarpsån is the receiving body for the
waste effluent, which flows into the river two kilometres downstream
from the factories. The Ybbarpsån soon widens to a long and narrow
reservoir, Storarydsdammen, built in 1960 to enhance the self-purification of the water. About 7 km south-west of the factories the river
meets the 5 m deep E. Sorrödssjön, from which a canal cut by monks in
the Middle Ages carries the water to the very shallow pond-like W.
Sorrödssjön. The Ybbarpsån flows into Rönneå at the castle Herrevadskloster.

Perstorp AB started just over a hundred years ago and by the end of
the 1940s had become a large industry. The original product was acetic
acid made by dry distillation of beech wood. Owing to rapid increase
in production, the industry became an environmental problem. It caused
the first great environmental debate in Skåne (Lillieroth, 1949). The
unfavourable effects on both water and air were discussed in the press
and journals. In the 1940s the authorities still gave priority to
economic interests. Although the water of the river and the lakes was
terribly polluted and directly poisonous due to phenol, it took some
decades to get the water restored. The industry was at last able to
reduce the discharge of phenol by internal measures, and in 1971 an
effective treatment plant was built. Only one year after the start of
this plant, it was possible to observe green algae in the water rather
than sewage-fungus, and a "normal" plankton developed in the two
eutrophic lakes. The earlier good fishing conditions have partially
been restored and the lakes are again attractive to birds. Many of
the substances produced are difficult to break down biologically and
often are also toxic. Therefore, there is still a risk of disturbance
in the active sludge process and reduced efficiency in treatment. In
such circumstances, purification is completed in the river and an
obvious oxygen decrease can occur, primarily in Storarydsdammen.

20.62   BÄLJANEÅ-PERSTORPSBÄCKEN

Bäljaneå, the outlet from the lake Bälingesjön, flows into the Rönneå
some kilometres west of the municipality of Klippan, and by that time
has received two smaller tributaries, Perstorpsbäcken and Smålarps-
bäcken. Bälingesjön is an example of the results of the numerous
lake-lowering projects carried out in Sweden at the end of the nineteenth and the beginning of the twentieth century. Some of these
projects were done without permission from the authorities. The
lowering was not successful. The ground gained was rich in boulders

partly covered with bog ore and unsuitable for the farmers. The only
benefit was a football ground on a sandy shore. The important lasting
result of the lowering appears to be a slight eutrophication, with
increased reed beds and plankton development. However, a recent
comparison of plankton composition with that in the 1940s, when
detailed studies were made (Lillieroth, 1950) shows that the eutrophication process has remained moderate. The lake played a valuable role
in increasing swimming possibilities for the people living there, far
from the coast, during the car-free part of this century. At present
it serves as a place for swimming instruction, boating and fishing.
Perch, pike and pike-perch can be caught.

The Bäljaneå runs for most of its course through bog-rich forested
districts, including large spruce plantations. Cultivated land is only
sparse. The river water is therefore very brown, especially in rainy
periods. Current speeds are very high in some places and small rapids
are formed. In the eastern part of Klippan the bottom is stony and
shaded by alder and willow; the fern *Osmunda regalis* occurs here. The
shaded stones are overgrown with aquatic mosses (e.g. *Fontinalis
antipyretica*) and liverworts.

Via Perstorpsbäcken, the Bäljaneå is the receiving body of water for
the municipality of Perstorp, and at Klippan it receives sewage
effluent following biological purification. The waste water from the
tannery at Klippan carries elevated chromium, leading to chromium
accumulation in the sediments. The purification of the polluted water
in Klippan is unsatisfactory, leading at times to marked decreases in
dissolved oxygen in the river and, at low dilution, also to very high
nitrogen and phosphorus contents.

Current speeds are less in the lowermost part of the river near its
entry to the Rönneå and there is a rich vegetation of, for example,
Scirpus lacustris, *Sparganium* and *Nuphar luteum*. Salmon migration
occurs to a certain extent in the Bäljaneå.

20.63 PINNÅN

Pinnån, the most attractive tributary, is the outlet of Åsljungasjön
and flows via Hjälmsjön down to the Rönneå. Åsljungasjön has been
lowered twice, both times without success. Hjälmsjön is a well-
frequented lake, with many large rowing and canoe competitions in
summer. The river Pinnån in its uppermost course goes through vast
forests with cultivated fields only close to the river. Downstream
of Örkelljunga it flows through large glacial, fluvial deposits
including ridges, plateaus and terraces - a suitable recreation area.
At Stidsvig the river is dammed up to form a small lake, Kopparmöllesjön
(see below). The Pinnån then comes to a fertile agricultural region,
where it winds between borders of alder and ash. There are several
rapids, previously utilized for milling and nowadays for hydro-electric
power. At Skvattemölla the water is used for fish ponds. The largest

waterfall is at Stora Mölla (power plant). At times of high flow the
excess water cascades over the rocks in the ravine and the spray glit-
ters like a rainbow. This waterfall is unique in Skåne.

The course of Pinnån through grazing land down to the Rönneå is tran-
quil. From the high valley slope at Stora Mölla there is a wonderful
panorama of the mouth area of Pinnån. At spring high water there is an
awe-inspiring stretch of lake caused by the extensive flooding of the
two rivers over fields and grazing area. The downstream part of Pinnån
is idyllic scenery,reminiscent of an English "park landscape". It is a
pity, therefore, that the river is exposed to industrial waste effluent
at Stidsvig from a large factory producing glue, fats, edible gelatin
and bone meal (Extraco AB). The effluent has a high content of organic
matter, nitrogen and phosphorus. It has led to a complicated river protection problem for a long time, since low flows in Pinnån can extend
for long periods. Before 1966 the effluent was untreated, resulting in
a saprobic vegetation (*Sphaerotilus natans* and *Carchesium polypinum*) and
anaerobic conditions in Kopparmöllesjön and the river downstream, with
accompanying fish mortality. The problem was compounded by discharges
of cooling water. Thick layers of black anaerobic sediment accumulated
in the lake and in the river downstream large quantities of *Vaucheria*
grew. Since 1966 the effluent has been treated in an activated sludge
plant and this has partially improved the state of the river. The
growth of saprobic organisms has decreased and oxygen conditions improved. Nevertheless, during summer the oxygen concentrations can still
become critical, because of the nitrification of ammonium (4 - 6 mg l^{-1}
N). Fish kills have still occurred. After installing equipment for
reduction of nitrogen discharge in 1982, conditions in the river have
stabilized and the future development looks promising.

Since 1967 Kopparmöllesjön has been invaded by an extremely dense
vegetation of *Ceratophyllum submersum*, *Potamogeton obtusifolius* and
Callitriche. Downstream from the lake the river is filled with *Callitriche* and filamentous algae (*Vaucheria, Microspora, Tribonema, Oedogonium*). A luxuriant growth of algae, often thickly covered with small
diatoms and mosses (*Fontinalis, Drepanocladus*), is characteristic of
Pinnån down to its mouth. Salmon migrate up to Stora Mölla, where
the fall causes a natural hindrance for further movement upstream.

20.64 RÖSSJÖHOLMSÅN-KÄGLEÅN

Rössjöholmsån has the largest drainage basin of the tributaries of the
Rönneå. The source area consists of the south side of the ridge
Hallandsåsen, including the lakes Västersjön and Rössjön, situated
along a fault which once gave rise to the horst. Rössjön (3.6 km^2) is
the eastern lake. Västersjön is more extensive, but not as deep. The
shores are mostly minerogenic, often rich in boulders. The littoral
vegetion is scant and that characteristic of oligotrophic *Lobelia*
lakes. In Västersjön there are more or less dense beds of reeds in
shallow parts, with *Scirpus lacustris* the dominant. These lakes occur

in a district which is very sparsely populated and have so far not been
subject to serious pollution. The water has a relatively low colour
(20 - 40 mg l^{-1} Pt) and Secchi disc transparency is about 2.5 m in
Västersjön and 3.5 m in Rössjön; alkalinity is below 0.1 mmol, so the
lakes may be susceptible to increasing acidification in the immediate
future.

Västersjön is strongly exploited for recreational activities (Fig.
20.4). Boating is especially disturbing and too many summerhouses have
been built in city-like aggregates. The idyllic bathing-places where
the "natives" could sit on the grass are gone and the risk of serious
pollution has become real.

Fig. 20.4 Västersjön on a summer Sunday: sailing, waterski-ing and
sometimes car-washing! August 1978.

While Västersjön in summer time echoes with the sound of motor
engines, Rössjön is still quiet, since all boat traffic is prohibited.
This lake has for a long time been protected by the owner of the Röss-
jöholm castle, and is now a "protected area for birds". Apart from
the castle the shores are quite free from settlement. They are well-
frequented by nature lovers, especially in spring and autumn. The
northern shores of Rössjön are more lobated than those of Västersjön
and offer suitable picnic sites. The rivulets from the ridge sometimes
flow in impressive ravines, which served as effective hiding places for
guerrillas during the Swedish-Danish war in the seventeenth century.

Västersjön and Rössjön can be classified as oligotrophic-mesotrophic
lakes and total phosphorus does not exceed 0.015 mg l^{-1} P. The dominant phytoplankton species are *Gomphosphaeria naegeliana, Tabellaria*

and *Asterionella*. The blue-green alga brings about a plankton turbidity of long duration. In Västersjön one can catch pike, perch, trout, bream, vandace and eel. A fish pond system south of Rössjön, is used for breeding pike spawn and to some extent for growing carp. The water discharged to the outlet from the two lakes is only slightly humic and on the whole of good quality. The supply of water from the bogs south of the lakes makes the river more brown and iron-rich, which is unfavourable for fish production (rainbow trout) in the large fish ponds downstream (Munka Ljungby). The low water level in summer is also detrimental.

Rössjöholmsån is at present the source of water supply for Ängelholm community. The intake is situated in the lower part of the river on the plain. The drinking water is taken from artificial ground water. The varying quality of the river water (at times high contents of iron, suspended matter and nitrate) creates problems in the filtration plant. Ängelholm would therefore like to take water directly from Rössjön, at about 12 m depth. The lake is already regulated with an amplitude of 1.4 m, which is rather wide with regard to the shore configuration. The planned regulation would lead to an elevation of the present minimum water level. This reduction ought to mean a positive ecological effect. The proposed regulation would also affect the discharge variation in the river. It seems that it might be advantageous to the fish ponds in Munka Ljungby in extremely dry periods, since the release from Rössjön will be limited under such conditions. According to the present stipulation zero release is possible. Permission for the abstraction of water from Rössjön has recently been approved by the authorities. Rössjöholmsån is also essential as a reproduction area for the salmon and trout populations in the water system of Rönneå.

During rainy seasons the lower part of Rössjöholmsån overflows as is the case in the main river and in Pinnån. About 2 km from the mouth the river is supplied with water from Kägleån, which changes the water quality of Rössjöholmsån considerably. It is a lowland river with high concentrations of nitrogen and phosphorus.

20.7 CONCLUDING COMMENTS

The River Rönneå has become "a problem child", according to an author in Skånes Natur (Holm, 1976). Unfortunately a similar fate has happened to other rivers in Skåne. However, it has now become apparent to the authorities that drastic preservation measures must be taken. Areas which are still fairly intact will be protected. Often the problem is one of "wear" caused by too many people moving around within a limited area. The Västersjön-Rössjön district is one such example which requires protection.

Pollution connected with densely populated areas seemed at one time to pose very complicated problems, but these have now been attacked

Fig. 20.5 Rönneå. *Upper* Billinge, 10 km from W. Ringsjön, where the shallow river is fringed by intensively grazed meadows and canoeing usually starts. *Lower* South of Djupadal, 17 km from W. Ringsjön, where the river meanders through marshy meadows. (Scandia Photopress AB)

with success by constructing highly effective waste treatment plants.
Diffuse pollution, however, still remains to be tackled. The crucial
situation for water protection is during long dry periods, when the
effects of eutrophication are particularly obvious on many of the
rivers and lakes of the Rönneå system. Water levels must be increased.
Steps required include the reduction of the present rate of abstraction
for drinking water, regulation to retain more water in rainy periods
and the transfer of water from other drainage basins.

There is reason to set great hopes on the "Bolmen project". This is
the plan by nine communities in Skåne to take "hygienically not
objectionable" water from Bolmen, a large pure lake in Småland. A
tunnel from Bolmen, more than half-finished, is to connect with the catchments of Ringsjön and Vombsjön. This would make it possible to increase
the quantity of water in Ringsjön and also to raise low flows in the
Rönneå by direct input of Bolmen water. Whether this project will be
brought to fruition seems to be a question of "profitability".

REFERENCES

1.   Published

Almestrand A. & Lundh A. (1951) Studies on the vegetation and hydro-
     chemistry of Scanian lakes. I-II. *Bot. Not., Suppl.* 2, 3.
Bunte C. (1976) Dokument från det förgångna. *Skånes Natur* 63, 27-40.
     (Documents from the past.)
Cronberg G. (1983) Förändringar i Ringsjöarnas växtplankton under 1900-
     talet. CODEN LUNBDS/(NBLI-3053)/1-57 (1983). ISSN 0348-0798.
     (Changes in the phytoplankton of Ringsjön during the twentieth
     century.)
Eskerod A. (1976) Glimtar från Ringsjön. *Skånes Natur* 63, 7-16.
     (Glimpses from Ringsjön.)
Holm H. (1976) Naturvårdssyn på Ringsjöbygd. *Skånes Natur* 63, 117-26.
     (Aspects on environment protection in the countryside of Ringsjön.)
(Kristianstads och Malmöhus) (1979) *Vägen till naturen i Skåne.* ISBN
     91-38-04171-5. Utg. av länsstyrelserna i Kristianstads och
     Malmöhus län. (*The Way to Nature in Skåne.* Published by Counties of
     Kiristianstad and Malmöhus.)
Lillieroth S. (1949) Om ogynnsamma följder av sjösänkning och vatten-
     förorening i nordvästra Skåne. *Skånes Natur* 36, 6-48. (Unfavourable
     consequences of the lowering of lakes and water pollution in N-W.
     Skåne.)
Lillieroth S. (1950) Über Folgen kulturbedingter Wasserstandsenkungen
     für Macrophyten- und Planktongemeinschaften in seichten Seen des
     südschwedischen Oligotrophiegebietes. *Acta Limnologica* 3. (English
     summary.)
Lundegårdh P.H., Lundqvist J. & Lindström M. (1964) *Berg och jord i
     Sverige.* Stockholm. (*Rock and Soil in Sweden.)*
Lundh A. (1951a) Some aspects of the higher aquatic vegetation in the
     lake Ringsjön in Scania. *Bot. Not.* 1951, 21-31.

Lundh A. (1951b) Studies on the vegetation and hydrochemistry of
    Scanian lakes. III. Distribution of macrophytes and some algal
    groups. *Bot. Not. Suppl.* 3, 1.
Lundh-Almestrand A. (1954) *Melosira islandica* and *M. granulata* in the
    Scanian lake Ringsjön. *Svensk bot. Tidskr.* 48, 491-95.
Statens Naturvårdsverk (1980) Undersökningar i Rönneåns avrinning-
    område 1977. PM 1336. (Investigations in the catchment of the
    Rönneå 1977. Published by National Swedish Environmental Protection
    Board.)
Trybom F. (1893) Ringsjön i Malmöhus län, dess naturförhållanden och
    fiske. *Medd. Kongl. Landtbruksstyr. N:4.* Stockholm. (Ringsjön in
    Malmöhus county, its nature and fisheries.)

2. Duplicated reports

Almestrand A. (1976) EXTRACO AB. Redogörelse för undersökning av sedi-
    mentförhållandena i Kopparmöllesjön jämte biologiska undersökningar
    i Pinnån 1976. (EXTRACO AB. An investigation of the sediments in
    Kopparmöllesjön and biological studies in Pinnån in 1976.)
Almestrand A. (1977) ÄNGELHOLMS KOMMUN. Limnologiska förhållanden
    inom Västersjön-Rössjön-Rössjöholmsåns avrinningsområde under
    tidsperioden 1946-1977. (Limnological conditions in the catchment
    of Västersjön-Rössjön-Rössjöholmsån during 1946-1977.)
Berzins B. & Lundqvist I. (1973) Limnologisk undersökning av Ybbarps-
    ån 1972. (A limnological study in Ybbarpsån in 1972.)
Björk S. & Lettevall U. (1968) BOLMEN-LAGAN-RINGSJÖN.
Franzén E. & Olofsson S. (1971) Vattendragsutredning för Rönneå.
    Etapp. 1. (General planning for the Rönneå.)
Länsstyrelsen i Kristianstads län och Malmöhus lan (1975) Naturvårds-
    plan Skåne. (Kristianstad and Malmöhus Counties 1975. Environmental
    protection planning in Skåne.)
Länsstyrelsen i Kristianstads län och Malmöhus lan (1976) Rönneån.
    Förslag till samordnat kontrollprogram. (Kristianstad and Malmöhus
    Counties, 1976. The Rönneå Scheme for a co-ordinated control
    programme.)
Mårtensson B., Sandberg R., Andersson Th. & Alkne H. (1979) Ringsjön
    som naturresurs. (Ringsjön as a natural resource.)
Mellanskånes Planeringskommitté (1980) Vattenundersökningar i Ringsjön
    1975-1979. (The Planning Committee of Central Skåne, 1980. Investi-
    gations on Ringsjön water 1975-1979.)
Nilsson B.M. (1974) Fritidsfiske inventering för Malmöhus län 1973.
    (Inventory of amateur angling in Malmöhus County 1973.)
Ryding S.-O. (1983a) Redovisning och utvärdering av limnologiska under-
    sökningar. I: Ringsjöområdet, ekosystem i förändring. Mellanskånes
    Planeringskommitté. (Account and evaluation of limnological investi-
    gations. In: Ringsjön, a changing ecosystem. The Planning Committee
    of Central Skåne.)
Ryding S.-O. (1983b) Vattenkvalitet och ämnestransport. Ringsjön och
    dess tillflöden 1975-1980. I: Ringsjoområdet, ekosystem i förändring.
    Bilaga I. Mellanskånes Planeringskommitté. (Water quality and

transport of nutrients in Ringsjön and its tributaries 1975-1980.
In: Ringsjön, an ecosystem in change. App. I. The Planning Committee
of Central Skåne.)

Appendix A: Guide to Further Literature on European Rivers

The following bibliography is intended to help the reader start to find
the way into the literature; it is very far from comprehensive, especially
in the case of the eastern and southern countries. Articles are not
included if they are listed by Whitton (1975) or at the end of chapters
in the present volume.

GENERAL PAPERS BASED LARGELY ON EUROPEAN STUDIES OR WITH EXTENSIVE BIBLIOGRAPHIES

Armitage *et al.*	1983	quality score based on macroinvertebrates
Armitage	1980	regulation in the U.K.
Backhaus	1973	algae as indicators
Breitag & von Tümpling	1982	D.D.R. biological methods
Brooker	1981	impoundment on fisheries, with extensive bibliography
Dawson & Haslam	1983	shading and marginal vegetation
Ellington & Burke	1981	environmental policy
Haslam	in press	macrophytes in E.E.C. rivers
Haslam & Wolseley	1981	assessment and management of river vegetation
Hellawell	1978	surveillance methods
Logan & Brooker	1983	macroinvertebrates (mainly U.K.)
Milner *et al.*	1981	extensive bibliography includes papers on fish not listed here
Müller	1974	drift
Newbold *et al.*	1983	semi-popular account of conservation and river engineering
Schmitz & Illies	1980	biological methods
Schmitz & Schuman	1982	temperature and salmonid zone
Sládeček	1973	saprobic system
Tarak	1981	law and pollution control

AUSTRIA

Ritrodat-Lunz study area	Bretschko	1981	zoobenthos
various montane	Kann	1978a	benthic algae
various montane	Kann	1978b	benthic algae
Danube (Donau)	Kann	1983	benthic algae
Finstertaler Bach, Gurgler Ache	Kawecka	1974b	organic pollution and algae

Pitburger Bach	Kownacka & Margreiter	1978	benthic macro-invertebrates
Limmat	Schanz & Betschart	1979	algal bioassay

BELGIUM

Vesdre	Houba & Remacle	1983	heavy metals
Helle, Roer Superieure	Leclercq	1977	chemistry, vegetation
near Boom	Maes	1982	organic C
Meuse	Mouvet & Bourg	1983	heavy metals
Flavion	Stroot	1984	Trichoptera ecology in trout river
Lieve	Willem & Frans	1972	general survey
Samson	Maquet	1983	
Ely	Esho & Benson-Evans	1983	sewage fungus

BULGARIA

Maljovica (Rila)	Kawecka	1974a	benthic algae
Rila Mts various	Kawecka	1981	benthic algae

CZECHOSLOVAKIA

Danube	Ertl *et al.*	1972	benthic algae
Danube	Holčík & Bastl	1977	fish yield
Schwarza	Kubicek	1968	drift
Vltava	Sládeček & Miskovsky	1976	bacterial periphyton

DENMARK

various	Andersen *et al.*	1984	modification of Trent Index
Fjederholt Å, Granslev Å	Dawson *et al.*	1982	macrophytes on chemistry
Gudenå	Ernst & Nielsen	1983	grayling biology
Odense Kanal	Fjerdingstad	1977	benthic bacteria
Skravad	Hvitred-Jacobson	1982	sewer overflows on dissolved oxygen
Rold Kilde spring	Iversen & Jessen	1977	*Gammarus pulex* biology
various	Iversen & Mortensen	1978	assessment of pollution by Danish and English methods
three streams	Iversen	1975	law of leaf break-down
various streams	Kern-Hansen & Dawson	1978	macrophyte biomass
four streams	Kern-Hansen	1978	*Gammarus*
Ravnkilde spring	Lindegaard *et al.*	1975	macroinvertebrates

Sønderup Å	Madsen *et al.*	1977	Ephemeroptera
	Ministry of the Environment, Denmark	1978	environmental
Bisballe Baek	Mortensen	1981	energy flow in population
S-W., various	Sand-Jensen & Rasmussen	1978	acidic stream macrophytes
Himmerland springs	Thorup & Linde-gaard	1977	macroinvertebrates

ENGLAND

E. & W. Allen	Abel & Green	1981	zinc on invertebrates
Nent	Armitage	1980	zinc, invertebrates
Trent	Aston & Milner	1980	power stations on Asellus
(Bristol) Avon	Aykulu	1978	phytoplankton
(Bristol) Avon	Aykulu	1982	epipelic algae
Wensum	Baker *et al.*	1978	macrophytes
Wensum	Baker & Lumbley	1980	general survey
Frome, Tadnoll Brook	Baldock *et al.*	1983	chalk stream protozoa
N. Tyne	Boon	1978	current speed and Tri-choptera
N. Tyne	Brennan *et al.*	1978	particulate material
N. Tyne	Brennan *et al.*	1978	particulate matter, chironomids
Holme	Brown *et al.*	1982	chemistry
Hayle	Brown	1977	Cu, Zn, invertebrates
Holme	Brown	1980	*Hydrobia* and toxic dis-charges
Derwent (N.-E. England)	Burrows & Whitton	1983	heavy metals, invertebrates
Bere	Casey	1981	macrophytes and discharge
Eden (Cumbria)	Crisp & Cubby	1978	fish
Piddle	Dawson *et al.*	1978	plant succession
Bere	Dawson	1976	leaf litter in chalk streams
Bere	Dawson	1980	transport of plant material
Duddon	Elliott & Tullett	1977	larval drift
Ruding	Extence	1981	drought on invertebrates
Irwell	Eyres *et al.*	1978	oligochaete ecology
Hull	Goulder	1981	phytoplankton
Moat Brook	Gower & Buckland	1978	sewage effluent on *Chirono-mus riparius*
Kennet, Lambourn	Ham *et al.*	1981/ 1982	macrophytes
various	Hanbury *et al.*	1981b	herbicide effects
(Dorset) Stour	Hansford	1978	*Simulium austeri*
various	Hargreaves *et al.*	1975	acidic: chemistry, plants
Nene	Harper & Davies	1977	macrophytes
Swale	Holmes & Whitton	1977	macrophytes

Wear	Holmes & Whitton	1977b	macrophyte survey
N. Tyne	Holmes & Whitton	1981	macrophyte survey
various	Holmes N. T. H.	1983	macrophyte communities
Twilfin Beck	Jones	1978	epilithic algae
Wye	Jones	1984	phytoplankton
Thames	Lack *et al.*	1978	diatoms
Thames	Lack *et al.*	1978	phytoplankton
Frome	Ladle *et al.*	1977	Simuliidae
various	Learner *et al.*	1978	Naididae biology
Coln	Mackey *et al.*	1982	general survey
Thames, Kennet	Mackey	1976a,b 1977a,b	chironomids
Thames	Kackey	1979	chironomids and fish
(Dorset) Stour	Mann	1978	perch biology
(Dorset) Stour	Mann	1979	coarse fish biology
borehole water	Marker & Casey	1982	benthic algae
Bere, Frome	Marker & Gunn	1977	benthic algae
Bere, Frome	Marker	1976	benthic algae
Wandle	McCrow	1974	invertebrates
(Bristol) Avon	McGill *et al.*	1979	chironomid exuviae in surveillance
Wye	Merry *et al.*	1981	macrophytes
Frome	Mills *et al.*	1983	stone loach biology
Duddon	Minshall & Minshall	1978	invertebrates
(Bristol) Avon	Moore	1976	seasonal succession of algae
Highland Water (stream)	Moore	1977a	seasonal succession of algae
Wellow Brook	Moore	1977b	seasonal succession of algae
Wylye	Moore	1978	seasonal succession of algae
Wey	Moss	1977	diatom biology
various	Murphy *et al.*	1981a	berbicide
Frome	Pinder	1980	chironomid distribution
(Buckinghamshire) Ouse	Pratten	1980	crayfish biology
Wandle	Price & Price	1983	macrophytes
Lowther	Saunders & Eaton	1976	phytobenthos measurement
Pennine, various	Say & Whitton	1981	zinc on plants
Bere	Thommen & Westlake	1981	macrophyte distribution
Medway,(Sussex) Ouse	Townsend *et al.*	1983	invertebrate community structure
Team	Wehr *et al.*	1981	zinc pollution
Frome	Westlake & Dawson	1982	weed-cutting
Tadnoll Brook (Dorset)	Wetton	1980	organic detritus
Chew	Wilson	1977	chironomid exuviae
Coquet	Wise	1976	Ephemeroptera distribution
Lambourn	Wright *et al.*	1982	macrophytes
Lambourn	Wright *et al.*	1983	macroinvertebrates

FINLAND

central region	Elorantas Kunnas	1979	benthic algae
Kalajoki, Pyhäjoki, Siikajoki	Kainua & Valtonen	1980	lamprey

FRANCE

Haut-Rhône	Amoros & Mathieu	1984	cyclopoid ecology
Haut-Rhône	Amoros	1980	Cladocera, Copepoda
Estaragne	Backhaus	1976	benthic algae
Ain	Bournaud & Keck	1980	benthic macroinvertebrates
Haut-Rhône	Bournaud *et al.*	1978	macroinvertebrate colonization
Haut-Rhône	Bournaud *et al.*	1982	Hydropsychidae
Aston	Chauvet	1983	reduced flow in mountain stream
Seine	Couté	1979	planktonic blue-greens
Somme	Descy	1976	benthic algae
Haut-Rhône	Dessaix & Roux	1980	2ndy production of gammarids
Estibère	Elliott & Tullett	1977	larval drift
Haut-Rhône	Fontaine	1982	Ephemeroptera
Riou Mort	Giani	1983	heavy metals on fauna
Haut-Rhône	Gibert *et al.*	1977	macroinvertebrates
Haut-Rhône	Ginet	1982	amphopods of interstitial water
Verdon	Gregoire & Champeau	1981	Ephemeroptera
Moselle	Hinzelin & Lectard	1978	yeasts
Haut-Rhône	Juget *et al.*	1976	hydrology, macroinvertebrates
Ardière	Khalaf & Tachet	1977	macroinvertebrate colonization
Ardière	Khalaf & Tachet	1980	macroinvertebrate colonization
Loire	Lécureuil *et al.*	1983	*Hydropsyche* biology
Haut-Rhône	Merle	1980	bacteria
S. Juva streams	Nelva	1979	invertebrate community structure
general	Neveu	1983	zoobenthos review
Haut-Rhône	Perrin & Roux	1978	benthic fauna
Haut-Rhône	Richardot-Coulet *et al.*	1983	benthic macroinvertebrates
Lunain	Rofes	1975	influence of chalk
Haut-Rhône	Roux *et al.*	1976	benthic macroinvertebrates
Haut-Rhône	Roux	1982	introduction to project
Haut-Rhône	Roux	1983	review, esp., invertebrates
Massif Central, various	Say & Whitton	1982a	heavy metals, algae

Pyrenees, various	Say & Whitton	1982b	heavy metals, algae
Haut-Rhône	Seyed-Reihani *et al.*	1982a	interstitial macro-invertebrates
Haut-Rhône	Seyed-Reihani *et al.*	1982b	interstitial macro-invertebrates
general	Vernaux	1983	quality assessment

GERMANY (B.R.D.)

Rheinisches Schiefergebirge streams	Caspers	1975	invertebrate pro-ductivity
Elbe freshwater estuary	Caspers	1981	nitrogen cycle
Ruhr system	Dietz	1973	heavy metals in macrophytes
general	Federal Ministry of the Interior	1977	management
S. Lower Saxony various	Grube	1975	macrophtes
Ruhr	Hartmann	1983	diatoms and water chemistry
Breitenbach	Illies	1975	invertebrate pro-ductivity
Oberpfalzer Wald	Kohler & Zeltner	1974	macrophytes
Friedberger Au various	Kohler *et al.*	1974	macrophytes
various	Kohler	1978	macrophytes as indicators
Bavaria, various	Kohler	1981	macrophyte-rotifiers
Hase	Koste	1976	rotifiers
various lowland	Kohmeyer & Krause	1975	shade and macrophytes
Munchen region	Kutscher & Kohler	1976	macrophytes
various	Lorris	1979	macrophytes as indicators
Danube	Maas & Kohler	1983	macrophytes
eastern Hesse	Marxsen	1980 a,b	suspended bacteria
Nagold, Alb	Monschau-Dudenhausen	1980	macrophytes as indicators
upper Fulda catch-ment stream	Pieper *et al.*	1982	*Gammarus* and chemistry
Elbe, Stor, Trave, Schwentine	Rheinheimer	1977 a,b	bacteria
Ibach, Schwarzen-bachle	Rosset *et al.*	1982	leaf decomposition
Schierenseebrooks	Statzner	1981	hydraulic stress on benthic macro-invertebrates
Moosach, Regen, artificial	Steinberg & Melzer	1983	C physiology

Eder	Steubing *et al.*	1983	chemistry, algae
Niedersachsen, various	Weder-Oldecop	1971	macrophytes
Niedersachsen	Wiegleb	1979a	macrophytes and chemistry
Niedersachsen	Wiegleb	1979b	macrophyte communities
N.-W. and S.	Wiegleb	1981a	discriminant analysis: water, macrophytes
Niedersachsen	Wiegleb	1981b	macrophyte communities
Netzbach (Saarland)	Zaiss	1981	methane oxidation
Silbergragen, Kernwasser	Ziemann	1975	chemistry

GERMANY (D.D.R.)

Saale	Braune	1972	algal crops
various	Jorga & Weise	1979	macrophytes as indicators
Havel	Klose	1968	plankton
Saale	Krausch	1976	macrophytes
various	Pietsch	1972	macrophytes as indicators
various	Stolz & Weise	1976	macrophytes and N
Zschopau	Werner & Weise	1982	macrophytes and P

GREENLAND

Narssaq streams	Jóhansson	1980	benthic algae
Ûntartoq	Pedersen	1977	hot springs

HUNGARY

Danube	Dvilhally-Tamás	1975	primary production
Backony Mts, various	Erzsebét	1973	benthic algae
Daugava	Kumsare	1972	phytoplankton
Tisza	Uherkovich	1966	plankton

ICELAND

south-west	Castenholz	1976	hot springs
Laxá	Jóhansson	1979	dimensions and animals
Kráká, Laxá	Lindegaard	1979	macroinvertebrates
by R. Grendalsa	Sandbeck & Ward	1982	decomposition in hot springs
south	Sperling	1976	hot springs in winter

IRELAND

Slaney catchment	An Foras Forbartha	1981a	management plan

Barrow catchment	An Foras Forbartha	1981a	management plan
general	An Foras Forbartha	1983	detailed bibliography
general	An Foras Forbartha	1983	includes comprehensive bibliography
Liffey catchment	An Foras Forbartha	1985	management plan
Lough Leane catchment	Casey et al.	1981	P
general	Clabby	1981	biological monitoring
general	McCumiskey	1981	review of resources
general	Toner & Lennox	1980	nitrate

ITALY

Montegrotto Terme	Andreoli & Tascio	1975	hot spring
Arno	Balloni et al.	1979	algae and eutrophication
Lambro	De Simone & Damiani Damiani	1983	PCBs, pesticides, mercury
rivers of Piacenza district	Ghetti et al.	1982	biological monitoring
Tevere (Tiber)	IRSA	1978	pollution survey
Ticino	Marchetti & De Paolis	1983	water quality
tributaries of Lago Maggiore	Mosello et al.	1982	water quality
Adige	Regione Veneto	1977	pollution
stream in Boschi di Carrega Park	Zullini & Ricci	1980	rotifers

NETHERLANDS

Renkumse Beck	Bovee-Meijn	1970	invertebrates
Hierden stream	Higler & Repko	1981	pollution in animals
Meuse, Rhine	Sloof et al.	1983	monitoring for toxicity

NORTHERN IRELAND

Lough Neagh tributaries	Smith et al.	1982	nitrate
Main	Stevens & Smith	1978	N, P

NORWAY

various	Braekke	1976	acid precipitation effects
Sognsvannsbekken	Brettum	1974	diatom drift
Tovdel	Leivestad & Muniz	1976	low pH and fish
Brurskardsbekken	Tangen et al.	1978	periphyton, drift algae
various	Wright et al.	1976	acid precipitation effects

POLAND

Oder	Boroweic & Tarnawski	1982	birds waste water on ciliate
Biała Przemsza	Czapik	1982	waste water on ciliate communities
Płoska	Czecuzuga et al.		phytoplankton
Drwinka + tributaries	Dratnal & Dumnicka	1982	benthic invertebrates
Unieść, Skotawa	Korzeniewski et al.	1982	influence of trout culture on rivers
Tatra streams	Kownacka & Kownacki	1972	zoocoenoses
Grabia	Niesiolowski	1980	Similuliidae on plants
Waksmundzki, Pysznianski, Koscieliski streams	Olechowska	1982	mayfly zonation
Biała Przemsza catchment	Pasternak	1973	heavy metals
Pilica	Penczak et al.	1976	roach biology
Bzura, Ultrata	Penczak et al.	1981	fish population parameters
Bzura, Ultrata	Penczak et al.	1984	fish diet
Vistula (Wisła)	Skoczén	1982	lead
Utrata	Zalewski & Penczak	1981	fish community

PORTUGAL

Azores, general	Johansson	1976	algae

RUMANIA

Danube	Botnariuc & Bolder	1975	primary production
Nera	Botosaneanu & Negrea	1976	gastropod association
Fagaras various	Kawecka	1981	benthic algae
Jiu	Mihăită	1970	saprobic survey of invertebrates
Jiu	Mihăită	1971	saprobic survey of invertebrates

SCOTLAND

Duchray Water	Harriman & Morrison	1982	acid precipitation invertebrates, fish
Tweed	Holmes & Whitton	1975 a,b	macrophytes
streams near Aberfeldy	Smith et al.	1983	mine impact assessment

SPAIN

Besós, Llobregat	Lucena *et al.*	1982	viruses
Llobregat	Prat *et al.*	1983	macroinvertebrates and water quality
Avenco	Sabater	1983	benthic algae

SWEDEN

general	Ahl	1980	chemistry
Kaltisjokk	Backlund & Renburg	1973	drift algae
Lapland various	Bjarnborg	1983	acidification
Kaltisjokk, Ångerån	Harmanen	1980	Ephemeroptera, Plecoptera
various	Henricson & Müller	1980	regulation
N. Jämtland, various	Johansson	1979a	benthic chlorophyll
N. W. Jämtland, various	Johansson	1979b	benthic algal biomass
Björbäck, Bakvattsbäck	Johansson	1980	benthic algal biomass + productivity
Jämtland, various	Johansson	1982a	benthic algae
Jämtland, various	Johansson	1982b	benthic algae
Rickleån	Johnson	1978	moss and detritus
Rickleån	Karlström & Backlund	1977	organic matter and chlorophyll
Rickleån	Karlström	1978	detrital metabolism
Kebnekaise Mts, various	Kawecka	1981	benthic algae
various	Kohler	1978	vegetation
Stampen stream	Malmqvist	1980	habitat selection by lamprey
Stampen stream	Malmqvist	1973	lamprey
Abisko region	Mendl & Müller	1978	*Amphinemura*
Vetkeån, Tolångaån	Müller	1978	algal productivity
Ulmeälven	Nilsson	1978	hydro-electric plant on flora
south, various	Otto & Svensson	1983	brown waters and invertebrates
far north	Steffan	1971	glacier brook chironomids
Tjulån	Wotton	1979	blackfly distributions

SWITZERLAND

Lützel	Bärlocher	1983	standing crop
Aare, Rhine	Bloesch	1977	benthic macroinvertebrates
artificial	Eichenberger *et al.*	1981	experiments on stream benthos
various	Gujer	1976	nitrification
Lussel, Lützel	Rosset *et al.*	1982	leaf decomposition

Rhine + tributaries	Schanz & Juon	1983	periphyton bioassay
artificial	Wuhrmann & Eichenberger	1980	experimental studies

U.S.S.R

Volga	Butorin *et al.*	1981	impact of reservoirs
Neris	Knabikas *et al.*	1973	mutagenicity of water
Azerbaijan, Caucasus streams	Kownacka & Kownacki	1972	zoocoenoses
Kola Peninsula	Kruglova	1973	zooplankton
Dnieper	Sivko *et al.*	1972	standing crop + productivity

WALES

Rheidol, Ystwyth	Brooker & Morris	1980	macroinvertebrates
Ely	Burton *et al.*	1982	plasmids
Ely	Esho & Benson-Evans	1983	primary productivity
Teifi	Jenkins & Wade	1981	macroinvertebrates
Teifi	Jenkins *et al.*	1984	macroinvertebrates
Clarach, Ystwyth	McLean & Jones	1975	algae and heavy metals
Elan, Tywi	Scullion	1983	impoundments on ecology
Usk	Thirb & Benson-Evans	1983	substrata for *Lemanea*
Emral Brook	Wilkinson & Jones	1977	dace biology

YUGOSLAVIA

Velika Morawa	Jankovic	1977	chironomid communities
Drava	Markič	1983	rotifiers
Pek	Nedeljković	1978	benthic fauna
Crnojevića	Smith *et al.*	1978	P in various components
Savinja	Vrhovšek	1981	macrophytes

REFERENCES FOR APPENDIX A

Abel P.D. & Green D.W.J. (1981) Ecological and toxicological studies
    on invertebrate fauna of two rivers in the Northern Pennine Orefield.
    In: Say P.J. & Whitton B.A. (Eds) *Heavy Metals in Northern England:*
    Environmental and Biological Aspects. 198 pp., pp. 109-122.
    Department of Botany, University of Durham.
Ahl T. (1980) Variability in ionic composition of Swedish lakes and
    rivers. *Arch. Hydrobiol.* 89, 5-16.
Amoros C. (1980) Structure et fonctionnement des écosystèmes du Haut-
    Rhône francaise. 15. Comparison des peuplementes Cladocères et
    Copepodes de deux anciens méandres d'âge différent. *Acta Oecol.*
    (Oecol. gen.) 1, 193-208.
Amoros C. & Mathieu J. (1984) Structure et fonctionnement des
    écosystèmes du Haut-Rhône français. 35. Relations entre les eaux
    interstitielles et les eaux superficielles; influence sur les
    peuplements de Copepodes Cyclopoëdes (Crustacées). *Hydrobiologia*
    108, 273-280.
Anderson M.M., Riget F.F. & Sparholt H. (1984) A modification of the
    Trent index for use in Denmark. *Water Res.* 18, 145-151.
Andreoli C. & Rascio N. (1975) The algal flora in the thermal baths
    of Montegrott Terme (Padua). Its distribution over one-year period.
    Int. Rev. ges. Hydrobiol. 60, 857-871.
An Foras Forbartha (1981a) *Draft Water Quality Management Plan for the*
    River Slaney Catchment (5 Vols). WR/C60, An Foras Forbartha, Dublin.
An Foras Forbartha (1981b) *Draft Water Quality Management Plan for the*
    River Barrow Catchment (5 Vols). WR/C62, An Foras Forbartha, Dublin.
An Foras Forbartha (1982) *Draft Water Quality Management Plan for the*
    River Barrow Catchment (5 Vols). WR/C61, An Foras Forbartha, Dublin.
An Foras Forbartha (1983) *A Review of Water Pollution in Ireland. A*
    Report to the Council. 152 pp. Water Pollution Advisory Council,
    Dublin.
An Foras Forbartha (1984) *Draft Water Quality Management Plan for the*
    River Liffey catchment (5 Vols). (In preparation), An Foras
    Forbartha, Dublin.
Armitage P.D. (1980) The effects of mine drainage and organic enrich-
    ment on benthos in the River Nent system, Northern Pennines.
    Hydrobiologia 74, 119-128.
Armitage P.D. (1980) Stream regulation in Great Britain. In: Ward
    J.V. & Stanford J.A. (Eds) *The Ecology of Regulated Streams.*
    pp. 165-181. Plenum Press, N.Y.
Armitage P.D., Moss D., Wright J.F. & Furse M.T. (1983) The perfor-
    mance of a new biological water quality score system based on macro-
    invertebrates over a wide range of unpolluted running-water sites.
    Water Res. 17, 333-347.

Aston R.J. & Milner A.G.P. (1980)   A comparison of populations of the
     isopod *Asellus aquaticus* above and below power stations in organi-
     cally polluted reaches of the River Trent. *Freshwat. Biol.* 10, 1-14.
Aykulu G. (1978)   A quantitative study of the phytoplankton of the
     River Avon, Bristol. *Br. phycol. J.* 13, 91-102.
Aykulu G. (1982)   The epipelic algal flora of the River Avon. *Br.
     phycol. J.* 17, 27-38.
Backhaus D. (1976)   Beiträge zur Ökologie der benthischen Algen des
     Hochgebirges in den Pyrenäen.   II. Cyanophyceen und übrige Algen-
     gruppes. *Int. Rev. ges. Hydrobiol.* 61, 471-516.
Backlund S. & Renberg I. (1973)   Driftens och kolonisationens
     utreckling under våren hos encelliga alger i Kaltisjokk. *Ökol.
     stat. Messawe Ber.* 26, 1-9.
Baker R., Driscol R.J. & Lambley P. (1978)   The River Wensum (1).
     Trans. Norfolk & Norwich Naturalists' Soc. 24 (4), 197-219.
Baker R. & Lambley P. (1980)   The River Wensum (2). *Trans. Norfolk
     & Norwich Naturalists' Soc.* 25, 65-81.
Baldock B.M., Baker J.H. & Sleigh M.A. (1983)   Abundance and producti-
     vity of protozoa in chalk streams. *Holarctic Ecology* 6, 238-246.
Balloni W., Vincenzini M. & Florenzano G. (1979)   Algae and eutrophi-
     cation of Arno River.   In: *Symposium on Eutrophication in Italy.*
     Roma 3-4 October 1978.   pp. 309-328.   (In Italian)
Barlocher F. (1983)   Seasonal variation of standing crop and digesti-
     bility of CPOM in a Swiss Jura stream. *Ecology* 64, 1266-1272.
Bjärnborg B. (1983)   Dilution and acidification effects during the
     spring flood of four Swedish mountain brooks. *Hydrobiologia* 101,
     19-26.
Bloesch J. (1977)   Bodenfaunistische Untersuchungen in Aare and Rhein.
     Schweiz. Z. Hydrol. 39, 46-68.
Boon P.J. (1978)   The pre-impoundment distribution of certain Tricho-
     ptera larvae in the north Tyne river system (northern England),
     with particular reference to current speed. *Hydrobiologia* 57,
     167-174.
Boroweic L. & Tarnawski D. (1982)   Przeloty i zimonwanit ptakow na
     Ordze pod Brzegiem.   The passage and wintering of birds on the River
     Oder near Brzeg. *Acta Zool. Cracov* 26, 3-30.   (In Polish, English
     summary)
Botnariuc N. & Boldor O. (1975)   Evolution de la production primaire
     dans les eaux de la zone inondable du Danube. *Verh. int. Verein.
     theor. angew. Limnol.* 19, 1823-1827.
Botosaneanu L. & Negrea A. (1976)   Sur quelques aspects biologiques
     d'une rivière des monts du Banat – Roumanie: la Nera (et surtout)
     sur une remarquable association de gastropodes. *Riv. Idrobiol.* 15
     (3), 403-432.   (In French, with Italian and English summaries)
Bournaud M., Chavanon G. & Tachet H. (1978)   Structure et fonctionne-
     ment des écosystèmes du Haut-Rhône français.   5. Colonisation par
     les macroinvertébrés de substrats artificiels suspendus en pleine
     eau ou posés sur le fond. *Verh. int. Verein. theor. angew. Limnol.*
     20, 1485-1493.
Bournaud M. & Keck G. (1980)   Diversité spécifique et structure des
     peuplements de macro-invertébrés benthiques au long d'un cours
     d'eau: le Furans (Ain). *Acta Oecologica, Oecol. génér.* 1, 131-150.

Bournaud M., Tachet H. & Perrin J.F. (1982) Structure et fonctionne-
    ment des écosystèmes du Haut-Rhône français. 32. Les Hydropsychi-
    dae (Trichoptera) du Haut-Rhône entre Genéve et Lyon. *Annls Limnol.*
    18, 61-80.

Bovée-Meijn N.P. (1970) Een hydrobiologisch onderzoek van de Renkumse
    Beek. *Overdruk uit De Levende Natuur* 73, 7-11.

Braekke F.H. (Ed.) (1976) *Impact of Acid Precipitation on Forest and
    Freshwater Ecosystems.* 111 pp. Fragrapport FR 6/76. Norges Land-
    bruksvitenskapelige forskningsrad.

Braune W. (1972) Experimentelle Untersuchungen *in situ* zur Biomass-
    bildung von Mikroalgen und zur Entwicklung naturlicher Algen-
    Biozönosen im Fliessgewässer. *Int. Rev. ges. Hydrobiol.* 57, 227-256.

Breitag G. & von Tumpling W. (1982) *Ausgewählte Methoden der Wasser-
    suchung. Band II. Biologische, Mikrobiologische und Toxikologische
    Methoden.* Gustav Fischer Verlag, Jena.

Brennan A., McLachlan A.J. & Wotton R.S. (1978) Particulate material
    and midge larvae (Chironomidae: Diptera) in an upland river.
    Hydrobiologia 59, 67-73.

Bretschko G. (1981) Vertical distribution of zoobenthos in an alpine
    brook of the RITRODAT-LUNZ study area. *Verh. int. Verein. theor.
    angew. Limnol.* 21, 873-876.

Brettum P. (1974) The relation between the new colonization and drift
    of periphytic diatoms in a small stream in Oslo, Norway. *Norw. J.
    Bot.* 21, 277-284.

Brooker M.P. (1981) The impact of impoundments of the downstream
    fisheries and general ecology of rivers. In: Coaker T.H. (Ed.)
    Advances in Applied Biology III. pp. 91-152. Academic Press,
    London.

Brooker M.P. & Morris D.L. (1980) A survey of the macro-invertebrate
    riffle fauna of the rivers Ystwyth and Rheidol, Wales. *Freshwat.
    Biol.* 10, 459-474.

Brown B.E. (1977) Effects of mine drainage on the River Hayle, Corn-
    wall. A) Factors affecting concentrations of copper, zinc and iron
    in water, sediments and dominant invertebrate fauna. *Hydrobiologia*
    52, 221-233.

Brown L. (1980) The use of *Hydrobia jenkinsi* to detect intermittent
    toxic discharges to a river. *Water Res.* 14, 941-947.

Brown L., Bellinger E.G. & Day J.P. (1982) A case study of nutrients
    in the R. Holme, West Yorkshire, England. *Environ. Pollut. B.* 3,
    81-100.

Burton N.F., Day M.J. & Bull A.J. (1982) Distribution of bacterial
    plasmids in clean and polluted sites in a South Wales river. *Appl.
    environ. Microbiol.* 44, 1026-1029.

Butorin N.V., Ziminova N.A., Romanenko V.I. & Ekzertsev E.A. (1981)
    Ecological changes in the river Volga due to creation of reservoirs.
    Verh. int. Verein. theor. angew. Limnol. 21, 877-879.

Casey H. (1977) Origin and variation of nitrate nitrogen in the chalk
    springs, streams and rivers in Dorset and its utilisation by higher
    plants. *Progr. Wat. Technol.* 8, 225-235.

Casey H. (1981) Discharge and chemical changes in a chalk stream head-
    water affected by the outflow of a commercial watercress-bed.
    Environ. Pollut. B. 2, 383-385.

Casey H. & Farr I.S. (1982) The influence of within-stream disturbance on dissolved nutrient levels during spates. *Hydrobiologia* 92, 447-462.

Casey T.J., O'Connor P.E. & Greene R.G. (1981) A survey of phosphorus inputs to Lough Leane. *Irish J. environ. Sci.* 1, 21-34.

Caspers H. (1981) Seasonal effects on the nitrogen cycle in the freshwater section of the Elbe estuary. *Verh. int. Verein. theor. angew. Limnol.* 21, 866-870.

Caspers N. (1975) Productivity and trophic structure of some West German woodland brooklets. *Verh. int. Verein. theor. angew. Limnol.* 19, 1712-1716.

Castenholz R.W. (1976) The effect of sulfide on the blue-green algae of hot springs. I. New Zealand and Iceland. *J. Phycol.* 12, 54-68.

Chauret E. (1983) Influence d'une réduction de débit sur un torrent de montagne: l'Aston (Ariège). *Annls Limnol.* 19, 45-50.

Clabby K.J. (1981) *The National Survey of Irish Rivers. A Review of Biological Monitoring.* 322 pp. An Foras Forbartha, Dublin.

Couté A. (1979) Cyanophycées planctoniques du bassin de la Seine. *Bull. Mus. natn. Hist. nat., Paris, 4 ser.* 1, 267-283.

Crisp D.T. & Cubby P.R. (1978) The populations of fish in tributaries of the River Eden on the Moor House National Nature Reserve, Northern England. *Hydrobiologia* 57, 85-93.

Czapik A. (1982) The effect of waste water on ciliate communities in the Biala Przemsza River. *Acta Hydrobiol.* 24, 29-37.

Czeczuga B., Gradzki F. & Bobiatyńska-Ksok E. (1968) Produkcja pier pierwotna na wybranym stowisku rzeki Ploski Cześć I. Prodikcja fitoplanktonu. Primary production in a chosen site of the River Ploska. Part I. Phytoplankton production. *Acta Hydrobiol.* 10, 85-94.

Dawson F.H. (1976) Organic contribution of stream edge forest litter fall to the chalk stream ecosystem. *Oikos* 27, 13-18.

Dawson F.H., Castellano E. & Ladle M. (1978) Concept of species succession in relation to river vegetation and management. *Verh. int. Verein. theor. angew. Limnol.* 20, 1429-1434.

Dawson F.H. & Haslam S.M. (1983) The management of river vegetation with particular reference to shading effects of marginal vegetation. *Landscape Planning* 10, 147-169.

Dawson F.H., Kern-Hansen U. & Westlake D.F. (1982) Water plants and the oxygen and temperature regimes of lowland streams. In: Symoens J.J., Hooper S.S. & Compère P. (Eds) *Studies on Aquatic Plants.* pp. 214-221. Royal Botanic Society of Belgium, Brussels.

Descy J.-P. (1976) Estimation de la qualité des eaux de la Somme par l'étude des peuplements algaux benthiques. *Bull. fr. Pisciculture* 260, 143-147.

De Simone R. & Damiani V. (1983) The Lambro River sediments: PCBs, pesticides and mercury. *Acqua e Aria* 9, 965-971. (In Italian)

Dessaix J. & Roux A.L. (1980) Structure and dynamics of the secondary production of Gammarids in the main stream. *Crustaceana, Suppl.* 6, 242-252.

Dietz F. (1973) The enrichment of heavy metals in submerged plants. In: Jenkins S.H. (Ed.) *Advances in Water Pollution Research*. 6th International Conference, Jerusalem, June 1972. pp. 53-60. Pergamon Press, Oxford.

Dratnal E. & Dunnicka E. (1982) Compositjon and zonation of benthic invertebrate communities in some chemically stressed aquatic habits of Niepolomice Forest (South Poland). *Acta Hydrobiol.* 24, 151-165.

Dvihally-Tamás S. (1975) Primary production of the Hungarian Danube. *Verh. int. Verein. theor. angew. Limnol.* 19, 1717-1722.

Edwards R.W. & Crisp D.T. (1982) Ecological implications of river regulation in the United Kingdom. In: Hey R.D., Bathurst J.C. & Thorne C.R. (Eds) *Gravel-bed Rivers.* pp. 843-865. Wiley, Chichester.

Eichenberger E., Schlatter F., Weilenmann H. & Wuhrmann K. (1981) Toxic and autrophicating effects of Co, Cu and Zn on algal benthic communities in rivers. *Verh. int. Verein. theor. angew. Limnol.* 21, 1131-1134.

Ellington A. & Burke T. (1981) *Europe: Environment. The European Communities' Environmental Policy.* 23 pp. Ecobooks, 16 Strutton Ground, London S.W.1. (Published with the assistance of the Commission of the European Communittis.)

Elliott J.M. & Tullett P.A. (1977) The downstream drift of larvae of *Dixa* (Diptera: Dixidae) in two stony streams. *Freshwat. Biol.* 7, 403-408.

Eloranta P. & Kumas S. (1979) The growth and species communities of the attached algae in a river system in Central Finland. *Arch. Hydrobiol.* 86, 27-44.

Ernst M.E. & Nielsen J. (1983) Gådenåstallingens (*Thymallus thymallus* L.) gydebiologi. *Meddelelser fra Ferskvandsfiskerilaboratoriet* 1983/1, 30 pp. (The spawning biology of the grayling (*Thymallus thymallus* (L.)) from the river Gudenå)

Ertl M., Juris S. & Tomajka J. (1972) Vorläufige Angaben über jahrsezeitliche Veränderungen und die Vertikale Verteilung des Periphytons in mittleren Abschnitt der Donau. *Arch. Hydrobiol., Suppl.* 44, 34-48.

Erzsebét K. (1973) Az Északi- (Öreg-) Bakony területén vegzett algólogiai és hidrobiologiai kutatások rövid ismertetése. *Veszprem Megyei Múzeumok Közleményei* 12, 143-163. (Hungarian, with German and English summaries)

Esho R.T. & Benson-Evans K. (1983) Studies on the sewage fungus complex in the River Ely, South Wales, U.K. *Nova Hedwigia* 37, 519-534.

Extence C.A. (1981) The effect of drought on benthic invertebrate communities in a lowland river. *Hydrobiologia* 23, 217-224.

Eyres J.P., Williams N.V. & Pugh-Thomas M. (1978) Ecological studies on Oligochaeta inhabiting depositing substrata in the Irwell, a polluted English river. *Freshwat. Biol.* 8, 25-32.

Federal Ministry of the Interior, Germany (1977) *Report on Water Management in the Federal Republic of Germany.* 47 pp. (Published on the occasion of the United Nations' International Water Conference in March 1977 in Mar del Plata, Argentina)

Fjerdingstad E. (1977) Effect of warm water discharges from power plants on the bacterial decomposition in the bottom mud of a stream. *Arch. Hydrobiol. Beih.* 9, 223-237.

Fontaine J. (1982) Le piégage lumineux, moyen d'approche de la faune entomologique d'un grand fleuve (Ephéméroptères, en particulier). *Bull. mens. Soc. Linn. Lyon* 51, 81-89.

Ghetti P.F., Bernina F., Bonazzi G., Cunzola A. & Ravanetti U. (1982) *Biological Monitoring of the Piacenza District Rivers.* 29 pp. Pubb. Amm. Prov. Piacenza e P. F. Ambients. (In Italian)

Giani N. (1983) Le Riou Mort, affluent du Lot pollué par les metaux lourds. III. Etude faunistique générale. *Annls Limnol.* 19, 29-44.

Gibert J., Ginet R., Mathieu J., Reygrobellet J.L. & Seyed-Reihani A. (1977) Structure et fonctionnement des écosystèmes du Haut-Rhône français. IV. Le peuplement des eaux phréatiques; premiers résultats. *Annls Limnol.* 13, 83-97.

Ginet R. (1982) Structure et fonctionnement des écosystèmes du Haut-Rhône francais. Les Amphipodes des eaux interstitielles en amont de Lyon. *Polskie Archwm Hydrobiol.* 29, 231-237.

Goulder R. (1982) A note on phytoplankton in the River Hull at Hempholme, North Humberside. *Naturalist, Hull* 106, 153-156.

Gower A.M. & Buckland P.J. (1978) Water quality and the occurrence of *Chironomus riparius* Meigen (Diptera: Chironomidae) in a stream receiving sewage effluent. *Freshwat. Biol.* 8, 153-164.

Gregoir A. & Champeau A. (1981) Les Ephéméroptères d'une rivière à debit regulé: le Verdon. *Verh. int. Verein. theor. angew. Limnol.* 21, 857-865.

Grube H.-J. (1975) Die Macrophytenvegetation der Fliessgewässer in Süd-Niedersachsen und ihre Beziehungen zur Gewässerverschmutzung. *Arch. Hydrobiol., Suppl.* 45, 376-475.

Gujer W. (1976) Nitrifikation in Fliessgewässern - Fallstudie Glatt. *Schweiz. Z. Hydrol.* 38, 171-189.

Ham S.F., Wright J.F. & Berrie A.D. (1981) Growth and recession of aquatic macrophytes in an unshaded section of the River Lambourn, England, from 1971 to 1976. *Freshwat. Biol.* 11, 381-390.

Ham S.F., Wright J.F. & Borie A.D. (1982) The effect of cutting on the growth and recession of the freshwater macrophyte *Ranunculus penicillatus* (Dumort.) Bab. var. *calcareus* (R. W. Butcher) C.D.K. Cook. *J. environ. Mgmt* 15, 283-271.

Ham S.F., Cooling D.A., Hiley P.D., McLeish P.R., Scorgie H.R.A. & Berrie A.D. (1982) Growth and recession of aquatic macrophytes on a shaded section of the River Lambourn, England, from 1971 to 1976. *Freshwat. Biol.* 11, 381-392.

Hansford R.G. (1978) Life-history and distribution of *Simulium austeni* (Diptera: Simuliidae) in relation to phytoplankton in some English rivers. *Freshwat. Biol.* 8, 521-532.

Hargreaves J.W., Lloyd E.J.H. & Whitton B.A. (1975) Chemistry and vegetation of highly acidic streams. *Freshwat. Biol.* 5, 563-576.

Harmanen M. (1980) Der Einfluss saurer Gewässer auf den Bestand der Ephemeriden - und Plecopterenfauna. *Gewässer Abwässer* 66/67, 130-136.

Harper D.M. & Davies C.E. (1977) Aquatic plants of the River Nene. *J. Northamptonshire Natural History Society* 37, 142-147.

Harriman R. & Morrison B.R.S. (1982) Ecology of streams draining forested and non-forested catchments in an area of central Scotland subject to acid precipitation. *Hydrobiologia* 88, 251-263.

Hartman D. (1983) Beziehungen zwischen der Diatomeen-Flora und dem Wasserchemismus in Fliessgewässern des Sauerlandes. 1. Die Rhur. *Ber. naturhist. Ges. Hannover* 126, 91-135.

Haslam S.M. (in press) *River Plants of Western Europe. The Macro-phytic Vegetation of Watercourses of the European Economic Commu-nity.* Cambridge U.P., Cambridge.

Haslam S.M. & Wolseley P.A. (1981) *River Vegetation: Its Identifi-cation, Assessment and Management. A Field Guide to the Macro-phytic Vegetation of British Watercourses.* 154 pp. Cambridge U.P., Cambridge.

Henricson J. & Müller K. (1980) Stream regulation in Sweden with some examples from central Europe. In: Ward V. & Stanford J.A. (Eds) *The Ecology of Regulated Streams.* pp. 183-199. Plenum Press, N.Y.

Higler L.W.G. & Repko F.F. (1981) The effects of pollution in the drainage area of a Dutch lowland stream on fish and macro-inverte-brates. *Verh. int. Verein. theor. angew. Limnol.* 21, 1077-1082.

Hinzelin F. & Lectard P. (1978) Les levures dans les eaux de la Moselle. *Hydrobiologia* 61, 209-224.

Holcík J. & Bastl I. (1977) Predicting fish yield in the Czechoslo-vakian section of the Danube river based on the hydrological regime. *Int. Rev. ges. Hydrobiol.* 62, 523-532.

Holmes N.T.H. (1983) Focus on Nature Conservation No. 4. *Typing British Rivers According to their Flora.* Nature Conservancy Council, Shrewsbury.

Holmes N.T.H. & Whitton B.A. (1975a) Macrophytes of the River Tweed. *Trans. bot. Soc. Edinb.* 42, 369-381.

Holmes N.T.H. & Whitton B.A. (1975b) Submerged macrophytes and angio-sperms of the River Tweed and its tributaries. *Trans. bot. Soc. Edinb.* 42, 383-395.

Holmes N.T.H. & Whitton B.A. (1977a) Macrophytic vegetation of the River Swale, Yorkshire. *Freshwat. Biol.* 8, 545-558.

Holmes N.T.H. & Whitton B.A. (1977b) Macrophytes of the River Wear: 1966-1976. *Naturalist, Hull* 106, 53-73.

Holmes N.T.H. & Whitton B.A. (1981) Plants of the River Tyne system before the Kielder Water Scheme. *Naturalist, Hull* 106, 97-108.

Houba C. & Remacle J. (1982) Distribution of heavy metals in sedi-ments of the River Vesdre (Belgium). *Environ. Technol. Lett.* 3, 237-240.

Huitved-Jacobsen T. (1982) The impact of combined sewer overflows on the dissolved oxygen concentration of a river. *Water Res.* 16, 1099-1105.

Illies J. (1975) A new attempt to estimate production in running waters (Schlitz studies on productivity, No. 12). *Verh. int. Verein. theor. angew. Limnol.* 19, 1705-1711.

IRSA (Istituto di Recerca sulle Acque, Consiglio Nazionale delle Ricerca) (1978) Indagine sulla 'inquinamento del fiume Tevere. *Quaderni IRSA* 27, 482 pp. (Researches on pollution of the River Tiber)

Iversen T.M. (1975) Disappearance of autumn shed beech leaves placed in bags in small streams. *Verh. int. Verein. theor. angew. Limnol.* 19, 1687-1692.

Iversen T.M. & Jessen J. (1977) Life-cycle, drift and production of *Gammarus pulex* L. (Amphipoda) in a Danish spring. *Freshwat. Biol.* 7, 287-300.

Iversen T.M. & Mortensen E. (1978) Sammerligningen af danske bedømmelse af vandrecipeinters forureningsgrad og det engelske Trent Index. *Vand* 3, 65-67. (Danish = A comparison between the method used in Denmark for assessment of the degree of pollution and the English Trent Index.)

Jankovic M.J. (1979) Communities on chironomid larvae in the Velika Morawa river. *Hydrobiologia* 64, 167-173.

Jenkins R.A. & Wade K.R. (1981) *A Survey of the Macroinvertebrate Fauna of the River Teifi*. Biological Report No. 13/81. Welsh Water Authority Laboratory, Llanelli.

Jenkins R.A., Wade K.R. & Pugh E. (1984) Macroinvertebrate - habitat relationships in the River Teifi catchment and the significance to conversion. *Freshwat. Biol.* 14, 23-42.

Johansson C. (1976) Freshwater algal vegetation in the Azores. *Bot. Soc. Broteriana* 50 (2nd series), 117-141.

Johansson C. (1979a) Chlorophyll content and the periphytic algal vegetation in six streams in Northern Jämtland, Sweden 1977. *Medd. Växtbiologiska Institutionen, Uppsala* 1979/2, 1-53.

Johansson C. (1979b) Biomass of the periphytic algal vegetation in some stony mountain streams in N.W. Jämtland, Sweden. A preliminary report. *Medd. Växtbiologiska Institutionen, Uppsala* 1979/3, 1-16.

Johansson C. (1980) Attached algal vegetation in some streams from the Narssaq area, South Greenland. *Acta Phytogeogr. Suec.* 68, 89-96.

Johansson C. (1982a) Attached algal vegetation in two stony streams in N.W. Jämtland, Sweden. Ecology, volume and primary production. *Medd. Växtbiologiska Institutionen, Uppsala* 1980/1, 1-30.

Johansson C. (1982b) The ecological characteristics of 314 algal taxa found in Jämtland Streams, Sweden. *Medd. Växtbiologiska Institutionen, Uppsala* 1982/2.

Jóhansson P.M. (1979) The River Laxá ecosystem, Iceland. *Oikos* 32, 306-309.

Johnston T. (1978) Aquatic mosses and stream metabolism in a North Swedish river. *Verh. int. Verein. theor. angew. Limnol.* 20, 1471-1477.

Jones F.H. (1984) The dynamics of suspended algal populations in the lower Wye catchment. *Water Res.* 18, 25-36.

Jones J.G. (1978) Spatial distribution in epilithic algae in a stony stream (Wilfin Beck) with particular reference to *Cocconeis placentula*. *Freshwat. Biol.* 8, 539-546.

Jorga W. & Weise G. (1979) Zum Bioindikationswert submerser Makrophyten und zur Rucklhaltung von Wasserinhaltsstoffen durch Unterwasserpflanzen in langsam fliessenden Gewässern. *Acta hydrochim. hydrobiol.* 7, 43-76.

Jorga W., Heym W.-D. & Weise G. (1982) Shading as a measure to prevent mass development of submersed macrophytes. *Int. Rev. ges. Hydrobiol.* 67, 271-281.

Juget J., Amoros C., Gamulin D., Reygrobellet J.L., Richardot M., Richoux P. & Roux C. (1976) Structure et fonctionnement des écosystèmes du Haut-Rhône français. II. Etude hydrologique et écologique de quelques bras morts. Premiers resultats. *Bull. Ecol. fr.* 7, 479-492.

Kainua K. & Valtomen T. (1980) Distribution and abundance in European river lamprey (*Lampetra fluviatilis*) larvae in three rivers running in to Bothnian Bay, Finland. *Can. J. Fish. aquatic. Sci.* 37, 1960–1966.

Kann E. (1978a) Systematik und Ökologie der algen österreichischer Bergbache. *Arch. Hydrobiol., Suppl.* 53, 405–643.

Kann E. (1978b) Typification of Austrian streams concerning algae. *Verh. int. Verein. theor. angew. Limnol.* 20, 1523–1526.

Kann E. (1983) Die benthischen Algen der Donau im Raum von Wien. *Arch. Hydrobiol., Suppl.* 68, 15–36.

Karlström U. & Backlund S. (1977) Relationship between algal cell number, chlorophyll and fine particulate organic matter in a river in Northern Sweden. *Arch. Hydrobiol.* 80, 192–199.

Karlström U. (1978) Role of the organic layer on stones in detrital metabolism in streams. *Verh. int. Verein. theor. angew. Limnol.* 20, 1463–1470.

Kawecka B. (1974a) Vertical distribution of algae communities in Maljovica stream (Rila - Bulgaria). *Polskie Archwm Hydrobiol.* 21, 211–228.

Kawecka B. (1974b) Effect of organic pollution on the development of diatom communities in the alpine streams Finstertaler Bach and Gurgler Ache (northern Tyrol, Austria). *Ber. nat.-med. Ver. Innsbruck* 61, 71–82.

Kawecka B. (1981) Sessile algae in European mountain streams. 2. Taxonomy and autecology. *Acta Hydrobiol.* 23, 17–46.

Kawecka B. (1982) Stream ecosystems in mountain grassland (West Carpathians). *Acta Hydrobiol.* 24, 357–365.

Kern-Hansen U. (1973) Drift of *Gammarus pulex* L. in relation to macrophyte cutting in four small Danish lowland-streams. *Verh. int. Verein. theor. angew. Limnol.* 20, 1440–1445.

Kern-Hansen U. & Dawson F.H. (1978) The standing crop of aquatic plants of lowland streams in Denmark and the inter-relationships of nutrients in plant, sediment and water. In: *5th International Symposium on Aquatic Weeds, September 1978,* pp. 143–150. European Weed Research Society.

Khalaf G. & Tachet H. (1977) La dynamique de colonisation des substrats artificiels par les macro-invertébrés d'un cours d'eau. *Annls Limnol.* 13, 169–190.

Khalaf G. & Tachet H. (1980) Colonization of artificial substrata by macroinvertebrates in a stream and variations according to stone size. *Freshwat. Biol.* 10, 475–482.

Klasvik B. (1974) *Computerized Analysis of Stream Algae.* 100 pp. Växtekologiska Studier 5.

Klose H. (1968) Untersuchungen uber den Indikationswert des Potamoplanktons. *Int. Rev. ges. Hydrobiol.* 53, 781–805.

Knabikas A., Lekevičius R. & Adomonyte L. (1983) Evaluation of mutageneity of water in the Neris River. *Acta Hydrobiologica Lituanica* 4, 59–68. (In Russian, English summary)

Kohler A. (1978) *Bericht uber die Forschungsreisen nach Südengland und Südschweden im Sommer 1978.* Institut für Landeskultur und Pflanzenökologie, Univ. Hohenheim, Stuttgart.

Kohler A. (1981) *Fliessgewässer in Bayern.* 18 pp. Akademie für Naturschutz und Landschaftspflege, Tagungsbericht 5/81. (ISSN 0173-7724)

Kohler A., Brinkmeier R. & Vollrath H. (1974) Verbreitung und Indikatorwert der submersen Makrophyten in den Fliessgewässern der Friedberger Au. *Ber. bayer. bot. Ges.* 45, 5-36.

Korzeniewski K., Banat Z. & Moczulska A. (1982) Changes in water of the Unieść and Skotawa Rivers, caused by intensive trout culture. *Polskie Archwm Hydrobiol.* 29, 683-691.

Koste W. (1976) Über die Rädertierbestände (Rotatoria) der oberen und mittleren Hase in den Jahren 1966-1969. *Osnabrucker Naturw. Mitteilungen* 4, 191-263.

Kownacka M. & Margreiter G. (1978) Die Zoobenthos-Gesellschaften des Piburger Baches (Ötztal, Tirol). *Int. Rev. ges. Hydrobiol.* 63, 213-232.

Krausch H.-D. (1976) Die Makrophyten der mittlerer Saale und ihre Biomasse. *Limnologica, Berl.* 10, 57-72.

Kristjansson J.K. & Alfredsson G.A. (1983) Distribution of *Themus* spp. in Icelandic hot springs and a thermal gradient. *Appl. environ. Microbiol.* 45, 1785-1789.

Kruglova A.N. (1983) Zooplankton in small rivers of the Kola peninsula. *Gibrobiol. Zh. Kiev* 19, 56-58. (In Russian, English summary)

Kumsare A. (1972) Zur hydrobiologischen Rayonierung der Daugava. *Verh. int. Verein. theor. angew. Limnol.* 18, 751-755.

Lack T.J., Youngman R.E. & Collingwood R.W. (1978) Observations on a spring bloom in the River Thames. *Verh. int. Verein. theor. angew. Limnol.* 20, 1435-1439.

Ladle M., Bass J.A.B., Philpott F.R. & Jeffrey A. (1977) Observations on the ecology of Simuliidae from the River Frome, Dorset. *Ecological Entomology* 2, 197-204.

Learner M.A., Lochhead G. & Hughes B.D. (1978) A review of the biology of British Naididae (Oligochaeta) with emphasis on the lotic environment. *Freshwat. Biol.* 8, 357-376.

Leclercq L. (1977) Végétation et caractéristiques physico-chimiques de deux rivières de Haute-Ardenne (Belgique): la Helle et la Roer Supérieure. *Lejeunia* 88, 1-42.

Lécureuil J.Y., Choret M., Bournault M. & Tachet H. (1983) Description, répartition et cycle biologique de la larve d'*Hydropsyche bulgaromanorum* Malicky 1977 (Trichoptera) in the lower Loire. *Annls Limnol.* 19, 17-24.

Leivestad H. & Muniz I.P. (1976) Fish kill at low pH in a Norwegian river. *Nature, Lond.* 259, 391-392.

Lindegaard C., Thorup J. & Bahn M. (1975) The invertebrate fauna of the moss carpet in the Danish spring Raunkilde and its seasonal, vertical, and horizontal distribution. *Arch. Hydrobiol.* 75, 109-139.

Lindengaard C. (1979) A survey of the macroinvertebrate fauna with special reference to Chironomidae (Diptera) in the rivers Laxá and Kráká, northern Iceland. *Oikos* 32, 281-288.

Lohmeyer W. & Krause A. (1975) Über die Auswirkungen des Gehölzbewuchses an kleinen Wasserläufen des Münsterlandes auf die Vegetation im Wasser und an den Böschungen im Hinblick auf die Unterhaltung der Gewässer. *Schriftenr. Vegetationskr.* 9, 1-105.

Logan P. & Brooker N.P. (1983) The macroinvertebrate faunas of riffles and pools. *Water Res.* 17, 263-270.

Loris B. (1979) Bio-Indikatoren: Wasserpflanzen uberwachen unsere Flüsse. *Bild der Wissenschaft* 1, 44-52.

Lucena F., Fiance C., Jofre J., Sancho J. & Schwartzbrod L. (1982) Viral pollution determination of superficial waters (river water and sea-water) from the urban area of Barcelona (Spain). *Water Res.* 16, 173-177.

Maas D. & Kohler A. (1983) Die Makrophytenbestände der Donau im Raum Tuttlingen. *Landschaft + Stadt* 15 (2), 49-60.

Mackey A.P. (1976-77) Quantitative studies in the Chironomidae (Diptera) of the Rivers Thames and Kennet. (1976a). I. The *Acorus* zone. *Arch. Hydrobiol.* 78, 240-267. (1976b). II. The Thames flint zone. *Arch. Hydrobiol.* 78, 310-318. (1977a). III. The *Nuphar* zone. *Arch. Hydrobiol.* 79, 62-102. (1977b). IV. Production. *Arch. Hydrobiol.* 80, 327-348.

Mackey A.P., Ham S.F., Cooling D.A. & Berrie A.D. (1982) An ecological survey of a limestone stream, the River Coln, Gloucestershire, England, in comparison with some chalk streams. *Arch. Hydrobiol., Suppl.* 64, 307-340.

Madsen B.L., Bengtsson J. & Butz I. (1977) Upstream movement of some Ephemeroptera species. *Arch. Hydrobiol.* 81, 119-127.

Maes L. (1982) Dissolved organic carbon and UV-absorption in a polluted lowland brook - pond system. Influence of water aeration by the Phallus technique. *Hydrobiologia* 89, 269-276.

Malmqvist B. (1980) Habitat selection of larval brook lampreys (*Lampetra planeri*, Blach) in a South Swedish stream. *Oecologia, Berl.* 45, 35-38.

Malmqvist B. (1983) Growth, dynamics, and distribution of a population of the lamprey *Lampetra planeri* in a South Swedish stream. *Holarctic Ecology* 6, 404-412.

Mann R.H.K. (1978) Observations on the biology of the perch, *Perca fluviatilis*, in the River Stour, Dorset. *Freshwat. Biol.* 8, 229-240.

Mann R.H.K. (1979) Aspects of the biology of coarse fish in the Dorset Stour. *Freshwater Biological Association, U.K., Annual Report* 47, 51-59.

Marchetti R. & De Paolis A. (1983) Water quality of the River Ticino. *Parco Ticino Notizie, Suppl.* 17-18, 32 pp.

Marker A.F.H. (1976) The benthic algae of some streams in southern England. I. Biomass of the epilithon in some streams. *J. Ecol.* 64, 343-358.

Marker A.F.H. & Gunn R.J.M. (1977) The benthic algae of some streams in southern England. III. Seasonal variations in chlorophyll a in the seston. *J. Ecol.* 65, 223-234.

Marker A.F.H. & Casey H. (1982) The population and production dynamics of benthic algae in an artificial recirculating hard-water stream. *Phil. Trans. R. Soc. Lond. B* 298, 265-308.

Markic M. (1983) The Rotatoria-Monogononta of the River Drava in Slovenia, Yugoslavia. *Hydrobiologia* 104, 229-330.

Marxsen J. (1980a, b) Untersuchungen zue Ökologie der Bakterien in der
    fliessenden Welle von Bachen. I. Chemismus, Primarproduktion, CO_2-
    Dunkelfixierung und Eintrag von partikularen organischen Material.
    Arch. Hydrobiol. Suppl. 4, 461-533. II. Die Zahl der Bakterien im
    Jahreslauf. *Arch. Hydrobiol., Suppl.* 57, 461-533.
Maquet B. (1983) Caractéristiques chimiques et biologiques des eaux
    de la vallée du Samson. *Annls Soc. r. Belg.* 113, 3-18.
McCumiskey L.M. (1981) A review of Ireland's water resources - rivers
    and lakes. *Irish J. environ. Sci.* 1, No. 2, 1-13.
McGill J.D., Wilson R.S. & Brake A.M. (1979) The use of chironomid
    pupal exuviae in the surveillance of sewage pollution within a
    drainage system. *Water Res.* 13, 887-894.
McLean R.O. & Jones A.K. (1975) Studies of tolerance to heavy metals
    in the flora of the rivers Ystwyth and Clarach, Wales. *Freshwat.
    Biol.* 5, 431-444.
Mendl H. & Müller K. (1978) The colonization cycle of *Amphinemura
    standfussi* Ris (Ins.: Plecoptera) in the Abisku area. *Hydrobio-
    logia* 60, 109-111.
Merle G. (1980) Analyses bactériologiques de l'eau du Rhône au
    voisinage de la centrale thermique du Bugey. *Cah. Lab. Hydrobiol.
    Montereau* 10, 63-76.
Merry D.G., Slater F.M. & Randerson P.F. (1981) The riparian and
    aquatic vegetation of the River Wye. *J. Biogeog.* 8, 313-327.
Mihăiţă A. (1970) Cercetaria saprobiologice în Bazinul Riului Jiu.
    Annal Analele Universitatii Burcuresti. *Biologie Animala* 19,
    101-107. (Rumanian, with Russian and French summaries)
Mihăiţă A. (1971) *Cercetări Saprobiologice în Bazinul Riului Jiu.*
    42 pp. Rezumatul tezei de doctorat, Universitatea din Bucuresti.
Mills C.A., Welton J.S. & Rendle E.L. (1983) The age, growth and
    reproduction of the stone loach *Noemacheilus barbatulus* (L.) in
    a Dorset chalk stream. *Freshwat. Biol.* 13, 283-292.
Milner N.H., Scullion J., Carling P.A. & Crisp D.T. (1981) The effects
    of discharge on sediment dynamics and consequent effects on inverte-
    brates and salmonids in upland rivers. In: Coaker T.H. (Ed.)
    Advances in Applied Biology Vol. VI. pp. 154-220. Academic Press,
    London.
Ministry of the Environment, Denmark (1978) *Environmental Protection
    Act.* 49 pp. National Agency of Environmental Protection, 29
    Strandgade, Copenhagen.
Monschau-Dudenhausen K. (1983) Wasserpflanzen als Belastungs-Indika-
    toren in Fliessgewässern dergestellt am Beispiel der Schwarzwald-
    flüsse Nagold und Alb. *Beih. Veröff. Naturschutz Landschaftspflege
    Bad.-Württ.* 28, 1-118.
Moore J.W. (1976) Seasonal succession of algae in rivers. I. Examples
    from the Avon, a large slow-flowing river. *J. Phycol.* 12, 342-349.
Moore J.W. (1977a) Seasonal succession of algae in rivers. II.
    Examples from Highland Water, a small woodland stream. *Arch.
    Hydrobiol.* 80, 160-171.
Moore J.W. (1977b) Some factors affecting algal densities in a
    eutrophic farmland stream. *Oecologia, Berl.* 29, 257-267.

Moore J.W. (1978) Seasonal succession of algae in rivers. III.
    Examples from the Wylye, a eutrophic farmland river. *Arch. Hydro-
    biol.* 83, 367-376.
Mortensen E. (1981) Energy flow in trout (*Salmo trutta* L.) population
    and the benthic invertebrate community of Bisballe baek, Denmark.
    I Jornadas de Ichtyologia Iberica. Leon, 25-30 mayo 1981.
Mosello R., Guidetti L., Calderoni A., Cominazzini C., De Giuli E. &
    Battioli M.T. (1982) *Water Quality of Tributary Streams of Lago
    Maggiore.* Symposium A.I.O.L., 1-3 December 1980, Chiavar. (In
    Italian)
Moss M.O. (1977) The temporal and spatial distribution of the valves
    of the diatom *Rhoicosphenia curvata* in the River Wye. *Freshwat.
    Biol.* 7, 573-578.
Mouvet C. & Bourg A.C.M. (1983) Speciation (including adsorbed species)
    of copper, lead, nickel and zinc in the Meuse River. Observed
    results compared to values calculated with a chemical equilibrium
    computer program. *Water Res.* 17, 641-649.
Müller C. (1978) On the productivity and chemical composition of some
    benthic algae in hard-water streams. *Verh. int. Verein. theor.
    angew. Limnol.* 20, 1457-1462.
Müller K. (1974) Stream drift as a chronobiological phenomenon in
    running water ecosystems. *A. Rev. Ecol. Systematics* 5, 309-323.
Nedeljkovic R. (1978) Einflus der Anschwemmungen des Pek-Flusses auf
    die Bodenfauna der Donau. *Verh. int. Verein. theor. angew. Limnol.*
    20, 1531-1535.
Nelva A. (1979) Structure écologique de trois rivières du Jura
    méridional d'après l'analyse multivariée de la macrofaune benthique.
    Arch. Hydrobiol. 87, 327-346.
Neveu A. (1983) Le zoobenthos des eaux courantes/The zoobenthos of
    running waters. In: Association Française Limnologie, *Traveaux
    Française en Limnologie.* pp. 48-52. XXe Congres SIL, Lyon,
    August 1983. (French and English)
Newbold C., Purseglove J. & Holmes N. (1983) *Nature Conservation and
    River Engineering.* 36 pp. Nature Conservancy Council, Shrewsbury.
Niesiolowski S. (1980) Studies on the abundance, biomass and vertical
    distribution of larvae and pupae of black flies (Simuliidae, Diptera)
    on plants of the Grabia river, Poland. *Hydrobiologia* 75, 149-156.
Olechowska M. (1982) Zonation of mayflies (Ephemeroptera) in several
    streams of the Tatra Mts and the Podhale region. *Acta Hydrobiol.*
    24, 63-71.
Otto C. & Svensson B.S. (1983) Properties of acid brown water streams
    in South Sweden. *Arch. Hydrobiol.* 99, 15-36.
Pasternak K. (1973) Rozprzestrzenienie metali ciezkich w wodach
    plynacych w rejonie wystepowania naturalnych złóż oraz przemyslu
    cynku i olowiu. The spreading of heavy metals in flowing waters
    in the region of occurrence of natural deposits and of the zinc
    and lead industry. *Acta Hydrobiol.* 14, 145-166.
Pedersen P.M. (1977) Blue-green algae from the thermal springs of
    Ûnartoq, Southwest Greenland. *Bot. Tidsskrift* 71, 80-83.
Penczak T., Moliński M., Kusto E., Palusiak H., Panusz H. & Zalewski,
    M. (1976) The ecology of roach (Rutilus rutilus L.) in the barbel
    region of the polluted Pilica River. 2. Dry weight, ash and
    contents of some elements. *Ekol. pol.* 24, 623-638.

Penczak T., Zalewski M., Suszycka R. & Moliński M. (1981) Estimates of density, biomass and growth of fish populations in two small lowland rivers. *Ekol. pol.* <u>29</u>, 233-255.

Penczak T., Kusto E., Krzyżanowska D., Moliński M. & Suszycka E. (1984) Food consumption and energy transformations by fish populations in two small lowland rivers in Poland. *Hydrobiologia* <u>108</u>, 135-144.

Perrin J.F. & Roux A.L. (1978) Structure et fonctionnement des écosystèmes du Haut-Rhône français. 6. La macrofaune benthique du fleuve. *Verh. int. Verein. theor. angew. Limnol.* <u>20</u>, 1494-1502.

Pieper H.-G., Meijering M.P.D. (1982) *Gammarus* occurrence as an indication for stable conditions in Hessian woodland brooks and rivers. *Polskie Archwm Hydrobiol.* <u>29</u>, 283-288.

Pietsch W. (1974) Ökologische Untersuchwung und Bewertung von Fliessgewässern mit Hilfe hoherer Wasserpflanzen - Ein Beitrag zur Belastung aquatischer Ökosysteme. *Mitt. Sekt. Geobot. Phytotax. Biol. Ges. DDR*, 13-29.

Pinder L.C.V. (1980) Spatial distribution of Chironomidae in an English chalk stream. In: Murray D.A. (1980). *Chironomidae: Ecology, Systematics, Cytology and Physiology.* pp. 153-161. Pergamon, Oxford.

Prat N., Rug M.A. & Gonzalez M.A. (1983) *Predicció i Control de la Qualitat de les Aigües dels Rius Besòs i Llobregat, II. El Poblament faunistic i la Seva Relació amb la Qualitat de les Aigües.* 185 pp. Estudis monografies 9, Servei del Medi Ambient. Diputació de Barcelona.

Pratten D.J. (1980) Growth in the crayfish *Austropotamobius pallipes* (Crustacea Astacidae). *Freshwat. Biol.* <u>10</u>, 401-412.

Price S.M. & Price J.H. (1983) The River Wandle: studies on the distribution of aquatic plants. *London Naturalist* <u>62</u>, 26-58.

Regione Veneto (1977) *Symposium on Pollution of the River Adige.* 15 January 1977. Verona. (In Italian)

Reid J.M., Cresser M.S. & MacLeod D.A. (1980) Observations on the estimation of total organic carbon from u.v. absorbance for an unpolluted stream. *Water Res.* <u>14</u>, 525-529.

Reinheimer G. (1977a, b) Mikrobiologische Untersuchungen in Flussen. I. Fluoreszenzmikroskopische Analyse der Bakterienflora einiger norddeutscher Flüsse. *Arch. Hydrobiol.* <u>81</u>, 106-118. II. Die Bakterienbiomasse in einigen norddeutschen Flüssen. *Arch Hydrobiol.* <u>81</u>, 259-267.

Richardot-Coulet M., Richoux P. & Roux C. (1983) Structure et fonctionnement des écosystèmes du Haut-Rhône francais. 29. Structure des peuplements de macroinvertébrés benthiques d'un ancien méandre. *Arch. Hydrobiol.* <u>96</u>, 363-383.

Rofes B. (1975) Influence de la craie en poudre sur les fonds envasés en eau courante. *Verh. int. Verein. theor. angew. Limnol.* <u>19</u>, 1693-1704.

Rosset J., Bärlocher F. & Oertli J.J. (1982) Decomposition of conifer needles and deciduous leaves in two Black Forest and two Swiss Jura streams. *Int. Rev. ges. Hydrobiol.* <u>67</u>, 695-712.

Roux A.L. (1976) Structure et fonctionnement des écosystèmes du Haut-Rhône français. 1. Presentation de l'étude. *Bull. Ecol.* 7, 475-478.

Roux A.L. (1983) Un hydrosysteme fluviatile: le Haute-Rhône et sa vallée - A fluvial hydrosystem: the French Upper Rhône and its valley. In: Association Française de Limnologie, *Travaux Français en Limnologie*, pp. 252-257. XX Congres SIL, August 1983 (French and English)

Roux A.L., Tachet H. & Neyron M. (1976) Structure et fonctionnement des écosystèmes du Haut-Rhône français. III. Une technique simple et peu onéreuse pour l'etude des macro-invertébrés benthiques des grands fleuves. *Bull. Ecol.* 7, 493-496.

Russev B. (1972) Über die Migration der Rheobionten in Fliessgewässern. *Verh. int. Verein. theor. angew. Limnol.* 18, 730-734.

Sabater S. (1983) Distribucion espacio-temporal de las poblaciones de algal del arroyo de l'Avenco (Barcelona). *Actas del Primer Congreso de Limnologia* (1983). pp. 159-166.

Sandbeck K.A. & Ward D.M. (1982) Temperature adaptations in the terminal processes of anaerobic decomposition of Yellowstone National Park and Icelandic Hot Spring microbial mats. *Appl. environ. Microbiol.* 44, 844-851.

Sand-Jensen K. & Rasmussen L. (1978) Macrophytes and chemistry of acidic streams from lignite mining areas. *Bot. Tidsskrift* 72, 105-112.

Saunders M.J. & Eaton J.W. (1976) A method for estimating the standing crop and nutrient content of the phytobenthos of stony rivers. *Arch. Hydrobiol.* 78, 86-101.

Say P.J. & Whitton B.A. (1980) Changes in flora down a stream showing a zinc gradient. *Hydrobiologia* 76, 255-262.

Say P.J. & Whitton B.A. (1981) Chemistry and plant ecology of zinc-rich streams in the Northern Pennines. In: Say P.J. & Whitton B.A. (Eds) *Heavy Metals in Northern England: Environmental and Biological Aspects.* 198 pp., pp. 55-64.

Say P.J. & Whitton B.A. (1982a) Chimie et ecologie de la vegetation de cours d'eau en France à fortes teneurs en zinc. Massif Central. *Annls Limnol.* 18, 3-18.

Say P.J. & Whitton B.A. (1982b) Chemistry and plant ecology of zinc-rich streams in France. 2. The Pyrenees. *Annls Limnol.* 18, 19-31.

Scharz F. & Betschart B. (1979) The use of periphyton from Lake Zurich to estimate the algal growth potential in River Limmat Water. *Schweiz. Z. Hydrol.* 41, 141-149.

Schmitz W. & Illies J. (1980) Die Verfahren der biologischen Beurteilung des Gütezustandes der Fliessgewässer. *Studien zum Gewässerschutz* 5.

Schmitz W. & Schuman G.O. (1982) Die sommerlichen Wassertemperaturen der Äschenzone mitteleuropäischer Fliessgewässer. *Arch. Hydrobiol.* 95, 435-443.

Scullion J. (1983) Effects of impoundments on downstream bed materials of two upland rivers in mid-Wales and some ecological implications of such effects. *Arch. Hydrobiol.* 96, 329-344.

Seyed-Reihani A., Gilbert J. & Ginet R. (1982a) Structure et fonctionnement des écosystèmes du Haut-Rhône français. 23. Ecologie de deux stations interstitielles; influence de la pluviosité sur leur peuplement. *Pol. Archwm Hydrobiol.* 29, 501-511.

Seyed-Reihani A., Ginet R. & Reygrobellet J.L. (1982b) Structure et fonctionnement des écosystèmes du Haut-Rhône français. 30. Le peuplement de trois stations interstitielles dans la plaine de Miribel-Jonage (Vallée du Rhône en amont de Lyon) en relation avec leur alimentation hydrogéologique. *Rev. fr. Sci. Eau* 1, 163-174.

Sivko T.N., Kovalevskaja R.Z., Mikheeva T.M. & Ostapenia A.P. (1972) Biological transformation of organic matter in a river. *Verh. int. Verein. theor. angew. Limnol.* 18, 756-760.

Skoczén I. (1982) The effect of rain waters on lead levels in the Vistula in the region of Cracow agglomeration. *Acta Hydrobiol.* 24, 95-107.

Sládeček V. & Miskovsky O. (1976) Bacterial periphyton of the Vltava river. *Hydrobiologia* 51, 181-187.

Sloor W., de Zwart D. & van de Kerkhoff J.F.J. (1983) Monitoring the rivers Rhine and Meuse in the Netherlands for toxicity. *Aquat. Toxicol.* 4, 189-198.

Smith B.D., Lyle A.A. & Maitland P.S. (1983) The ecology of running waters near Aberfeldy, Scotland, in relation to a proposed barytes mine: an impact assessment. *Environ. Pollut. (Ser. A)* 32, 269-306.

Smith C.S., Adams R.S., Schmitt M.R. & Adams S.S. (1978) Phosphorus in the water, sediment, and vegetation of the Crnojevica River, Montenegro, Yugoslavia. *Verh. int. Verein. theor. angew. Limnol.* 20, 1536-1542.

Smith R.V., Stevens R.J., Foy R.H. & Gibson C.E. (1982) Upward trend in nitrate concentration in rivers discharging into Lough Neagh for the period 1969-1979. *Water Res.* 16, 182-188.

Sperling J.A. (1975) Algal ecology of southern Icelandic hot springs in winter. *Ecology* 56, 183-190.

Stazner B. (1981) The relation between "hydraulic stress" and microdistribution of benthic macroinvertebrates in a lowland running water system, the Schierenseebrooks (North Germany). *Arch. Hydrobiol.* 91, 192-218.

Steffan A.W. (1971) Chironomid (Diptera) biocoenoses in Scandinavian glacier brooks. *Canadian Entomologist* 103, 478-486.

Steinberg C. & Melzer A. (1983) Aufnahme, Transport und Abgabe von Kohlenstoff durch submerse Makrophyten von Fliesswasserstandorten. *Schweiz. Z. Hydrol.* 45, 333-344.

Steubing L., Fricke G. & Jehn H. (1983) Veranderung des Algenspektrums der Eder im Verlauf von vier Jahrzehnten. *Arch. Hydrobiol.* 205-222.

Stevens R.J. & Smith R.V. (1978) A comparison of discrete and intensive sampling for measuring the loads of nitrogen and phosphorus in the River Main, County Antrim. *Water Res.* 12, 828-830.

Stolz L. & Weise G. (1976) Einflus des Nährstoffgehaltes auf die Stoffproduktion submerser Macrophyten in Fliessgewässern, dargestellt an *Fontinalis antipyretica*. *Limnologica, Berl.* 10, 405-417.

Stroot P. (1984) Faunistique et répartition longitudinale des Trichoptères dans une rivière salmonicole de basse montagne, en Belgique. *Hydrobiologia* 108, 245-258.

Tangen K., Müller C. & Brettum P. (1978) Periphytic and drifting
microalgae in a tributary stream of Øvre Heimdalsvattnet. *Holarctic Ecol.* 1, 148-154.

Tarak P. (1981) Environmental legislation and pollution control
regarding water resources. In: *Water for Human Consumption. Man
and his Environment.* A selection of papers prepared for the IVth
World Congress of the International Water Resources Association
(IWRA). pp. 494-506. Tycooly International Publishing Ltd, Dublin.

Thirb H.H. & Benson-Evans K. (1983) Studies on the life cycle of
Lemanea using natural artificial substrates. *Hydrobiologia* 98,
110-124.

Thomas E.A. (1976) Der flutende Hahnesfuss (*R. fluitans*), ein neues
limnologisches Problem am Rhine. *Z. Wasser-Energie-Luft* 10,
230-234.

Thommen G.H. & Westlake D.F. (1981) Factors affecting the distribution
of populations of *Apium nodiflorum* and *Nasturtium officinale* in
small chalk streams. *Aquatic Botany* 11, 21-36.

Thorup J. & Lindegaard C. (1977) Studies on Danish springs. *Folia
Limnol. Scand.* 17, 7-15.

Toner P.F. & Lennox L.J. (1980) Nitrate content of Irish river waters.
Irish J. environ. Sci. 1, 75-76.

Townsend C.R., Hildrew A.G. & Francis J. (1983) Community structure
in some southern English streams: the influence of physicochemical
factors. *Freshwat. Biol.* 13, 521-544.

Uherkovich G. (1966) Übersicht über das Potamophytoplankton der
Tisza (Theiss) in Ungarn. *Hydrobiologia* 28, 252-280.

Vernaux J. (1983) Les methodes biologiques pratiques d'appreciation
de la qualité. Practical biological methods used in France for the
assessment of the quality of running waters. In: Association
Francaise de Limnologie, Travaux Français en Limnologie, pp. 178-191.
XX Congres SIL, Lyon, August 1983. (French and English)

Vrhovsek D., Martinincic & Kralj M. (1981) Evaluation of the polluted
river Savinja with the help of macrophytes. *Hydrobiologia* 80,
97-110.

Weber-Oldecop D.W. (1971) Wasserpflanzengesellschaften im östlichen
Niedersachsen (II.) *Int. Rev. ges. Hydrobiol.* 56, 79-122.

Wehr J.D., Say P.J. & Whitton B.A. (1981) Heavy metals in an indus-
trially polluted river, the Team. In: Say P.J. & Whitton B.A. (Eds)
*Heavy Metals in Northern England: Environmental and Biological
Aspects.* 198 pp. pp. 99-109. Department of Botany, University of
Durham.

Welton J.S. (1980) Dynamics of sediment and organic detritus in a
small chalk stream. *Arch. Hydrobiol.* 90, 162-181.

Werner I. & Weise G. (1982) Biomass production of submersed macrophytes
in a selected stretch of the River Zschopau (South GDR) with special
regard to orthophosphate incorporation. *Int. Rev. ges. Hydrobiol.*
67, 45-62.

Westlake D.F. & Dawson F.H. (1982) Thirty years of weed cutting on a
chalk-stream. In: *Proceedings, European Weed Research Society 6th
Symposium on Aquatic Weeds.* pp. 132-140.

Whitton B.A. (1975) *River Ecology*. 725 pp. Blackwell Scientific Publications, Oxford.

Wiegleb G. (1979a) Der Zusammenhang zwischen Gewässergute und Macrophytenvegetation in niedersachsicher Fliessgewässern. *Landschaft + Stadt* 11, 32-35.

Wiegleb G. (1979b) Vorläufige Übersicht über die Pflanzengesellschaften der niedersächsischen Fliessgewässer. *Natursch. u. Landschaftspflege Niedersachsen* 11, 85-116.

Wiegleb G. (1981a) Application of multiple discriminant analysis on the analysis of the correlation between macrophyte vegetation and water quality in running waters of Central Europe. *Hydrobiologia* 79, 90-100.

Wiegleb G. (1981b) Struktur, Verbreitung und Bewertung von Makrophytengesellschaft niedersächsischer Fliessgewässer. *Limnologica (Berlin)* 13, 427-448.

Wilkinson D.R. & Jones J.W. (1977) The fecundity of dace, *Leuciscus leuciscus* (L.) in Emral Brook, Clwyd, North Wales. *Freshwat. Biol.* 7, 135-145.

Willis M. (1983) A comparative survey of *Ancylus fluviatilis* (MÜLLER) populations in the Afon Crafnant, N. Wales, above and below an input of zinc from mine-waste. *Arch. Hydrobiol.* 98, 198-214.

Wilson R.S. (1977) Chironomid pupal exuviae in the River Chew. *Freshwat. Biol.* 7, 9-18.

Wise E.J. (1976) Studies on the Ephemeroptera of a Northumbrian river system. I. Serial dilution and relative abundance. *Freshwat. Biol.* 6, 363-372.

Wotton R.S. (1979) The influence of a lake on the distribution of black-fly species (Diptera: Simuliidae) along a river. *Oikos* 22, 368-372.

Wright J.F., Cameron A.C., Hiley P.D. & Berrie A.D. (1982) Seasonal changes in biomass of macrophytes on shaded and unshaded sections of the River Lambourn, England. *Freshwat. Biol.* 12, 271-283.

Wright J.F., Hiley P.D., Cameron A.C., Wigham M.E. & Berrie A.D. (1983) A quantitative study of the macroinvertebrate fauna of five biotopes in the River Lambourn, Berkshire, England. *Arch. Hydrobiol.* 96, 271-292.

Wright R.F., Dale T., Gjessing E.T., Hendrey G.R., Henriksen A., Johanhessen M. & Muniz I.P. (1976) Impact of acid precipitation on freshwater ecosystems in Norway. *Water Air Soil Pollut.* 6, 483-499.

Wurhmann K. & Eichenberger D. (1980) Künstliche Bäche als Hilfsmittel der experimentellen Fliessgewässer-Ökologie. *Vom Wasser* 54, 1-18.

Zaiss U. (1981) Natural ebullition of mine gas and its microbial oxidation in the Netzbach brook, Saarland. *Verh. int. Verein. theor. angew. Limnol.* 21, 1381-1385.

Zalewski M. & Penczak T. (1981) Characterization of the fish community of the Utrata River drainage-basin, and evaluation of the efficiency of catching methods. *Pol. Archwm Hydrobiol.* 28, 385-396.

Zullin A. & Ricci C. (1980) Bdelloid rotifers and nematodes in a small Italian stream. *Freshwat. Biol.* 10, 67-72.

Appendix B: Authorities, Synonyms and English Names

Latin names, authorities, synonyms and English common names for selected vascular plants and fish. (The synonyms given are those used in some chapters or which were in original manuscripts from authors.)

VASCULAR PLANTS

PTERIDOPHYTES

Isoetes
Osmunda regalis L.

DICOTYLEDONS

Apium graveolens L.	wild celery
Apium nodiflorum (L.) Lag.	fool's watercress
Berula erecta (Huds.) Coville	lesser water-parsnip
(= *Sium erectum* Huds.)	
Callitriche	water-starwort
Ceratophyllum demersum L.	hornwort
Filipendula ulmaria (L.) Maxim.	meadowsweet
Hottonia palustris L.	water violet
Impatiens glandulifera Royle	Himalayan balsam
Limosella aquatica L.	mudwort
Littorella uniflora (L.) Aschers.	shoreweed
Lobelia dortmanna L.	water Lobelia
Mentha aquatica L.	water mint
Myriophyllum	water milfoil
Nuphar lutea (L.) Sm.	yellow water lily
Nuphar pumila (Timm.) DC	least water lily
Nymphaea alba L.	white water lily
Nymphoides peltata (S.G. Gmel.) O. Kuntze	fringed water lily
Oenanthe	water dropwort
Ranunculus fluitans Lam., *R. penicillatus*	water buttercup
(Dumort.) Bab.	
Rorippa amphibia (L.) Besser	great yellow-cress
Rorippa islandica (Oder) Borbas	marsh yellow-cress
Rorippa nasturtium-aquaticum (L.) Hayek	watercress
(= *Nasturtium officinale* R. Br.)	
Rorippa sylvestris (L.) Besser	creeping yellow-cress
Utricularia	bladder-wort
Veronica beccabunga L.	brooklime

MONOCOTYLEDONS

Acorus calamus L.	sweet fly
Alisma plantago-aquatica	water-plantain
Butomus umbellatus L.	flowering-rush
Catabrosa aquatica (L.) Beauv.	whorl-grass
Cladium mariscus (L.) Pohl	great fen-sedge
Elodea canadensis Michx.	Canadian pondweed
Elodea ernstiae St John	
Eleocharis palustris (L.) Roem & Schult.	common spike-rush
Glyceria maxima (Hartm.) Holmberg	reed sweet-grass
Hydrocharus morsus-ranae L.	frog-bit
Juncus	rush
Lemna gibba L.	fat duckweed
Lemna minor L.	common duckweed
Lemna trisulca L.	ivy-leaved duckweed
Phalaris arundinacea L.	reed canary-grass
Phragmites australis (Cav.) Trin. ex. Steud.	common reed
Potamogeton lucens L.	shining pondweed
Potamogeton pectinatus L.	fennel pondweed
Scirpus fluitans L.	floating club-rush
Scirpus lacustris L. (= *Schoenoplectus lacustris* (L.) Palla)	common club-rush
Scirpus tabernaemontani (= *Schoenoplectus tabernaemontani* (C.C. Gmel.) Palla)	
Sparganium emersum (= *S. simplex* Huds.)	unbranched bur-reed
Sparganium erectum L.	branched bur-reed
Sparganium minimum Wallr.	least bur-reed
Stratrotes aloides L.	water-soldier
Typha angustifolia L.	lesser bulrush
Typha latifolia L.	bulrush, great reedmace
Wolffia arrhiza (L.) Wimm.	rootless duckweed
Zannichellia palustris	horned pondweed

FISH

Abramis brama (L.)	bream
Acipenser naccarii Bp.	
Acipenser sturio L.	sturgeon
Alburnus alburnus (L.)	bleak
Alburnus bipunctatus (Bloch) = *Alburnoides bipunctatus*	bystryanka
Alosa alosa (L.)	allis shad, shad, mayfish
Alosa fallax (Lacepede)	twaite shad, twait
Anguilla anguilla (L.)	eel
Aspius aspius (L.)	asp, rapfen
Barbus barbus (L.)	barbel
Barbus meridionalis Risso	southern barbel
Carassius auratus (L.)	goldfish

Carassius carassius (L.)	crucian carp
Chrondrostoma nasus (L.)	nase
Chrondrostoma toxostoma (Vallot)	soiffe
Cobitis taenia (L.)	spined loach
Coregonus oxyrhynchus (L.)	houting
Cottus gobio (L.)	bullhead
Cottus poecilopus Heck	alpine bullhead
Ctenopharyngodon idella (Val.)	grass carp
Cyprinus carpio (L.)	carp
Esox lucius (L.)	pike
Gambusia affinis (Baird & Girard)	mosquito-fish
Gasterosteus aculeatus (L.)	three-spined stickleback
Gobio gobio (L.)	gudgen
Gymnocephalus cernua (L.)	ruffe
Huso huso (L.)	beluga
Ictalurus melas (Rafinesque)	black bullhead
Ictalurus nebulosus (Lesueur)	brown bullhead
Lampetra fluviatilis (L.)	river lamprey, lampern
Lampetra planeri (Bloch)	brook lamprey
Lepomis gibbosus (L.)	pumpkinseed
Leucaspius delineatus (Heck)	
Leuciscus cephalus (L.)	chub
Leuciscus idus (L.)	ide
Leuciscus leuciscus (L.)	dace
Leuciscus souffia	
Liza ramada (Risso)	thin-lipped grey mullet
= *Chelon ramada* (Risso)	
Liza saliens (Risso)	
Lota lota (L.)	burbot
Micropterus salmoides (Lacep.)	large-mouthed bass
Noemacheilus barbatulus (L.)	stone loach
Oncorhynchus tshawytscha (Walbaum)	Chinook salmon
Padogabius martensi (Gthr)	
Perca fluviatilis (L.)	perch
Petromyzon marinus (L.)	sea lamprey
Phoxinus phoxinus (L.)	minnow
Platichthys flesus (L.)	flounder
Pungitius pungitius (L.)	ten-spined stickleback, nine-spined stickleback
Rutilus pigus (Lacep.)	Danubian roach
Rutilus rubilia (Bp.)	
Rutilus rutilus (L.)	roach
Sabanejewia larvata (De Fil.)	
Salmo gairdneri	rainbow trout
Salmo salar	salmon
Salmo trutta (L.) (*fario*)	brown trout, brook trout
Salmo trutta (L.) (*marmoratus*)	
Salmo trutta (L.) (*trutta*)	sea trout
Salvelinus alpinus (L.)	Arctic charr
Salvelinus fontinalis (Mitch.)	brook charr
Scardinius erythrophthalmus (L.)	rudd
Silurus glanis (L.)	wels

Stizostedion lucioperca (L.)	zander, pike-perch
Tinca tinca (L.)	tench
Thymallus thymallus (L.)	grayling

Organism Index

Subject Index

644